Cabinetmaking Patternmaking and Millwork

This book is dedicated to
my wife, Lorraine,
for her enthusiasm, help, and encouragement.

Cabinetmaking Patternmaking and Millwork

Gaspar J. Lewis

10 9 8 7 6 5 4 3

LIBRARY OF CONGRESS CATALOG CARD NUMBER: 79-50917

ISBN: 0-8273-1816-2

Printed in the United States of America
Published simultaneously in Canada
by Nelson Canada,
A Division of International Thomson Limited

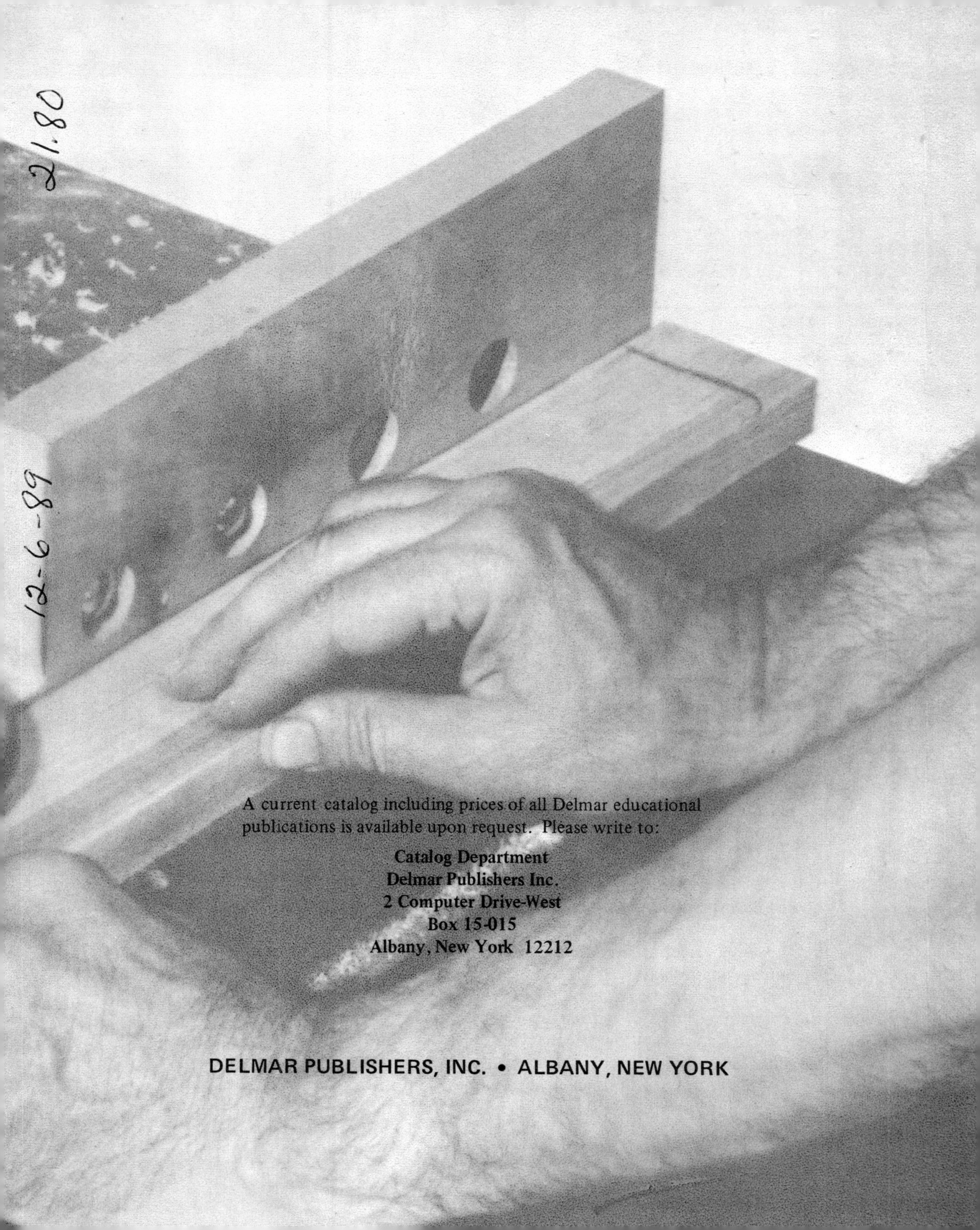

A current catalog including prices of all Delmar educational publications is available upon request. Please write to:

Catalog Department
Delmar Publishers Inc.
2 Computer Drive-West
Box 15-015
Albany, New York 12212

DELMAR PUBLISHERS, INC. • ALBANY, NEW YORK

PREFACE

Wood is a vital part of our lives. No other material can be used to make so many goods and is so readily available. Those who have mastered the skills required to make wood products have many opportunities for employment in our society. Even greater opportunities are open to woodworkers with advanced and specialized woodworking skills.

CABINETMAKING, PATTERNMAKING, AND MILLWORK is designed to develop skills in these three areas of specialization. It is intended to be used in advanced woodworking courses at the senior high school level, in vocational and technical schools, and in apprenticeship programs.

The first three sections of the text deal with elements of woodworking that are common to all three fields. Section 1 discusses the nature of wood and its uses. Section 2 explains how to use hand tools, portable power tools, and stationary woodworking machines. Section 3 describes methods of joinery. The last three sections deal with each particular trade, stressing the skills unique to each. Through hands-on experience, the student learns how to make the products produced by the cabinetmaker, patternmaker, and millworker.

Each unit of the text begins with a list of objectives that explains exactly what the student is expected to learn. These objectives help the student and instructor evaluate the student's progress through the text. In order to meet the objectives, the student must effectively learn the skills presented in the unit.

Material is presented through information, illustrations, and step-by-step procedures. This threefold approach gives the student many opportunities to grasp the material. After reading the information, the student gains practical, hands-on experience by following the step-by-step procedures to produce typical woodworking products. The text highly illustrates both the information and the procedures. New terms are italicized and defined in the unit and in the glossary at the end of the text.

There are activities and unit review questions at the end of each unit. The activities range from simple observation and reporting to producing actual projects. This wide range lets the instructor adapt the activities to the individual and class setting. The unit review questions further test the student's knowledge of the material. By successfully completing the procedures in the unit, the activities, and the unit review, the student will meet the objectives of the unit.

A note about metrics. Though the woodworking industries will eventually adopt the metric system of measurement, lumber and other wood products are not yet being manufactured in metric sizes. Since this is the case, most of CABINETMAKING, PATTERNMAKING, AND MILLWORK uses the English system of measurement. Only *Section 5 Patternmaking* includes metric equivalents because of this trade's close association with the metalworking industries which use metrics extensively. Metric conversion charts are listed in the appendix at the end of the text.

Other woodworking texts from Delmar Publishers include:

BASIC CONSTRUCTION BLUEPRINT READING
EXTERIOR AND INTERIOR TRIM
HAND WOODWORKING TOOLS
PORTABLE POWER TOOLS
PRACTICAL PROBLEMS IN MATHEMATICS FOR CARPENTERS
SIMPLIFIED STAIR LAYOUT
UNITS IN WOODWORKING

ABOUT THE AUTHOR

Gaspar J. Lewis is a journeyman carpenter and cabinetmaker with over thirty years experience in the teaching profession. He is a member of many professional organizations including the American Vocational Association, Florida Vocational Association, and the Vocational Industrial Clubs of America. Mr. Lewis is currently a cabinetmaking instructor at Pinellas Vocational-Technical Institute in Clearwater, Florida.

ACKNOWLEDGMENTS

The author wishes to thank the school board of Pinellas County, Florida, and the administration, faculty, and students of Pinellas Vocational-Technical Institute, Construction Trades Technology, for their encouragement and permission to use and photograph their classrooms and equipment.

The author also wishes to express special thanks to the following contributors.

Lorraine Lewis, his wife, for typing and proofreading the original manuscript
Frank Ballard of the American Plywood Association for providing technical assistance
George Henderson of Custom Cabinets
Bill Hermann, patternmaker, Schmidt Aluminum Castings
Russ Lisk of Stanley Power Tools

Reviewers

George Bedell
Patternmaking Reviewer
Industrial Arts Chairman
Averill Park Central School
Averill Park, New York

Kenneth L. Desjardins
Cabinetmaking and Millwork Instructor
Los Angeles Trade Technical College
Los Angeles, California

C. LeRoy Michaelis
Consulting Editor

Companies and Associations

Adhesive Machinery Corporation,
Division of Ornsteen Chemicals, Inc.
Adjustable Clamp Company
American Plywood Association
Amerock Corporation
Andersen Corporation
Anderson Lumber Company
Berkshire Pattern and Woodworking Shop
Black and Decker Manufacturing Company
Boise-Cascade Corporation
Bostich Division of Textron, Inc.
C.E. Morgan, Division of Combustion Engineering
Custom Cabinets
Disston, Inc.
Duraflake Division, Williamette Industries Inc.
Ekstrom Carlson
Greenlee Tool Company
Hardwood Plywood Manufacturers Association
Ingersoll-Rand, Millers Falls Division
Iroquois Millwork Corporation
Irvington Moore
Jim & Slim's Tool Supply
Kane's Furniture
Makita, U.S.A.
Masonite Corporation
Mattison Machine Works
Modern Casting
Mohawk Finishing Products, Inc.
Moisture Register Company
National Association of Pattern Manufacturers Inc.
Newman Machine Company Inc.
Oliver Machinery Company
Parten Machinery Company
Rockwell International, Power Tool Division
B.M. Root Company
S & W Saw and Tool Company, Inc.
Schmidt Aluminum Castings
The Stanley Works
L.S. Starrett Company
James L. Taylor Manufacturing Company
Timesavers, Inc.
Tyler Machinery Company, Inc.
U.S. Dept. of Commerce,
National Bureau of Standards
Western Wood Products Association
Weyerhauser Company
Woodworking and Furniture Digest

Delmar Staff

Mark W. Huth – Sponsoring Editor
Kathleen E. Beiswenger –Associate Editor
John A. Foley –Editorial Intern

CONTENTS

section 1

Wood and Woodworking

unit 1 The Cabinetmaker, Patternmaker, and Millworker

OBJECTIVES

After studying this unit, the student will be able to:

- describe the skills needed by the cabinetmaker, patternmaker, and millworker.
- explain how these woodworking trades are similar.
- discuss the job opportunities for these woodworking trades.
- explain how woodworking skills are acquired.

WOOD AND WOODWORKERS

Wood is a remarkable substance. It comes in a variety of strengths, colors, grains, and textures. It has a natural beauty and warmth that is valued for building homes, furniture, and thousands of other products. It can be shaped to just about any form, finished easily, and with the proper care, last indefinitely.

Wood is one of the world's few renewable resources. Even though it has been used since the beginning of man, wood is still in abundant supply because it can be replenished. A new tree can be planted to replace those cut down. With a steady supply of wood to rely on, there is a constant demand for those skilled in the woodworking trades.

Cabinetmaking, patternmaking, and millwork are trades that work primarily with wood. Much of the same knowledge and many of the same skills are required of each. People in these trades are all woodworkers who lay out, cut, fit, smooth, assemble, and finish wood to complete a project.

Fig. 1-1 The cabinetmaker is installing drawer guides on a cabinet.

Fig. 1-2 The cabinetmaker must know how to use power tools.

Fig. 1-3 The cabinetmaker must select the proper joints and fastening devices. *(American Plywood Association)*

CABINETMAKERS

Cabinetmakers make and repair furniture, store fixtures, office equipment, kitchen cabinets, and many other articles, figure 1-1. They work mainly with wood but also use other materials, such as plastics, metals, and glass.

To be good in their craft, cabinetmakers must have a wide range of skills. They must be able to:

- use a variety of hand tools, portable power tools, and stationary power tools, figure 1-2.
- read and interpret blueprints and drawings of the article to be made.
- lay out dimensions of the parts to be cut.
- estimate materials needed for the job.
- develop a plan of procedure.
- cut out and assemble the parts using proper joints and fastening devices, figure 1-3.
- sand and scrape surfaces to prepare them for finishing.

Fig. 1-4 Cabinetmakers often custom design their projects. Here a decorative groove is being cut in a cabinet door panel. *(Rockwell International)*

- coat assembled pieces with fillers, stains, and finishes by dipping, brushing, or spraying.
- install hardware such as hinges, drawer guides, pulls, and catches.

There are many opportunities for the cabinetmaker today. Although many wood products are now mass-produced by millworkers in large shops, there are still custom shops where cabinetmakers do all the work from start to finish, figure 1-4. Some shops make only kitchen cabinets; others make store fixtures or office equipment. The boatbuilding, automotive, and aeronautic industries all employ cabinetmakers. These speciality shops value the tradeworker skilled in all phases of cabinetmaking.

PATTERNMAKERS

Patternmakers make and repair one-piece or multipiece wood units called *patterns.* Patterns are used to form sand molds to make castings of different kinds of metal or plastic, figure 1-5. To make these patterns, patternmakers work with wood, such as pine and mahogany, and other materials, such as leather, wax, and plastics.

Patternmaking is a highly specialized trade. Patternmakers need basic woodworking skills as well as a knowledge of casting procedures. Precision is important because a mar in the pattern will be reproduced in the castings made from it, figure 1-6.

Patternmakers must know how to:

- use a variety of hand tools, portable power tools, and stationary power tools.
- read blueprints and decide which material to use, estimate the amount, and develop a plan of procedure.
- determine the parting line and lay out the pattern on the stock, dividing it if necessary so that the pattern can be removed from the mold cleanly and easily.
- cut, shape, and assemble the parts using proper joints and fastening devices and making allowances for draft and finishing, figure 1-7.

Fig. 1-5 A pattern (left) and the casting made from it *(Berkshire Pattern & Woodworking Shop)*

- smooth surfaces and apply fillets to interior angles.
- check dimensions of the completed pattern with the blueprint.
- color the pattern according to the color code and finish the pattern with shellac, lacquer, or wax.
- repair or rebuild worn or broken patterns.

Fig. 1-6 The patternmaker must be able to read plans to make a pattern.

Fig. 1-7 The patternmaker must know how to cut wood precisely using a variety of tools. *(Modern Casting)*

There is a great demand for patternmakers today. Patterns are needed for such diverse products as engine blocks and cooking utensils. The experienced journeyman can work in a variety of fields and command a good salary.

Foundries are the main employer of patternmakers. The automobile and equipment manufacturing industries especially use many patterns. Many patternmakers also work in independent pattern shops. These job shops make patterns for companies who do not employ their own patternmakers.

Another field open to the patternmaker is modelmaking. Modelmakers make precise models of new products from engineering designs. These models may be the forerunners of a new car that will eventually need patterns in order to be mass-produced. They may also be a one-of-a-kind model developed for a specific purpose.

Fig. 1-8 Millworkers produce wood parts in large quantities. *(Reprinted from the August 1978 issue of WOODWORKING & FURNITURE DIGEST © 1979 Hitchcock Publishing Company, ALL RIGHTS RESERVED)*

MILLWORKERS

Millworkers make moldings and parts for doors, windows, furniture, sashes, mantels, stair rails, posts, treads, balusters, and many other wood products. They are employed in plants where large quantities of these products are made in standard designs, sizes, grades, and types, figure 1-8.

A millworker may be a machine operator or a skilled machine woodworker, figure 1-9. Some millworkers do only one part of the job, such as setting up, assembling, or finishing. Others spend many years learning a specific craft, such as the millworker who lays out and builds staircases.

The well-rounded millworker must know how to:

- read blueprints and specifications and estimate materials needed for the job.
- set up, adjust, and maintain woodworking machines.
- grind, sharpen, and balance cutter blades to a specified shape and weight, figure 1-10.
- operate woodworking machines to saw, shape, and smooth lumber to specifications.
- cut mortises and tenons, bore holes, and cut grooves, dadoes, and designs in wood stock.
- verify dimensions of parts using gauges and templates.
- assemble fabricated parts to make millwork products.
- coat the finished product by brushing or spraying.

With so much specialization in the woodworking field, the demand for millworkers exceeds the supply. Many new and different woodworking machines are developed each year. Some of these machines are very expensive. Though there may be an ample supply of machine operators available, employers will always seek out the millworker who can set up, operate, and maintain these machines efficiently.

TRAINING

There are many ways to learn the skills needed to enter the woodworking trades. Depending on the job, these skills may be acquired at a trade

Fig. 1-9 This machine operator is feeding stock into a drum sander.

Fig. 1-10 This millworker is gumming a circular saw blade.

school, through on-the-job training, or in an apprenticeship program.

Vocational schools, junior colleges, and technical institutes offer programs in the woodworking fields. These programs provide courses in math and blueprint reading and introduce the student to woodworking tools, machines, and procedures. While they develop basic skills needed by the woodworker, on-the-job training is still necessary.

Many woodworkers learn their skills on the job. Their training usually begins by observing the skilled operator at work. When trainees first operate a machine, they are closely supervised by an experienced worker. Trainees gradually learn the skills needed to perform the job on their own.

Certain woodworking careers require a high degree of skill and a wide range of knowledge. Patternmaking, for instance, requires more than on-the-job training. The best way to qualify for these jobs is to complete an *apprenticeship*.

Apprenticeships range from 3 to 5 years. Apprentices attend classes a few hours each week to learn the theory of their trade. The rest of the week is spent working under an experienced craftsperson, called a *journeyman*. Apprentices begin by helping the journeyman in routine duties and performing basic tasks under close supervision. As they progress, the work becomes increasingly more complex and the supervision more general.

A high school education is usually needed to enter an apprenticeship. Apprentices receive a percentage of the regular salary for that trade. This percentage increases as the apprentice gains experience.

ACTIVITIES

1. Visit a cabinetmaking shop, an independent job shop or foundry pattern shop, and a wood mill to observe the work of these trades. Find out what on-the-job training is available.
2. Study the history and development of cabinetmaking, patternmaking, and millwork.
3. List and describe branches of the cabinetmaking industry.
4. List and describe millwork specialties.
5. List and describe other woodworking occupations.
6. Visit a technical institute, junior college, or vocational school and find out what courses and programs are offered in the woodworking fields.
7. Contact your state's Apprenticeship Council and find out what woodworking apprenticeships are available in your area and what they require.

UNIT REVIEW

Questions

1. What does it mean to say that wood is a renewable resource?
2. What skills do the cabinetmaker, patternmaker, and millworker have in common?
3. What will happen if there is a mar in the pattern made by the patternmaker?
4. Where are cabinetmakers employed?
5. Where are patternmakers employed?
6. Where are millworkers employed?

7. How are woodworking skills acquired?
8. How does an apprenticeship differ from on-the-job training?
9. What is a journeyman?
10. Why might a student who has completed a technical course in woodworking still need on-the-job training?

Multiple Choice

1. The cabinetmaker makes and repairs
   a. sand molds.
   b. machines.
   c. patterns.
   d. furniture.
2. The patternmaker makes and repairs patterns used to form
   a. models.
   b. sand molds.
   c. fillets.
   d. moldings.
3. The millworker uses woodworking machines to make
   a. moldings.
   b. models.
   c. patterns.
   d. castings.
4. All cabinetmakers, patternmakers, and millworkers must be able to
   a. assemble furniture.
   b. make stair parts.
   c. do foundry work.
   d. use woodworking machines.
5. The millworker is more concerned than other woodworkers with
   a. using hand tools.
   b. making wood parts.
   c. making store fixtures.
   d. making patterns.
6. All woodworkers must know
   a. casting procedures.
   b. the kinds and nature of metals.
   c. how to apply fillets.
   d. the kinds and nature of wood.
7. Who mass-produces wood products?
   a. The cabinetmaker
   b. The patternmaker
   c. The millworker
   d. All woodworkers
8. Who must determine a parting line?
   a. The cabinetmaker
   b. The patternmaker
   c. The millworker
   d. All woodworkers
9. Who must grind, sharpen, and balance cutter blades?
   a. The cabinetmaker
   b. The patternmaker
   c. The millworker
   d. All woodworkers
10. A woodworker who makes one-of-a-kind examples of new products from engineering designs is called
    a. a modelmaker.
    b. a journeyman.
    c. an apprentice.
    d. a foundry worker.

unit 2 Wood

OBJECTIVES

After studying this unit, the student will be able to:

- name the parts of a tree and state their function.
- describe how lumber is cut.
- explain moisture content at various stages of seasoning.
- identify and describe the properties, characteristics, and uses of common hardwoods and softwoods.
- compute board feet.

There are many different kinds of wood. They differ in color, strength, durability, workability, and in other ways. These differences give a wide choice for selecting the most suitable wood for a job. Anyone who works with wood must understand the characteristics of wood in order to use it intelligently.

STRUCTURE OF WOOD

Wood is made up of many hollow cells held together by a natural substance called *lignin.* The size, shape, and arrangement of these cells determine the strength, weight, and other properties of wood.

Wood grows outward from the center of the tree, called the *pith.* New wood cells are formed in the *cambium layer* just inside the bark of the tree. The outer bark forms a protective shield for the tree. *Medullary rays* are bands of cells that carry food from the inner bark toward the center of the tree, figure 2-1.

Heartwood and Sapwood

The new wood nearest the bark is called *sapwood* and contains living cells. As the tree grows, the inner cells become inactive and turn to heartwood. *Heartwood* is the central part of the tree and is usually darker in color. Although it is dead, it does not decay or lose its strength. It is more durable than sapwood and is preferred for exterior use. The heart of redwood, for instance, is often used for outdoor furniture, porches, and siding. If used for the same purpose, its sapwood would decay more quickly.

Annular Rings

Each spring and summer a tree adds new layers to its trunk. Wood growth in the spring is fast and lighter in color. In summer, the growth is slower and darker forming *annular rings.*

The age of a tree is determined by counting the number of annular rings. The annular rings show the effect of dry spells, periods of abundant

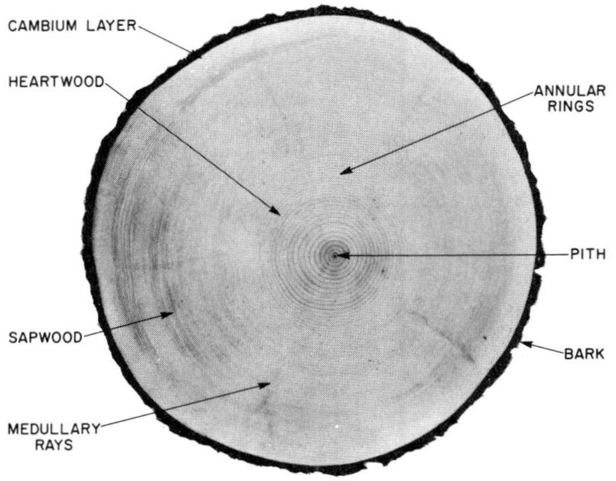

Fig. 2-1 Cross section of a tree *(Western Wood Products Assn.)*

rainfall and sunshine, and damage from fire, storms, and insects. In tropical climates, where the weather and growth of the tree is more constant, annular rings are not so prominent and sometimes are not visible. This is one way of identifying a tropical wood from those grown in temperate climates.

METHODS OF CUTTING LUMBER

There are two common ways of cutting lumber: plain-sawing and quartersawing, figure 2-2. The *plain* or *flat-sawed method* is the most common. In this method, the log is cut tangent to the annular rings. It is cheaper and less wasteful. Greater widths are obtained and kiln drying is easier. Plain-sawed lumber shrinks more and warps easily, however.

Quartersawing is more expensive and wasteful than plain-sawing, but results in less shrinkage and warping. This method cuts the log at right angles to the annular rings. It has a more durable surface and produces a more desirable wood grain pattern.

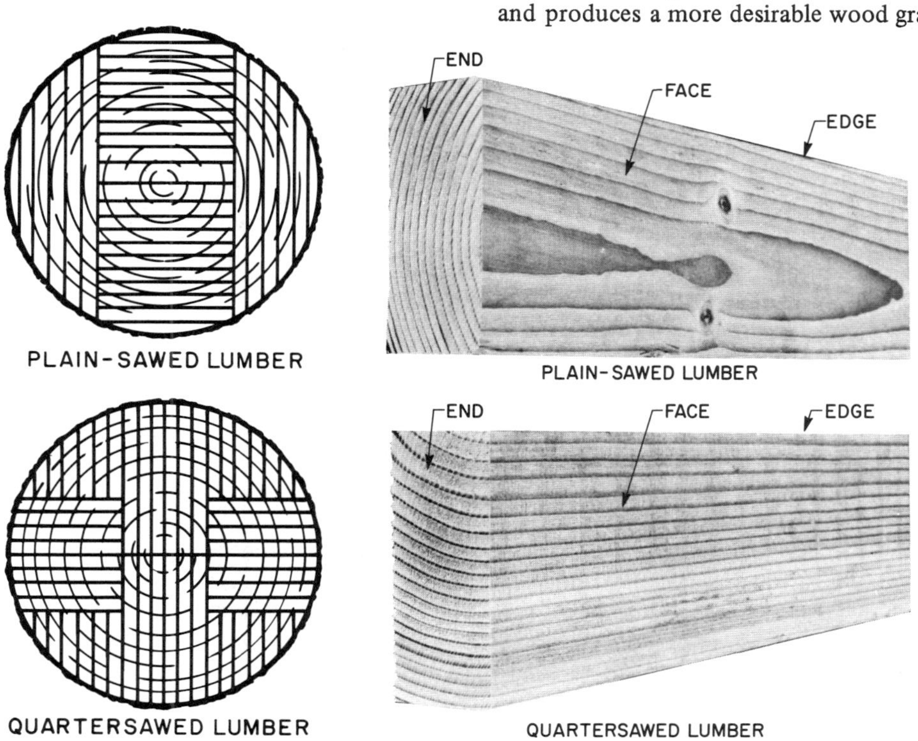

Fig. 2-2 Methods of cutting lumber *(Weyerhaeuser)*

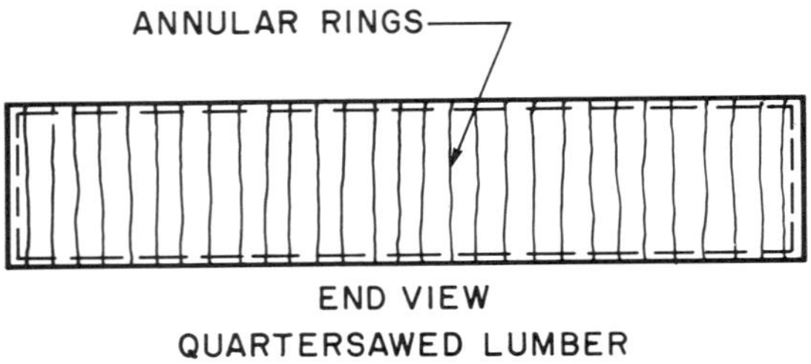

Fig. 2-3 There is little shrinkage and no warping as quartersawed lumber dries out.

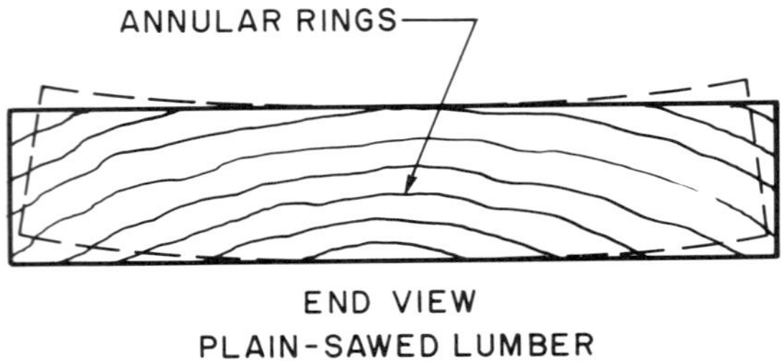

Fig. 2-4 Plain-sawed lumber shrinks unequally because the rings are unequal in length.

MOISTURE CONTENT

When a tree is first cut down, it contains a great amount of water. In this state it is called *green*. Much of this water must be removed before the wood can be used satisfactorily. The wood is then called *dry*.

Green lumber may be as much as three times heavier than dry lumber because of the water contained in it. Green wood has water in the hollow part of the wood cells and also in the cell walls. When wood starts to dry, the water in the hollow part of the cells is first removed. When all of this free water is gone, the wood has reached its *fiber saturation point*. About one-third of the wood's weight is now water. This is expressed as a percentage, usually about 30 percent moisture content.

As wood continues drying, the water in the cell walls is removed. The wood starts to *shrink* or become smaller. Lumber must be dried to a moisture content of less than 10 percent to be used for cabinetwork or furniture. It must therefore shrink considerably from its fiber saturation point of 30 percent moisture content.

When wood dries out, most of the shrinkage is in the direction of the annular rings. This explains why quartersawed lumber does not warp as it dries, figure 2-3. Wood warps as it dries according to the way it has been sawed from the log. Looking at the end of a piece of lumber, the part that has the longest rings shrinks the most. Therefore, a piece with rings of unequal length shrinks unequally, figure 2-4.

Dry wood swells if moisture is absorbed. It also swells unequally if the annular rings in the lumber are of unequal length. This is why the unprotected undersides of floors and tables warp during periods of high humidity.

Fig. 2-5 Air drying lumber *(Western Wood Products Assn.)*

Lumber shrinks mostly in width and thickness. So little shrinkage takes place along its length, end to end, that it is not considered. Although shrinking and swelling cannot be absolutely prevented, it can be held to a minimum by sealing all surfaces of the finished product.

SEASONING

Two common methods of *seasoning* or drying wood are air drying and kiln drying.

Air-dried lumber is stacked in piles with spacers between each layer. This is called *sticking*. Sticking permits air to circulate through the pile, figure 2-5.

Kiln-dried lumber is stacked in piles similar to air-dried lumber and then placed in an oven

Fig. 2-6 Kiln drying lumber *(Western Wood Products Assn.)*

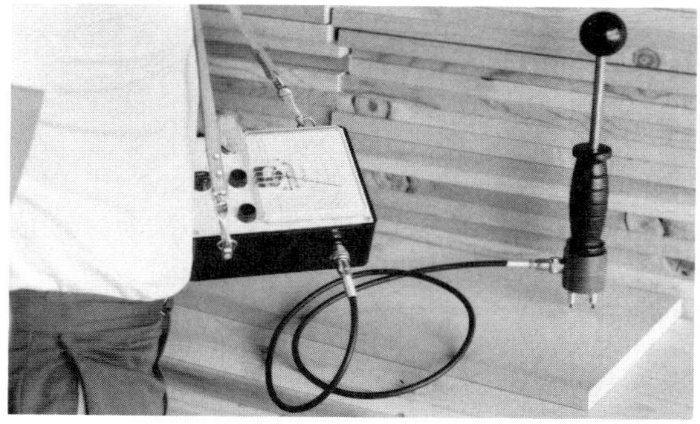

Fig. 2-7 Moisture meter *(Moisture Register Co.)*

called a *kiln.* The drying time and temperature is carefully controlled inside a kiln, figure 2-6.

Kiln drying has many advantages over air drying. Although it is more costly than air drying, it takes much less time. Moisture content is controlled and reduced to a definite and much lower amount than through air drying. Kiln-dried lumber can be dried to a moisture content of less than 10 percent. Air-dried lumber often cannot be dried to much less than 20 percent, and only after a considerable period of time depending on temperature, humidity, wind, and the thickness of the lumber.

DETERMINING MOISTURE CONTENT

The easiest and most common method of determining moisture content is with a *moisture meter,* figure 2-7. The points of the meter are inserted in the wood and the moisture content is read on the scale.

The moisture content of wood is also determined by the oven-drying method. Small samples of wood are weighed before baking. They are then placed in an oven at about 220 degrees until they stop losing weight. The amount of moisture originally contained in the samples can then be determined by finding the difference in weights, dividing by the dry weight, and multiplying by 100.

Example: Before baking, a sample piece of wood weighs 10 ounces. After baking, it weighs 8 ounces. The moisture content is therefore:

$$\frac{(10-8) \times 100}{8} = 25\% \text{ moisture content}$$

LUMBER DEFECTS

A defect in lumber is any fault that detracts from its appearance, use, or strength, figure 2-8.

Warping is caused when wood is dried too fast, handled or stored carelessly, or milled before it is thoroughly dry. Warps are classified as cups, bows, crooks, and winds, figure 2-9.

Splits in the end of a board running lengthwise are called *checks.* These occur when the end dries faster than the rest of the stock. Checks can be prevented to some degree by sealing the lumber ends with paint, wax, or other sealers.

Cracks that run along and between the annular rings are called *shakes.* They are often caused by storm damage.

Pith is the spongy center of the tree and should not be used. *Knots* are parts of branches of a tree. They are not necessarily defects unless they are loose or weaken the piece. *Pitch pockets* are small openings that hold pitch. A *wane* is bark or missing wood on the edge of lumber, figure 2-10. *Pecky wood* has small grooves or channels running with the grain. Other defects are stains, decay, and wormholes.

KINDS OF WOOD

Hardwood and Softwood

Woods are classified as either hardwood or softwood. *Hardwood* comes from trees, called *deciduous,* that shed their leaves annually. Common hardwoods are ash, basswood, beech, birch, cherry,

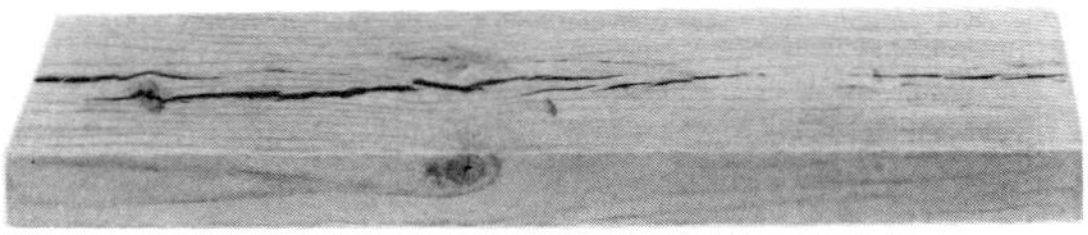

Fig. 2-8 Checks, shakes, and other defects *(Western Wood Products Assn.)*

hickory, lauan, mahogany, maple, oak, poplar, teak, and walnut. *Softwood* is cut from cone-bearing trees, called *coniferous,* that are commonly known as evergreens. Some of the more common softwoods are cedar, cypress, fir, pine, redwood, and spruce. Some softwoods may actually be harder than some hardwoods. For instance, a softwood like fir is harder and stronger than basswood.

There are other methods of classifying hardwoods and softwoods, but this is the method that is widely used.

Open and Close-Grained Wood

Wood is also divided into two groups according to its cell structure. *Open-grained lumber* has large cells. When cut, it shows tiny openings or pores in the surface. To obtain a good finish these pores must be filled. Examples of open-grained wood are ash, hickory, mahogany, oak, and walnut.

Close-grained woods have cells so small that, when cut, they can hardly be seen. Wood from coniferous trees is close-grained. It is not necessary to fill close-grained lumber. Some close-grained woods are cherry, maple, and poplar.

PROPERTIES, CHARACTERISTICS, AND USES

To select the best wood for a given job, the woodworker must know the properties and characteristics of wood. Wood for furniture must be hard enough to resist denting and present a good appearance. Basswood, although classified as a hardwood, is therefore not suitable. Because of their characteristics some softwoods, like fir and pine, are often used for millwork products such as windows and doors. Cypress, redwood, and cedar are often selected for exterior use because of their resistance to decay. Ash and hickory are used when woods of excellent bending qualities are desired.

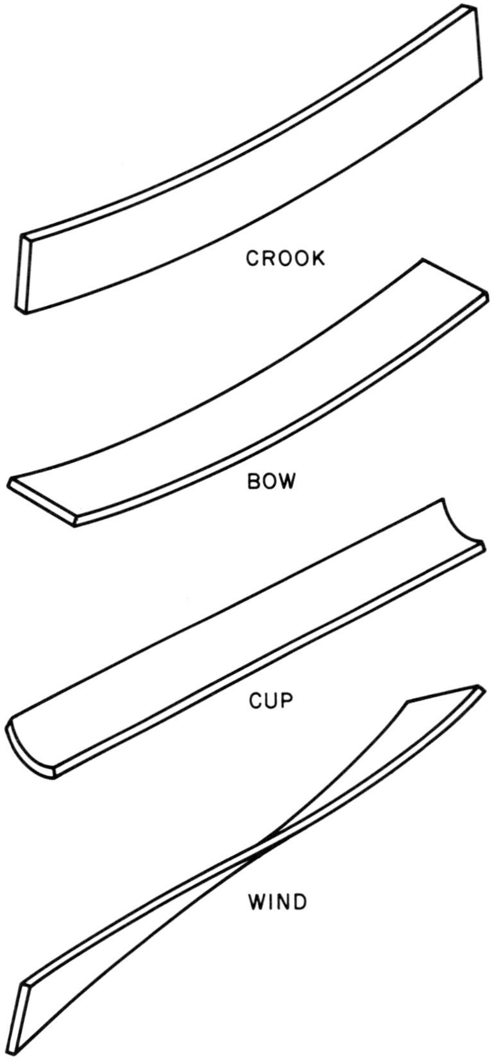

Fig. 2-9 Kinds of warps

Fig. 2-10 Wane *(Western Wood Products Assn.)*

KIND	COLOR	GRAIN	HARDNESS	STRENGTH	WORK-ABILITY	ELASTICITY	DECAY RESISTANCE	USES	OTHER
Ash	Light Tan	Open Coarse	Hard	High	Hard	Very High	Low	Tool Handles Oars Baseball Bats	
Basswood	Creamy White	Close Fine	Soft	Low	Easy	Low	Low	Drawing Bds Veneer Core	Imparts No Taste Or Odor
Beech	Light Brown	Close Medium	Hard	High	Medium	Medium	Low	Food Containers Furniture	
Birch	Light Brown	Close Fine	Hard	High	Medium	Medium	Low	Furniture Veneers	
Cherry	Lt. Reddish Brown	Close Fine	Medium	High	Medium	High	Medium	Furniture	
Hickory	Light Tan	Open Medium	Hard	High	Hard	Very High	Low	Tool Handles Diving Boards	
Lauan	Lt. Reddish Brown	Open Medium	Soft	Low	Easy	Low	Low	Veneers Paneling	
Mahogany	Russet Brown	Open Fine	Medium	Medium	Excellent	Medium	High	Quality Furniture	
Maple	Light Tan	Close Medium	Hard	High	Hard	Medium	Low	Furniture Flooring	
Oak	Light Brown	Open Coarse	Hard	High	Hard	Very High	Medium	Flooring Boats	
Poplar	Greenish Yellow	Close Fine	Medium Soft	Medium Low	Easy	Low	Low	Furniture Veneer Core	
Teak	Honey	Open Medium	Medium	High	Excellent	High	Very High	Furniture Boat Trim	Heavy Oily
Walnut	Dark Brown	Open Fine	Medium	High	Excellent	High	High	High Quality Furniture	

Fig. 2-11 Common Hardwoods

KIND	COLOR	GRAIN	HARDNESS	STRENGTH	WORK-ABILITY	ELASTICITY	DECAY RESISTANCE	USES	OTHER
Red Cedar	Dark Reddish Brown	Close Medium	Soft	Low	Easy	Poor	Very High	Exterior	Cedar Odor
Cypress	Orange Tan	Close Medium	Soft to Medium	Medium	Medium	Medium	Very High	Exterior	
Fir	Yellow to Orange Brown	Close Coarse	Medium to Hard	High	Hard	Medium	Medium	Framing Millwork Plywood	
Ponderosa Pine	White with Brown Grain	Close Coarse	Medium	Medium	Medium	Poor	Low	Millwork Trim	Pine Odor
Sugar Pine	Creamy White	Close Fine	Soft	Low	Easy	Poor	Low	Patternmaking Millwork	Large Clear Pieces
Western White Pine	Brownish White	Close Medium	Soft to Medium	Low	Medium	Poor	Low	Millwork Trim	
Southern Yellow Pine	Yellow Brown	Close Coarse	Soft to Hard	High	Hard	Medium	Medium	Framing Plywood	Much Pitch
Redwood	Reddish Brown	Close Medium	Soft	Low	Easy	Poor	Very High	Exterior	Light Sapwood
Spruce	Cream to Tan	Close Medium	Medium	Medium	Medium	Poor	Low	Siding Subflooring	Spruce Odor

Fig. 2-12 Common softwoods

The tables in figures 2-11 and 2-12 give some of the properties, characteristics, and uses of common hardwoods and softwoods. Remember that there are many different species of each wood. Each species differs in appearance and properties. For example, there are many varieties of pine, such as Pond, Red, Virginia, Western White, Jack, Lodgepole, Ponderosa, Eastern White, Sugar, Loblolly, Longleaf, Shortleaf, and Slash.

THICKNESS (INCHES)		WIDTH (INCHES)	
NOMINAL	ACTUAL	NOMINAL	ACTUAL
1	3/4	2	1 1/2
5/4	1	3	2 1/2
6/4	1 1/4	4	3 1/2
2	1 1/2	5	4 1/2
2 1/2	2	6	5 1/2
3	2 1/2	8	7 1/4
3 1/2	3	10	9 1/4
4	3 1/2	12	11 1/4

Fig. 2-13 Softwood Lumber Sizes

THICKNESS (INCHES)	
NOMINAL	ACTUAL
1/2	5/16
3/4	9/16
1	13/16
1 1/4	1 1/16
1 1/2	1 5/16
2	1 3/4
3	2 3/4
4	3 3/4

Fig. 2-14 Hardwood Lumber Sizes

GRADES

Softwood Grading

Softwood lumber is divided into three main groups for grading. *Yard lumber* is used for general building purposes. *Structural lumber* is used for heavy construction. *Factory* and *shop lumber* are used for millwork items such as windows, doors and moldings.

Yard lumber is further divided into *select* and *common grades.* Select grades run from A to D and common grades from 1 to 5. Each of the other main groups of softwoods are also subdivided to identify their quality. More complete information on grading softwoods can be obtained from the National Bureau of Standards.

Hardwood Grading

First and seconds (FAS) is the best grade of hardwood and yields about 85 percent clear cuttings. Each piece must be at least 6 inches wide and 8 feet long. The next best grade is called *select.* The minimum width is 4 inches and the length must be 6 feet or more. *No. 1 common grade* allows even narrower widths and shorter lengths with about 65 percent clear cuttings. Hardwood grades are established by the National Hardwood Lumber Association.

SIZES OF LUMBER

Lumber that comes directly from the sawmill and has not been surfaced is called *rough lumber.* Rough lumber comes in nominal sizes, a size that is fairly close to what it actually measures. For instance, a 1 x 6 piece of rough softwood measures approximately 1 inch in thickness and 6 inches in width. This size may vary because of the heavy machinery used to cut the log. When this piece is *planed* (surfaced), it is reduced in thickness and width. Its nominal size does not change, but its actual size does. Therefore, when surfaced, although it is still called a 1 x 6, its actual size is 3/4 inch by 5 1/2 inches. The same applies to all surfaced lumber. The nominal size and the actual size are not the same.

Softwoods are available in standard thickness, width, and length. Hardwoods come surfaced in standard thickness only. Widths and lengths are not sized and are called *random* widths and lengths.

Figure 2-13 shows the nominal and actual sizes of the thicknesses and widths of softwoods. Figure 2-14 shows only the nominal and actual thicknesses of hardwoods.

BOARD MEASURE

Lumber is measured by the board foot. A *board foot* is a piece of lumber 1 inch thick by 12 inches wide by 12 inches long, or its equal. For instance, a piece of lumber 1 inch thick by 6 inches wide by 2 feet long is still one board foot. A piece 1 inch thick by 4 inches wide must be 3 feet long to equal 1 board foot.

Two formulas are used to determine board feet. When the length of the pieces are in feet, the formula is:

$$\text{Board Feet} = \frac{\text{no. of pieces x thickness x width x length}}{12}$$

When the length of the pieces are in inches, the formula is:

$$\text{Board Feet} = \frac{\text{no. of pieces x thickness x width x length}}{144}$$

Lumber less than 1 inch thick is figured the same as if it were 1 inch. Lumber that is thicker than 1 inch is figured to the next quarter inch. For example, a piece that is 1 1/8 inches thick is figured as 1 1/4 inches.

ACTIVITIES

1. Examine a cross section of a tree. Determine its age, identify its parts, and describe periods of slow and fast growth.
2. Expose a piece of heartwood and sapwood from the same log to the weather to compare their durability.
3. Visit a sawmill to observe methods of cutting lumber.
4. Determine the moisture content of lumber through the oven-drying method.
5. Visit a kiln to observe the method of kiln drying lumber.
6. Use a moisture meter to determine the moisture content of sample pieces of lumber.
7. Gather as many different kinds of wood as possible and make a display identifying each kind.
8. Obtain samples and make a display showing different kinds of defects in lumber.

UNIT REVIEW

Completion

1. The center of the tree is called the ________.
2. New wood cells are formed in the ________.
3. The wood nearest the bark is called ________. The wood in the central part of the tree is called ________.
4. The age of a tree is determined by counting its ________.
5. Two common ways of cutting lumber are ________ and ________.
6. Lumber will not shrink until it has passed its ________.
7. Lumber that is to be used for cabinetwork must have a moisture content of less than ________ percent.
8. Quartersawed lumber will not warp because the annular rings are of ________ length.
9. Two methods of seasoning lumber are ________ and ________.
10. An easy method of determining the moisture content of lumber is with ________.

Multiple Choice

1. A check is a
   a. split in the end of a board.
   b. separation of the annular rings.
   c. spongy center part.
   d. small groove or channel.

2. An example of hardwood is
   a. fir.
   b. cedar.
   c. birch.
   d. spruce.
3. An example of softwood is
   a. pine.
   b. mahogany.
   c. oak.
   d. ash.
4. An example of open-grained wood is
   a. pine.
   b. spruce.
   c. cedar.
   d. oak.
5. An example of close-grained wood is
   a. hickory.
   b. walnut.
   c. mahogany.
   d. maple.
6. A wood with excellent bending qualities is
   a. basswood.
   b. hickory.
   c. poplar.
   d. redwood.
7. A wood exceptionally high in decay resistance is
   a. beech.
   b. cypress.
   c. hickory.
   d. maple.
8. A wood that is very strong is
   a. cedar.
   b. redwood.
   c. poplar.
   d. oak.
9. A wood that is used to make high-quality furniture is
   a. poplar.
   b. fir.
   c. walnut.
   d. spruce.
10. A wood that is often used to make patterns because of its fine grain and ease of workability is
    a. fir.
    b. oak.
    c. sugar pine.
    d. ash.
11. Each piece of first and second (FAS) hardwood must be at least
    a. 4 inches wide and 4 feet long.
    b. 5 inches wide and 5 feet long.
    c. 6 inches wide and 6 feet long.
    d. 6 inches wide and 8 feet long.
12. When a 1 x 6 piece of softwood is surfaced, it actually measures
    a. 7/8″ x 5 3/4″.
    b. 13/16″ x 5 3/4″.
    c. 3/4″ x 5 1/2″.
    d. 3/4″ x 5 3/4″.
13. A piece of 1-inch hardwood when surfaced to standard thickness is
    a. 7/8 inch thick.
    b. 13/16 inch thick.
    c. 3/4 inch thick.
    d. 11/16 inch thick.
14. A 1″ x 4″ x 15′ piece of lumber contains
    a. 4 board feet.
    b. 5 board feet.
    c. 10 board feet.
    d. 15 board feet.
15. Sixteen 2″ x 4″ x 12′ pieces of lumber contain
    a. 64 board feet.
    b. 96 board feet.
    c. 118 board feet.
    d. 128 board feet.

unit 3 Plywood, Hardboard, and Particleboard

OBJECTIVES

After studying this unit, the student will be able to:

- describe the kinds, sizes, and grades of plywood, hardboard, and particleboard.
- explain how plywood, hardboard, and particleboard are made.
- discuss the uses of plywood, hardboard, and particleboard.

PLYWOOD

Plywood is a sandwich of thin layers or *plies* of wood. The plies are glued together with the grain of each ply at right angle to the next one, figure 3-1. This cross-graining results in a panel that is stronger than a solid wood board.

Plywood panels always contain an odd number of plies, usually 3, 5, or 7 layers. This insures that the grain on both sides of the plywood sheet runs in the same direction. The center layer or *core* is sometimes thicker than the outer layers. The core can be veneer, lumber, or sometimes particleboard.

Because of its construction, plywood resists shrinkage more than solid wood. It maintains its size and shape better with changes in humidity. It is less likely to warp and does not split or check like solid wood. Plywood comes in large sheets which may make it easier to work with than smaller pieces of solid wood which would require gluing.

MANUFACTURING PLYWOOD

To make plywood, thin sheets of material, called *veneers,* must be cut from the log. First the outer bark is removed by a barker lathe. Then the log is soaked in a hot water vat or steamed to make cutting easier. Veneers are then cut on either a rotary lathe or a veneer slicer.

Rotary Lathe

Most softwoods are cut by the *rotary lathe* method. The logs are mounted on a huge rotary lathe and rotated against a sharp knife, figure 3-2. As the log turns, the knife peels off a continuous layer like unwinding paper from a roll, figure 3-3. The cut follows the log's annular ring. This produces a bold grain marking.

The lathe continues to cut until the entire log is used. The small core spindles and limited trimmings that are left are used for making hardboard and paper pulp.

The long ribbon of veneer is automatically cut to desired lengths, sorted, and dried to a moisture content of 5 percent. After drying, the sheets are fed through glue spreaders and coated with adhesive. The crossbands and faces are then glued and bonded to the core and assembled to make panels of 3, 5, or 7 plies, figure 3-4.

Large presses bond the assembly under controlled heat and pressure, figure 3-5. After they are trimmed and sanded to exact size, the panels are inspected and stamped to certify their grade and type.

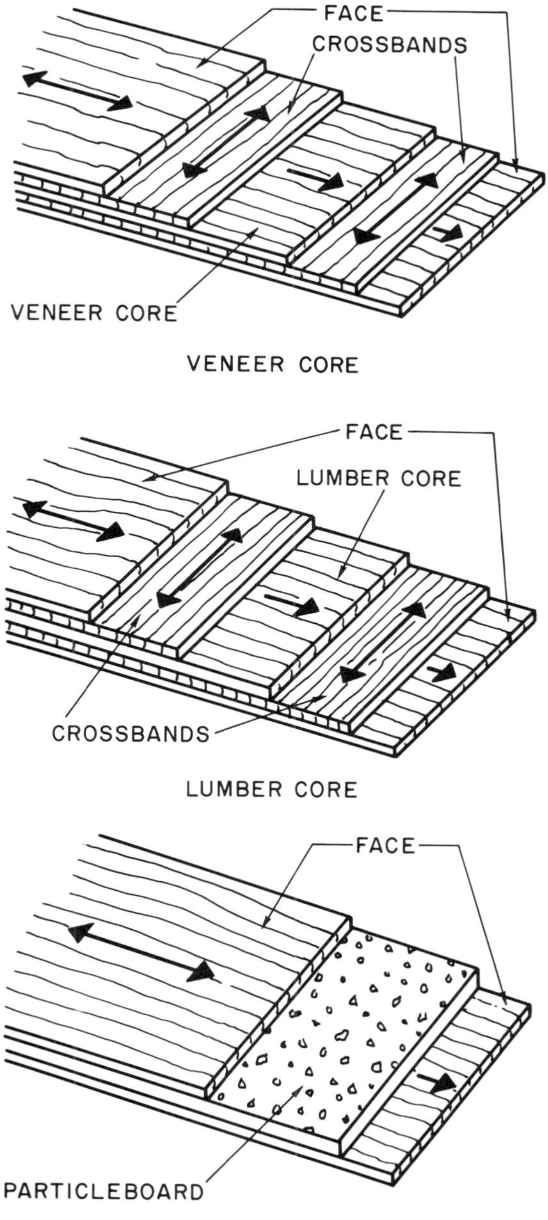

Fig. 3-1 Plywood construction

Veneer Slicer

The *veneer slicer* is used for cutting face veneers, figure 3-6. Hardwood plywoods are usually manufactured in this way. In this method, the log is moved down against a razor-sharp knife that slices off the veneer. The two most common ways of cutting face veneers on the veneer slicer are flat slicing and quarter slicing, figure 3-7.

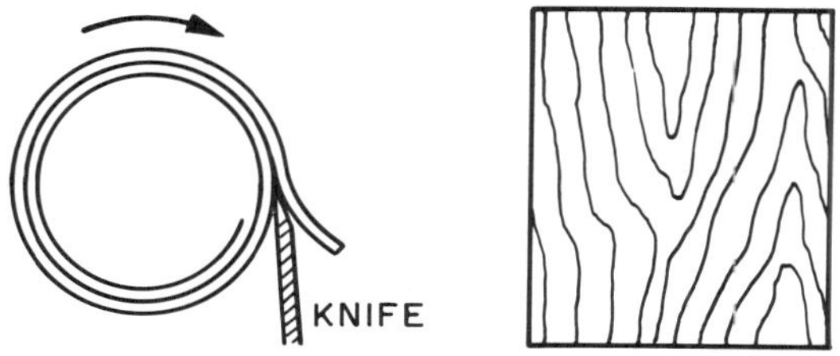

Fig. 3-2 Cutting veneer on a rotary lathe

Fig. 3-3 Peeling veneer from a log on a rotary lathe *(American Plywood Assn.)*

Fig. 3-4 Gluing and assembling plywood veneers into panels *(American Plywood Assn.)*

In *flat slicing,* the log is first sawed to a six-sided hexagon and then split in half. Each half is called a *flitch.* The flitch is held firmly in the slicer clamps and moved up and down against a sharp blade. The veneer is sliced off on the downstroke. Flat slicing produces veneers with a close-grained pattern at the edges and wider grain pattern towards the center.

Quarter slicing requires that the hexagonal log be cut in quarters rather than halves. This produces a more desirable wavy-striped pattern since the cut is parallel to the meduallary rays. Quarter slicing is done the same way as flat slicing, but it is more costly.

Because face veneers are cut thin, adjacent sheets contain similar grain markings. These markings are matched in large sheets to make interesting designs. The pile of veneers may be opened up like a book and taped together edge to edge for a *book-match* design. An *end match* is similar except it is taped end to end. Other matches are the *slip, herringbone, diamond,* and *reverse diamond,* figure 3-8.

After the veneers are matched, the crossbands and faces are bonded to the core. The American standard size for face veneers is 1/28 inch. The thickness may vary for certain types of work.

Core Construction

Plywood panels may have a core of veneer, lumber, or particleboard, see figure 3-1. *Veneer-core plywood* is made by gluing together 3, 5, 7, or 9 plies of thin veneer.

Lumber-core plywood has a thick, solid wood core that is made by edge-gluing narrow strips of

Fig. 3-5 Large presses bond the assembly under heat and pressure. *(Hardwood Plywood Manufacturers Assn.)*

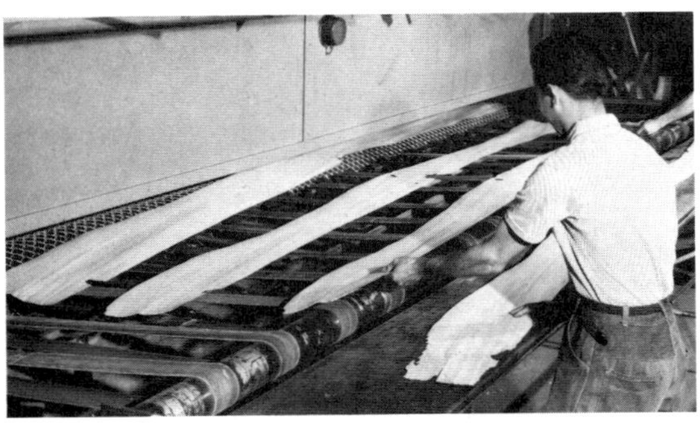

Fig. 3-6 Cutting veneers on the veneer slicer *(Hardwood Plywood Manufacturers Assn.)*

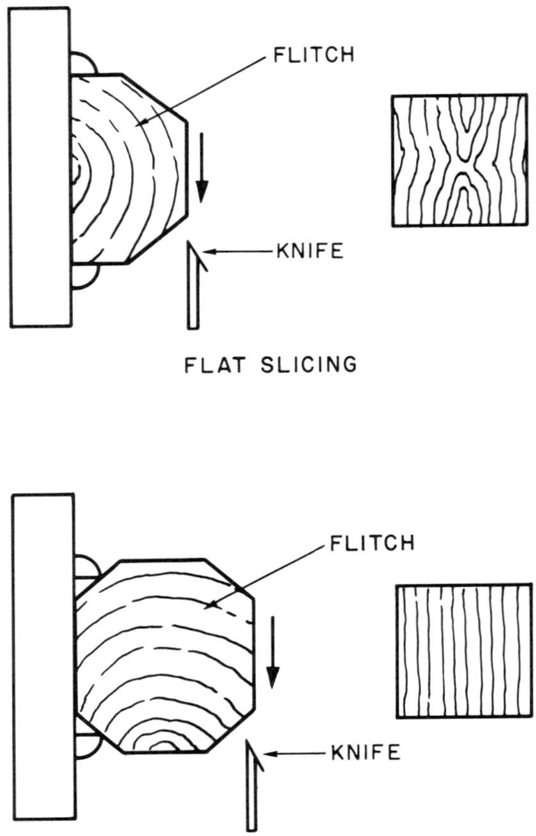

Fig. 3-7 Methods of cutting on the veneer slicer

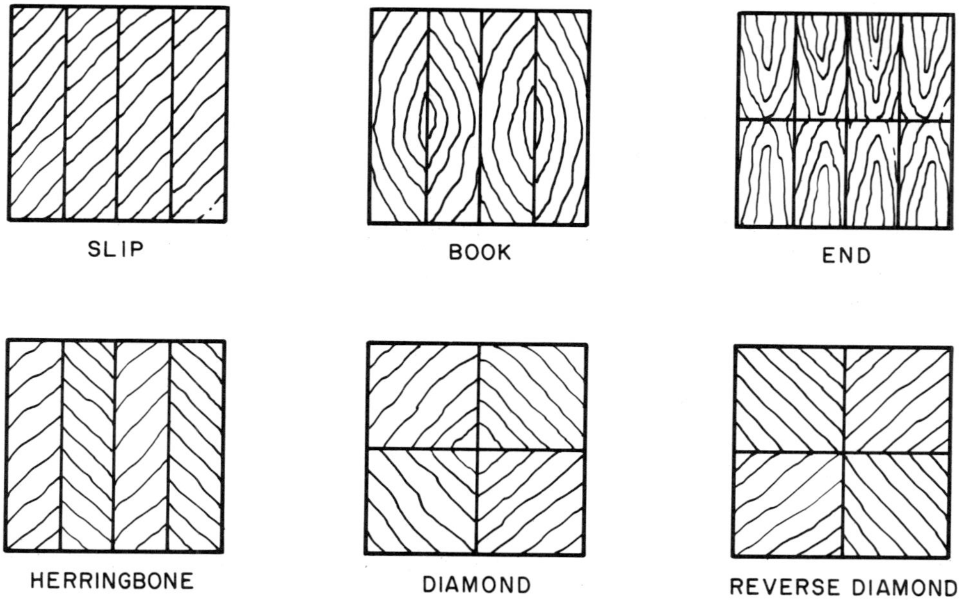

Fig. 3-8 Veneer matching

quality, kiln-dried lumber together. Inexpensive poplar or basswood is often used. The lumber core is then planed and trimmed to desired size. The crossbanding veneers in lumber-core plywood are cut slightly larger than needed.

Another type of panel uses a core of particleboard. *Particleboard-core plywood* gives the plywood panel maximum dimensional stability.

SOFTWOOD PLYWOOD

Most softwood plywood is made of Douglas fir, though other kinds of wood are also used. Softwood is grouped according to its strength, figure 3-9. Douglas fir, for instance, is in Group 1 which is the strongest.

Common plywood sheets measure 4′ x 8′, but greater lengths are available. Widths range from 2 to 5 feet. Regular stock thicknesses are from 1/4 to 1 1/8 inches, although special-purpose panels are available as thin as 1/8 inch.

There are two types of softwood plywood – exterior and interior. They are classified according to the type of glue used to make them. *Exterior plywood* is made with waterproof glue and retains its shape and strength when moisture is present. *Interior plywood* is not suited to construction that is constantly exposed to moisture. Both exterior and interior plywood come in several grades.

Plywood veneer is graded according to its quality. In declining order, the veneer grades are N, A, B, C plugged, C, and D, figure 3-10. Plywood panels are stamped with two letters. For instance, a sheet with AC stamped on it means that one side has an A face and the other side has a C face.

To select the right size, type, and grade of softwood plywood for a particular use, consult the chart in figure 3-11.

Special Plywoods

Overlaid panels are plywood sheets surfaced with another kind of material. Some of these have resin fiber sheets heat-fused to one or both sides. Another type is faced with hardboard. Both furnish an excellent base for paint by concealing all traces of the wood grain. These panels are ideal for exterior use, such as outdoor signs and displays.

Another type of overlay is a film of clear plastic. It has a highly protective finish that still lets the beautiful grain of the natural wood show through.

Group 1	Group 2	Group 3	Group 4	Group 5
Apitong[a, b]	Cedar, Port Orford	Alder, Red	Aspen	Basswood
Beech, American	Cypress	Birch, Paper	Bigtooth	Fir, Balsam
Birch	Douglas Fir 2[c]	Cedar, Alaska	Quaking	Poplar, Balsam
Sweet	Fir	Fir, Subalpine	Cativo	
Yellow	California Red	Hemlock, Eastern	Cedar	
Douglas Fir 1[c]	Grand	Maple, Bigleaf	Incense	
Kapur[a]	Noble	Pine	Western Red	
Keruing[a, b]	Pacific Silver	Jack	Cottonwood	
Larch, Western	White	Lodgepole	Eastern	
Maple, Sugar	Hemlock, Western	Ponderosa	Black (Western Poplar)	
Pine	Lauan	Spruce	Pine	
Caribbean	Almon	Redwood	Eastern White	
Ocote	Bagtikan	Spruce	Sugar	
Pine, Southern	Mayapis	Black		
Loblolly	Red Lauan	Engelmann		
Longleaf	Tangile	White		
Shortleaf	White Lauan			
Slash	Maple, Black			
Tanoak	Mengkulang[a]			
	Meranti, Red[a, d]			
	Mersawa[a]			
	Pine			
	Pond			
	Red			
	Virginia			
	Western White			
	Spruce			
	Red			
	Sitka			
	Sweetgum			
	Tamarack			
	Yellow-poplar			

a. *Represents trade group of closely related species.*

b. *Species of same genus collectively: Apitong if originating in Philippines; Keruing if in Malasia or Indonesia.*

c. *Douglar fir 1 grown in WA, OR, CA, ID, MT, WY, and Alberta and British Columbia, Canada. Douglas fir 2 grown in NV, UT, CO, AZ and NM.*

d. *Limited to species having specific gravity of 0.41 or more based on green volume and oven dry weight.*

Fig. 3-9 Species group classifications *(American Plywood Assn.)*

N Smooth surface "natural finish" veneer. Select, all heartwood or all sapwood. Free of open defects. Allows not more than 6 repairs, wood only, per 4 x 8 panel, made parallel to grain and well matched for grain and color.

A Smooth, paintable. Not more than 18 neatly made repairs, boat, sled, or router type, and parallel to grain, permitted. May be used for natural finish in less demanding applications.

B Solid surface. Shims, circular repair plugs and tight knots to 1 inch across grain permitted. Some minor splits permitted.

C *Plugged* Improved C veneer with splits limited to 1/8 inch width and knotholes and borer holes limited to 1/4 x 1/2 inch. Admits some broken grain. Synthetic repairs permitted.

C Tight knots to 1-1/2 inch. Knotholes to 1 inch across grain and some to 1-1/2 inch if total width of knots and knotholes is within specified limits. Synthetic or wood repairs. Discoloration and sanding defects that do not impair strength permitted. Limited splits allowed.

D Knots and knotholes to 2-1/2 inch width across grain and 1/2 inch larger within specified limits. Limited splits are permitted. Limited to interior grades of plywood.

Fig. 3-10 Veneer grades for softwood plywood *(American Plywood Assn.)*

SPECIFIC GRADES AND THICKNESSES MAY BE IN LOCALLY LIMITED SUPPLY. SEE YOUR DEALER BEFORE SPECIFYING.

Type	Grade Designation (2)	Description and Most Common Uses	Typical (3) Grade-trademarks	Veneer Grade: Face	Veneer Grade: Inner Plies	Veneer Grade: Back	Most Common Thicknesses (inch)					
Interior Type	N-N, N-A N-B INT-APA	Cabinet quality. For natural finish furniture, cabinet doors, built-ins, etc. Special order items.	N-N G-1 INT-APA PS 1-74 000 N-A G-2 INT-APA PS 1-74 000	N	C	N,A, or B						3/4
Interior Type	N-D-INT-APA	For natural finish paneling. Special order item.	N-D G-2 INT-APA PS 1-74 000	N	D	D	1/4					
Interior Type	A-A INT-APA	For applications with both sides on view, built-ins, cabinets, furniture, partitions. Smooth face; suitable for painting.	A-A G-1 INT-APA PS 1-74 000	A	D	A	1/4		3/8	1/2	5/8	3/4
Interior Type	A-B INT-APA	Use where appearance of one side is less important but where two solid surfaces are necessary.	A-B G-1 INT-APA PS 1-74 000	A	D	B	1/4		3/8	1/2	5/8	3/4
Interior Type	A-D INT-APA	Use where appearance of only one side is important. Paneling, built-ins, shelving, partitions, flow racks.	A-D GROUP 1 INTERIOR PS 1-74 000 APA	A	D	D	1/4		3/8	1/2	5/8	3/4
Interior Type	B-B INT-APA	Utility panel with two solid sides. Permits circular plugs.	B-B G-2 INT-APA PS 1-74 000	B	D	B	1/4		3/8	1/2	5/8	3/4
Interior Type	B-D INT-APA	Utility panel with one solid side. Good for backing, sides of built-ins, industry shelving, slip sheets, separator boards, bins.	B-D GROUP 2 INTERIOR PS 1-74 000 APA	B	D	D	1/4		3/8	1/2	5/8	3/4
Interior Type	DECORATIVE PANELS—APA	Rough-sawn, brushed, grooved, or striated faces. For paneling, interior accent walls, built-ins, counter facing, displays, exhibits.	DECORATIVE B-D G-1 INT-APA PS 1-74	C or btr.	D	D		5/16	3/8	1/2	5/8	
Interior Type	PLYRON INT-APA	Hardboard face on both sides. For counter tops, shelving, cabinet doors, flooring. Faces tempered, untempered, smooth, or screened.	PLYRON INT-APA 000		C & D					1/2	5/8	3/4
Exterior Type	A-A EXT-APA	Use where appearance of both sides is important. Fences, built-ins, signs, boats, cabinets, commercial refrigerators, shipping containers, tote boxes, tanks, ducts. (4)	A-A G-1 EXT-APA PS 1-74 000	A	C	A	1/4		3/8	1/2	5/8	3/4
Exterior Type	A-B EXT-APA	Use where the appearance of one side is less important. (4)	A-B G-1 EXT-APA PS 1-74 000	A	C	B	1/4		3/8	1/2	5/8	3/4
Exterior Type	A-C EXT-APA	Use where the appearance of only one side is important. Soffits, fences, structural uses, boxcar and truck lining, farm buildings. Tanks, trays, commercial refrigerators. (4)	A-C GROUP 1 EXTERIOR PS 1-74 000 APA	A	C	C	1/4		3/8	1/2	5/8	3/4
Exterior Type	B-B EXT-APA	Utility panel with solid faces. (4)	B-B G-2 EXT-APA PS 1-74 000	B	C	B	1/4		3/8	1/2	5/8	3/4
Exterior Type	B-C EXT-APA	Utility panel for farm service and work buildings, boxcar and truck lining, containers, tanks, agricultural equipment. Also as base for exterior coatings for walls, roofs. (4)	B-C GROUP 2 EXTERIOR PS 1-74 000 APA	B	C	C	1/4		3/8	1/2	5/8	3/4
Exterior Type	HDO EXT-APA	High Density Overlay plywood. Has a hard, semi-opaque resin-fiber overlay both faces. Abrasion resistant. For concrete forms, cabinets, counter tops, signs, tanks. (4)	HDO A-A G-1 EXT-APA PS 1-74	A or B	C or C plgd	A or B			3/8	1/2	5/8	3/4
Exterior Type	MDO EXT-APA	Medium Density Overlay with smooth, opaque, resin-fiber overlay one or both panel faces. Highly recommended for siding and other outdoor applications, built-ins, signs, displays. Ideal base for paint. (4)(6)	MDO B-B G-2 EXT-APA PS 1-74 000	B	C	B or C			3/8	1/2	5/8	3/4
Exterior Type	303 SIDING EXT-APA	Proprietary plywood products for exterior siding, fencing, etc. Special surface treatment such as V-groove, channel groove, striated, brushed, rough-sawn and texture-embossed MDO. Stud spacing (Span Index) and face grade classification indicated on grade stamp.	303 SIDING 6-S GROUP 1 24 oc SPAN EXTERIOR PS 1-74 000 APA	(5)	C	C			3/8	1/2	5/8	
Exterior Type	T 1-11 EXT-APA	Special 303 panel having grooves 1/4" deep, 3/8" wide, spaced 4" or 8" o.c. Other spacing optional. Edges shiplapped. Available unsanded, textured and MDO.	303 SIDING 6-S/W T 1-11 GROUP 2 16 oc SPAN EXTERIOR PS 1-74 000 APA	C or btr.	C	C				19/32	5/8	
Exterior Type	PLYRON EXT-APA	Hardboard faces both sides, tempered, smooth or screened.	PLYRON EXT-APA 000		C					1/2	5/8	3/4
Exterior Type	MARINE EXT-APA	Ideal for boat hulls. Made only with Douglas fir or western larch. Special solid jointed core construction. Subject to special limitations on core gaps and number of face repairs. Also available with HDO or MDO faces.	MARINE A-A EXT-APA PS 1-74 000	A or B	B	A or B	1/4		3/8	1/2	5/8	3/4

(1) Sanded both sides except where decorative or other surfaces specified.
(2) Can be manufactured in Group 1, 2, 3, 4 or 5.
(3) The species groups, Identification Indexes and Span Indexes shown in the typical grade-trademarks are examples only. See "Group," "Identification Index" and "Span Index" for explanations and availability.
(4) Can also be manufactured in Structural I (all plies limited to Group 1 species) and Structural II (all plies limited to Group 1, 2, or 3 species).
(5) C or better for 5 plies. C Plugged or better for 3 plies.
(6) Also available as a 303 siding.

Fig. 3-11 Guide for using plywood *(American Plywood Assn.)*

SPECIFIC GRADES AND THICKNESSES MAY BE IN LOCALLY LIMITED SUPPLY. SEE YOUR DEALER BEFORE SPECIFYING.

	Grade Designation	Description and Most Common Uses	Typical Grade-trademarks (1)	Veneer Grade: Face	Veneer Grade: Inner Plies	Veneer Grade: Back	Most Common Thicknesses (inch)				
Interior Type	C-D INT-APA	For wall and roof sheathing, subflooring, industrial uses such as pallets. Most commonly available with exterior glue (CDX). Specify exterior glue where construction delays are anticipated and for treated-wood foundations. (7)	C-D 32/16 APA INTERIOR PS 1-74 000 C-D 24/0 APA INTERIOR PS 1-74 000 EXTERIOR GLUE	C	D	D	5/16	3/8	1/2	5/8	3/4
Interior Type	STRUCTURAL I C-D INT-APA and STRUCTURAL II C-D INT-APA	Unsanded structural grades where plywood strength properties are of maximum importance: structural diaphragms, box beams, gusset plates, stressed-skin panels, containers, pallet bins. Made only with exterior glue. See (6) for species group requirements. Structural I more commonly available. (7)	STRUCTURAL I C-D 24/0 APA INTERIOR PS 1-74 000 EXTERIOR GLUE	C(3)	D(3)	D(3)	5/16	3/8	1/2	5/8	3/4
Interior Type	STURD-I-FLOOR INT-APA	For combination subfloor-underlayment. Provides smooth surface for application of resilient floor covering. Possesses high concentrated- and impact-load resistance during construction and occupancy. Manufactured with exterior glue only. Touch-sanded. Available square edge or tongue-and-groove. (7)	STURD-I-FLOOR 24oc T&G 23/32 INCH INTERIOR APA 000 EXTERIOR GLUE NRB-108	C Plugged	(4)	D				19/32 5/8	23/32 3/4
Interior Type	STURD-I-FLOOR 48 O.C. (2-4-1) INT-APA	For combination subfloor-underlayment on 32- and 48-inch spans. Provides smooth surface for application of resilient floor coverings. Possesses high concentrated- and impact-load resistance during construction and occupancy. Manufactured with exterior glue only. Unsanded or touch-sanded. Available square edge or tongue-and-groove. (7)	STURD-I-FLOOR 48oc 2-4-1 T&G 1 1/8 INCH INTERIOR APA 000 EXTERIOR GLUE NRB-108	C Plugged	C(5) & D	D	1-1/8				
Interior Type	UNDERLAYMENT INT-APA	For application over structural subfloor. Provides smooth surface for application of resilient floor coverings. Touch-sanded. Also available with exterior glue. (2)(6)	UNDERLAYMENT GROUP 1 APA INTERIOR PS 1-74 000	Plugged C	C(5) & D	D		3/8	1/2	19/32 5/8	23/32 3/4
Interior Type	C-D PLUGGED INT-APA	For built-ins, wall and ceiling tile backing, cable reels, walkways, separator boards. Not a substitute for Underlayment or Sturd-I-Floor as it lacks their indentation resistance. Touch-sanded. Also made with exterior glue. (2) (6)	C-D PLUGGED GROUP 2 APA INTERIOR PS 1-74 000	Plugged C	D	D		3/8	1/2	19/32 5/8	23/32 3/4
Exterior Type	C-C EXT-APA	Unsanded grade with waterproof bond for subflooring and roof decking, siding on service and farm buildings, crating, pallets, pallet bins, cable reels, treated-wood foundations. (7)	C-C 42/20 APA EXTERIOR PS 1-74 000	C	C	C	5/16	3/8	1/2	5/8	3/4
Exterior Type	STRUCTURAL I C-C EXT-APA and STRUCTURAL II C-C EXT-APA	For engineered applications in construction and industry where full Exterior type panels are required. Unsanded. See (6) for species group requirements. (7)	STRUCTURAL I C-C 32/16 APA EXTERIOR PS 1-74 000	C	C	C	5/16	3/8	1/2	5/8	3/4
Exterior Type	STURD-I-FLOOR EXT-APA	For combination subfloor-underlayment under resilient floor coverings where severe moisture conditions may be present, as in balcony decks. Possesses high concentrated-and impact-load resistance during construction and occupancy. Touch-sanded. Available square edge or tongue-and-groove. (7)	STURD-I-FLOOR 20oc 5/8 INCH EXTERIOR APA 000 NRB-108	C Plugged	C(5)	C				19/32 5/8	23/32 3/4
Exterior Type	UNDERLAYMENT C-C PLUGGED EXT-APA	For application over structural subfloor. Provides smooth surface for application of resilient floor coverings where severe moisture conditions may be present. Touch-sanded. (2)(6)	UNDERLAYMENT C-C PLUGGED GROUP 2 APA EXTERIOR PS 1-74 000	Plugged C	C(5)	C		3/8	1/2	19/32 5/8	23/32 3/4
Exterior Type	C-C PLUGGED EXT-APA	For use as tile backing where severe moisture conditions exist. For refrigerated or controlled atmosphere rooms, pallet fruit bins, tanks, box car and truck floors and linings, open soffits. Touch-sanded. (2)(6)	C-C PLUGGED GROUP 2 APA EXTERIOR PS 1-74 000	Plugged C	C	C		3/8	1/2	19/32 5/8	23/32 3/4
Exterior Type	B-B PLYFORM CLASS I & CLASS II EXT-APA	Concrete form grades with high reuse factor. Sanded both sides. Mill-oiled unless otherwise specified. Special restrictions on species. Available in HDO and Structural I. Class I most commonly available. (8)	B-B PLYFORM CLASS I APA EXTERIOR PS 1-74 000	B	C	B				5/8	3/4

(1) The species groups, Identification Indexes and Span Indexes shown in the typical grade-trademarks are examples only. See "Group," "Identification Index" and "Span Index" for explanations and availability.
(2) Can be manufactured in Group 1, 2, 3, 4, or 5.
(3) Special improved grade for structural panels.
(4) Special veneer construction to resist indentation from concentrated loads, or other solid wood-base materials.
(5) Special construction to resist indentation from concentrated loads.
(6) Can also be manufactured in Structural I (all plies limited to Group 1 species) and Structural II (all plies limited to Group 1, 2, or 3 species).
(7) Specify by Identification Index for sheathing and Span Index for Sturd-I-Floor panels.
(8) Made only from certain wood species to conform to APA specifications.

Fig. 3-11 Guide for using plywood (Cont'd.) *(American Plywood Assn.)*

PREMIUM GRADE #1	Select, matched face veneers with little contrasting color
GOOD GRADE #1	Avoids sharp contrasts in color and grain.
SOUND GRADE #2	Allows imperfect matching of color and grain. Does not have open defects.
UTILITY GRADE #3	Allows mismatching of veneers, tight knots, streaks. May show slight splits.
BACKING GRADE #4	Allows larger defects that do not impair the strength of the panel.
SPECIALTY GRADE SP	Custom-made of select, matched veneers for special needs.

Fig. 3-12 Veneer grades for hardwood plywoods

Plywood faces can also be rough-sawn, grooved, brushed, or striated (combed surface) to create a special effect.

HARDWOOD PLYWOOD

Hardwood plywood is ideal for making fine furniture. Many kinds of beautifully-grained hardwoods are used for face veneers. Exotic and rare woods, both domestic and foreign, are matched to produce interesting face designs.

Hardwood plywood is available in 2 to 4-foot widths. Four-foot widths are the most common. Lengths range from 3 to 12 feet. Wider and longer panels are also available on special orders. Thicknesses generally range from 1/8 to 3/4 inch.

There are four types of interior and exterior hardwood plywood. *Type I* is waterproof. *Type II* is water-resistant and made for interior use only. *Type III* is only moisture-resistant and can be damaged by water. It will resist indoor dampness however. The fourth type, called *Technical Type I*, is waterproof but differs in construction from Type I.

Hardwood plywood is graded according to the quality of the plies. In descending order, there are six grades: *premium, good, sound, utility, backing,* and *specialty*, figure 3-12.

Fig. 3-13 Handsaw plywood with the face side up. *(American Plywood Assn.)*

WORKING WITH PLYWOOD

The woodworking tools used to work solid lumber are also used on plywood. However, precautions must be taken to protect the thin face veneers.

Sanding must be done carefully to avoid cutting through the veneer. This may expose the glue or the inner ply whose grain runs at right angles to the face veneer. When sawing, care must be taken to keep the thin veneer from splintering, figure 3-13. Using a fine-tooth saw and sawing with the face side up will help avoid this. Planing must be done with shallow cuts to avoid splintering out the inner plies.

Because of the construction of plywood, the end grain of the inner plies are exposed at the edges. For a pleasing appearance, these edges must be covered. A number of ways to finish plywood edges are discussed in *Unit 24 Cabinet Doors.*

HARDBOARD

The woodworking industries uses every bit of wood from a log. Wood chips and board trimming, once considered waste, are now used to make hardboard.

Hardboard consists of refined wood chips that are manufactured into sturdy boards and panels. It is grainless and has a uniform thickness, density, and appearance. It resists marring, scuffing, and abrasion.

Hardboard can be used wherever a dense, hard panel is needed. It is widely used for cabinet backs and drawer bottoms. It is also widely used for displays, signs, toys, games, sports equipments, and in the construction, automotive, and transportation industries.

MANUFACTURING HARDBOARD

Hardboard is made from wood chips that are converted to fibers. These wood fibers are coated with adhesives and permanently bonded under heat and pressure to form panels. The makeup of the fibers determine the properties of different hardboards.

Defibering and Refining

To manufacture hardboard, wood is first reduced to uniform wood chips in a chipper. Next the wood chips are reduced to fibers by either a mechanical or explosion method.

In the *mechanical process,* the chips are first steamed under pressure in a *digester.* This softens the chips so they can be reduced by friction to unbroken, flexible fibers. Water is introduced at this stage to form a soupy pulp.

In the *explosion method* of defibering, chips are loaded into a high-pressure cylinder. Steam pressure is applied gradually and then suddenly released. The resulting explosion produces a mass of wood fibers. If a wet process is used, water is then added.

After defibering, the soupy pulp is screened for size and undersirable residues. Small amounts of chemicals may be added during refining to obtain desired properties.

Forming Hardboard

Forming hardboard is done by either a wet or dry process. In the *wet process,* the soupy pulp flows onto a traveling mesh screen. Water is drawn off through the screen and also by pressing the mat through a series of press rolls.

In the *dry process,* the fibers are laid out in a mat using air as the medium instead of water. Air-formed mats are thicker and softer than wet-formed ones and require more care in loading into the presses.

Pressing

From this point on, wet and dry production is virtually the same. The mats are pressed under heat (380-550 degrees Fahrenheit) and pressure (500-1500 psi). This combination of heat and pressure welds the wood fibers back together and produces properties not found in natural wood. All panels emerge from the hot presses with an extremely low moisture content.

Some panels are *tempered* next. They are coated with oil and baked to increase the hardboard's hardness, strength, and water resistance.

Humidifying

To prevent warping, the boards are sent to a humidifier. This raises the moisture content to approximately atmospheric humidity. Although hardboards are humidified, they must adjust to local atmospheric conditions before being installed.

Trimming

Carbide-tipped saws trim the panels to standard sizes. Sheets are cut to practically any size a customer wants. Hardboard is also fabricated and finished in a variety of ways. After final inspection, the sheets are wrapped and shipped out.

CLASSES OF HARDBOARD

Hardboard is classified, in descending order, as *tempered, standard, service-tempered, service,* and *industrialite,* figure 3-14. Tempered hardboard is processed with special additives to give it more stiffness, strength, hardness, and resistance to water and abrasion as compared to standard hardboard. Each class diminishes in the quality of these properties.

Hardboard also comes in smooth-one-side (S1S) or smooth-two-sides (S2S). This refers to how many sides of the panel have a smooth surface.

Hardboard is perforated for decorative effects, figure 3-15. It is sometimes laminated to obtain greater thicknesses. It is also available with primed or coated surfaces or with decorative overlays.

WORKING WITH HARDBOARD

Because it is a wood-base product, hardboard can be sawed, routed, shaped, and drilled with standard woodworking tools. It can be securely glued or fastened with screws, staples, or nails. Hardboard can also be laminated with veneers, plastic films, and laminates. Depending on the type of board and the radius of the curve desired, hardboard can also be bent. Stains, sealers, shellac, lacquer, and water or oil-base paints may be applied to it by brush, roller, or spray gun.

PARTICLEBOARD

Particleboard is a sheet material made of wood flakes, chips, sawdust, and planer shavings, figure 3-16. These wood particles are mixed with an adhesive, formed in a mat, and pressed into sheet form. The kind, size, and arrangement of these particles help determine the quality of the board.

Class	Surface	Nominal thickness	Water resistance (max av per panel): Water absorption based on weight, S1S	Water absorption based on weight, S2S	Thickness swelling, S1S	Thickness swelling, S2S	Modulus of rupture (min av per panel)	Tensile strength (min av per panel): Parallel to surface	Perpendicular to surface
		inch	percent	percent	percent	percent	psi	psi	psi
1 Tempered	S1S	1/12	30	—	25	—	7000	3500	150
	S1S and S2S	1/10	20	25	16	20			
		1/8	15	20	11	16			
		3/16	12	18	10	15			
		1/4	10	12	8	11			
		5/16	8	11	8	10			
		3/8	8	10	8	9			
2 Standard	S1S and S2S	1/12	40	40	30	30	5000	2500	100
		1/10	25	30	22	25			
		1/8	20	25	16	18			
		3/16	18	25	14	18			
		1/4	16	20	12	14			
		5/16	14	15	10	12			
		3/8	12	12	10	10			
3 Service-tempered	S1S and S2S	1/8	20	25	15	22	4500	2000	100
		3/16	18	20	13	18			
		1/4	15	20	13	14			
		3/8	14	18	11	14			
4 Service	S1S and S2S	1/8	30	30	25	25	3000	1500	75
		3/16	25	27	15	22			
		1/4	25	27	15	22			
		3/8	25	27	15	22			
		7/16	25	27	15	22			
		1/2	25	18	15	14			
	S2S	5/8	—	15	—	12			
		11/16	—	15	—	12			
		3/4	—	12	—	9			
		13/16	—	12	—	9			
		7/8	—	12	—	9			
		1	—	12	—	9			
		1-1/8	—	12	—	9			
5 Industrialite	S1S and S2S	3/8	25	25	20	20	2000	1000	35
		7/16	25	25	20	20			
		1/2	25	25	20	20			
	S2S	5/8	—	22	—	18			
		11/16	—	22	—	18			
		3/4	—	20	—	16			
		13/16	—	20	—	16			
		7/8	—	20	—	16			
		1	—	20	—	16			
		1-1/8	—	20	—	16			

Fig. 3-14 Classes of hardboard *(Masonite Corp.)*

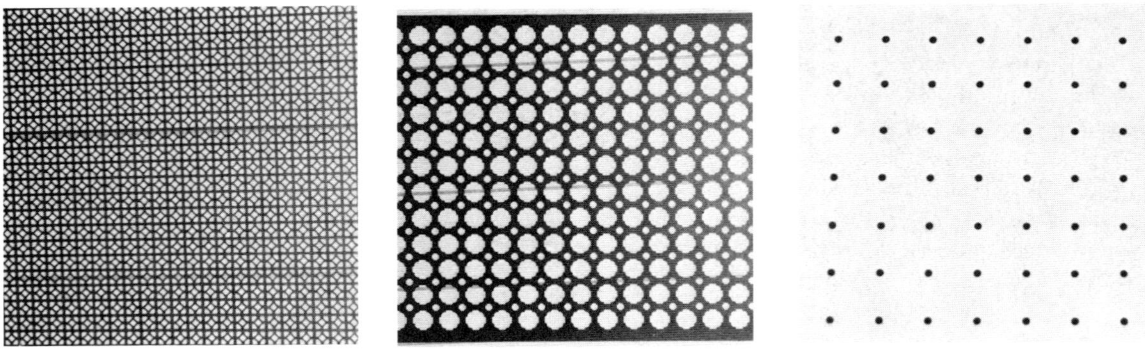

Fig. 3-15 Types of perforated hardboard

Particleboard is widely used in the furniture industry as a core material for plastic laminates and wood veneers, figure 3-17. Because it has no grain, it is uniformly strong in every direction, retains its shape, and will not warp. It is used to make toys and games and upholstered furniture. Particleboard is also used in the construction of kitchen cabinets and countertops. Lower qualities are used as underlayment for floors and for cores of veneered doors, figure 3-18.

MANUFACTURING PARTICLEBOARD

The quality of particleboard is also determined by the manufacturing method. In the most popular method, called the *Bison method,* the wood is first turned into flakes on special machines, then dried to a carefully controlled moisture content. An air separator removes metal or stone from the dried flakes. An air classifier sifts out the particles and sorts them into fine, medium, and coarse bins.

The selected particles are blended with a *resin* (adhesive) and formed into mats. Fine particles are laid on both outer surfaces of the mat, with coarser particles in the center. The mat is then run through heated presses and formed into boards.

The highest quality particleboard has larger wood flakes in the center and finer particles near the surfaces. This results in an extremely hard board with a very smooth surface. Softer and lower quality boards contain the same size particles throughout. These boards have a rougher surface texture.

Fig. 3-16 Particleboard is made of wood flakes, shavings, and sawdust. *(Duraflake Div., Willamette Industries Inc.)*

Besides wood, particleboard is also made commercially of such things as cotton stalks, flax, coconut shells, and cacao bean husks. A combination of 80 percent municipal trash and 20 percent wood may result in a particleboard with commercial potential that uses some of society's waste.

GRADES AND SIZES OF PARTICLEBOARD

The quality of particleboard is indicated by its hardness (density) that ranges from 28 to 55 pounds. Sheets are available in 4 1/2′ x 18′, 5′ x 16′ and 4′ x 24′ sizes. Smaller sizes are available with 4′ x 8′ being the most common. Thicknesses range from ¼ to 2 inches. In addition to the plain surface, particleboard is available in colors, primed, filled, wood-banded, or can be custom made for special needs.

Fig. 3-17 Particleboard is an excellent core material for plastic laminated furniture. *(Duraflake Div., Willamette Industries, Inc.)*

Fig. 3-18 Particleboard is used as floor underlayment. *(Duraflake Div., Willamette Industries, Inc.)*

WORKING WITH PARTICLEBOARD

Because particleboard is a wood product, it can be cut with the same woodworking tools used to cut solid wood. However, ordinary tools will dull quickly when used to cut particleboard. Carbide-tipped cutting tools are the best to use. Particleboard edges may also be filled and painted, covered with veneer strips, plastic laminates, or wood bands.

ACTIVITIES

1. Submerge pieces of exterior and interior plywood in water to compare their resistance to moisture.
2. Submerge pieces of hardboard and particleboard in water to compare their resistance to moisture.
3. Visit lumber yards and distributors and examine their stock of plywood, hardboard, and particleboard for kinds, sizes, and grades.
4. Make a display identifying different types of plywood, hardboard, and particleboard.

UNIT REVIEW

Multiple Choice

1. Plywood always contains
   a. an even number of layers.
   b. four layers.
   c. three layers.
   d. an odd number of layers.

2. The best looking face veneer of a softwood plywood panel is indicated by which grade classification?
   a. A
   b. B
   c. E
   d. N

3. Most softwood plywood is made of
   a. pine.
   b. spruce.
   c. fir.
   d. cedar.

4. Hardwood plywood face veneers are usually cut on a
   a. rotary lathe.
   b. large bandsaw.
   c. veneer slicer.
   d. flitch.

5. The standard thickness for hardboard plywood face veneers is
   a. 1/4 inch.
   b. 1/8 inch.
   c. 1/16 inch.
   d. 1/28 inch.

6. When sanding plywood faces,
   a. always use coarse sandpaper.
   b. never use a power sander.
   c. avoid sanding through the face veneer.
   d. always sand across the grain.

7. Most softwood plywood veneers are cut on a
   a. rotary lathe.
   b. large bandsaw.
   c. veneer slicer.
   d. flitch.

8. To increase the strength of some hardboard panels, they are
   a. pressed tighter.
   b. tempered.
   c. flitched.
   d. defibered.

9. The quality of particleboard is indicated by its
   a. appearance.
   b. strength.
   c. density.
   d. kind of wood.

10. Particleboard is widely used for
   a. outdoor furniture.
   b. cores for plastic laminates.
   c. boat decks and hulls.
   d. shipping crates.

Questions

1. How are the plies assembled to make plywood sheets?
2. What are some of the advantages of using plywood instead of solid wood?
3. What are the three types of plywood core construction?
4. What does it mean when a softwood plywood panel is stamped *AC?*
5. What is the difference between interior and exterior plywood?
6. What are some uses of hardboard?
7. Why must hardboard be humidified?
8. How does hardboard differ from particleboard?
9. How are the particles arranged in high-quality particleboard?
10. What are some of the characteristics of particleboard?

section 2

Woodworking Tools

unit 4 Hand Tools

OBJECTIVES

After studying this unit, the student will be able to:

- identify and describe commonly used hand tools.
- use hand tools properly and safely.
- recondition cutting tools and handsaws.

HAND TOOLS

Hand tools are very important in the woodworking trades. Though power tools are widely used, there are still many occasions when hand tools are the better choice for a job.

Hand tools may be easier and faster to use. They are often easier to control, reducing the chance of injury to the operator or damage to the workpiece. Certain jobs, such as finishing, often require the finer applications of hand tools. The successful woodworker must know how to choose the proper hand tool, use it skillfully, and keep it in good condition.

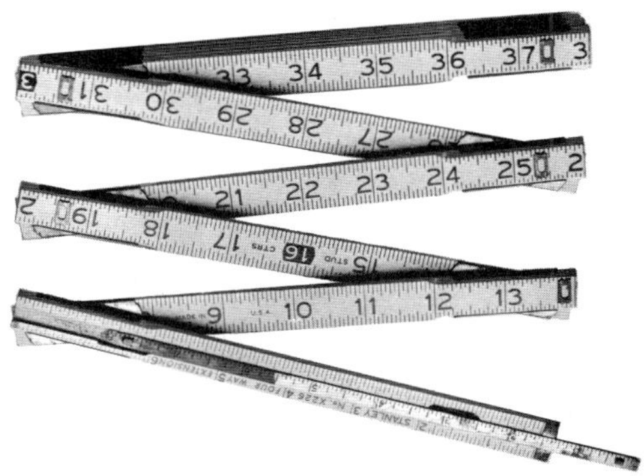

Fig. 4-1 Folding rule

LAYOUT TOOLS

A *rule* is used for measuring actual units of length. Rules used in woodworking are divided into feet, inches, and usually sixteenths of an inch. The ability to read a rule quickly and accurately is important.

The most commonly used rules are the *folding rule,* figure 4-1, and the *pocket tape,* figure 4-2. The folding rule sometimes has a metal extension on one end for taking inside measurements. Pocket

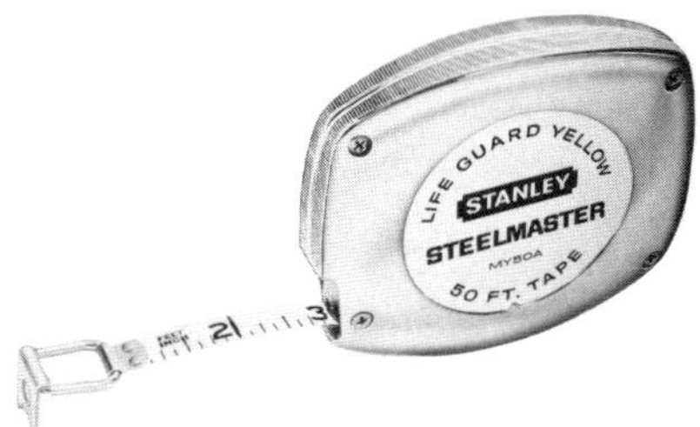

Fig. 4-2 Steel tape *(The Stanley Works)*

Fig. 4-3 Marking gauge *(The Stanley Works)*

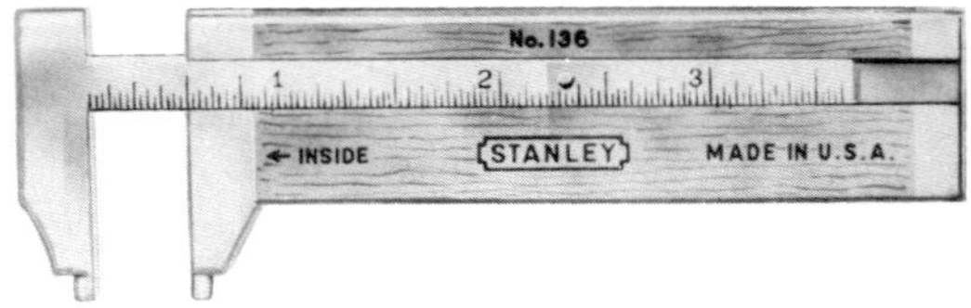

Fig. 4-4 Caliper rule *(The Stanley Works)*

tapes are available in lengths of 6, 8, 10, and 12 feet or longer. They have an adjustable clip on the end for taking outside and inside measurements. Steel tapes in lengths of 50 and 100 feet are also commonly used.

The *marking gauge,* figure 4-3, is used for laying out lines parallel to an edge. The face of the gauge is placed against the edge with the pin resting on the stock. As the gauge is moved along the stick, the pin scribes a line parallel to the edge.

The *caliper rule,* figure 4-4, is used for gauging the outside and inside measurements of turnings and other round objects.

The *steel square,* figure 4-5, is used to square across stock, to check the straightness of a line, and to lay out a 45-degree or 90-degree angle. The steel square is often used in carpentry to lay out rafters and stairs.

The *combination square,* figure 4-6, is used to lay out and check 90 and 45 degree angles on smaller pieces of stock. It is also used as a marking gauge to draw lines parallel to the edge of a board or as a depth gauge to test the depth of mortises, grooves, and dadoes.

The *sliding T bevel,* figure 4-7, often called a *bevel square,* is used to lay out, transfer, and check any desired angle.

Wing dividers, figure 4-8, are used to lay out circles, to divide a distance into equal spaces, and to scribe or mark irregular lines.

Trammel points, figure 4-9, can be attached to a strip of wood and adjusted to lay out circles with diameters too large for the ordinary compass.

TOOTH-CUTTING TOOLS

Handsaws used in woodworking refer to either a crosscut saw or a ripsaw, figure 4-10. The *crosscut saw* is used for cutting across the grain.

Fig. 4-5 Steel square *(Millers Falls)*

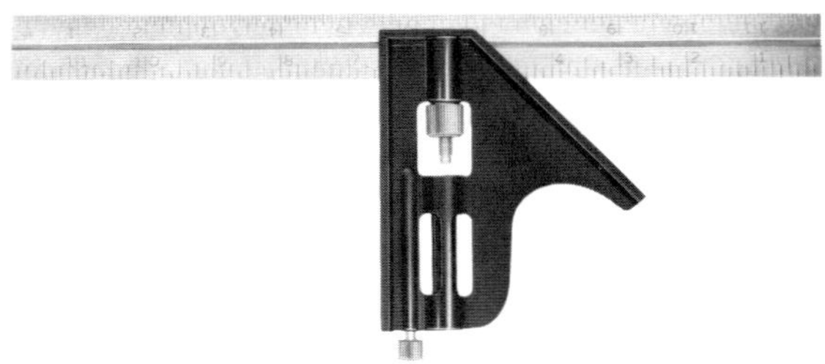

Fig. 4-6 Combination square *(The Stanley Works)*

Fig. 4-7 Sliding T bevel

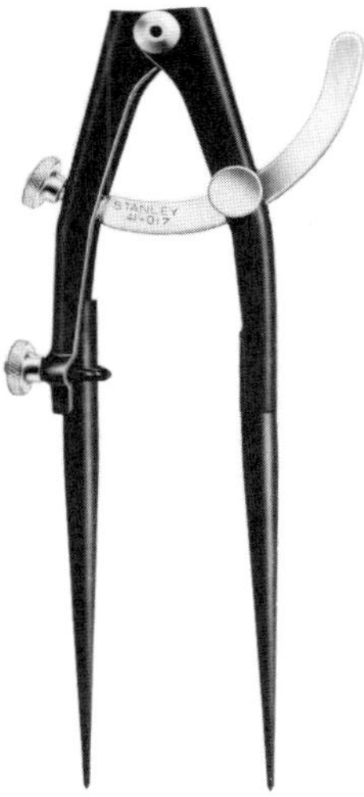

◁
Fig. 4-8 Wing dividers *(The Stanley Works)*

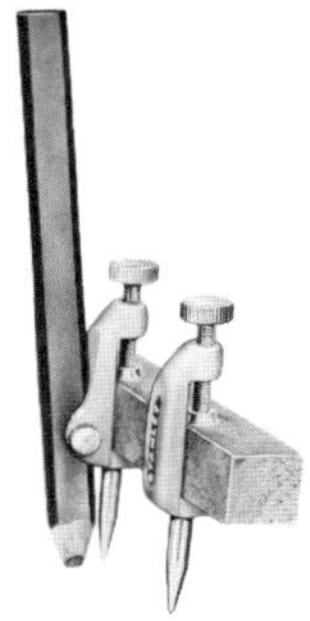

△
Fig. 4-9 Trammel points *(The Stanley Works)*

△
Fig. 4-10 Handsaws *(The Stanley Works) (Millers Falls)*

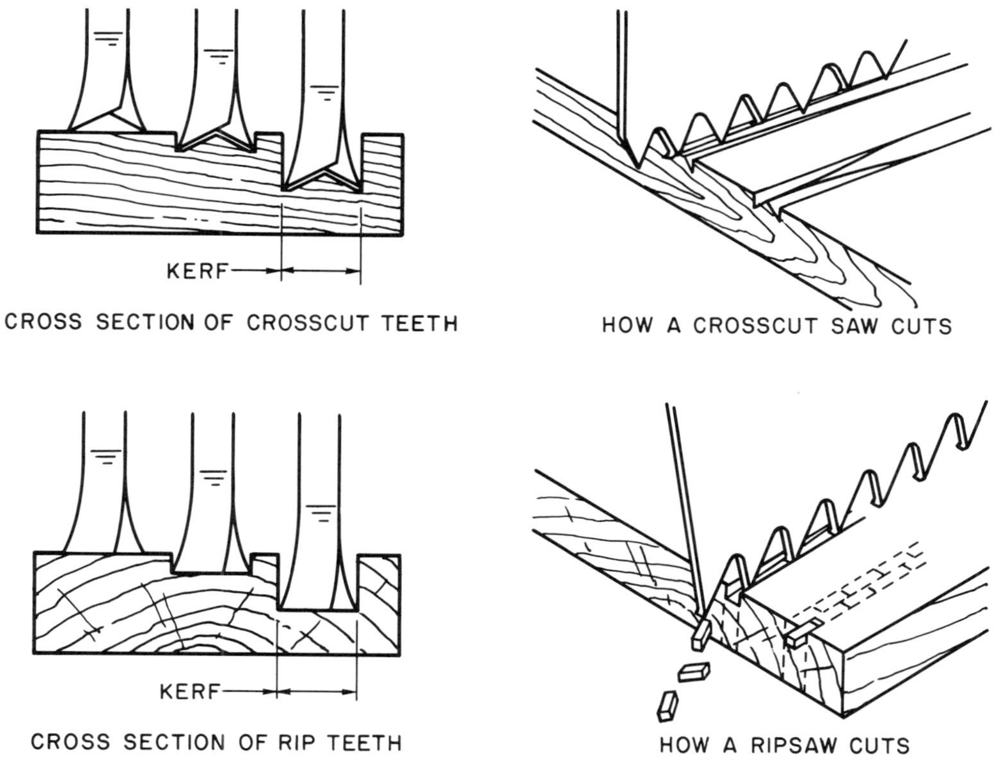

Fig. 4-11 Cutting action of handsaws *(Disston Inc.)*

The *ripsaw* cuts with the grain. The shape of the teeth determines their cutting action, figure 4-11.

Handsaws are designated by the number of points of teeth per inch. Common sizes for crosscut saws are 7, 8, 10, 11, and 12 points. The coarser saws are used for rough work while those with more points are used for finish work. A 10 to 12-point crosscut saw, for instance, is used to cut the thin face veneer of plywood to keep it from splintering. Ripsaws usually have 5 1/2 to 7 points per inch. The number of points is usually stamped near the heel of the saw. The standard length of handsaws from heel to toe is 26 inches.

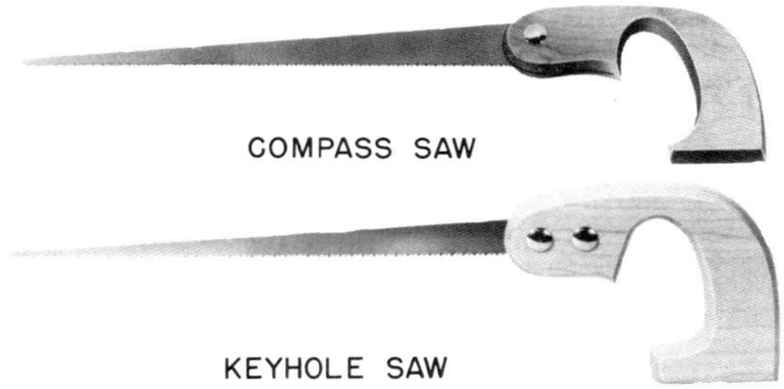

Fig. 4-12 Compass and keyhole saws *(Disston Inc.)*

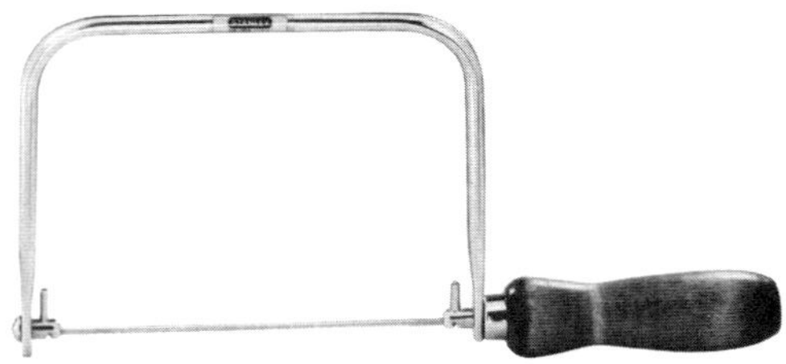

Fig. 4-13 Coping saw *(The Stanley Works)*

Fig. 4-14 Backsaw *(The Stanley Works)*

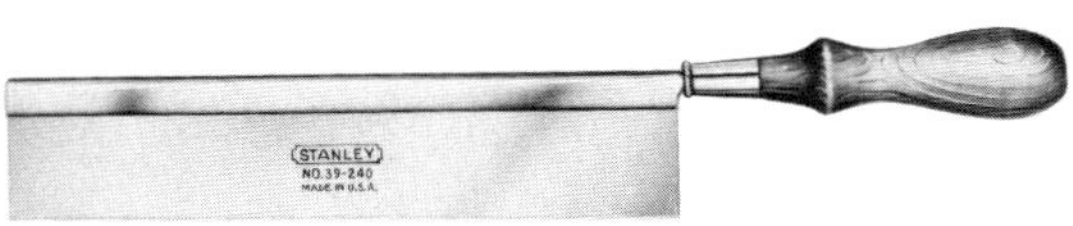

Fig. 4-15 Dovetail saw *(The Stanley Works)*

Compass and *keyhole saws,* figure 4-12, have tapered blades with toes that come almost to a point. The keyhole saw blade is narrower and has fine teeth. These saws are used to cut out circles or curved work. The blades vary from 10 to 14 inches in length. The teeth are shaped so that the saws may be used for crosscutting or ripping.

The *coping saw,* figure 4-13, has a U-shaped steel frame, a very fine removable blade, and usually a wood handle which, when turned, tightens the blade. The coping saw is used for cutting the ends of molding to make coped joints and for other small curved cutting. The blade is inserted in the frame with the teeth pointing away from the handle. Cuts are made as the saw is pushed, not pulled.

The *backsaw,* figure 4-14, is a crosscut saw with very fine teeth, usually 10 to 16 points per inch. The blade has a reinforced back for rigidity and better control. The backsaw is used for making very fine joints. It is particularly useful when making dadoed and rabbeted joints by hand.

The *dovetail saw,* figure 4-15, is similar to the backsaw except it has a straight handle and is lighter in weight. As its name implies, it is used to make dovetail joints, but it is also used to make fine cuts for other joints. The blade may be 8, 10, or 12 inches long with 15 to 17 points to the inch.

The *hacksaw,* figure 4-16, is used to cut metal. Blades are removable, and points range from 18 to 32 teeth per inch. Coarse teeth blades are used for cutting heavy metals, while fine teeth blades are used for cutting thin metals. Blades are inserted with the teeth pointing away from the handle. Cutting is done on the push stroke.

Refitting Handsaws

Refitting handsaws involves three steps: jointing, setting, and filing, figure 4-17. To begin, clamp the saw in a saw vise.

Fig. 4-16 Hacksaw *(Greenlee Tool Co.)*

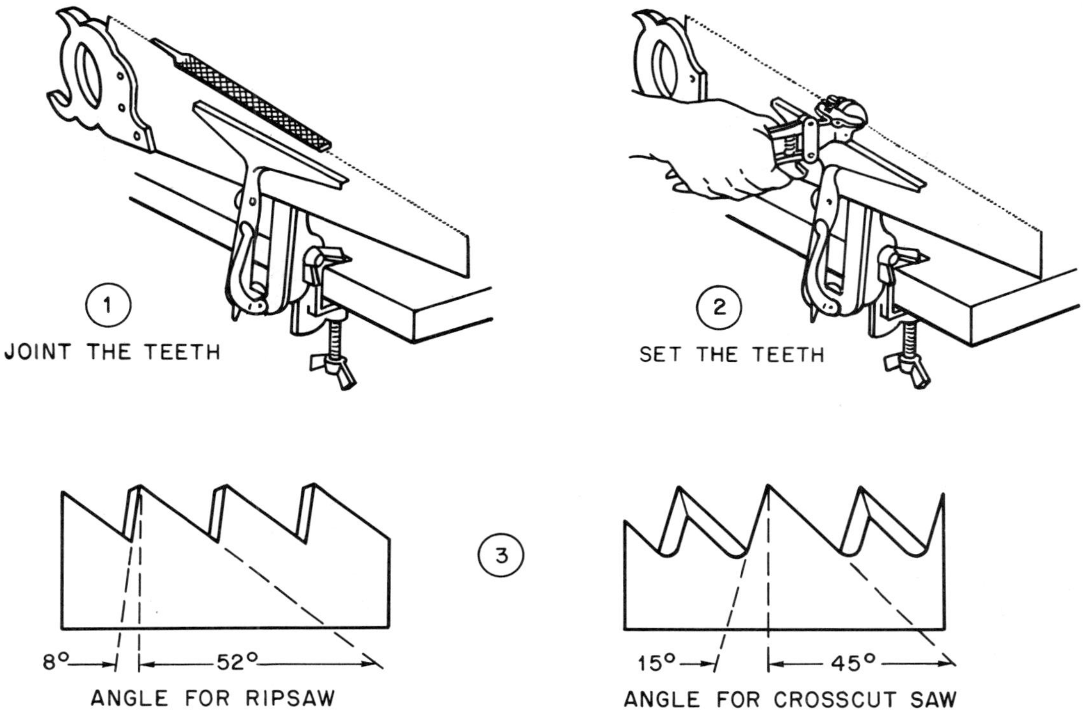

Fig. 4-17 Refitting handsaws

1. *Joint* the teeth to make them a uniform height. This is done by running a flat-mill file over the tops of the teeth.
2. *Set* the teeth with a saw set, figure 4-18. Setting bends the teeth alternately left and right to provide clearance for the blade so it does not bind in the cut *(kerf)*.
3. *File* the teeth.
   a. *Ripsaw:* Hold the file 90 degrees to the face of the blade. File straight across the same side of each tooth while maintaining the same angle. File until the shiny flat tip caused by the jointing disappears.
   b. *Crosscut saw:* Hold the file at a 60-degree angle to the face of the blade with the top of the file tilted toward the toe at about 15 degrees. File every other tooth from one side and those in between from the other. File until the shiny flat tip caused by the jointing disappears.

Fig. 4-18 Saw set

EDGE-CUTTING TOOLS

Bench planes remove a small amount of stock from a surface. They are used for smoothing rough surfaces and bringing work down to the desired finished size. Bench planes differ in length, weight, and type of blade. These differences determine the type of work they are able to do.

In order from largest to smallest, bench planes include the jointer, fore, jack, and smooth

Fig. 4-19 A jack plane *(The Stanley Works)*

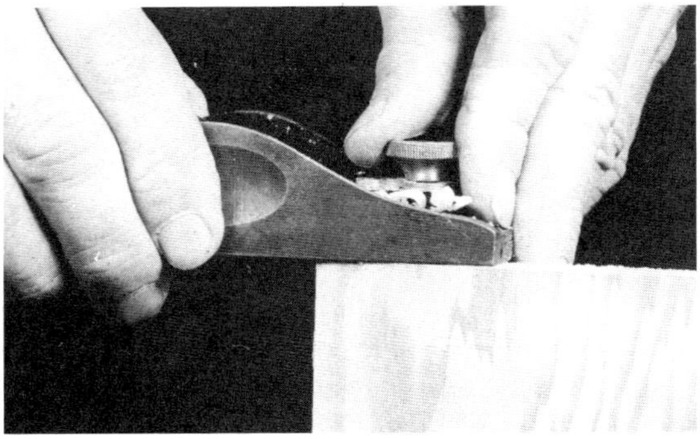

Fig. 4-20 Planing an end with a block plane *(American Plywood Assn.)*

planes. The *jointer* and *fore planes* are used mainly for producing flat surfaces and to straighten edges on long stock. The *jack plane* is an all-purpose plane, figure 4-19. The *smooth plane* is ideal for smoothing rough boards. The *block plane* is the smallest commonly used plane and can easily be held in one hand. Its low blade angle makes it ideal for planing end grains and smoothing short surfaces, figure 4-20.

When planing the face or edge of a piece of wood, hold the wood securely in a vise. Always plane with the grain. Push forward while applying pressure downward on the knob, figure 4-21. As the front of the plane clears the end of the wood, relax the pressure on the knob but maintain pressure downward on the handle. If the plane is sharp and used properly, a long, smooth, curled shaving should form. Check the edge for squareness with a steel square.

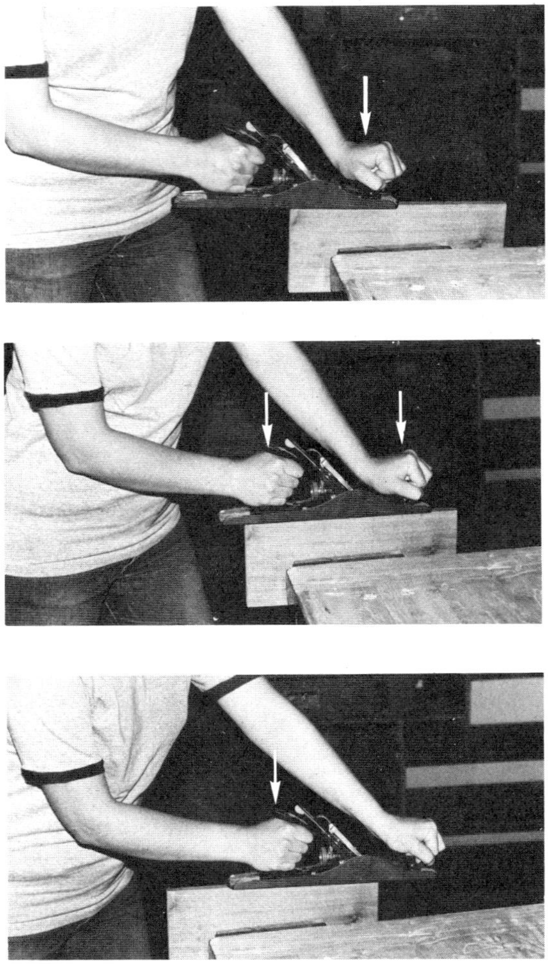

Fig. 4-21 Planing an edge with a jack plane

The *spokeshave,* figure 4-22, is a plane used to trim and smooth curved surfaces, both convex and concave. It is used like any other plane. Its short bottom allows it to follow most curves unless they are extremely tight. Some spokeshaves have adjusting nuts to regulate the depth of cut.

The *drawknife,* figure 4-23, is an open-beveled blade with handles on both ends. A handle is grasped in each hand, and with the bevel up, the drawknife is pulled through the stock. This tool is used to rough down the work by removing a lot of material in a short time.

Fig. 4-22 Spokeshave *(The Stanley Works)*

Fig. 4-23 Drawknife *(Greenlee Tool Co.,*

Fig. 4-24 Wood chisel *(Greenlee Tool Co*

Fig. 4-25 Wood gouge *(Greenlee Tool Co*

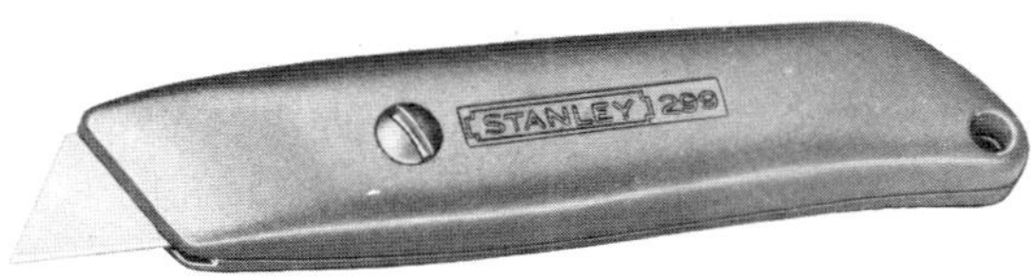

Fig. 4-26 Knife *(The Stanley Works)*

> **Caution:** Care should be taken with the drawknife because the cutting edge is being drawn toward the operator. Cutting should always be done with the grain of the wood.

Wood chisels, figure 4-24, are flat pieces of tempered steel with one end ground on a bevel. They are used to trim wood, to form mortises, and to smooth grooves and other surfaces. The depth of cut is controlled by placing the beveled edge against the wood and raising and lowering the handle.

Chisels are measured by the width of the blade, with sizes ranging from 1/8 to 2 inches. They also vary according to the shape, length, thickness, and method by which the blade is attached to the handle.

> **Caution:** Wood chisels are very sharp. When using a chisel, keep both hands behind the cutting edge at all times. Always cut away from the body.

Gouges, figure 4-25, are similar to chisels except that the blade is curved and may be ground with an inside or outside bevel. They come in various sizes and are used for cutting small, deep curves. The shank of the gouge may be bent or straight. The bent shank raises the handle clear of the work in order to make longer grooves. Gouges are used the same as chisels. When gouging, avoid taking heavy cuts.

In woodworking, the *knife,* figure 4-26, is used more frequently for marking than for cutting. The mark made by the knife is finer than one made by a pencil. Also, the knife mark scores the grain and prevents splitting beyond the mark when a cut is made. The bevel on the knife should have a long taper and be kept sharp by whetting on an oilstone. The knife should never be ground.

Reconditioning Edge-Cutting Tools

Worn or nicked cutting tools must be reconditioned. Tools are brought back to their original shape by grinding. Then a keen edge is restored by whetting (rubbing) it on a *whetstone,* see figure 4-46. Whetstones, also called *oilstones,* come in coarse, medium, and fine grits. The finer the grit, the sharper the edge it produces.

To recondition an edge-cutting tool, figure 4-27:

1. Shape the tool to its original form on the grinder. The angle should be between 25 and 30 degrees. Dip the tool in water frequently to keep it from overheating. Overheating will cause the tool to lose its temper (hardness). Hold the tool at the same angle to produce a hollow grind or concave surface to the bevel.
2. Oil the whetstone liberally. While holding the tool at the desired angle, whet the beveled

side by moving it across the stone in a back and forth, circular, or figure-eight pattern.

3. While holding the tool flat against the stone, whet the straight side in a similar manner.
4. Repeat steps 2 and 3 on a finer stone.
5. If an even finer edge is desired, hone the tool on a leather strap or a fine Lily stone.
6. Test the edge for keenness by stroking a piece of paper against the cutting edge.

Some cutting tools, such as the knife, should never be ground. The blade is fine enough so that whetting is enough to bring it back to its original keenness. Chisels are only ground if the edge is nicked or the bevel is rounded over. The edge of a chisel may be whetted many times before it needs grinding.

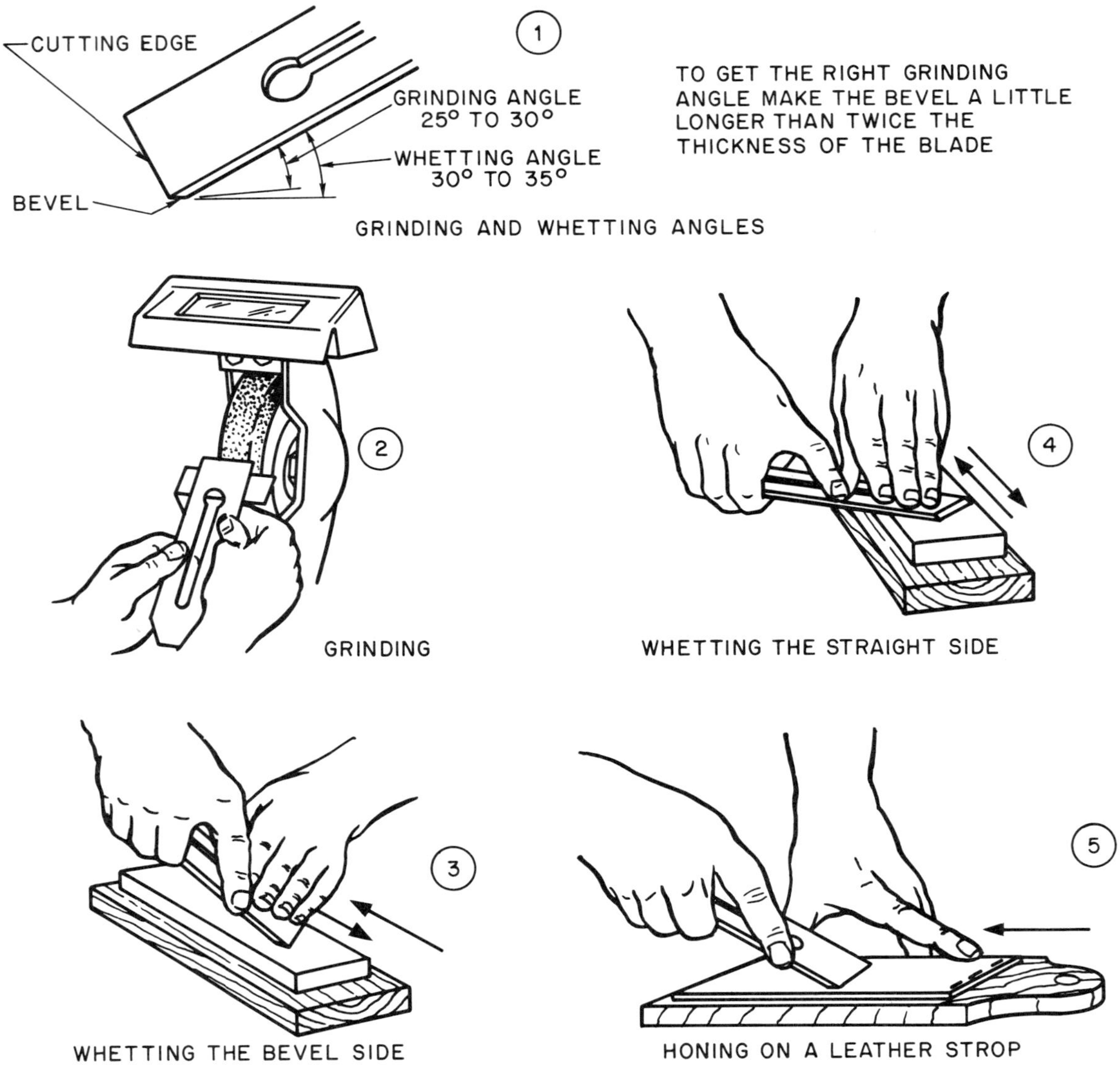

Fig. 4-27 Reconditioning edge-cutting tools

BORING TOOLS

The *bit brace,* figure 4-28, is a crank-shaped tool that holds and turns bits to bore holes. It also holds screwdriver bits to drive screws. Its size is determined by its *sweep,* or the diameter of a circle made by the handle. Sizes range from 8 to 12 inches. A ratchet is used when there is not enough room to make a complete turn of the handle.

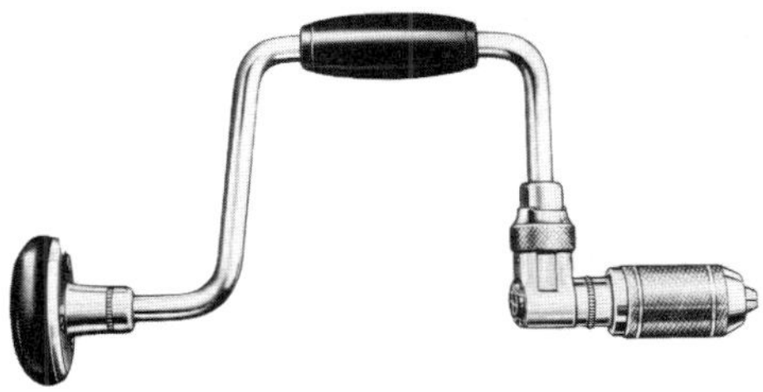

Fig. 4-28 Bit brace *(The Stanley Works)*

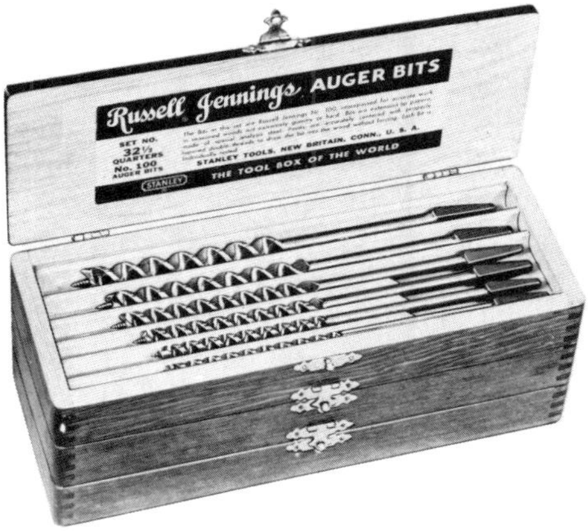

Fig. 4-29 Auger bits *(The Stanley Works)*

Auger bits, figure 4-29, are used in the bit brace for boring holes in wood. Sizes range from 1/4 to 1 inch graduated in sixteenths of an inch. The size is determined by the number of sixteenths of an inch. For instance, a #8 bit indicates a 8/16-inch or 1/2-inch bit. A #12 bit will bore a 3/4-inch hole. (12/16 = 3/4) Bits are available with fast, medium, or slow feed screws and in single or double twist.

Boring with an auger bit may splinter the back of a workpiece. To prevent this, bore just until the feed screw comes through the workpiece, then finish the hole from the other side. Another method is to clamp a piece of scrap wood to the back of the workpiece and then bore the hole.

Auger bits are sharpened with an auger bit file, figure 4-30. File across the inside of the spurs only enough to bring them to a sharp edge. Excessive filing will reduce their height making the bit useless. Never file the outside of the spur as this will change the diameter of the bit. File the cutting lips on the top side and on the original bevel until the edge is sharp.

The *expansive bit,* figure 4-31, is like an adjustable auger bit. It is used to bore holes larger than 1 inch in diameter. Several cutter blades are supplied that can be adjusted to bore holes from 1 to 5 inches. The expansive bit is sharpened like an auger bit.

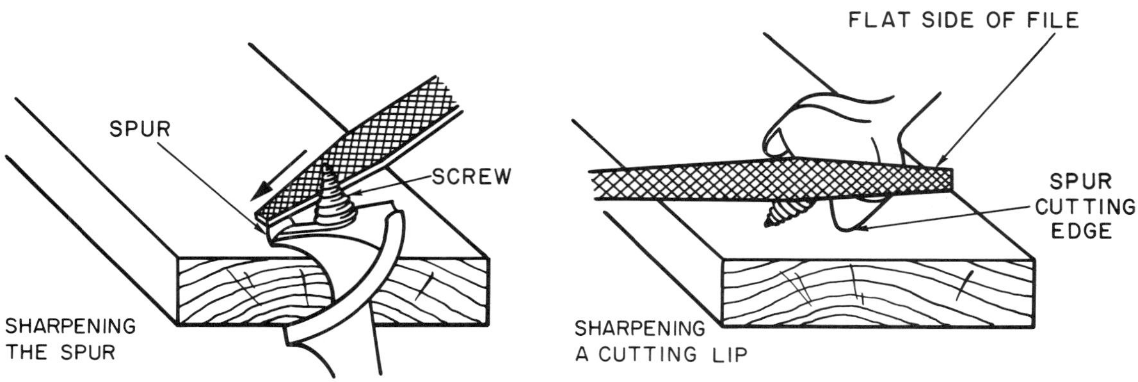

Fig. 4-30 Sharpening an auger bit

The *forstner bit,* figure 4-32, is used to bore holes which cannot be bored with other bits, such as a hole almost completely through the material. Other bits will pierce the entire thickness of the stock. Forstner bits are also useful for boring thin stock or near the end of stock because an auger bit might split the work. Sizes range from 1/4 to 2 inches and are numbered in the same manner as auger bits. The forstner bit has no feed screw and must be centered by its circular steel rim.

Fig. 4-31 Expansive bit *(The Stanley Works)*

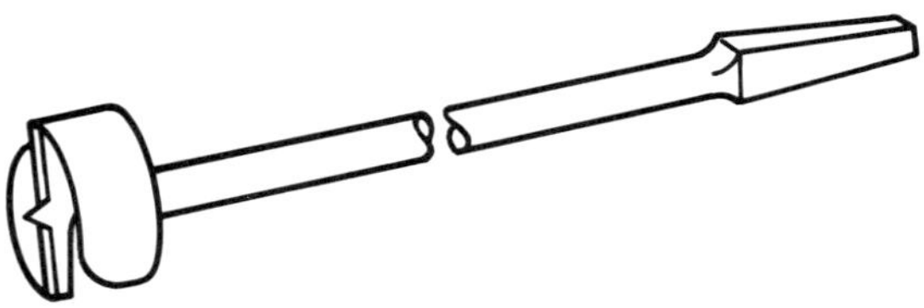

Fig. 4-32 Forstner bit

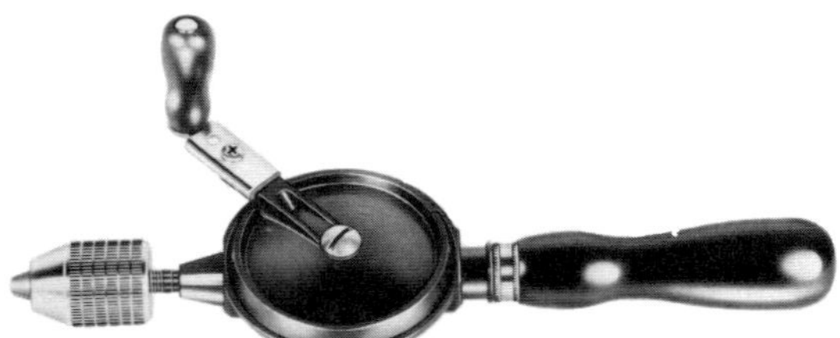

Fig. 4-33 Hand drill *(The Stanley Works)*

The *hand drill,* figure 4-33, holds and turns twist drills to drill small holes up to 1/4 inch. It also holds and turns countersinks. It is a very useful tool when only a few holes are to be drilled or no power is available.

Twist drills, figure 4-34, are used to make holes 1/6 to 1/2 inch in diameter. Increments are in 1/64ths. These drills are particularly useful for drilling holes for screws. They are held and turned in hand or electric drills.

The two cutting edges of twist drills are sharpened by grinding. The edges form a 59-degree angle with the center line of the drill. A drill point gauge is used to test this angle, figure 4-35. The surface in back of the cutting edge must be ground lower to make sure the cutting edge is the highest point. This clearance is called *relief.*

Countersinks, figure 4-36, are used in a bit brace or hand drill. They form recesses in which flathead screws can be set flush with the surface of the material.

Combination drills, figure 4-37, are available that will drill and countersink in one operation. They come in a variety of sizes to accommodate ordinary size wood screws.

ASSEMBLING TOOLS

The *hammer,* figure 4-38, is used to drive nails. The claw of the hammer is used to withdraw nails. Hammers are available in a number of styles and weights. The claw may be straight or curved; the face may be bell-shaped or plain, smooth, or

Fig. 4-34 Twist drills

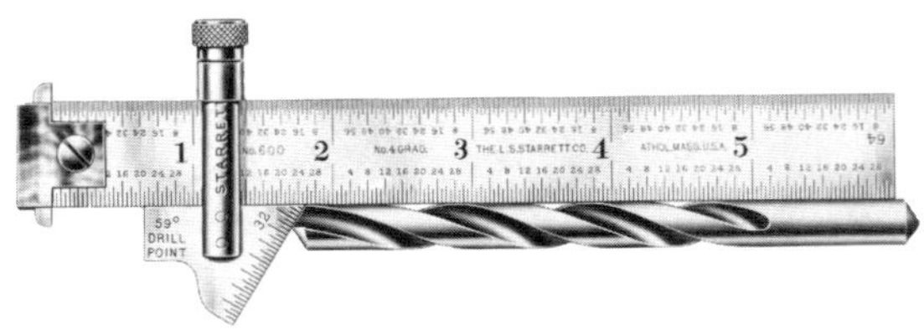

Fig. 4-35 Drill point gauge *(L.S. Starrett Co.)*

checkered. Weights range from 7 to 20 ounces. Handles may be of wood or other material. Hammers are selected to suit the type of work to be done.

To drive a nail, grasp the hammer near the end of the handle and swing it with the entire forearm, striking the nailhead squarely. If the nail bends, pull it out and start a new one.

Fig. 4-36 Countersink *(The Stanley Works)*

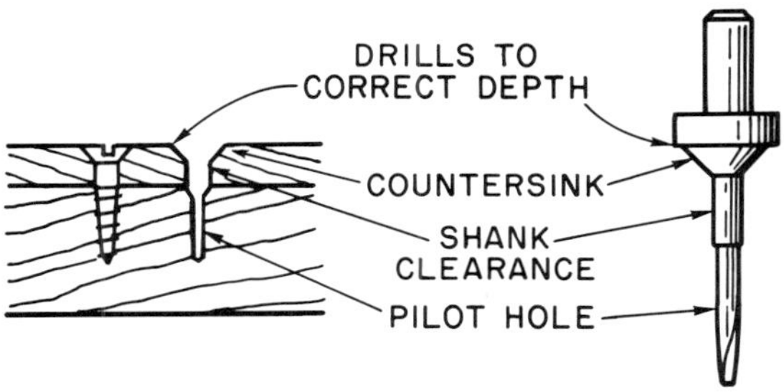

Fig. 4-37 Combination wood drill and countersink

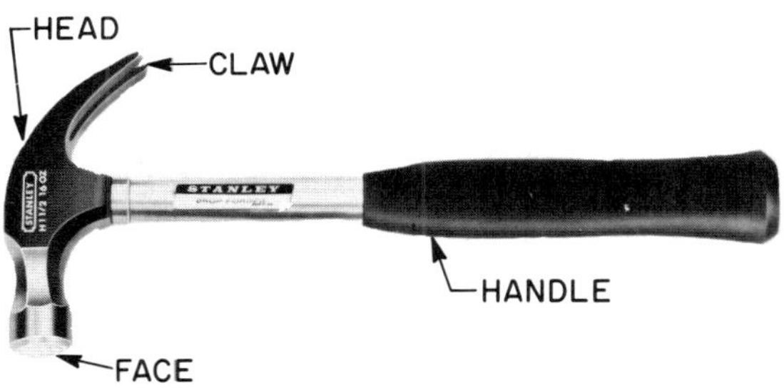

Fig. 4-38 Hammer *(The Stanley Works)*

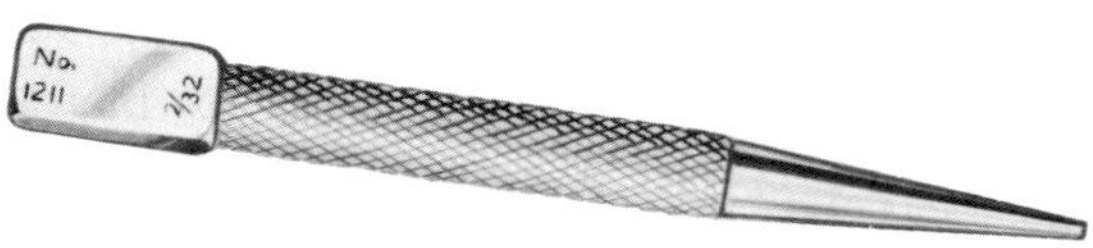

Fig. 4-39 Nail set *(The Stanley Works)*

Nail sets, figure 4-39, are used to set nailheads below the surface. The most common sizes are 1/32, 2/32, and 3/32 inch. The size refers to the small end which is hollowed out to prevent it from slipping off the nailhead. If this hollowed out point becomes flattened, the nail set is useless. Do not set hardened nails or hit the point of the nail set with a hammer as this will cause the point to flatten.

Screwdrivers, figure 4-40, come in many styles and sizes. Common screwdrivers have a straight tip. Phillips screwdrivers have a cross-shaped tip. Spiral screwdrivers with interchangeable bits can drive screws at a faster rate and with greater ease. A screwdriver should fit snugly without play in the screw being driven. Their sizes are determined by the length of the blade.

Screwdriver bits, figure 4-41, can be used with the bit brace to drive large screws with less effort. These bits are available in different sizes to accommodate a variety of screws.

SMOOTHING TOOLS

The *hand scraper,* figure 4-42, is a rectangula or curved piece of tool steel about 1/16 inch thick. It has a square or beveled edge that is sharpened and turned to produce a hook. This edge scrapes the wood to remove mill marks and other defects prior to sanding.

The *cabinet scraper,* figure 4-43, is a double-handled tool that holds a rectangular hand scraper blade. The thickness of the scraping can be adjusted

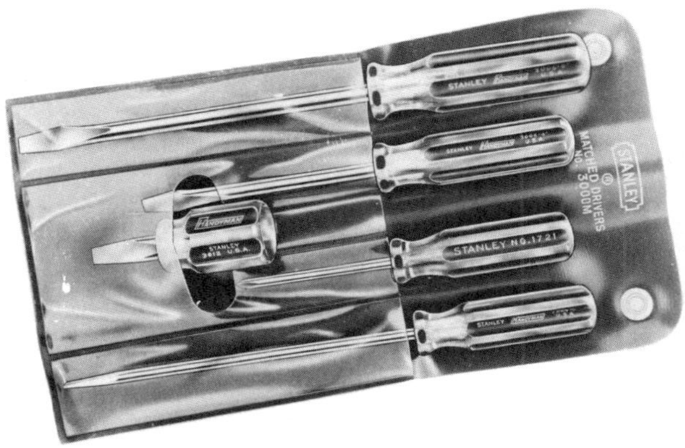

Fig. 4-40 Screwdrivers *(The Stanley Works)*

by means of a thumbscrew. It is used with a downward, pushing action and is particularly useful for smoothing large, flat areas after planing.

The *burnisher,* figure 4-44, is a round, pointed tool with a handle. Its steel is extremely hard and smooth. The burnisher is used to turn over the edge of a scraper blade to form a hook. This is called *burnishing.*

Files and *rasps,* figure 4-45, are used to smooth parts that are difficult to smooth with other tools. They are available in a variety of shapes, sizes, and cuts. Generally files are used

Fig. 4-41 Screwdriver bit *(The Stanley Works)*

Fig. 4-42 Hand scraper *(The Stanley Works)*

Fig. 4-43 Cabinet scraper *(The Stanley Works)*

Fig. 4-44 Burnisher

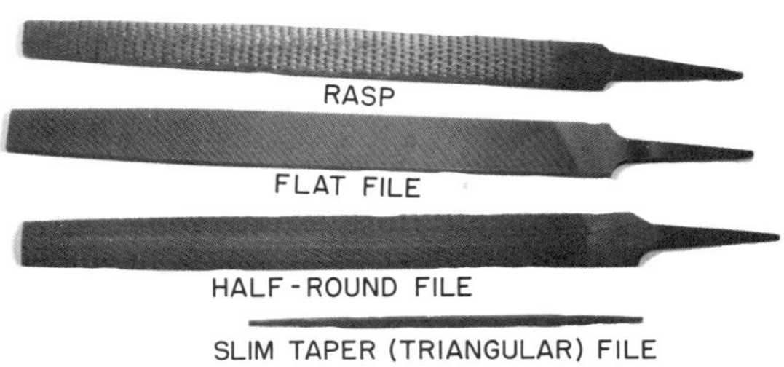

Fig. 4-45 Files and rasps

Fig. 4-46 Whetstone

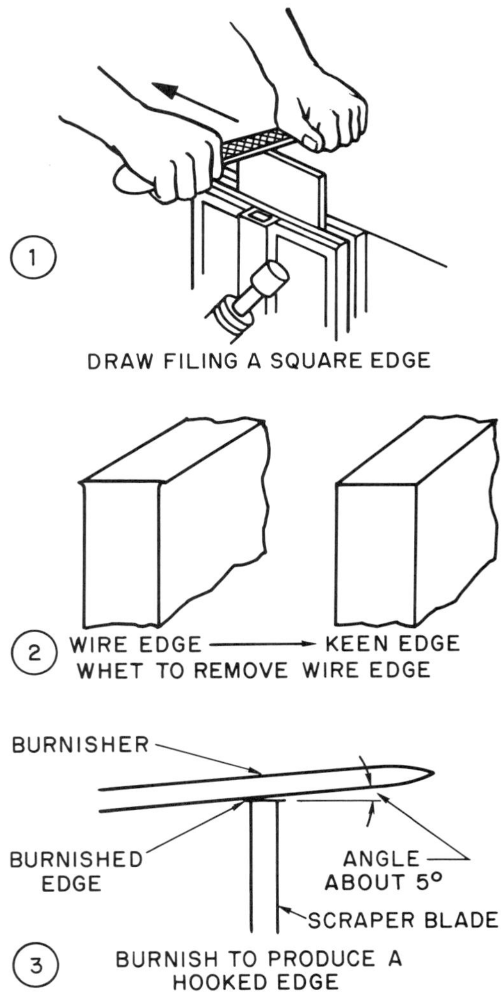

Fig. 4-47 Sharpening a scraper blade

when only a little stock is to be removed. Rasps are used when considerable stock is to be removed. Files are also used to sharpen handsaws. These are triangular in shape and are called slim taper, extra slim taper and double extra slim taper files according to their width. They range in lengths of 6 to 10 inches.

> **Caution**: To avoid injury, files and rasps should not be used without handles.

Attempting to file metal that is too hard will ruin a sharp file. To determine if metal can be filed, ride the file over the metal. If it does not cut, the metal cannot be filed. Such metal is sharpened by grinding and whetting.

Whetstones, figure 4-46, are used for whetting plane blades, chisels, and other cutting-edge tools. They come in coarse, medium, or fine grits. Combination stones have a coarse grit on one side and a finer grit on the other side. Whetstones, also called *oilstones,* are stored in wood boxes to prevent breakage. The stone is kept clean and sharp by lightly oiling every time it is used.

Sharpening Scraper Blades

Scraper blades may be sharpened either with a square or beveled edge. The square edge produces a cutting edge on both sides of the blade, while the beveled edge produces a cutting edge on one side. Sharpening is done in a similar manner for straight or beveled edges, figure 4-47.

1. Draw file the blade to produce a straight edge. Hold both ends of the file with each hand and pull across the edge. Sight the edge occasionally to make sure it is not hollowed out in the center. Continue draw filing until a wire edge appears.
2. Whet the blade, in the same manner as for chisels and plane irons, until a keen edge is obtained.
3. Turn the sharp edge with a burnisher to produce a hooked edge. Stroke the back of the edge with a burnisher, gradually raising the handle with increasing pressure to turn the edge.

ACTIVITIES

1. Measure distances specified by the instructor using the various layout tools described in this unit.
2. Lay out lines parallel to the edge of a piece of stock using a marking gauge and combination square. Rip the board closely to these lines using a ripsaw.
3. Practice squaring lines across a piece of stock. Cut along the lines using a crosscut saw.
4. Lay out a 5-inch circle and cut it out using dividers, bit brace, auger bit, and compass saw.
5. Square the end of a piece of stock using a block plane.
6. Plane the edge of a long piece of stock using a jack plane. Make sure it is straight and square.
7. Plane a curved surface smooth using a spokeshave.
8. Fasten two pieces of stock of equal width and length together with flathead screws using a hand drill, twist drills, countersink, and screwdriver. Keep the ends and edges of the boards flush.
9. Visit a tool sharpening shop and observe the equipment and methods used to sharpen hand tools.

10. Under the direction and supervision of your instructor:
    a. grind, whet, and hone a wood chisel or plane iron.
    b. sharpen a pocket knife to a keen edge.
    c. sharpen an auger bit.
    d. file, whet, and burnish a cabinet scraper blade.
    e. joint, set, and file a ripsaw or crosscut saw.

UNIT REVIEW

Matching

Match each tool in column I with its correct application in column II.

	Column I	*Column II*
____	1. Combination square	a. Used to cut irregular shapes
____	2. Trammel points	b. Used to smooth grooves
____	3. Caliper rule	c. Used for sharpening plane blades
____	4. Coping saw	d. Used to lay out 45-degree angles
____	5. Hacksaw	e. Used to remove mill marks
____	6. Block plane	f. Used to measure outside and inside measurements
____	7. Wood chisel	g. Used to trim wood end grains
____	8. Drawknife	h. Used to lay out circles
____	9. Hand scraper	i. Used to rough down stock
____	10. Whetstone	j. Used to cut metal

Multiple Choice

1. Burnishing means to
   a. burn wood to accent the grain.
   b. remove mill marks.
   c. turn the edge of a scraper blade.
   d. produce a very keen edge.

2. Countersinks are tools that are used to
   a. make sink cutouts in countertops.
   b. drive large screws.
   c. set nails below the surface.
   d. form recesses to set flathead screws flush.

3. The size of a screwdriver is determined by
   a. the diameter of the blade.
   b. its overall length.
   c. the length of the blade.
   d. the size of the tip.

4. To prevent a nail set from slipping off the nailhead, the point is
   a. ground smooth and clean.
   b. checkered.
   c. hollowed out.
   d. bell-shaped.

5. The forstner bit
   a. has no feed screw.
   b. holds and turns a countersink.
   c. is used to bore thick stock.
   d. drives flathead screws.

6. Twist drills are generally used to make holes
   a. 3/8 to 1/2 inch in diameter.
   b. 1/16 to 1/2 inch in diameter.
   c. 1/4 to 3/4 inch in diameter.
   d. 1/8 to 7/8 inch in diameter.

7. In woodworking, the knife is used more often to ________ than to cut.
   a. whittle
   b. sharpen pencils
   c. mark
   d. make pegs

8. A #14 auger bit will bore a hole
   a. 3/4 inch in diameter.
   b. 1/2 inch in diameter.
   c. 7/16 inch in diameter.
   d. 7/8 inch in diameter.

9. The spokeshave is a tool used to
   a. shave spokes.
   b. plane long surfaces.
   c. trim end grain.
   d. smooth curved surfaces.

10. The longest of the bench planes is the
    a. fore plane.
    b. smooth plane.
    c. jointer plane.
    d. jack plane.

11. Bending handsaw teeth alternately left and right to provide clearance for the blade is called
    a. setting.
    b. jointing.
    c. tempering.
    d. filing.

12. Overheating a cutting tool while grinding may cause the tool to lose its
    a. joint.
    b. edge.
    c. temper.
    d. shape.

13. Filing handsaw teeth to a uniform height is called
    a. gumming.
    b. jointing.
    c. setting.
    d. shaping.

14. In order to produce teeth of uniform height when filing handsaws,
    a. use the same number of strokes on each tooth.
    b. draw a straight line along the bottom of the gullets.
    c. use a depth gauge on each tooth.
    d. file until the shiny tip of the tooth disappears.

15. Never file the outside of the spurs of an auger bit because this will
    a. shorten the spurs.
    b. change the diameter of the bit.
    c. produce a burr on the inside.
    d. change the original bevel.

unit 5 Portable Power Tools

OBJECTIVES

After studying this unit, the student will be able to:

- identify and describe commonly used portable power tools.
- demonstrate the safety precautions to take when operating portable power tools.
- show how to use portable power tools.

Portable power tools make the woodworker's job easier. These tools work faster and require less effort to use than hand tools. However, portable power tools can cause serious injury if operated improperly. Always observe safety rules when using a portable power tool.

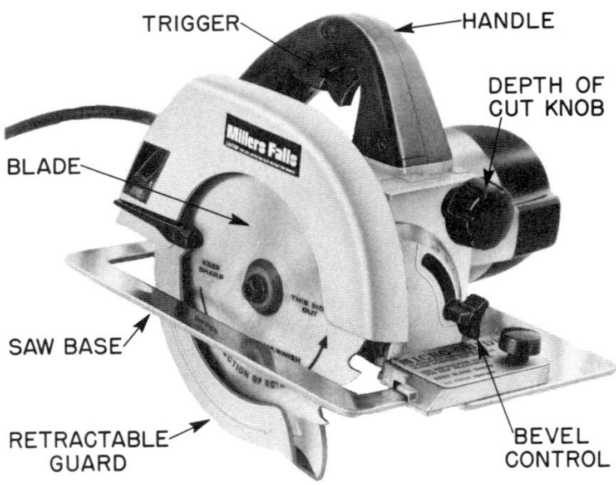

Fig. 5-1 Portable circular saw *(Millers Falls)*

SAFETY PRECAUTIONS FOR PORTABLE POWER TOOLS

- Understand how the tool works and how it is used *before* trying to operate it.
- Never allow your attention to wander when using power tools.
- Electrical shock is a frequent cause of accidents when using power tools. Tools that are not double-insulated should be connected to a grounded electrical outlet.
- Place extension cords so they will not be damaged or cut.
- Unplug tools when making adjustments or changing cutters.
- Do not wear loose clothing or jewelry which might get caught in the tool.
- Secure the workpiece.
- Use all safety guards supplied with the tool.
- Wear safety glasses when operating power tools.

CIRCULAR SAW

The portable *circular saw,* figure 5-1, has an electric motor which drives a circular saw blade. It is controlled by a trigger switch on the handle. There are adjustments for depth of cut and for cutting at an angle. A retractable safety guard covers the blade.

Circular saws are available in very light to heavy-duty models, with motor sizes ranging from 1/6 to 1 1/2 horsepowers (HP). The size, which is determined by the diameter of the blade, ranges from 4 1/2 to 12 inches. Maximum depth of cut is from 1 1/4 to 4 3/8 inches. The blade may be driven directly by the motor or through a worm gear. Portable circular saws cut on the upstroke and guides are provided to follow lines to be cut.

The circular saw blades used in the portable circular saw are the same types as those used in the stationary table saw and radial arm saw. More information about circular blades may be found in *Unit 7 Table Saw* and *Unit 8 Radial Arm Saw.*

Fig. 5-2 Using the portable circular saw

Using the Circular Saw

1. Mark the cutout pattern on the stock.
2. Adjust the depth of cut so that the blade pierces the entire thickness of the stock and does not extend more than 1/2 inch out the back side.
3. Make sure the guard works properly.

Caution: Never wedge the guard back in an open position.

4. With the blade clear of the stock, start the saw.

Caution: Always wear safety goggles when sawing.

5. Advance the saw into the work slowly but firmly, figure 5-2. Cut to the line for a short distance and then follow the line by the guide on the saw.
6. Near the end of the cut, watch the line at the saw blade to finish the cut. Release the switch. Keep the saw away from your body until it has completely stopped. Be aware that the guard can possibly stick in the open position.

Making Pocket Cuts With The Circular Saw

A *pocket cut* is an interior cut which removes a middle section of the stock, figure 5-3.

1. Hold the guard open. Tilt the saw forward on the front edge of the base so that the saw blade is over and in line with the cutout.
2. Make sure the teeth of the blade are clear of the work. Start the motor and lower the blade into the work until the base rests on it.
3. Advance the saw into the corner. Release the switch and wait until the saw stops before removing it from the cut.
4. Cut the other sides the same way. The corners may have to be finished with a handsaw.

SABER SAW

The saber saw, sometimes called the *jigsaw* or *bayonet saw,* is widely used for making curved cuts, figure 5-4. As with the circular saw, the saber

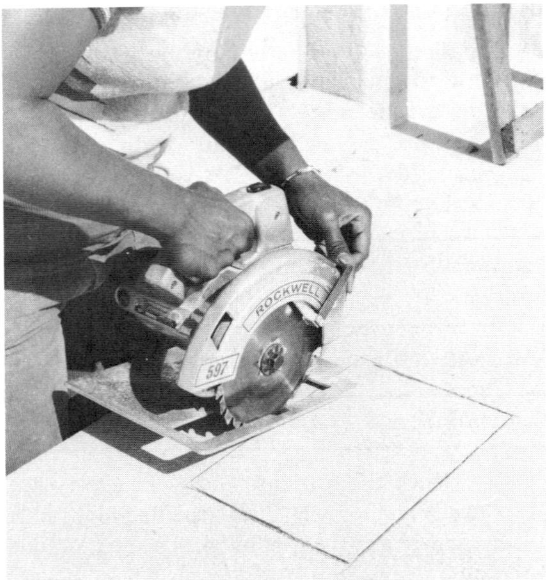

Fig. 5-3 Making pocket cuts with the circular saw

Fig. 5-4 Saber saw *(Millers Falls)*

saw cuts on the upstroke and, consequently, splinters out the top side of the work. To produce a splinter-free cut, cut with the face side down if possible.

There are many styles and varieties of saber saws. The length of the stroke and the capacity of the motor determines its size. Strokes range from 1/2 to 1 inch with the longest stroke being the best for faster and easier sawing. The base of the saw may be tilted to make bevel cuts.

Many blades are available for fine or coarse cutting in wood and for cutting metal. Wood-sawing blades have teeth from 6 to 12 points to the inch. Knife blades for cutting paper, leather, and other materials are also available.

Using the Saber Saw

1. Outline the cut to be made and secure the work.
2. Hold the base of the saw firmly on the work and, with the blade clear of the work, start the motor.
3. Push the saw forward into the work and follow the line closely as in handsawing. Feed the saw into the work as fast as possible, but without forcing it, to finish the cut. Keep the saw away from your body until it has stopped.

Making Pocket Cuts With the Saber Saw

1. Tilt the saw forward on its base with the blade in line and clear of the work.
2. Start the motor. Holding the base steady, gradually lower the saw until the blade penetrates the stock and the base rests firmly on the work, figure 5-5.
3. Continue the cut to the end. Back away from the end for a short distance.
4. Make a curved cut to the other side and continue until all sides are cut.
5. Saw in the opposite direction to cut out the corners.

ELECTRIC DRILL

Like other portable power tools, *electric drills* come in many sizes and styles, figure 5-6. The size of an electric drill is determined by the capacity of the chuck. The *chuck* holds the cutting tool. The most popular sizes are 1/4, 3/8, and 1/2 inch, although there are many other sizes.

The drill has an encased motor that drives the chuck. Lighter drills usually have a pistol-type handle, while the heavier drills have a double handle.

Using the Electric Drill

1. Select the proper size bit or twist drill. Insert it and tighten the chuck with the chuck key.
2. Place the bit on the center point of the hole to be drilled. Holding the drill at the correct angle, start the motor, figure 5-7.

Fig. 5-5 Making pocket cuts with the saber saw *(American Plywood Assn.)*

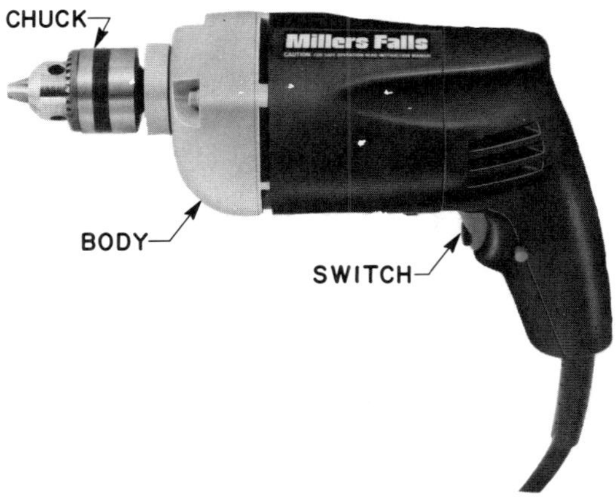

Fig. 5-6 Portable electric drill *(Millers Falls)*

3. Apply pressure when drilling into the stock. Do not wobble the drill. Remove the bit as necessary to clear the waste. When drilling is finished, shut off the motor.

> **Caution:** Do not wear loose clothing when operating electric drills. Loose clothing may wrap around a drill and injure the operator. Also, tie back long hair.

PORTABLE POWER PLANE

Power planes, figure 5-8, provide a smoother and more accurate cut than hand planes. They save time and eliminate a lot of hard work.

Power planes vary in length and weight, depth and width of cut, speed, and horsepower. The power block plane, for instance, is lightweight and can be held in one hand. An electric motor powers the spiral cutting blade of the power plane. The base has an adjustment for depth of cut and a side guide. The handle holds the trigger switch.

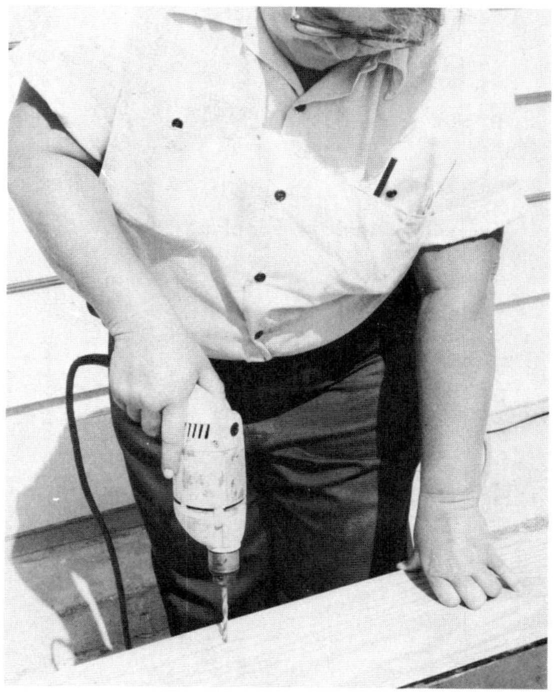

Fig. 5-7 Using the electric drill

Caution: Operate a power plane carefully. The high-speed cutting blade is unguarded.

Operating the Power Plane

1. Set the angle of the side fence and adjust the depth of cut.
2. Place the plane on the work with the cutting blade clear of the work.
3. Start the motor and with a steady, even pressure make the cut through the work, figure 5-9.
4. Keep the tool clear of the body until it has stopped.

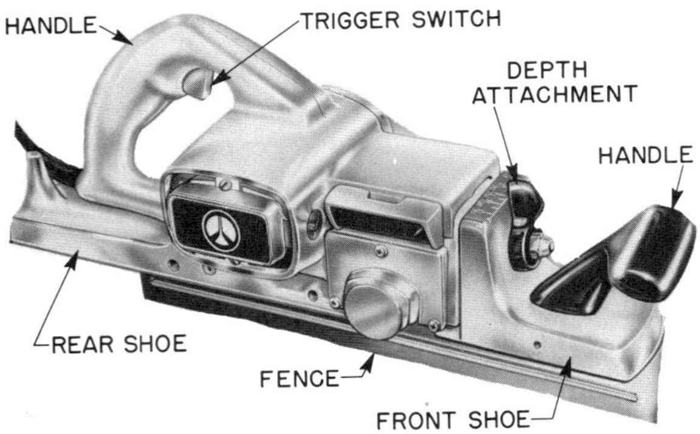

Fig. 5-8 Portable power plane *(Rockwell)*

Fig. 5-9 Using the power block plane

ELECTRIC ROUTER

The *router* is one of the most versatile portable tools in the woodworking trades, figure 5-10. With the proper accessories, the router can mold edges to different shapes; cut straight, curved, and V grooves; make dadoes, rabbets, and dovetails; and trim laminates, figure 5-11.

The size of a router is determined by its horsepower rating. Routers are available in models ranging from 1/4 to 2 1/2 horsepower with speeds of 18,000 to 27,000 rpm. The motor must be high speed to produce a clean, smoothly cut edge. The router's motor powers a chuck which holds the cutting bit. The base moves up and down on the motor to control the depth of cut. A trigger or toggle switch controls the motor.

The router can be guided in several ways:

- By using a router bit with a pilot, figure 5-12.
- By running the base of the router against a straightedge, figure 5-13.
- By using a guide attached to the base of the router, figure 5-14.

Fig. 5-10 Electric router *(Millers Falls)*

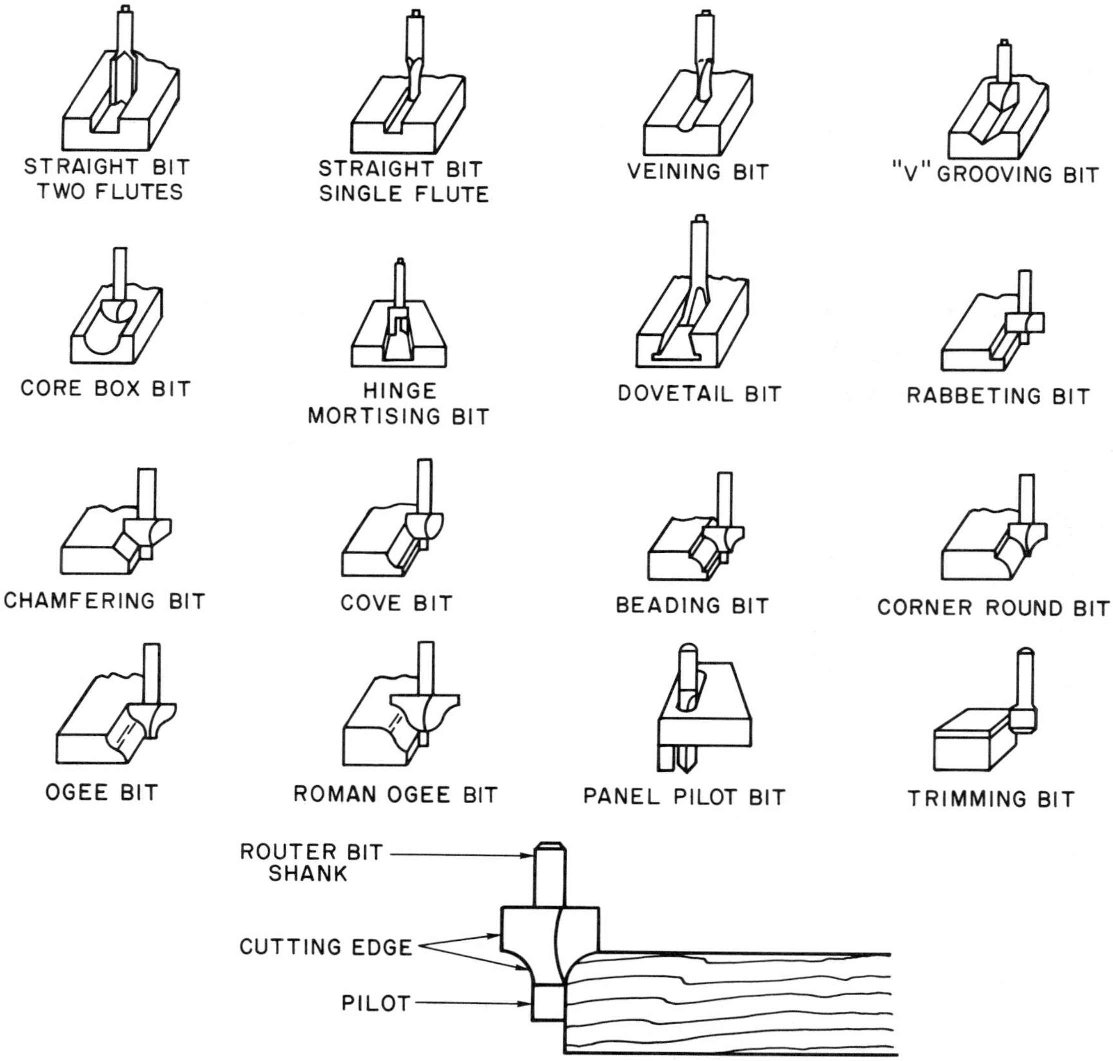

Fig. 5-11 Router bits

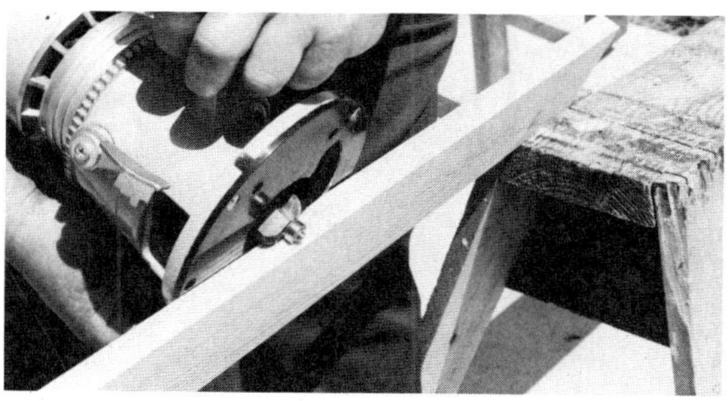

Fig. 5-12 Guiding the router with a pilot bit

Fig. 5-13 Guiding the router with a straightedge

- By using a template or a pattern with template guides attached to the base of the router, figure 5-15.
- Freehand routing in which the sideway motion is controlled by the operator, figure 5-16.

Using the Router

1. Select the correct bit for the job.

Fig. 5-14 Guiding the router with an attachment *(Black & Decker Mfg. Co.)*

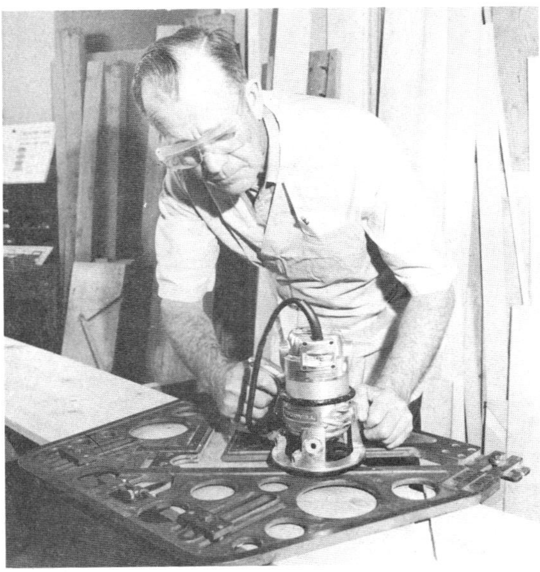

Fig. 5-15 Guiding a router with a template guide & pattern *(Rockwell)*

2. Insert the bit in the chuck. Make sure the chuck grabs at least 1/2 inch of the bit. Adjust the depth of cut.
3. Control the motion of the router by one of the methods described.
4. Lay the base of the router on the work with the cutting edge clear of the work. Start the motor.
5. Advance the bit into the cut while pulling the router in the correct direction, figure 5-17. Feed the router counterclockwise on outside edges and clockwise on inside edges.
6. Keep the router away from your body until it has stopped.

BELT SANDER

The *belt sander* is an excellent tool for sanding cabinetwork, figure 5-18. The size of the belt determines the size of the sander. Belt widths are from 2 to 4 inches, and belt lengths are from 21 to 27 inches. Grits are available from very fine to very coarse. Some sanders have a bag for collecting the sanding dust.

Fig. 5-16 Freehand routing *(Black & Decker Mfg. Co.)*

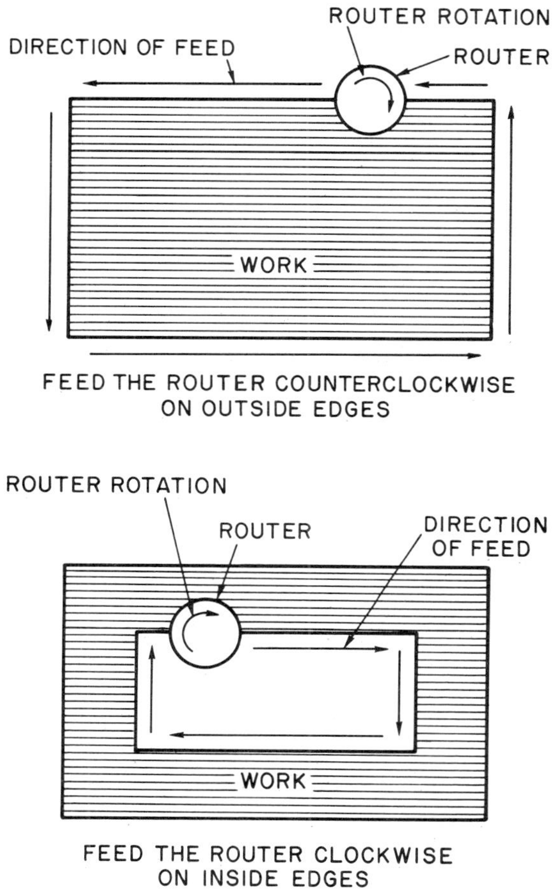

Fig. 5-17 Direction of feed

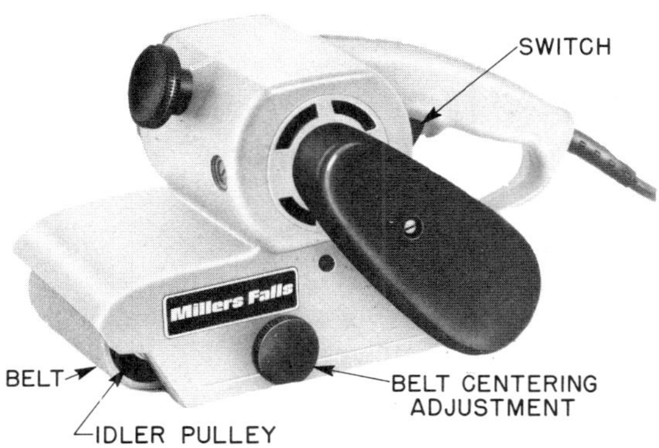

Fig. 5-18 Belt sander *(Millers Falls)*

Belts are usually installed by retracting the forward pulley (see the manufacturer's instructions). There is also an adjustment on the forward pulley for centering the belt. An arrow is stamped on the inside of the belt to indicate the direction of run. The belt should run so that the *lap* of the belt (where it is joined) will ride over the work, figure 5-19.

Using the Belt Sander

1. Secure the work to be sanded. Make sure the belt is tracking properly.
2. Start the machine and lower the sander so that the sanding pad lays flat on the work and pull the sander back, figure 5-20.
3. Sand the entire surface with a skimming motion by lifting the sander clear on the forward

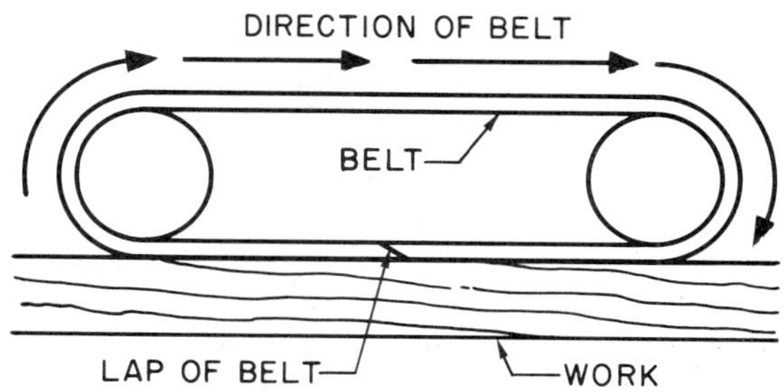

Figure 5-19 Lap and run of the sander belt

Fig. 5-20 Using the belt sander

stroke. Do not sand in one spot for long and be careful not to tilt the sander. Do not exert excessive pressure. The weight of the sander is enough. Always sand with the grain to produce a smooth finish.

FINISHING SANDER

The *finishing sander* is used for final sanding and for rubbing down finishing coats, figure 5-21. It is generally used to remove small amounts of material. Finishing sanders are made in many styles and sizes but are usually classified according to their motion.

Finishing sanders have either an orbital (circular) motion, a straight line (back and forth) motion, or a combination of motions controlled by a switch.

Fig. 5-21 Finishing sander *(Millers Falls)*

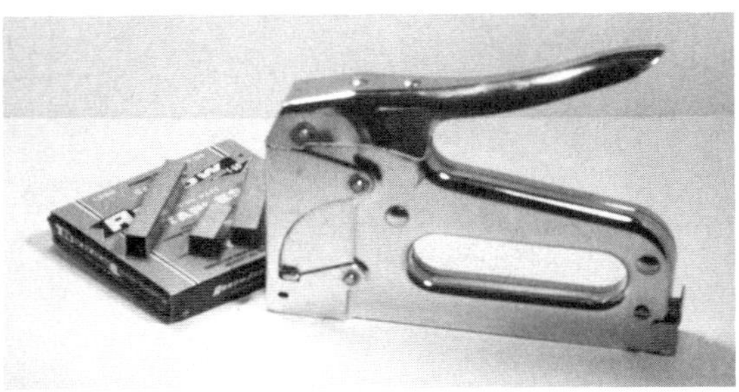

Fig. 5-22 Hand-powered spring stapler

The orbital motion has a faster cutting action but leaves scratches across the grain. The straight line motion cuts slower but produces a scratch-free surface.

Most sanders take one-third of a sheet of sandpaper. It is usually attached to the pad by friction or a spring device. Most kinds and grits of sandpaper can be used on finishing sanders.

Using the Finishing Sander

1. Select the correct sandpaper and attach it to the pad. Make sure it is tight because a loose sheet will tear easily.
2. Start the motor and hold the sander flat on its pad. Sand the surface evenly by slowly pushing and pulling the sander with the grain. Do not use extreme pressure because this may overload the machine and burn out the motor.

STAPLERS

Generally two kinds of staplers are used to speed up and reduce the cost of fastening. The *spring stapler,* figure 5-22, is a hand-powered tool that drives small staples by pressing the lever downward. The *pneumatic stapler,* figure 5-23, is operated by driving the stapler with compressed air. All power staplers are equipped with a cartridge or magazine which is loaded with specially designed fasteners. These fasteners come in many sizes.

Fig. 5-23 Pneumatic stapler *(Bostitch – Textron)*

Using Staplers

Because of many different designs of staplers, follow the manufacturer's operating directions carefully. Always keep the stapler pointed toward the work. Use the correct size and shape of fastener for a job. Use the recommended air pressure for the pneumatic stapler.

> **Caution:** Never point the stapler at other workers or fire a staple except into the work. A stapler can cause serious injury if misused. All new pneumatic staplers have a safety guard to prevent firing unless it has been depressed. Do not remove the guard.

ACTIVITIES

1. Study catalogs from tool suppliers and note the different makes, sizes, and kinds of portable power tools.
2. Select a portable router you believe is best for general work. Outline the reasons for the selection.
3. Compare the direct-drive and worm gear-driven portable electric circular saw. List the advantages and disadvantages of each.
4. Compare a handsaw and a circular saw. List the advantages and disadvantages of each.
5. Demonstrate how to use one portable power tool to your class. Emphasize proper procedure and safety precautions.
6. Make a chart of safety precautions to follow when using portable power tools.

UNIT REVIEW

Questions

1. What precaution must be taken to prevent electrical shock when using a portable power tool?
2. How is the size of a portable electric circular saw determined?
3. What is a pocket cut?
4. What are two other names for the saber saw?
5. How is a splinter-free cut obtained with a saber saw?
6. How is the size of a portable electric drill determined?
7. Why is it dangerous to wear loose clothing when using an electric drill?
8. Why would a woodworker choose a power plane instead of a hand plane?
9. What is one method of controlling the sideway motion of the portable router?
10. How is the pneumatic stapler operated?

Multiple Choice

1. When using portable power tools
   a. make sure extension cords are long and thin.
   b. wear loose clothing.
   c. remove all safety guards.
   d. make sure the tool is grounded.

2. The direction of feed on an electric router is
   a. counterclockwise on both inside and outside edges.
   b. clockwise on both inside and outside edges.
   c. counterclockwise on outside edges, clockwise on inside edges.
   d. clockwise on outside edges, counterclockwise on inside edges.

3. The portable circular saw and saber saw both
   a. cut on the downstroke.
   b. cut on the upstroke.
   c. cut in both directions.
   d. make curved cuts.

4. The best length of stroke in a saber saw for faster and easier cutting is
   a. 3/4 inch.
   b. 1 1/2 inches.
   c. 1 inch.
   d. 2 inches.

5. Wood-sawing blades for saber saws range from
   a. 5 to 10 points per inch.
   b. 6 to 12 points per inch.
   c. 8 to 16 points per inch.
   d. 7 to 14 points per inch.

6. The cutting tool of an electric drill is held securely by
   a. a collet.
   b. a chuck.
   c. a clamp.
   d. a clip.

7. Care must be taken when operating portable power planes because
   a. they have a tendency to jump.
   b. there is a danger of kickback.
   c. the cutting blades are unguarded.
   d. wood chips may fly out.

8. A router bit must be inserted in the chuck at least
   a. 1/2 inch.
   b. 3/4 inch.
   c. 1 inch.
   d. 1 1/4 inches.

9. Portable electric routers can be used to
   a. plane stock to thickness.
   b. rip lumber to width.
   c. cut stock to length.
   d. trim laminates.

10. A smooth finish is produced with a belt sander by
    a. pressing heavily against the stock.
    b. sanding with the grain.
    c. sanding with a circular motion.
    d. tilting the front of the sander upward.

unit 6 Jointer and Planer

OBJECTIVES

After studying this unit, the student will be able to:

- describe the parts and purpose of the jointer and the planer.
- adjust and operate the jointer to joint, face, bevel, rabbet, bevel rabbet, and taper stock.
- use the planer to surface stock to thickness and to plane a bevel and a taper.
- use a whetstone to sharpen jointer and planer knives.

The *jointer* is used to straighten the edges and faces of stock. This removes the warp from the stock before it is planed. The *planer* smooths the stock to a uniform thickness.

Though their purposes are different, the jointer and planer have similar cutterheads, figure 6-1. The cutterhead has three or more knives mounted in it and revolves at high speed. The stock is cut as it passes over or under the cutterhead. The number of knife cuts per second and the smoothness of the cut is determined by:

- the speed of the cutterhead.
- the rate of speed that the stock is fed through the machine.
- the number of knives in the cutterhead.

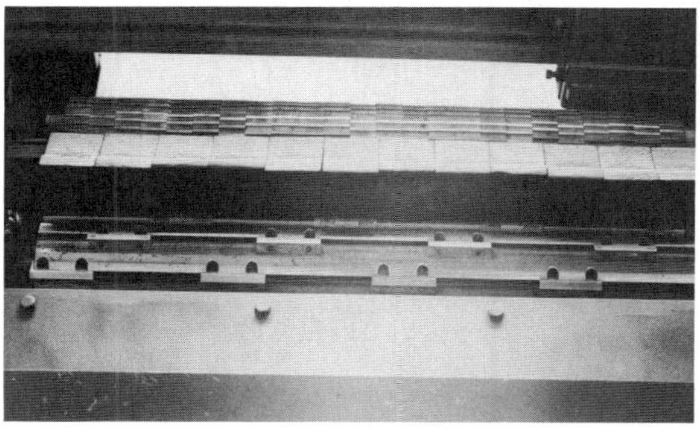

Fig. 6-1A The planer's cutterhead. This one contains short, staggered knives.

Fig. 6-1B The jointer's cutterhead. This one contains three knives.

THE JOINTER

Although other operations may be performed on the jointer, its main purpose is to straighten the edges and faces of stock. Straightening the edge of stock is called *jointing;* straightening the face of stock is called *facing.*

Warp must be removed from stock before it is used. Bows, cups, and twists are removed by facing. Crooks are straightened by jointing. Jointing and facing are easier to do if the stock is the shortest rough length possible. Surfaces of stock are smoothed by running it through a planer, but the planer will not remove any warp.

The jointer, figure 6-2, consists of two adjustable tables with the cutterhead mounted in between. The stock is placed on the infeed table, fed over the cutterhead, and passed to the outfeed table. A sliding and tilting fence is used to guide the stock as it is being cut. A swinging guard covers the cutterhead and should be used whenever possible.

The size of the jointer is determined by the length of the cutterhead or the width of the tables. In some jointers, the knives are the same length as the cutterhead. In newer jointers, the knives are shorter and staggered around the cutterhead to reduce operating noises.

ADJUSTING THE JOINTER

Adjusting the Outfeed Table

The height of the outfeed table must be exactly the same height as the jointer knives at their highest point, figure 6-3. The cut stock must come to rest exactly on the surface of the outfeed table.

If the outfeed table is excessively high, the stock may hit the outfeed table and not pass through the machine. If the outfeed table is slightly high, the stock may ride over it. This will raise the stock slightly off the infeed table and result in a tapered edge. If the outfeed table is low, the stock will drop down on the outfeed table near its tail end and *snipe* a cut out of the tail end. The outfeed table must be at the correct height to obtain a straight and true cut. Release the hand wheel and raise or lower the outfeed table to adjust it to the correct height.

Adjusting for Parallel Planes

To make accurate, straight cuts with the jointer, the outfeed table and infeed table must be in parallel planes.

1. Lay a straightedge as long as the jointer firmly on the outfeed table, figure 6-4.

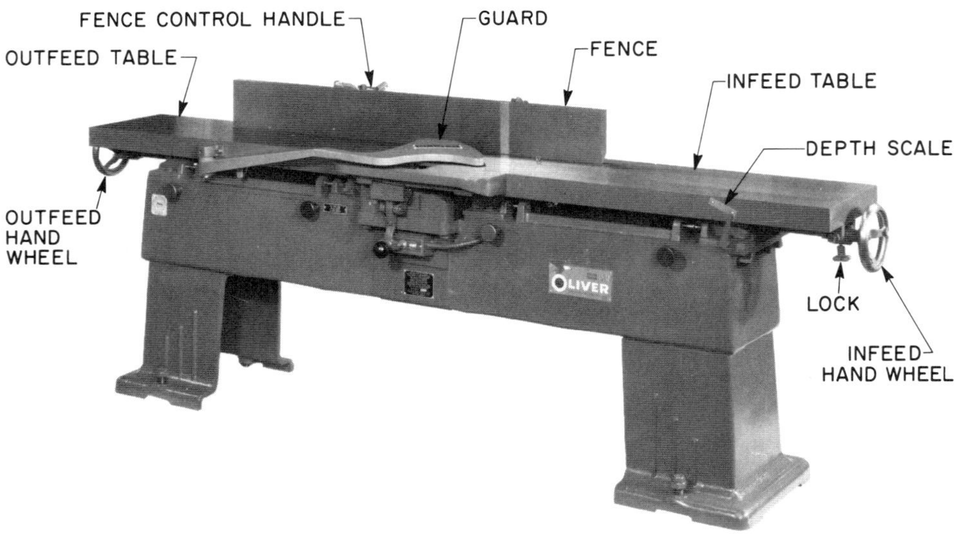

Fig. 6-2 The jointer *(Oliver Machinery Co.)*

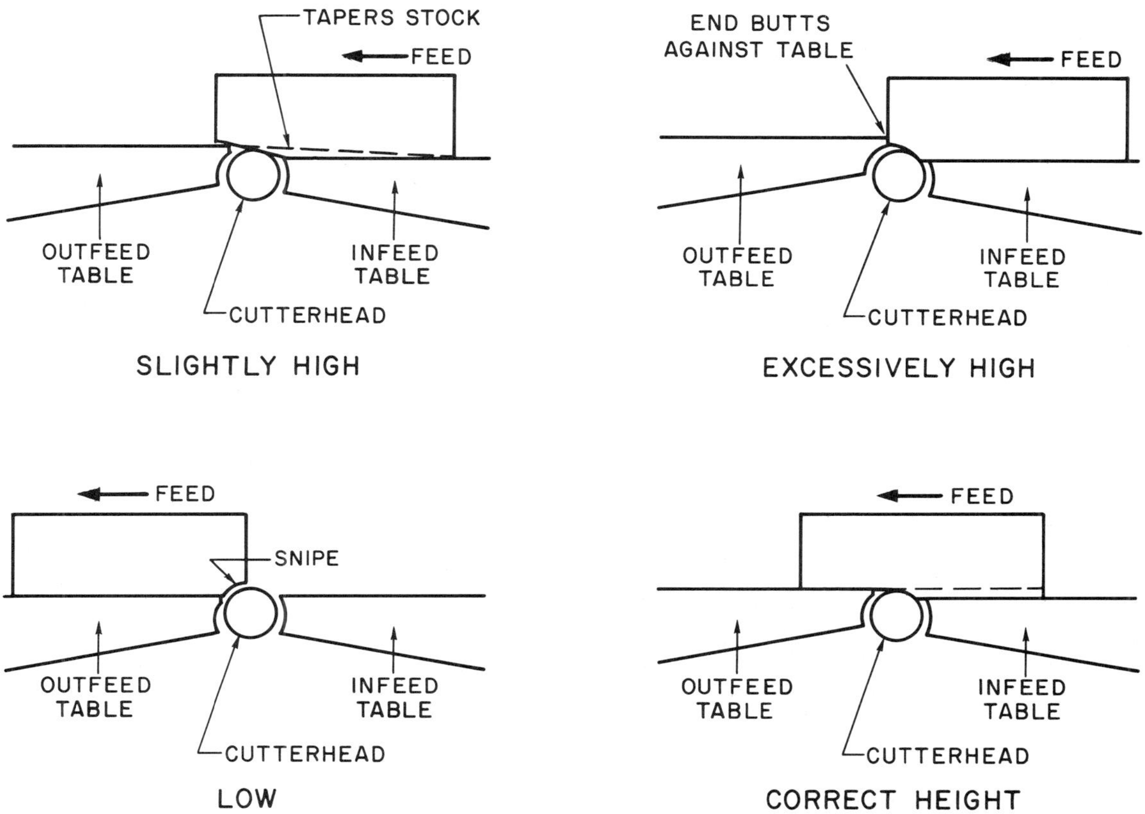

Fig. 6-3 The height of the outfeed table in relation to the cutterhead

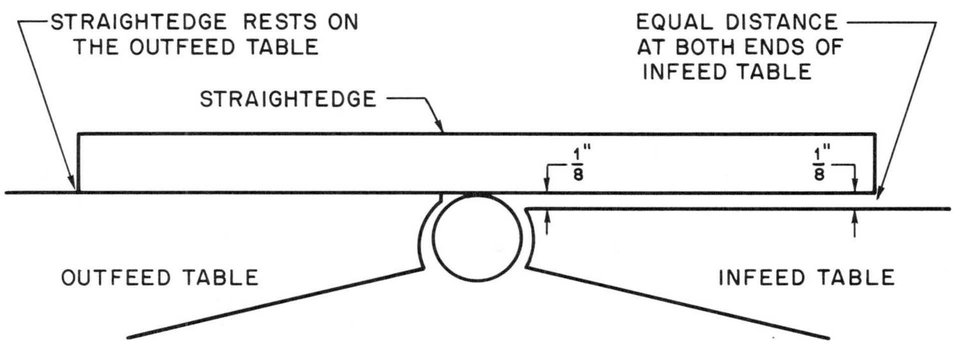

Fig. 6-4 The outfeed and infeed tables are in parallel planes.

2. Lower the infeed table about 1/8 inch.
3. Check the distance between the straightedge and both ends of the infeed table. If the distance is the same, both outfeed and infeed tables are in parallel planes.
4. If the distance is not the same, adjust the tables according to the manufacturer's directions until they are.

Adjusting for Depth of Cut

In order to make a cut, the infeed table must be lower than the outfeed table. The difference in the height of the two tables determines the depth of cut.

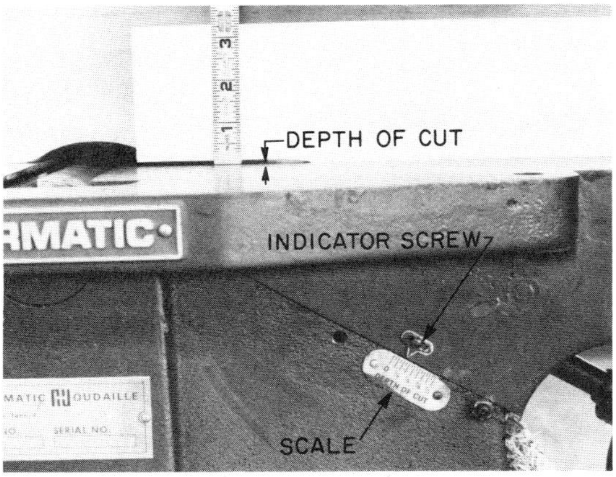

Fig. 6-5 Checking for depth of cut (guard is removed for clarity)

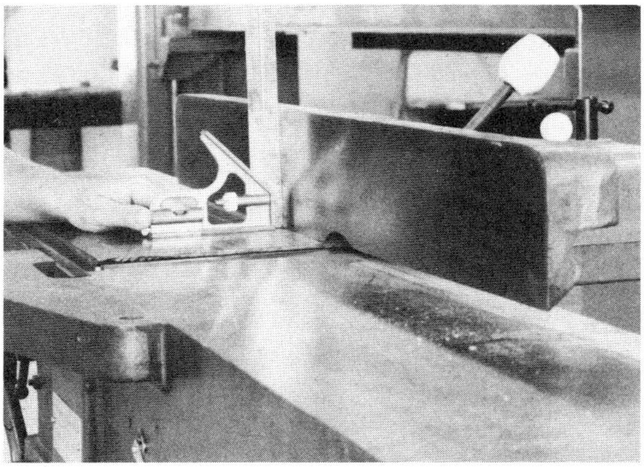

Fig. 6-6 Squaring the jointer fence (guard is removed for clarity)

On most jointers a scale indicates the depth of cut. For ordinary jointing a 1/8-inch cut is average. For facing operations the average depth of cut is 1/16 inch. For finishing cuts remove 1/32 inch.

To check the depth-of-cut scale for accuracy:

1. Set the depth of cut for 1/8 inch.
2. Joint the edge of a piece of stock two inches.
3. Move the stock back on the infeed table. Measure the distance from the cut edge to the surface of the infeed table, figure 6-5.
4. If the measurement does not match the scale, loosen the scale indicator and move to the correct measurement.

Adjusting the Fence

The fence must be at right angle to the table for most operations. Check this angle with a square, figure 6-6. The jointer fence may also be tilted in or out by 45 degrees. To tilt, loosen the fence control handle and adjust the angle of cut. Check the angle with a sliding T bevel, figure 6-7. To equalize wear on the jointer knives, slide the fence sideways in different positions.

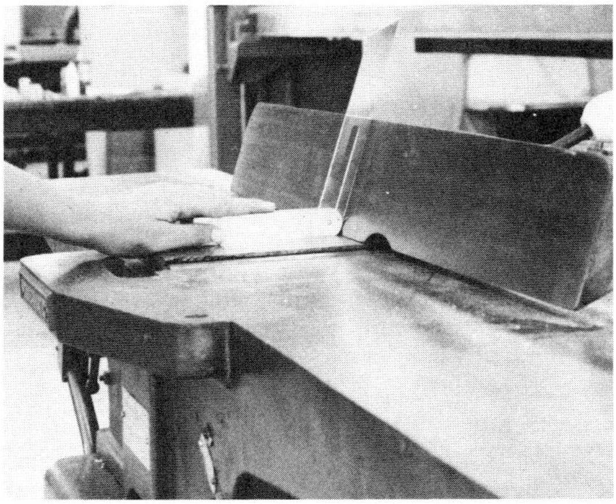

Fig. 6-7 Checking the angle of the fence (guard is removed for clarity)

OPERATING THE JOINTER

Facing

Rough stock should be faced before it is jointed in order to obtain two surfaces at right angles to each other. To face a surface:

1. Adjust the depth of cut to 1/16 inch.
2. Place the warped or hollowed out face against the infeed table and hold it firmly. Place the left hand near the front and the right hand toward the rear.

> **Caution:** Do not use extreme pressure to hold the stock. This may cause the hands to slip farther and faster if the stock catches or is thrown out of control.

3. Turn on the jointer.
4. Move the stock over the revolving cutterhead and onto the outfeed table. As the stock passes onto the outfeed table, apply pressure downward and forward on both tables.
5. As the end of the stock nears the cutterhead, hold the stock with the left hand. With the other, use a push stick to push the stock through the cutterhead, figure 6-8.
6. Check the surface to see that all warp has been removed. If not, make additional 1/16-inch cuts until the surface is faced.

> **Caution:** Do not face or joint pieces shorter than 12 inches. The forward end of short stock may dip down as it crosses over the space between the two tables. This will cause the cutterhead to throw the work backwards instead of cutting it. It will also draw the operator's fingers forward under the guard and into the cutterhead.

Square-Edge Jointing

Square-edge jointing is done in a similar manner to facing except the stock is held on its edge. The faced surface of the stock is held against the fence, figure 6-9. If a piece of stock has an extreme crook, draw a straight line as close to the edge as possible. Saw it to the line and then joint the edge.

Beveling

Cutting a bevel or chamfer on the jointer is done the same as square-edge jointing except that

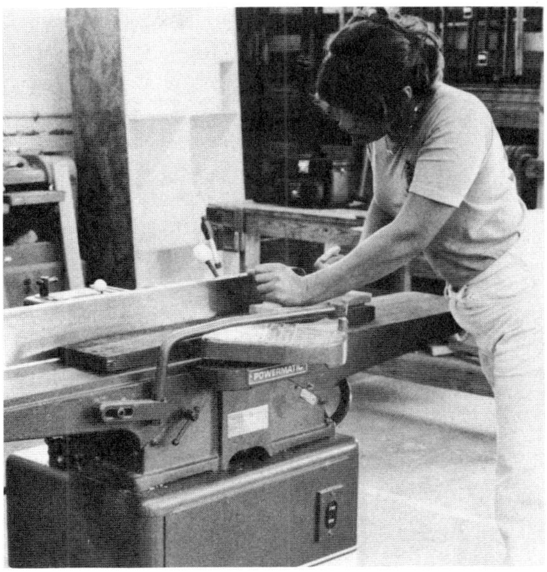

Fig. 6-8 Use a push stick when facing.

Fig. 6-9 Square-edge jointing.

Source

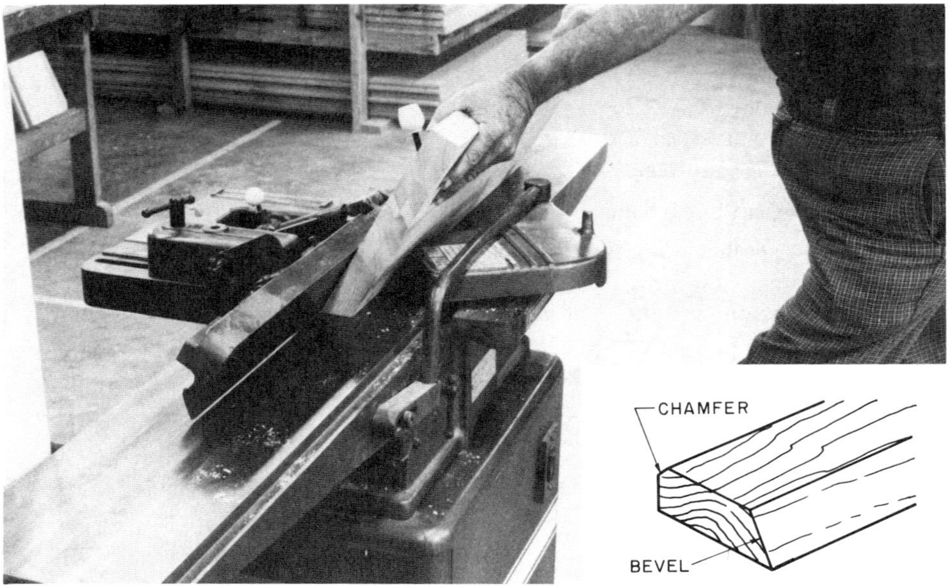

Fig. 6-10 Cutting a bevel or chamfer

the fence is tilted to produce the desired angle, figure 6-10. Care must be taken to hold the stock against the fence.

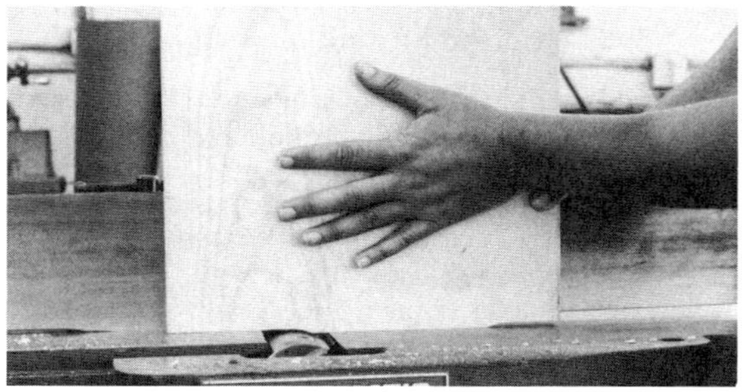

Fig. 6-11 End jointing (guard is removed for clarity)

End Jointing

The ends of stock may also be jointed when necessary. However, never joint the ends of stock unless it is at least 12 inches wide. Joint the end until the cut end rests on the outfeed table, figure 6-11. Turn the stock around and complete the cut, putting pressure on the outfeed table when enough of the end rests on it. This method of end jointing prevents the tail edge of the stock from splitting out.

Fig. 6-12 When rabbeting on the jointer, remove the guard.

Rabbeting

When rabbeting on the jointer, it is necessary to remove the guard.

1. Slide the fence over toward the end of the knives until the distance from the end of the knives to the fence is the desired width of the rabbet.
2. Lower the infeed table to make a 1/8-inch cut.

3. Turn on the jointer and make one cut.
4. Continue lowering the infeed table and making cuts until the desired depth of the rabbet is obtained, figure 6-12. Taking small cuts with each pass insures a much more accurate and smoother cut.

Bevel Rabbeting

Tilt strips are used to cut a beveled rabbet or a wide bevel on stock faces. Tilt strips are made in various thicknesses from 1/16 inch up. They are slightly longer than the infeed table of the jointer. A block fastened to the underside of the tilt strip will catch the end of the infeed table, figure 6-13. The other end of the tilt strip is just short of the cutterhead.

Fig. 6-13 Making a beveled rabbet on the jointer

To cut a beveled rabbet:

1. Slide the fence away from the ends of the knives for the desired width of the rabbet and lock in position.
2. Select a tilt strip of the thickness necessary to give the desired bevel to the rabbet. Lay the tilt strip on the infeed table and against the jointer fence.
3. Lower the infeed table to make a 1/8-inch cut. Move the stock over the cutterhead.
4. Continue lowering the infeed table and making passes until the desired depth of rabbet is obtained.

A bevel across part or all of the width of the stock is also made by using tilt strips. Simply slide the fence over so that all of the stock face rests on the jointer table. Make passes, lowering the table with each pass, until the desired width of bevel is obtained.

Tapering

To cut a long taper:

1. Mark where the taper begins on the face. Mark the amount of taper on the end of the stock.
2. Lower the infeed table about 1/8 inch below the outfeed table.
3. Position the stock so the start of the taper is directly above the jointer knives at its highest point. Clamp a stop block on the infeed table to mark where the stock ends.
4. Start the jointer.
5. Open the guard. Hold the back end of the stock against the stop block while holding the forward end above the jointer table surface.
6. Lower the forward end until it comes to rest on the outfeed table. Push the stock through holding the back end down on the table, figure 6-14.
7. Continue making cuts in this manner until the desired amount of taper is reached as indicated by the marks on the end of the stock.

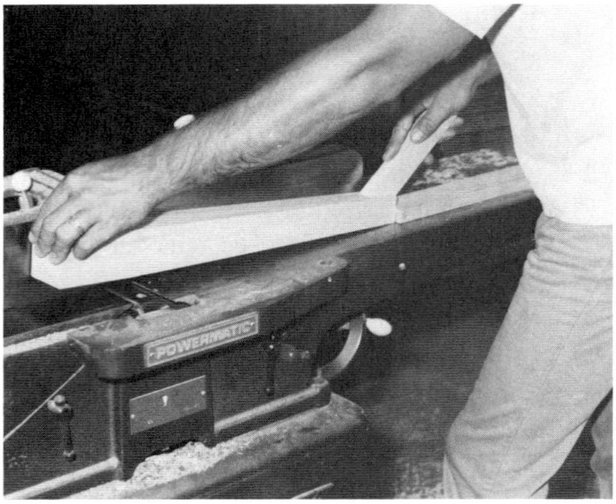

Fig. 6-14 Making a tapered leg on the jointer (guard is removed for clarity)

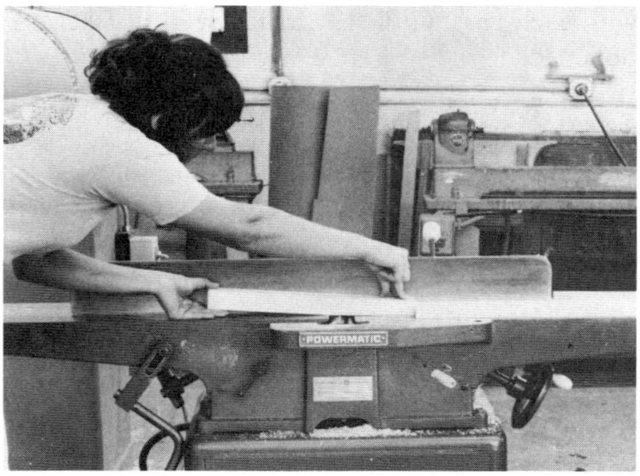

Fig. 6-15 Pull the stock through the cutterhead to make a short taper.

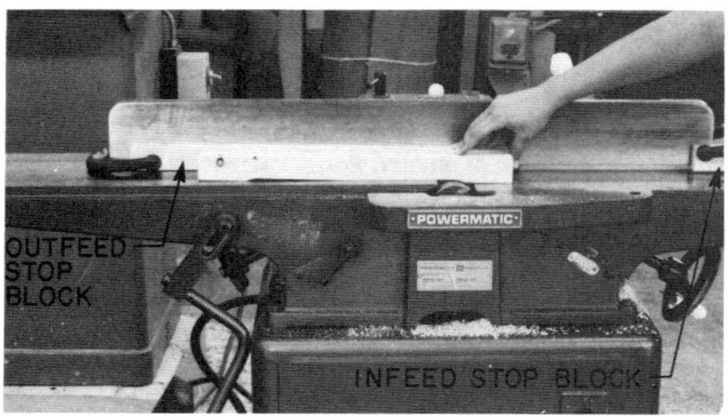

Fig. 6-16 Cutting a stop taper

Caution: Use a push stick to complete the cut if the tapered end is smaller than 1 1/2 inches.

A short taper is cut in a similar manner except the stock is pulled through the jointer instead of pushed, figure 6-15.

Cutting a stop taper is similar to cutting a through taper. However, a stop block must also be clamped to the outfeed table. It is positioned so the forward end of the stock stops where the taper must end, figure 6-16.

Making a Raised Panel

Raised panels are commonly found on doors and walls, figure 6-17. If a shaper is not available in the shop, a raised panel can be made on the jointer.

1. Clamp a straightedge to the outside edge of the jointer so that its edge is above and parallel to the jointer table surfaces, figure 6-18. The heights of the straightedge above the jointer table determines the angle of the beveled edge.
2. Slide the jointer fence toward the end of the jointer knives until the desired width of the bevel is obtained.
3. Lower the infeed table about 1/8 inch.
4. Cut the four sides. Hold the panel at an angle by riding the edge against the fence and down on the jointer table. It is better to cut the

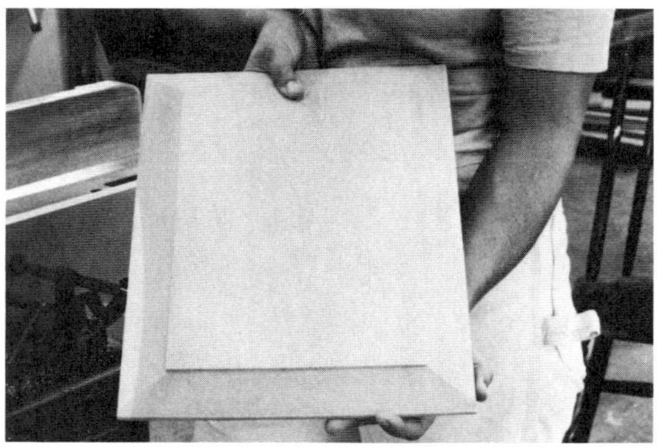

Fig. 6-17 A raised panel

Fig. 6-18 Making a raised panel on the jointer

ends first, then the edges. Splintered ends are then removed when cutting with the grain (the edges).

5. Continue cutting, lowering the table about 1/16 inch with each cut, until the desired depth of cut is reached.

Sharpening Jointer Knives

Maintaining a sharp edge on jointer knives is done by rubbing them with a whetstone at regular intervals, figure 6-19.

1. Turn off and unplug the jointer. Remove the guard.
2. Lower the infeed table about 1/4 inch.
3. Wrap cloth or heavy paper around the stone, leaving a 2-inch end exposed. Lay the covered part of the stone on the infeed table, and the exposed part on the jointer knife.
4. Rotate the cutterhead until the whetstone lays flat on the bevel of the knife. Hold the cutterhead in position by its pulley.
5. Wet the stone with a light oil. Rub it across each jointer knife until its edge is keen. Use the same number of strokes on each knife.

Badly worn or nicked jointer knives must be reground and sharpened. Some jointers are equipped with a knife grinding attachment. The manufacturer's directions should be closely followed when

Fig. 6-19 Sharpening jointer knives with an oilstone

using this attachment or when removing and replacing jointer knives.

THE PLANER

The main purpose of the planer, or *surfacer,* is to smooth stock and reduce it to a uniform thickness, figure 6-20. Most planers in small shops are single-surface planers. These cut the top surface with each pass through the machine. In large mills, double-surface planers cut both sides of stock with a single pass. Planers may remove as much as 1/2 inch with each pass, but 1/8 inch is average.

Like the jointer, the size of the planer is determined by the length of its cutterhead or the width of its tables. Planers usually range in size from 12 to 54 inches.

The stock to be planed rests on a table that can be adjusted closer to or farther away from the cutterhead. A gauge indicates the thickness of the material as it comes out of the machine. The stock must be long enough to reach between the outfeed rollers before it leaves the infeed rollers.

The stock is fed between two infeed rollers. On some models, the speed of the rollers is adjustable. The top infeed roller of a single-surface planer is corrugated and usually made in sections, figure 6-21. A sectional infeed roller allows pieces of slightly different thicknesses to be fed through the machine at the same time.

Fig. 6-20 Single-surface planer *(Oliver Machinery Co.)*

The stock then passes under a chip breaker, which also is usually made in sections. The chip breaker prevents splintering as stock is removed by the cutting knives that revolve against the direction of feed.

A pressure bar just beyond the cutterhead holds the stock firmly on the table to give a smoother cut. Finally, smooth outfeed rollers move the stock out of the machine. Each of these parts are adjusted to close tolerances for smooth operation. Refer to the manufacturer's maintenance manual to adjust these parts.

OPERATING THE PLANER

Some planers have separate motors for the cutterhead and rollers. Variable speed rollers cannot be adjusted unless the rollers are turning. For smoother surfaces on wide work, roller speed is set slower. For faster planing when the quality of the surface is not as important, roller speed is increased.

Caution: Wear ear protection if operating the planer for extended periods of time.

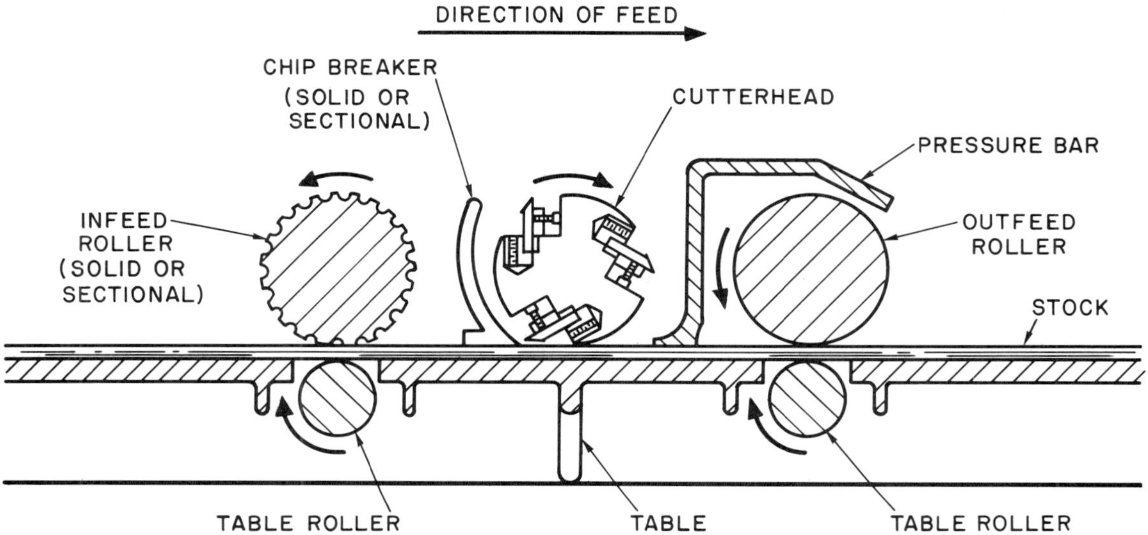

Fig. 6-21 Cross section of a single-surface planer

Fig. 6-22 Straight planing

To plane stock to a uniform thickness:

1. Measure the thickness of the stock and adjust the planer to remove about 1/8 inch. If the stock varies in thickness, use the thickest part as a guide.
2. Start both motors. Adjust the speed of the rollers.
3. Feed the stock into the planer, figure 6-22. If the stock has been faced, feed it in with the face side down. Support long, overhanging ends.
4. If the stock feeds in at an angle, straighten it by pushing on the end. Some planers have a reverse switch to turn the rollers backwards if stock gets stuck in the machine.

Caution: Do not look into the planer and watch the stock being planed. Chips may fly out.

5. As the stock feeds out of the planer, support the end until the cut is complete.
6. On the first pass, check the thickness of the stock with the thickness gauge. If there is a difference, adjust the gauge.
7. Run all pieces through on the last pass to insure the uniform thickness of all pieces.

The planer cannot be used to flatten warped stock. The pressure of the infeed roller will flatten the board as it moves under the cutterhead, but the board will spring back to its warped shape as soon as it leaves the outfeed roller. A warped board must be straightened on a jointer first.

Planing a Bevel

Stock can be beveled on a planer so one edge is thinner than the other edge.

1. Cut a narrow strip of wood the same length as the stock. Its thickness will determine the amount of bevel.
2. Fasten the strip flush to one edge on the underside of the stock with wire brads. Make sure fasteners are not so long they will be clipped off by the cutterhead.
3. Adjust the planer. Plane the stock with the narrow strip down on the table, figure 6-23.
4. Continue making passes until the desired thickness or width of bevel is obtained.

Planing a Taper

Stock can be tapered on a planer so one end is thinner than the other end.

Fig. 6-23 Planing a bevel

Fig. 6-24 Fasten wedge-shaped strips to the stock.

1. Cut two similar wedge-shaped strips to the desired amount of taper for the length of the stock.
2. Fasten the strips flush to each edge on the underside of the stock with wire brads, figure 6-24.
3. Adjust the planer. Plane in the usual manner with the strips down on the table, figure 6-25.
4. Continue making passes until the desired thickness and length of taper are obtained.

Sharpening Planer Knives

Planer knives are whetted in a manner similar to jointer knives. To whet planer knives:

1. Turn off and unplug the planer. Raise the hood clear to expose the cutterhead.
2. Hold the cutterhead steady and rub a lightly oiled whetstone on each knife to produce a keen edge.
3. Use the same number of strokes on each knife to produce equal edges.

Like jointers, planers sometimes are equipped with a knife grinding attachment. This attachment grinds and hones the planer knives without removing them from the cutterhead. Follow the manufacturer's instructions carefully when using the attachment.

Fig. 6-25 Planing a taper

ACTIVITIES

1. Visit a wood mill and observe the jointing, facing, and planing methods and machinery used.
2. Adjust the jointer to make true cuts.
3. Face stock and joint a square edge, beveled edge, and chamfered edge.
4. Joint an end.
5. Cut a rabbet to a specified size on the edge of stock using the jointer.
6. Make a beveled rabbet or a wide bevel of specified size and angle on the face of stock by using tilt sticks and the jointer.
7. Make a table leg with a short through taper and long stop taper on the jointer.
8. Design a raised panel and make it on the jointer.
9. Bevel plane and taper stock to specifications using the planer.
10. Bring jointer and planer knives to a keen edge using a whetstone.

UNIT REVIEW

Multiple Choice

1. Straightening the face of a board is called
   a. facing.
   b. planing.
   c. jointing.
   d. truing.
2. The size of a jointer is determined by
   a. its depth of cut.
   b. the length of its tables.
   c. the width of the knives.
   d. the length of its knives.
3. The main purpose of a jointer is to
   a. bevel the faces of lumber.
   b. smooth the edges and faces of lumber.
   c. rabbet the edges of lumber.
   d. straighten the edges and faces of lumber.
4. Facing and jointing operations are more easily accomplished by first
   a. removing all warp.
   b. bringing the stock to uniform width.
   c. bringing the stock to the shortest possible rough length.
   d. bringing the stock to uniform thickness.
5. The height of the jointer outfeed table must correspond with the height of the
   a. infeed table.
   b. jointer knives.
   c. bottom edge of the jointer fence.
   d. jointer cutterhead.
6. When using the jointer, the depth of cut is determined by the
   a. height of the outfeed table.
   b. difference in height of the two tables.
   c. material to be cut.
   d. depth of cut scale.

7. The jointer fence should occasionally be moved sideways to different positions to
   a. equalize wear on the knives.
   b. keep the guard operating properly.
   c. expose as little of the knives as possible.
   d. keep the fence sliding mechanism clear.

8. Jointing a board means
   a. removing all warp.
   b. straightening the face.
   c. straightening the edge.
   d. cutting the uniform thickness.

9. On a single-surface planer, the only roller that may be corrugated is the
   a. top infeed roller.
   b. bottom infeed roller.
   c. top outfeed roller.
   d. bottom outfeed roller.

10. Stock to be planed must be at least
    a. the length of the cutterhead.
    b. 16 inches.
    c. the center-to-center distance between the infeed and outfeed rollers.
    d. 12 inches.

Questions

1. How are the infeed and outfeed tables of the jointer checked for parallel planes?
2. How is the height of the outfeed table of the jointer adjusted?
3. Describe what happens when jointing an edge if the outfeed table is too high.
4. Describe what happens when jointing an edge if the outfeed table is too low.
5. If there is a difference between the actual depth of cut and the jointer depth-of-cut scale, how is this corrected?
6. Describe the danger of jointing a piece of stock that is too short.
7. What method is used for end jointing?
8. What is the advantage of an infeed roller that is made in sections?
9. What is the purpose of the planer chip breaker? The pressure bar?
10. Describe how a bevel may be planed on the side of stock.

unit 7 Table Saw

OBJECTIVES

After studying this unit, the student will be able to:

- describe and adjust the parts of the table saw for all operations.
- select, remove, sharpen, and replace saw blades.
- crosscut and rip on the table saw using proper attachments.
- cut grooves, dadoes, and rabbets using the dado head and shape stock using the molding head.

TABLE SAW

The table saw is frequently used in the woodworking shop because it is so versatile. Most table saws are the *tilting-arbor* type, figure 7-1. It turns a circular saw blade which is mounted on an arbor below the tabletop surface. The blade can be raised or lowered to adjust the depth of cut, and its axis can be tilted to 45 degrees. A scale shows the degree of tilt.

The size of the table saw is determined by the diameter of the circular saw blade. Blades can measure up to 16 inches or more. Since the blade cuts on the downturning side, the stock is placed faceup on the table. Any splintering therefore occurs on the bottom.

The table saw has three important features: the safety guard, the rip fence, and the miter gauge.

The *safety guard* covers the saw blade to protect the operator. There are many different kinds of safety guards, but most have antikickback features. These prevent the stock from being thrown back at the operator if the work binds in the cut. *A splitter* directly in back of the blade, for instance, keeps the saw cut open and prevents stock from binding. Although some operations cannot be done with the safety guard in place, it should be used whenever possible.

> **Caution:** Never stand directly in back of the saw blade. If the stock does kick back, it will hit you. Stand to the side of the stock. Make sure no one else is in line with the blade.

A *rip fence* is a metal straightedge that is clamped on the table parallel to the blade. The rip fence guides the work during ripping operations. Its distance from the saw blade can be adjusted. It is also aligned so there is about 1/64 inch more clearance from the blade at the back of the table than at the front.

With the rip fence, the table saw can rip, bevel rip, groove, bevel groove, rabbet, and tenon. When

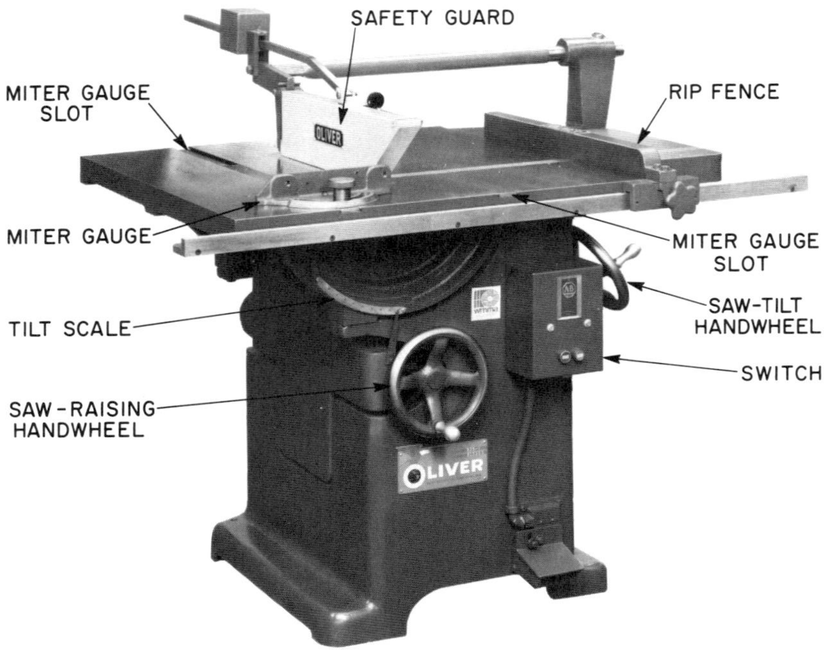

Fig. 7-1 The table saw *(Oliver Machinery Co.)*

ripping (cutting with the grain), the stock is held against the rip fence and fed past the blade.

A *miter gauge* guides the work for crosscutting operations. It slides across the table in one of the two grooved slots running parallel on either side of the blade. To check if slots are parallel to the blade, measure the distance from the groove to the front and back of the blade. Measure from the same saw tooth. If the groove is not parallel, loosen the tabletop bolts. Tap the tabletop until the grooves are parallel to the blade, then tighten the tabletop back down.

The miter gauge also angles left or right up to 45 degrees. It is checked for accuracy with a framing square. Raise the blade to its full height. Place one edge of the square against the miter gauge and the other edge against the side of the saw blade. Adjust the miter gauge as necessary. The miter gauge also has stops for 45 and 90-degree cuts. These should also be checked for accuracy.

When crosscutting (cutting across the grain), the stock is held firmly against the miter gauge and fed past the blade. With a miter gauge, the table saw can crosscut, bevel crosscut, dado, miter, bevel miter, and tenon.

Caution: It is very dangerous to make freehand cuts. A guide such as a rip fence, miter gauge, or jig should be used. Otherwise, the blade may bind in the cut and kick back the stock.

CIRCULAR SAW BLADES

The table saw, radial arm saw, and portable circular saw all use the same types of circular saw blades. Selecting the proper blade for the job is very important.

The more teeth in a circular saw blade, the smoother the cut edge will be, but the slower the feed. A coarse-tooth blade leaves a rough edge, but the stock may be fed faster. If the feed is too slow, the blade may overheat causing it to lose its shape and wobble. The saw will then start to bind in the cut. This is why a fine-tooth blade is not used to cut heavy or rough stock.

Saw blades are made of high-speed steel or tungsten carbide. Carbide-tipped blades retain their cutting edge longer than steel ones. They are used to cut material that contains adhesives or other foreign material, particleboard, hardboard, plywood, and plastics.

Regardless of the type, the number and shape of the saw teeth vary. The shape of the teeth determines a blade's cutting action.

High-Speed Steel Blades

Steel blades are usually classified as rip, crosscut, or combination saw blades, figure 7-2. Ripsaw blades usually have less teeth than crosscut or combination blades. Like the rip handsaw, the teeth are filed or ground at right angles to the face of the blade. This produces a cutting edge all the way across the tooth. Ripsaw blades are used to cut solid lumber with the grain when a smooth edge is not necessary. It is also used to cut green (unseasoned) lumber and heavy-dimension, dry lumber.

Like the crosscut handsaw, the sides of crosscut circular saw teeth are alternately shaped on a bevel to produce a point instead of an edge. These points slice the wood fibers smoothly. Crosscut blades are used to cut across the grain of solid lumber. They are also used to cut plywood, but they dull quicker than carbide blades when used for this purpose.

The combination blade has both types of teeth and is used for a variety of ripping and crosscutting operations. This blade eliminates changing blades for each operation.

All blades must clear the saw cut for the saw blade. Without this clearance, the saw blade will bind in the cut. The set of the teeth of most blades provides this clearance.

A *taper-ground blade,* also called a *planer blade,* is not set. It is thicker at the tips of the teeth and thinner towards the center to provide clearance. It usually has a combination of rip and crosscut teeth. Taper-ground blades cut a very smooth edge and are used on straight, dry lumber of relatively small dimension. They do not provide much clearance, however, and the slightest twist in the stock will cause it to bind against the blade.

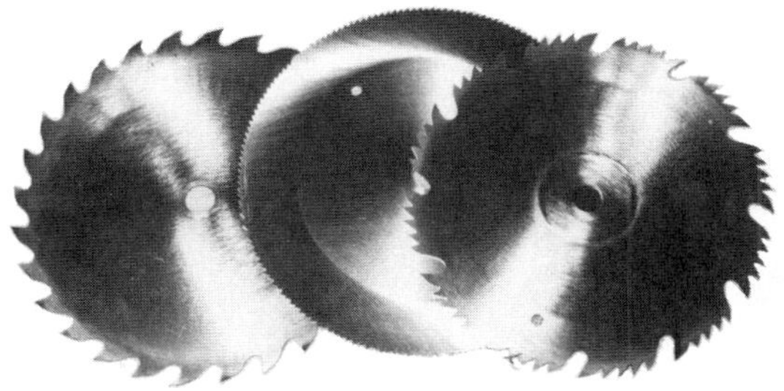

Fig. 7-2 High-speed steel circular saw blades

Carbide-Tipped Blades

Carbide-tipped blades maintain a sharp edge under conditions that cause other blades to dull quickly. They are very useful for high-production work and for cutting composition materials.

Generally 12-inch, carbide-tipped blades contain from 8 to 80 teeth or more. Larger diameters have even more teeth. Some carbide-tipped blades therefore produce smooth cuts, since the more teeth in a saw blade the smoother the cut.

Three main styles of carbide-tipped teeth are common, figure 7-3. These are the square grind, alternate top bevel, and triple-chip grind.

The *square grind* is similar to ripsaw teeth in a steel blade. It is used mainly to cut solid wood with the grain and to cut composition boards when the quality of the edge is not important.

The *alternate top bevel grind* is used for crosscutting solid lumber and for cutting plywood, hardboard, and particleboard.

A *triple-chip grind* can cut brittle material without chipping out the surface. It cuts an extremely smooth edge on plastic-laminated material. When cutting hardwood plywood, a 60-tooth, triple-chip-grind blade hardly splinters any of the thin face veneer. Like a planer blade, it produces a smooth edge on straight, dry lumber.

Carbide-tipped blades also come with a combination of teeth. The leading tooth in each set is square ground, while the following teeth are ground at alternate bevels. Carbide-tipped teeth are not set. The carbide tips are slightly thicker than the saw blade and therefore provide clearance in the saw cut for the blade.

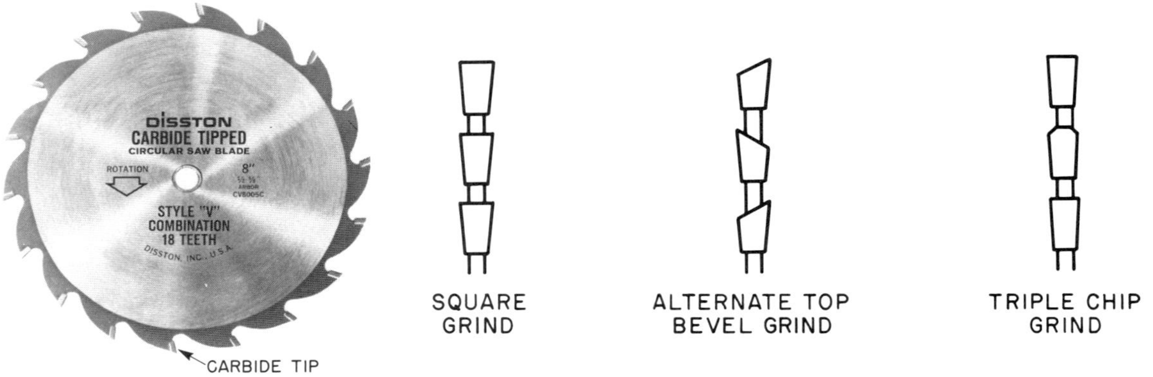

Fig. 7-3 Carbide-tipped circular saw blade *(Disston)*

Removing and Replacing Blades

Whether saw blades are steel or carbide-tipped, they should be replaced when dull. Always use a sharp saw blade.

Arbor shafts have a right or left-hand thread depending on the side it is located. In either case, the arbor nut is loosened in the same direction that the saw blade rotates. The arbor nut is tightened against, or opposite, the rotation of the saw blade. This design prevents the arbor nut from loosening during operation.

To remove the blade, turn off and unplug the table saw and remove the guard. Remove the insert plate that surrounds the blade and hold the blade steady with a strip of softwood, figure 7-4. With a wrench pull the arbor nut in the same direction as the blade turns. Remove the nut, collar, and saw blade.

To replace the blade, reverse the procedure. Do not put the blade on backwards. Make sure the teeth point toward the operator. Replace the collar, tighten the nut securely, and rotate the blade by hand to make sure it is running clear. Replace the metal insert.

Sharpening Circular Saw Blades

Circular saw blades are usually sent to a sharpening shop to be reconditioned with special equipment. It involves four steps: jointing, gumming, setting, and filing. These steps can also be done on the job.

Fig. 7-4 Removing and replacing circular saw blades

Jointing rounds teeth to an even height:

1. Lower the blade below the tabletop and place an oilstone over the teeth.
2. Start the machine and *slowly* raise the blade until the teeth touch the stone, figure 7-5.
3. Stop the machine and inspect the teeth. All should have touched the stone. Do not joint more than necessary.

Gumming grinds the gullets to the proper depth:

1. Raise the blade until a pencil held flat on the table comes just below the deepest *gullet*, or valley, between the teeth. Rotate the blade to draw a line around it. This will be the guide for gumming, figure 7-6.

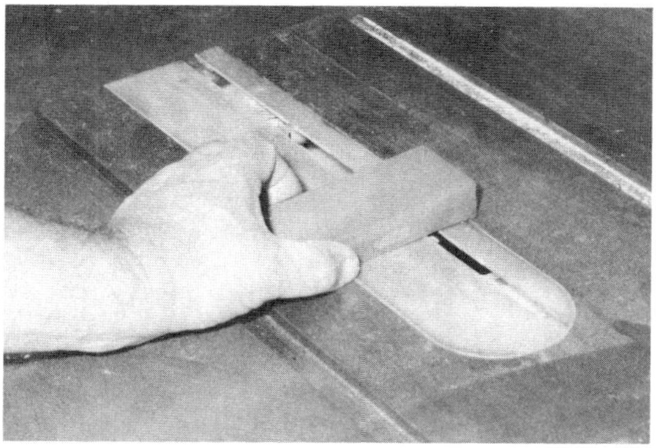

Fig. 7-5 Jointing a circular saw blade on the job

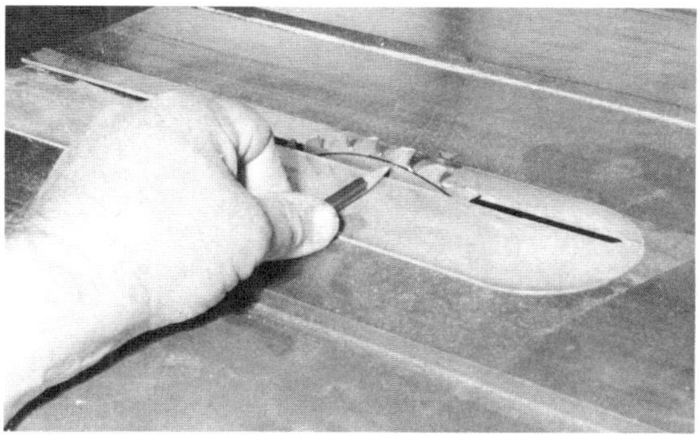

Fig. 7-6 Marking the blade for gumming

Fig. 7-7 Gumming the blade

2. Remove the blade. Grind out the gullets on the bench grinder with a narrow, round-edge grinding wheel, figure 7-7. Grind down to the pencil mark.
3. Maintain the correct shape. Do not overheat the teeth.

Setting bends the teeth right and left to provide clearance:

1. Place the blade on the beveled anvil of the saw set, figure 7-8.
2. Set the teeth outward by striking the top one-third of each tooth with a punch. Remember that taper-ground blades are not set.

Filing brings the teeth to a sharp point:

1. Secure the blade in a circular saw vise, figure 7-9.
2. File the teeth the same as handsaw teeth. Maintain a uniform shape and height. Use a flat or triangular file.
3. File until the shiny tip disappears. Do not file any more than necessary.

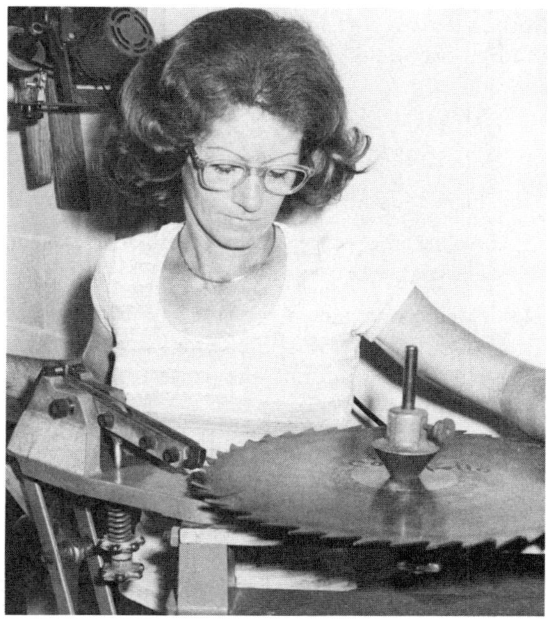

Fig. 7-8 Setting the blade

Fig. 7-9 Filing the blade

Fig. 7-10 Crosscutting on the table saw

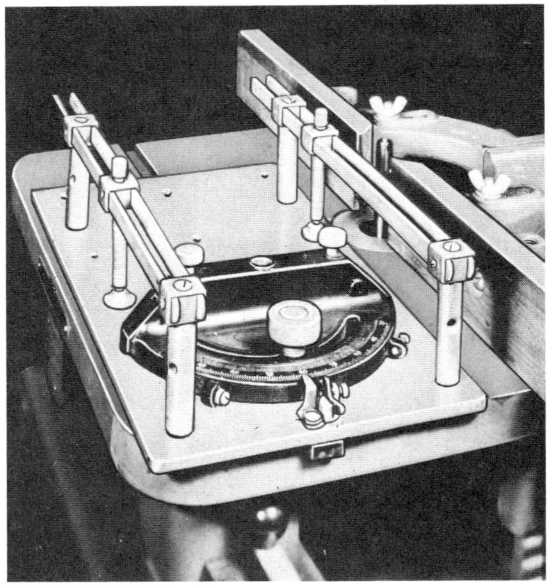

Fig. 7-11 Clamp attachment for the miter gauge *(Rockwell)*

CROSSCUTTING ON THE TABLE SAW

The miter gauge is used for most crosscutting work. Usually the miter gauge runs in the left-hand groove, unless the right-hand groove is more convenient.

To cut stock to length with square ends:

1. Check the miter gauge for accuracy.
2. Place the stock faceup against the miter gauge.
3. Adjust the blade so the teeth just project above the surface of the stock.
4. Square one end of the stock by holding the work firmly against the miter gauge with one hand while pushing the gauge forward with the other hand.
5. From the squared end, measure the desired distance and mark it on the front edge of the stock.
6. Cut the stock to the line, figure 7-10.

Caution: Eye protection should be worn during all table saw operations.

During crosscutting operations, the stock has a tendency to creep away from the saw blade. This results in an inaccurate cut, especially when cutting close to the end of stock. To prevent this, an *auxiliary miter fence* with sandpaper glued to its face is screwed to the miter gauge, see figure 7-12. This holds the stock to be cut more securely against the fence. The auxiliary fence may be as long as necessary to accommodate longer lengths of stock or to make duplicate lengths.

A clamp attachment on the miter gauge may also be used to hold stock while crosscutting, figure 7-11.

Crosscutting Duplicate Length

To cut a number of identical lengths:

1. Square one end of the pieces to be cut.
2. Clamp a stop block to the auxiliary miter fence the desired distance from the saw blade, figure 7-12.
3. Place the squared end of the stock against the stop block and make the cut. Remove the pieces.

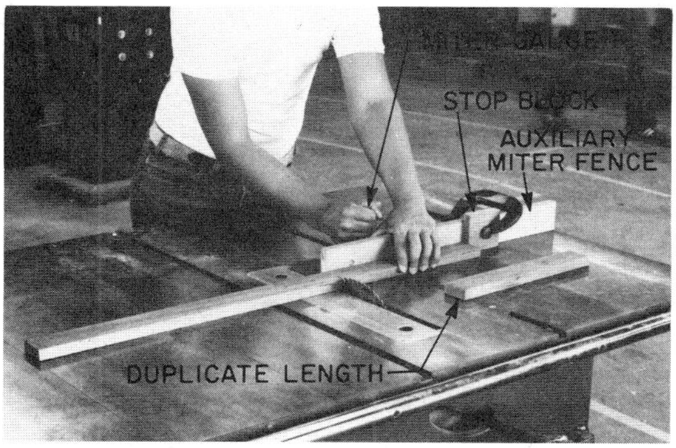

Fig. 7-12 Cutting duplicate lengths (guard is removed for clarity)

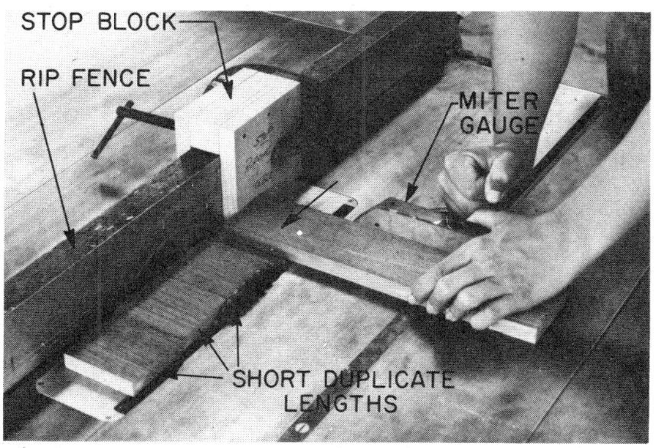

Fig. 7-13 Cutting short duplicate lengths (guard is removed for clarity)

4. Slide the remaining stock across the table until its end contacts with the stop block to cut the desired number of pieces.

To cut short pieces of identical length:

1. Clamp a stop block to the rip fence just in front of the blade. This allows clearance between the workpiece and the rip fence. It also prevents the work from binding between the saw blade and fence and eliminates kickback.
2. Adjust the fence so that the distance between the stop block and the blade is the desired length.
3. Square one end of the stock.
4. Slide the squared end against the stop block and make a cut, figure 7-13.
5. Repeat the procedure until the desired number of pieces are cut.

Mitering

Miter cuts are cuts made at an angle to the edge or side of the stock. Flat miters are cut like square ends except the miter gauge is turned to the desired angle, figure 7-14. Edge miters are cut by adjusting the miter gauge square with the blade and tilting the blade to the desired angle. Compound miters are cut with the miter gauge turned and the blade tilted to the required angles, figure 7-15.

A *mitering jig* is very useful for fast, efficient, and accurate miter cutting, figure 7-16. With the

Fig. 7-14 Cutting a flat miter (guard is removed for clarity)

Fig. 7-15 Making a compound miter cut (guard is removed for clarity)

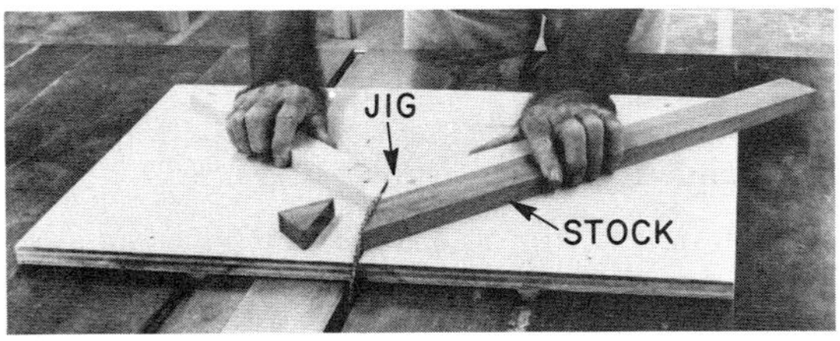

Fig. 7-16 Making miter cuts with a mitering jig (guard is removed for clarity)

Fig. 7-17 Wide stock may be cut by reversing the miter gauge. (guard is removed for clarity)

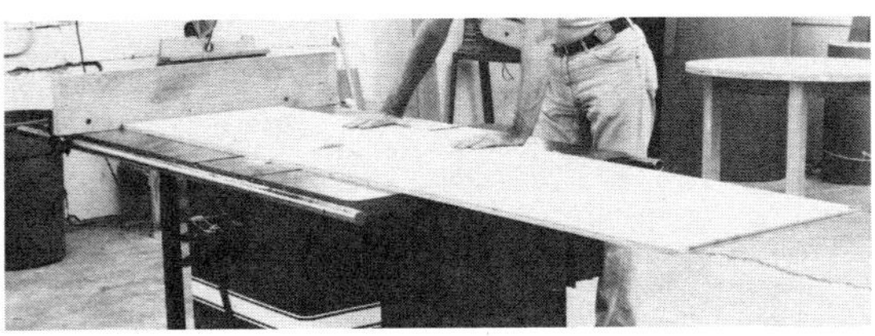

Fig. 7-18 The rip fence may be used as a guide to crosscut wide stock. (guard is removed for clarity)

mitering jig, the miter gauge does not have to be turned one way for left-hand miters and the other way for right-hand miters.

Crosscutting Wide Stock

Wide stock may be crosscut by guiding the stock with the miter gauge reversed in the slot, figure 7-17. Another method uses the rip fence as a guide, figure 7-18. The end of the stock is held against the fence to prevent wobbling which might cause the stock to bind against the blade.

When one end of wide stock is uneven, a straightedge clamped to the underside of the stock acts as a guide along the edge of the table, figure 7-19. It is clamped the necessary distance from the saw blades. The straightedge is held against the table edge when making the cut.

RIPPING ON THE TABLE SAW

All ripping operations use the rip fence as a guide. Usually it is placed to the right of the blade.

Another useful aid is an *auxiliary rip fence.* A straight piece of 3/4-inch plywood, about 12 inches wide, and as long as the metal fence is adequate. When cuts are made very close to the fence, an auxiliary fence prevents the saw blade from cutting into the metal fence. Also, the additional height gives a broader surface to steady wide work when its edge is being cut. Feather boards can also be clamped to it, figure 7-20.

Feather boards are used to hold work against the fence and down against the tabletop during ripping operations. They are wood pieces with one

Fig. 7-19 Use the table edge as a guide to cut uneven ends of wide stock. (guard is removed for clarity)

end shaped at a 45-degree angle. Saw cuts are made in this end about 1/4 inch apart, figure 7-21. This gives the end some spring and allows it to apply pressure to stock when clamped against it.

Caution: When ripping stock:

1. If the stock starts to bind in the cut when ripping, lift it clear of the blade immediately. If this cannot be done, hold the stock in position and shut off the power. Never try to back the work out of a cut with the saw blade running.
2. Do not pick small waste pieces from the table saw when the saw is running. Remove them with a stick or wait until the saw has stopped.
3. When ripping long stock, support overhanging ends.
4. Never leave a saw unattended when it is turned on.

Ripping Stock

To rip stock to width:

1. Measure the desired distance from the fence to the nearest saw tooth. Lock the fence in place.
2. Adjust the height of the blade. With the stock clear of the blade, turn on the power.
3. Hold the stock against the fence with the left hand. Push the stock forward with the right hand and let the stock slide through the left hand. Use a smooth, continuous motion and feed the stock as fast as possible, figure 7-22.
4. *For wide stock:* Remove the left hand as the end of the stock approaches it. Push the stock all the way through the blade with the right hand if the stock is at least 5 inches wide.

   For narrow stock: Grasp a push stick with the right hand. Use the stick to push the stock through the blade, figure 7-23.

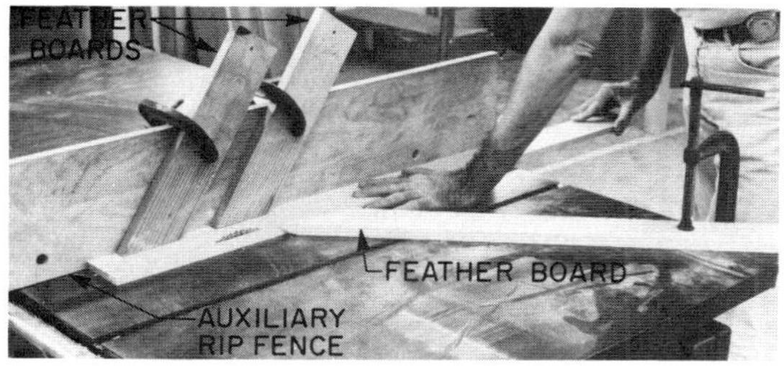

Fig. 7-20 Feather boards are clamped to an auxiliary rip fence. (guard is removed for clarity)

Caution: The left hand should never come up to or beyond the saw blade when ripping.

Make sure the stock is pushed all the way through the saw blade. Leaving the cut stock between the fence and a running saw blade may cause it to kick back and injure anyone in its path. Remember that ripping is done with a smooth, forward movement. This produces a smoother cut.

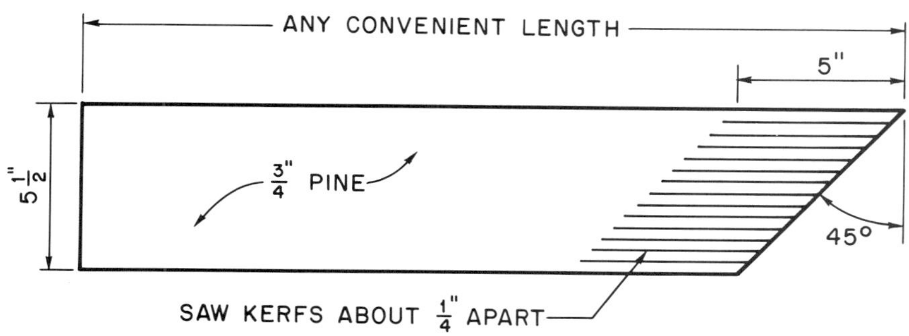

Fig. 7-21 Feather board design

The push stick is designed to be held at a 45-degree angle to the table surface. This is the best angle for holding down stock and pushing it ahead. The push stick must be thin enough to go between the fence and blade and still have clearance.

Caution: The push stick must not come in contact with the revolving blade. If it does, it may be forced out of the operator's hand and cause an injury.

Fig. 7-22 The correct position for ripping a board

Fig. 7-23 Use a push stick to rip narrow pieces. (guard is removed for clarity)

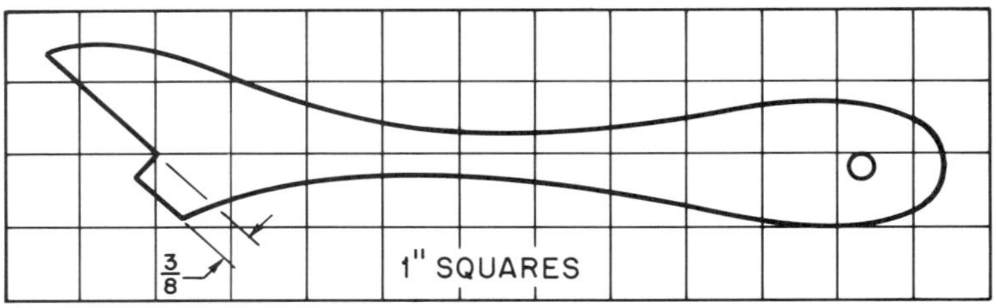

Fig. 7-24 Push stick design *(Powermatic)*

Fig. 7-25 When bevel ripping, the fence is placed on the side away from the tilt of the saw blade. (guard is removed for clarity)

Fig. 7-26 To resaw, cut half the width on the first pass. Reverse the stock and make the second cut. (guard is removed for clarity)

Before starting to rip narrow stock, make sure a push stick is handy. If one is not available, make one. A good design is shown in figure 7-24.

Bevel Ripping

Cutting a bevel with the grain is similar to straight ripping except that the blade is tilted to the desired angle. On some saws, the blade tilts to the right; on others, the blade tilts to the left. In either case, the fence is placed on the side that the blade tilts away from, figure 7-25.

Resawing

Cutting thick pieces to make two or more thinner pieces is called *resawing.* Resawing is generally done on the band saw because it has a thinner blade which removes less waste. Also, wider pieces can be resawed on the band saw than on the table saw.

To resaw on the table saw:

1. Raise the saw blade to cut slightly more than half the width of the piece.
2. Set the fence for the desired thickness and make one cut, figure 7-26.
3. Turn the piece over and make the second cut.

Ripping Thin Stock

When ripping thin stock, the cut pieces may slide between the rip fence and the tabletop. To avoid this, an auxiliary wood table of plywood is used.

1. Adjust the fence to the desired position.
2. Lower the blade below the table surface.
3. Clamp a thin panel of plywood on the table surface and against the fence.
4. Start the power and raise the blade so that it cuts through the plywood. This material can now be cut, figure 7-27.

Taper Ripping

Tapered pieces have one end narrower than the other. They are made on the table saw by using a *taper-ripping jig.* The jig consists of a wide board with the length and amount of taper cut out of one edge. The straight edge is held against the ripping fence. The stock is held in the cutout of the jig, figure 7-28. A handle on the jig makes it safer to use. If the jig is the same thickness as the stock, the jig cutout can be covered to prevent waste from flying out.

If equal tapers are needed on both edges, the jig is made with two notches. The piece is held in one notch to make the first cut. Then, to compensate for the tapered edge when the piece is turned over, the stock is held in the second notch to make the taper on the opposite edge, figure 7-29.

Fig. 7-27 An auxiliary wood table prevents thin strips from slipping under or through the table insert. (guard is removed for clarity)

Fig. 7-28 Cutting a taper with a taper-ripping jig (guard is removed for clarity)

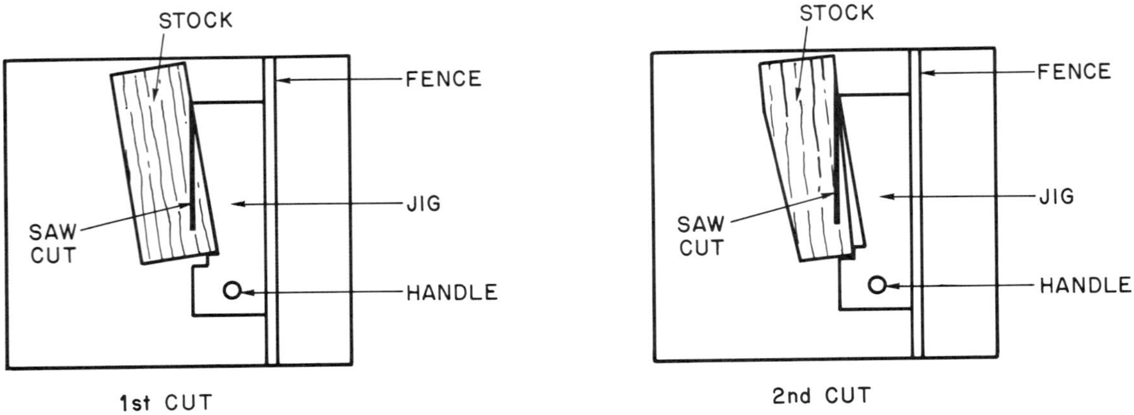

Fig. 7-29 Jig for cutting tapers on opposite edges

Making Cove Cuts

Cove cuts are concave cuts made by feeding stock across the saw blade at an angle. A number of light cuts gradually hollows out the stock. This reduces the side stress on the saw blade. A straightedge is clamped to the tabletop at an angle to guide the stock. The angle determines the radius of the cove.

To make a cove cut:

1. Raise the saw blade to the overall depth of cut.
2. Lay the straightedge on the table and bring it up against the teeth of the blade, figure 7-30.
3. Move the straightedge, keeping it against the teeth, until the angle it makes with the blade equals the width of the cut.
4. Clamp the straightedge in this position. Place the clamps on the side away from the blade.
5. Lower the blade to cut about 1/16 inch. Feed the stock through the blade. Make more passes by raising the blade by 1/16 inch with each cut until the desired depth and width of cut are obtained.

> **Caution:** Use a push stick if the blade comes close to the top surface.

Fig. 7-30 Making cove cuts (guard is removed for clarity)

THE DADO HEAD

The *dado head* is commonly used to make dado, groove, and rabbet cuts on the table saw. It is mounted on the saw arbor like a saw blade. One type consists of two outside circular blades and several inner chippers of different thicknesses, figure 7-31. By using different chippers, the width of the cut is changed. Paper washers may be inserted between the blades to widen the cut slightly. The dado head must be removed from the saw arbor to change settings.

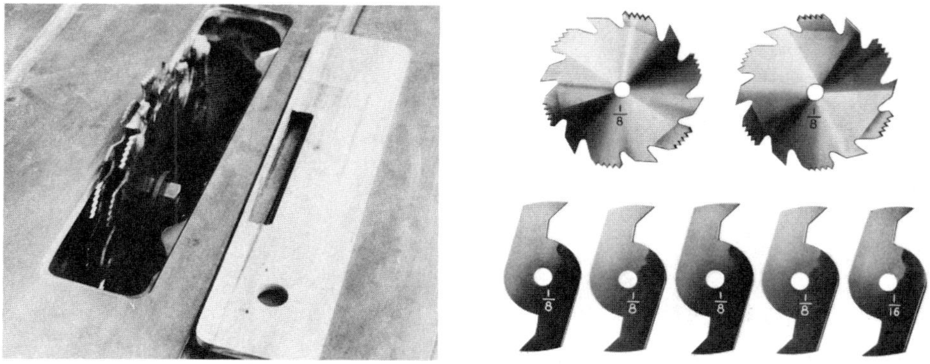

Fig. 7-31 The dado head has two outer blades and several inner chippers.

Most dado heads make cuts from 1/4 inch to 13/16 inch. When installing this type of dado head, make sure the tips of the chippers do not touch the sides of the outside blades. Chipper tips are thicker than the body of the chipper itself to insure a clean cut.

Another type of dado head is a one-unit adjustable dado head, figure 7-32. This type does not have to be removed from the saw arbor to change settings. In some models a calibrated dial is turned to adjust for the width of cut desired.

Both types of dado heads are available with high-quality steel or carbide-tipped blades.

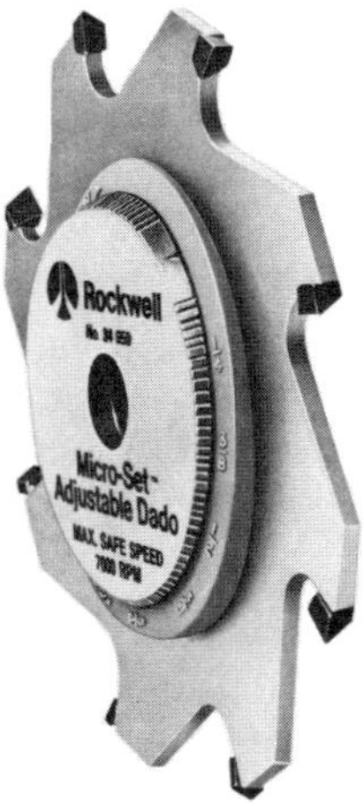

Fig. 7-32 One-unit adjustable dado head

Cutting Dadoes

A *dado* is a wide cut made part way through the stock across the grain, figure 7-33. To cut dadoes with a dado head, use the miter gauge as a guide. Adjust the dado head for height. Test the width and depth of cut on scrap stock. Then cut in the desired location as in crosscutting, figure 7-34.

Cutting Grooves

A *groove* is the same kind of cut as a dado except it is cut with the grain, see figure 7-33. To cut grooves with a dado head, use the rip fence as a guide. Adjust the rip fence to the desired location and the dado head to the correct height. Test the width and depth of cut on scrap stock. When all adjustments are made, cut the grooves as in straight ripping, figure 7-35.

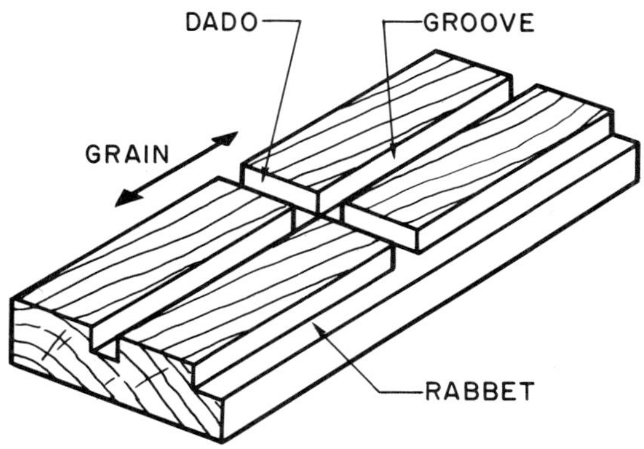

Fig. 7-33 Dado, groove, and rabbet cuts

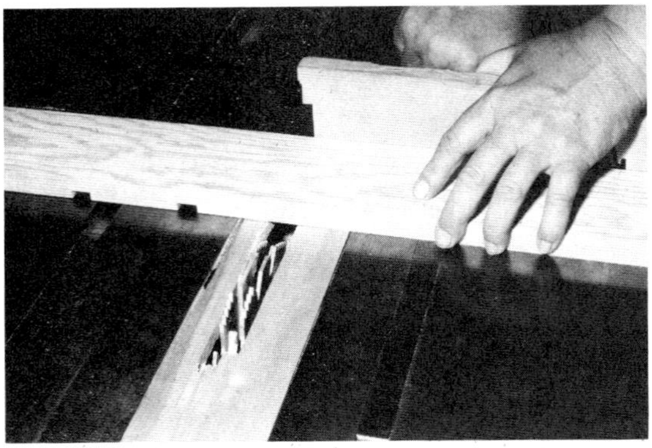

Fig. 7-34 Cutting dadoes with the dado head (guard is removed for clarity)

Fig. 7-35 Cutting a groove with the dado head (guard is removed for clarity)

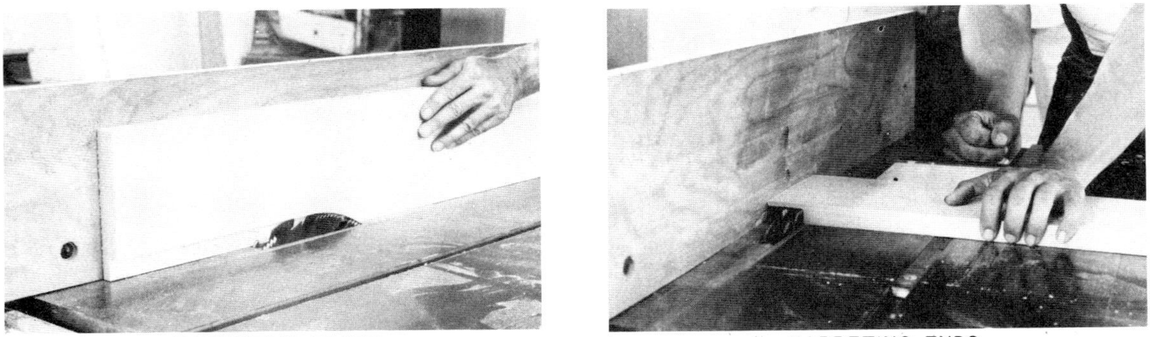

RABBETING EDGES

RABBETING ENDS

Fig. 7-36 Rabbeting edges and ends with a dado head (guard is removed for clarity)

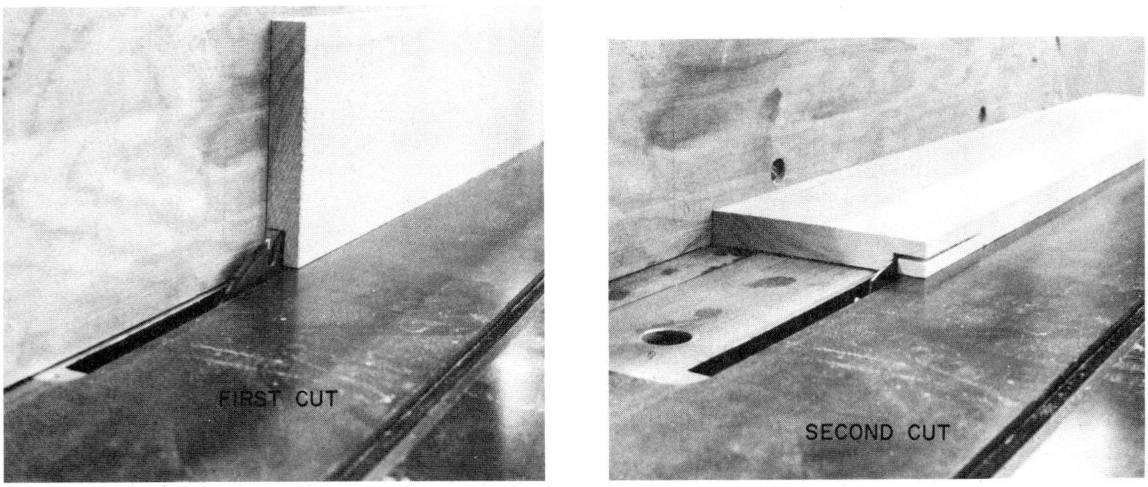

Fig. 7-37 Rabbeting an edge with a single saw blade (guard is removed for clarity)

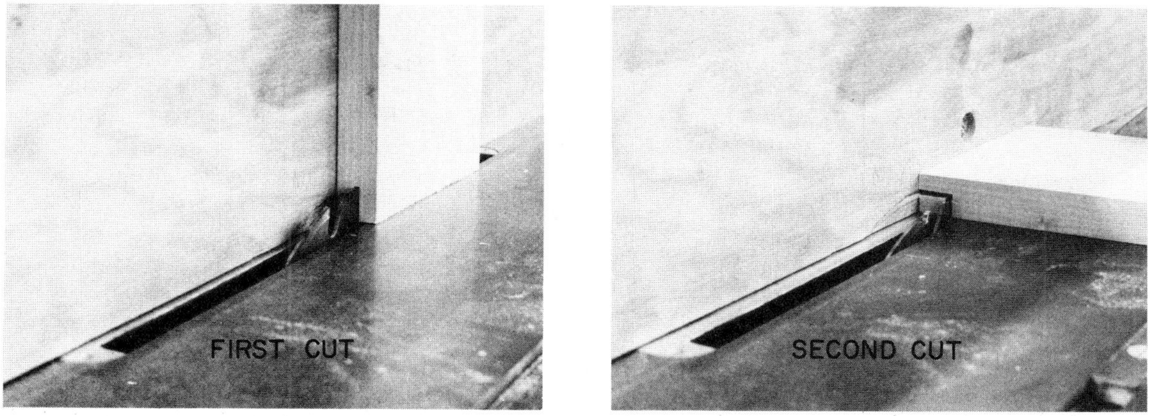

Fig. 7-38 Rabbeting an end with a single saw blade (guard is removed for clarity)

Cutting Rabbets

A *rabbet* is a cut in the corner of the edge or end of stock, see figure 7-33. If the rabbet does not go through from end to end or from edge to edge, it is called a *blind rabbet.* Rabbets can be made on the table saw using the dado head or a single saw blade.

To rabbet the edge of stock with the dado head, use the rip fence as a guide. To rabbet the end of stock with the dado head, use the miter gauge as a guide, figure 7-36.

Rabbets can also be made with a single circular blade. This method requires two passes over the saw blade. To rabbet edges, use the rip fence to guide both cuts, figure 7-37. Adjust the fence and blade for each cut. To rabbet ends, figure 7-38, the first cut is guided by the rip fence with the stock standing on its end. The second cut is guided by the miter gauge. The saw blade is adjusted to the proper height for each pass to insure a clean corner.

A *tenoning jig* can be used to hold the end for the first cut more securely. This jig slides in the miter gauge slot, and the work is clamped to it.

MAKING TENONS

Though the tenon for a mortise-and-tenon joint is usually made on a tenoner, a table saw may also be used.

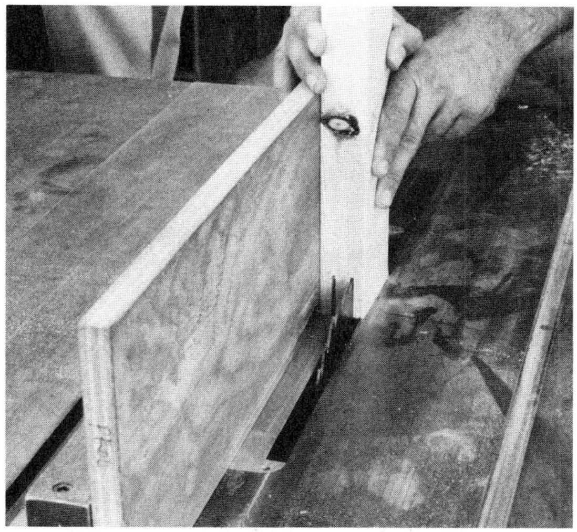

Fig. 7-39 Cutting a tenon on the table saw (guard is removed for clarity)

1. Raise the blade to the desired tenon length, adjust the rip fence, and stand the stock on end. Make one cut.
2. Reverse the stock and make another cut, figure 7-39. The distance between the two cuts should be the thickness of the tenon.
3. Adjust the fence so the distance to the blade is the desired tenon length. The blade height should equal the shoulder depth. Using the

fence as a stop, hold the stock against the miter gauge and make a shoulder cut on each side, figure 7-40.

4. *Relish* (cut back) the tenon to reduce it to width so it fits the mortise. Without moving the fence, adjust the blade for the amount of relish. Make saw cuts on each edge.
5. Finish the relish cuts. Adjust the fence and raise the blade to the height of the tenon. Hold the piece on end with its edge against the rip fence and cut, figure 7-41.

Tenons can also be cut using a dado head by adjusting the fence and blade to control the depth and length of the tenon. A tenoning jig can hold the end securely when cutting.

Fig. 7-40 Cutting the shoulders of the tenon (guard is removed for clarity)

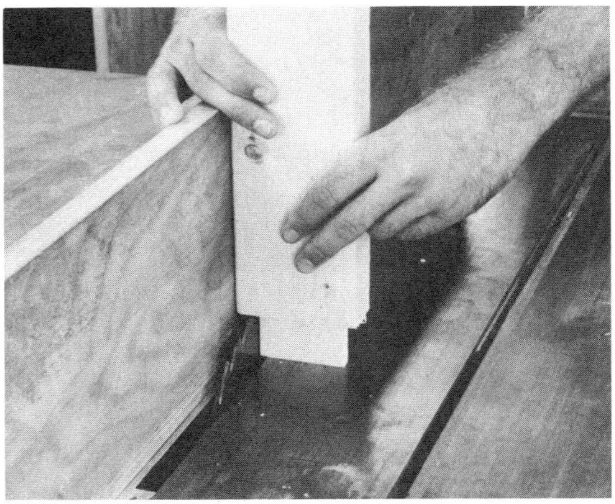

Fig. 7-41 Relishing the tenon (guard is removed for clarity)

THE MOLDING HEAD

A *molding head* may be used on the table saw to cut thousands of different shapes, figure 7-42. It usually has a round head into which three or more cutting knives are inserted. The molding head is mounted on the arbor shaft like a saw blade. Shapes are cut by guiding the stock with the ripping fence or the miter gauge. Feather boards are used whenever possible to insure a smooth cut.

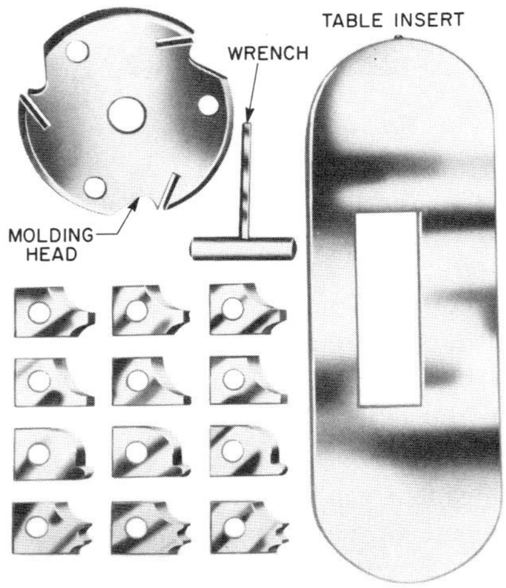

Fig. 7-42 Molding head with four standard cutting blades

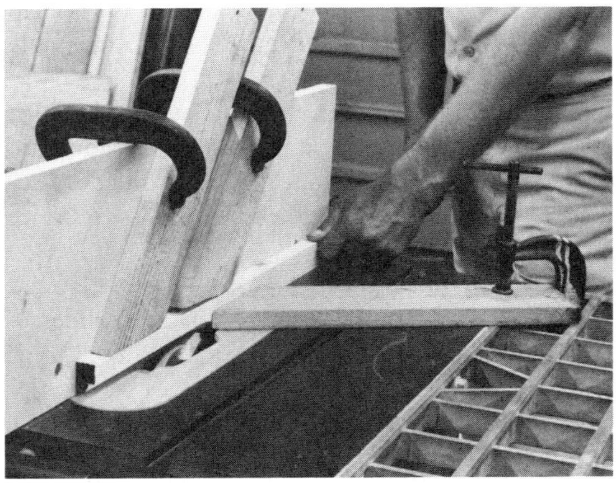

Fig. 7-43 Using the molding head to shape an edge (guard is removed for clarity)

ACTIVITIES

1. Show how to adjust the table saw for crosscutting and ripping. What attachments are available in the class shop to make table saw operations safer and easier to perform?
2. What type of safety guards are used on the table saws in the class shop? Describe the features and advantages of each. Choose which one you prefer and explain why.
3. Make a list of the circular saw blades available in the class shop. Note the type, size, number of teeth, and uses.
4. Make a list of safety rules to follow when using the table saw. Post it in the shop.
5. Make a number of wood blocks with the ends cut at various angles. Use these to test the tilt of the saw blade.
6. If the equipment is available, sharpen a circular saw blade in the class shop.
7. Practice the crosscutting and ripping operations described in this unit.
8. Remove and replace a circular saw blade.
9. Mount the molding head on the table saw. Make samples of the types of cutting knives available in the shop.
10. Practice making cove cuts and tenons on the table saw.

UNIT REVIEW

Multiple Choice

1. The size of a table saw is determined by
   a. the thickest stock it can cut.
   b. the widest cut it can make.
   c. the diameter of its blade.
   d. the width of its table.
2. The guide used for cutting with the grain is called a
   a. rip fence.
   b. miter gauge.
   c. tilting arbor.
   d. width jig.
3. The more teeth there are in a saw blade,
   a. the longer it will stay sharp.
   b. the faster the feed.
   c. the rougher the cut.
   d. the smoother the cut.
4. Carbide-tipped and taper-ground saw blades have no
   a. bevel.
   b. set.
   c. need of sharpening.
   d. points, only edges.
5. The triple-chip, carbide-tipped circular saw blade is designed for
   a. cutting brittle material.
   b. ripping rough stock.
   c. crosscutting dry lumber.
   d. general use.
6. Table saw arbor nuts are loosened by turning them
   a. clockwise.
   b. counterclockwise.
   c. in the same direction as the blade rotation.
   d. opposite to the blade rotation.

7. Stock should be at least _________ wide for pushing it through the blade by hand while ripping.
   a. 1 inch
   b. 3 inches
   c. 5 inches
   d. 10 inches

8. When bevel ripping, the fence is placed on the
   a. right side.
   b. side away from the tilt of the blade.
   c. left side.
   d. side toward the tilt of the blade.

9. A dado is a wide cut made part way through the stock and
   a. with the grain.
   b. on the end.
   c. across the grain.
   d. on the edge.

10. It is safer to make an end rabbet on a single saw blade if _________ is used.
    a. a tenoning jig
    b. a clamp attachment
    c. a feather board
    d. an auxiliary fence

11. Cutting thick pieces of stock to make two or more thinner pieces is called
    a. rabbeting.
    b. resawing.
    c. feathering.
    d. dadoing.

12. When cutting dadoes with a dado head, use the _________ as a guide.
    a. miter gauge
    b. rip fence
    c. table edge
    d. feather board

13. A cut in the corner of the edge or end of stock is called
    a. a groove.
    b. a dado.
    c. a rabbet.
    d. a cove.

14. A compound miter is cut by
    a. tilting the miter gauge to the desired angle.
    b. tilting the miter gauge and the blade to the desired angles.
    c. tilting the blade and adjusting the miter gauge square to the blade.
    d. adjusting both the miter gauge and the blade so they are parallel.

15. The _________ consists of two outside circular blades and several inner chippers of different thicknesses.
    a. molding head
    b. triple-chip grind blade
    c. alternate top-bevel grind blade
    d. dado head

Questions

1. How are the miter gauge grooves lined up with the saw blade?
2. How is the stock placed on the table for ripping and crosscutting operations? Why?
3. Which blade should be used to cut thin plywood? Why?
4. Which blade should be used to crosscut seasoned, solid lumber? Why?

5. What may be the result of using a fine-toothed crosscut blade for ripping thick, solid lumber?
6. Why are taper-ground circular saw blades used?
7. Explain how to check the miter gauge for accuracy in order to cut square ends.
8. How is the work prevented from creeping away from the saw blade when crosscutting?
9. Describe three methods of crosscutting wide stock.
10. What is the purpose of using an auxiliary rip fence when ripping?

unit 8 Radial Arm Saw

OBJECTIVES

After studying this unit, the student will be able to:

- describe and adjust the parts of the radial arm saw.
- crosscut and rip on the radial arm saw.
- make compound miters and determine angles for hoppers.
- perform special cutting and sanding operations using the dado head and other accessories.
- describe the purpose of the power miter box and show how to use it.

RADIAL ARM SAW

Like the table saw, the radial arm saw can perform many operations, figure 8-1. The advantage of this machine is that the stock can remain stationary while the saw moves across it. Another advantage is that the cut is made above the work and is easily seen. The saw blade is never hidden below the work, and layout lines are clearly visible.

The radial arm saw is particularly suited for crosscutting. Crosscutting long lengths on the table saw, for instance, is difficult because the stock tends to wobble as it moves across the saw blade. On the radial arm saw, it is not necessary to move the stock, and a better cut is achieved.

Although ripping operations may be done on the radial arm saw, the table saw is better suited for ripping. If both machines are available, use the table saw for ripping and the radial arm saw for crosscutting. When ripping on the radial arm saw, the saw is locked in place and the stock is fed through the saw.

The size of a radial arm saw is determined by the diameter of the largest blade it can hold. The same circular blades used on the table saw and portable circular saw are used on the radial arm saw.

The arm of the radial arm saw moves horizontally in a complete circle, figure 8-2. The motor unit, which holds the saw blade, tilts to any desired angle, figure 8-3. It also rotates in a complete circle, figure 8-4. The depth of cut is controlled by raising or lowering the arm, figure 8-5.

The flexibility of the radial arm saw allows many kinds of cut to be made. Practice adusting the radial arm saw to its many different positions to learn how to control the saw.

REMOVING AND REPLACING BLADES

To remove a saw blade from the radial arm saw, disconnect the saw and remove the guard. Hold the arbor steady and turn the nut in the same direction that the blade turns. Usually the arbor is held steady by inserting an Allen wrench in its end, figure

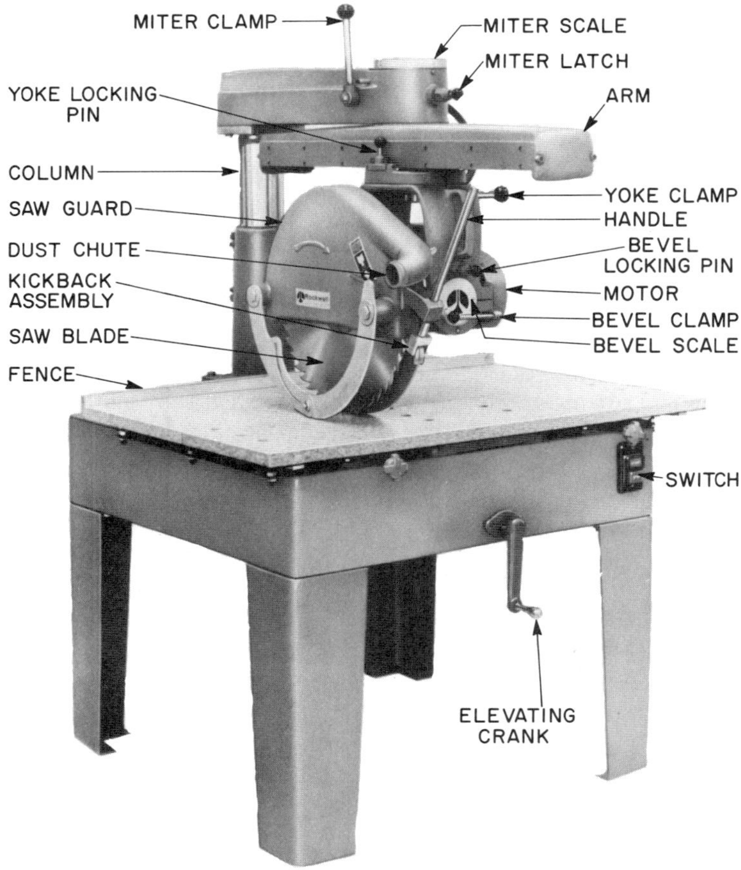

Fig. 8-1 Radial arm saw *(Rockwell International)*

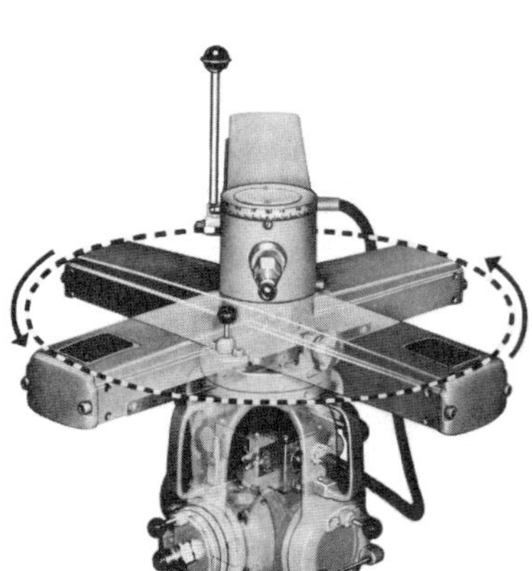

Fig. 8-2 The arm of the radial arm saw moves in a complete circle. *(Rockwell International)*

Fig. 8-3 The motor unit tilts to any angle.

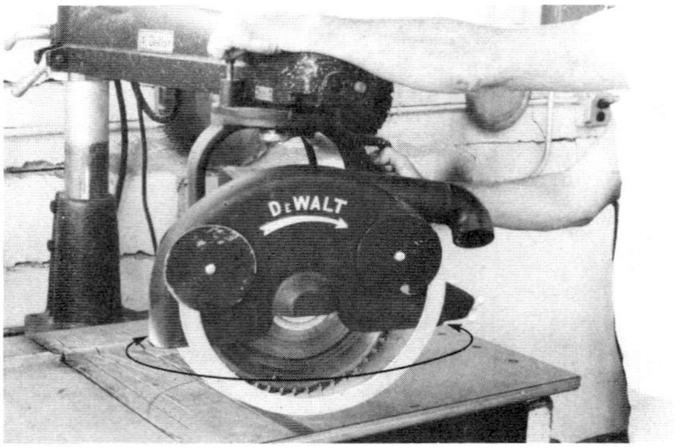

Fig. 8-4 The motor unit also rotates in a complete circle.

Fig. 8-5 Depth of cut is controlled by raising or lowering the arm.

Fig. 8-6 Removing the blade from the radial arm saw

8-6. If this is not possible, hold the saw blade with a strip of wood against its teeth while loosening the arbor nut.

To replace a blade, tighten the arbor nut opposite the direction the blades turns. Do not cross-thread the nut. Cross-threading will damage the brass arbor nut. Always replace and use the saw blade guard.

CROSSCUTTING

When crosscutting, usually most of the stock is held to the left of the blade. Hold the stock with the left hand and pull the saw with the right hand. When the stock is to the right, hold the stock with the right hand while pulling the saw with the left hand. Do not cross arms while operating the radial arm saw. When crosscutting stock thicker than the capacity of the saw, cut through half the thickness, turn the stock over, and make another cut.

Straight crosscutting:

1. Make sure the track is at a right angle to the fence. Check this with a steel square.
2. Adjust the depth of cut so the teeth of the saw blade is about 1/16 inch below the table surface.
3. Make sure the saw is all the way back. Hold the stock against the fence. Turn on the power.

Fig. 8-7 Straight crosscutting

4. Bring the saw forward and cut to the layout line, figure 8-7. When the cut is complete, return the saw to its starting position and turn off the power.

Caution: When crosscutting stock, do not pull the saw quickly through the work. Forcing the saw may cause it to jam in the work or ride over it, possibly causing injury to the operator. Little effort is required to pull the saw through the work because the cutting action tends to feed the blade into the stock.

Crosscutting Identical Lengths

To cut many pieces of the same length:

1. Clamp a stop block to the table the desired distance from the blade.
2. Square one end of the stock to be cut.
3. Slide the squared end against the stop block and cut the desired length, figure 8-8.
4. Repeat step 3 until the desired number of pieces are cut.

Do not force the stock against the stop block. This may move the stop block and result in pieces of unequal length. A pencil mark at the end of the stop block on the table will show if the stop block has moved. Check it when cutting many identical pieces. Also be careful that no sawdust or wood chips are trapped between the end of the stock and the stop block. This will result in pieces shorter than desired.

Fig. 8-8 Crosscutting identical lengths

If the pieces to be cut are so long that a stop block cannot be clamped to the table, clamp a strip of wood to the fence extending it beyond the table. Clamp a stop block to the strip the desired distance from the saw blade.

Bevel Crosscutting

Crosscutting at a bevel is similar to straight crosscutting except that the motor unit is tilted, figure 8-9.

To tilt the motor unit, raise the saw blade, loosen the bevel clamp, and pull the bevel locking pin. Tilt the motor to the desired angle and lock the bevel clamp. The bevel locking pin locks the motor unit into a 90, 45 or 0-degree angle position. The saw is held at any other angle by the bevel clamp only. To bevel cut an angle of 50 degrees or more, a temporary stop must be attached to the arm to prevent the saw blade from hitting the base of the column.

When the desired angle is obtained, start the motor and lower the saw until it cuts into the table surface about 1/16 inch.

Mitering

To cut a miter, the arm of the saw is rotated either to the right or the left to the desired angle,

Fig. 8-9 Bevel crosscutting

Fig. 8-10 Mitering *(Rockwell International)*

Fig. 8-11 Using a mitering jig

Fig. 8-12 Making a compound miter

figure 8-10. Usually the arm is swung to the right for most miter cuts. On this side, it is possible to cut wider stock.

To move the arm, raise the saw blade to clear the table surface. Loosen the miter clamp and lift the miter latch. Move the arm to the desired angle and tighten the miter clamp. The miter latch locates the arm in either a 90 or 45-degree angle position. At any other position the arm is held securely by the miter clamp only.

With the arm locked in the desired position, start the saw and lower it until it just cuts into the table surface. Make miter cuts in a manner similar to straight crosscutting.

Miters may also be cut by using a mitering jig. The mitering jig in figure 8-11 has strips of wood fastened to a plywood sheet at a 45-degree angle to its edge. The jig is clamped to the table surface. This holds the stock at a 45-degree angle so the radial saw arm can miter it while in a straight crosscutting position. Right and left-hand miters can then be cut without swinging the arm of the radial saw. Mitering jigs may also be constructed to cut miters other than 45 degrees.

Making Compound Miters

The *compound miter* cut is a miter cut at a bevel, figure 8-12. It is made by adjusting the arm to the correct angle and also tilting the motor the desired amount. Compound miters are sometimes called *hopper cuts* because they are made to build

Fig. 8-13 A hopper-type light fixture

hoppers. *Hoppers* are boxes which have four or more slanted sides. This makes the box wider at the top than at the bottom, figure 8-13.

Because of the slanting sides, butt joints require more than a 90-degree angle and a miter joint more than 45 degrees. When building hoppers, it is necessary to know at what angle to swing the arm and to tilt the motor of the radial saw to make the cuts.

The angles of the mitered joints for hopper boxes are found by a simple shop method.

1. Bevel the edge of a short piece of 2" x 4" stock to the angle of the desired hopper.
2. Hold the piece on the radial arm saw with the beveled edge resting firmly on the table surface.
3. With the piece against the fence, swing the arm 45 degrees and cut the end, figure 8-14.
4. Use the cut end to test the swing of the radial arm or the miter gauge and the tilt of the blade, figure 8-15.

For a six-sided hopper, make the cut at a 60-degree angle. For an eight-sided hopper make the cut at a 67 1/2-degree angle. These are the regular miter cuts for boxes with straight sides.

The angles of butt joints for hopper boxes are found in a similar manner:

1. Bevel the edge of a piece of 2" x 4" stock to the angle of the hopper.
2. Hold the stock on its beveled edge against the fence.
3. Tilt the blade to the angle of the hopper.
4. Cut the end with the arm at a right angle to the fence.
5. Use the cut end to test the tilt of the saw blade and the swing of the radial arm or the miter gauge.

POWER MITER BOX

The *power miter box* is a small, compact tool that produces accurate miters and square cuts, figure 8-16. A circular saw, similar to a portable circular saw, is mounted above the base and moves up and down. The saw and motor may be turned 45 degrees to the right or left. The table contains positive stops at usually 90, 67 1/2, and 45 degrees.

To turn the saw, unscrew the locking handle slightly, press the spring-loaded lever, and turn it to the desired angle. Angles other than the positive stops are locked by tightening the locking handle. The blade is removed and replaced just as the radial arm saw blade.

To make cuts, the stock is held firmly on the table and against the fence. The saw is adjusted to

Fig. 8-14 Cutting a test block for a hopper miter cut (guard is removed for clarity)

Fig. 8-15 Using the test block to adjust the saw to make mitered hopper joints (guard is removed for clarity)

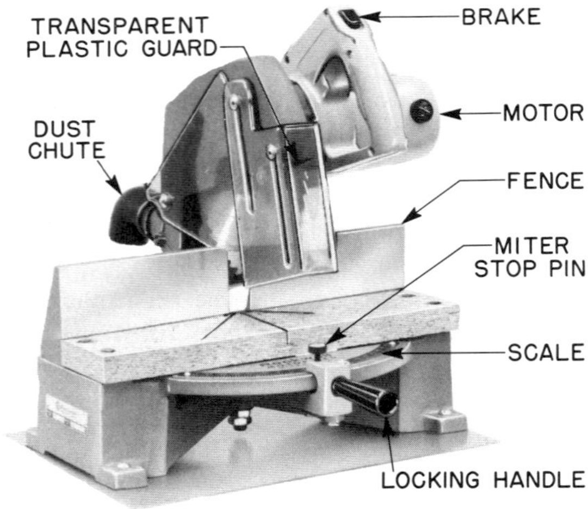

Fig. 8-16 Power miter box

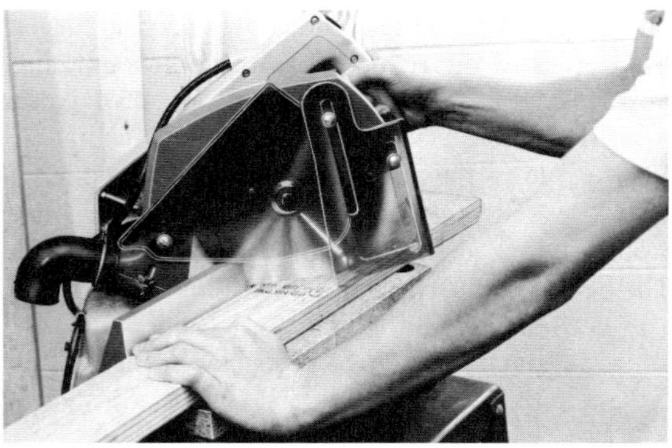

Fig. 8-17 Making miter cuts with the power miter box *(Rockwell)*

the desired angle, turned on, and brought down through the stock, figure 8-17.

A transparent guard lets the operator see layout lines and gives maximum protection. The guard should be in place at all times during operation. A crosscutting saw blade is ideally suited for this machine.

RIPPING ON THE RADIAL ARM SAW

When ripping on the radial arm saw, the saw is locked in place and the stock is fed through the saw. The arm is locked at a right angle to the fence. The saw is rotated to either an inrip or outrip position with the saw blade parallel to the fence. The inrip position is used to cut narrow stock, figure 8-18. In this position, the motor unit is swiveled so the saw blade end is closest to the fence. In the outrip position, the saw blade end is farthest away from the fence, figure 8-19.

To adjust the saw for ripping:

1. Raise the motor and pull it to the end of the arm.
2. Pull and loosen the yoke clamp handle and lift the yoke locking pin.
3. Revolve the motor to the inrip or outrip position until the yoke locking pin engages. Tighten the yoke clamp.
4. Locate the saw for the desired width of cut using the rip scale on the side of the arm as a guide, figure 8-20.

Fig. 8-18 The inrip position of the radial arm saw

Fig. 8-19 The outrip position of the radial arm saw

5. Lock the saw carriage by tightening the rip lock against the side of the arm.
6. Lower the motor until the saw teeth are just below the table surface.
7. Adjust the safety guard so the infeed end almost touches the material to be cut. This prevents sawdust from flying at the operator.
8. Lower the kickback assembly until the fingers are approximately 1/8 inch lower than the material.
9. With the stock against the fence, feed it evenly into the saw blade.

Caution: Do not feed stock from the kickback side of the guard. Feed the stock *against* the rotation of the blade, figure 8-21. Feeding stock in the wrong direction may pull it from the operator's hands and through the saw with great force. This can injure anyone in its path. Do not force the stock into the saw. Feed it with a continuous motion as fast as the saw will cut it. Use a push stick when ripping narrow stock.

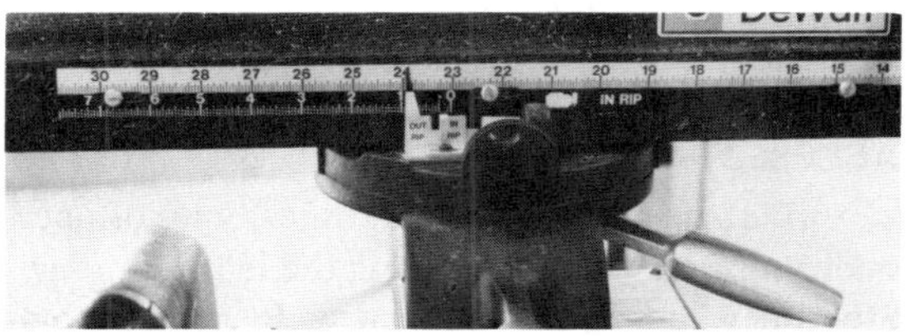

Fig. 8-20 The rip scale indicates the width of cut for both inrip and outrip positions.

Fig. 8-21 Correct feed for ripping (ring guard is removed for clarity)

Bevel Ripping

Bevel ripping is similar to straight ripping except the motor is tilted to the desired angle, figure 8-22.

Taper Ripping

Taper ripping uses the same type of jigs that are used on the table saw. The only difference in using the radial arm saw is that the saw blade is above the work.

Cove Cuts

Cove cuts may also be made on the radial arm saw:

1. Rotate the saw on its horizontal axis to obtain the desired radius of the cove. Lock it in position on the arm.
2. Adjust the motor to cut about 1/8 inch below the surface of the stock.
3. Feed the stock as in ripping.
4. Continue lowering the motor and making cuts of about 1/8 inch with each pass until the desired depth of cut is obtained, figure 8-23.

Caution: Do not attempt to take too deep a cut when making coves. This will place too much stress on the sides of the saw blade.

Fig. 8-22 Ripping stock at a bevel

USING THE DADO HEAD

The dado head may be used on the radial arm saw also. *Dadoing* is very similar to crosscutting. With the arm at right angles to the fence, adjust the height of the dado head to cut the desired depth. Allow the saw to come to full speed and make the cut smoothly and slowly, figure 8-24. Make a cut on a scrap piece first to test for depth and width of cut.

Fig. 8-23 Making cove cuts. The radius of the cove is determined by the angle of the blade.

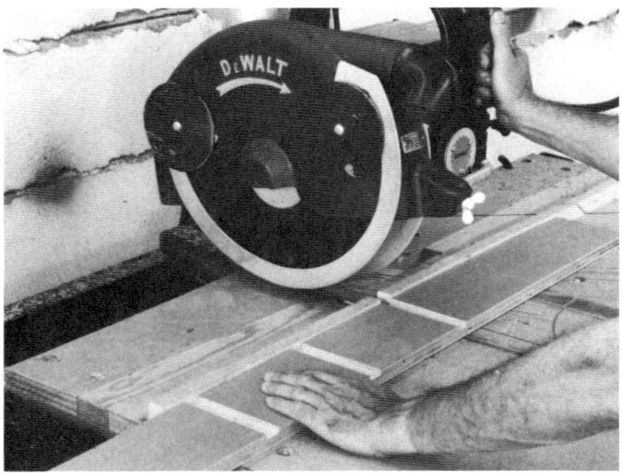

Fig. 8-24 Cutting dadoes with a dado head

Fig. 8-25 Making end rabbets with the dado head

End rabbets are made like dadoes except the cut is made at the end of the stock. When the end rabbet is made on both sides of the stock, a *tenon* is formed, figure 8-25.

A *blind dado* is cut by clamping a stop block on the arm of the saw. The stop block limits the amount of travel as desired, figure 8-26.

Grooving is done with the dado head in the rip position, figure 8-27. Adjust the height of the dado head and lock the motor unit in position on the arm to cut the groove where desired. Groove the stock as in ripping. Remember to lower the guard on the infeed side and adjust the kickback assembly. Making blind grooves is awkward and dangerous and should not be attempted with the radial arm saw.

Edge rabbeting is done the same way as grooving except that the motor unit is moved to cut the stock at the edge.

V-grooves are made with or across the grain by tilting the dado head to a 45-degree angle. Cuts are similar to crosscutting or ripping, figure 8-28.

ACCESSORIES

With special equipment, other types of operations may be performed using the radial arm saw. Stock may be surfaced and sanded. By using a molding head, wood can also be shaped in many different ways, figure 8-29.

Fig. 8-26 A stop block is clamped to the arm to make blind dadoes.

Fig. 8-27 Cutting a groove

Fig. 8-28 Making a V groove by tilting the dado head

SURFACING SANDING SHAPING

Fig. 8-29 Surfacing, sanding, and shaping with accessories *(Dewalt, Div. of Black & Decker)*

ACTIVITIES

1. Show how to adjust the radial arm saw for crosscutting and ripping.
2. What size circular saw blades do the radial arm saws in the class shop require?
3. Crosscut five idential lengths on the radial arm saw.
4. Cut a miter using a mitering jig on the radial arm saw.
5. Make a four-sided hopper whose sides slant at 60 degrees with mitered joints.
6. Make a four-sided hopper whose sides slant at 60 degrees with butt joints.
7. Make a cove cut using the radial arm saw.
8. Mount the dado head on the radial arm saw. Practice cutting dadoes, grooves, V grooves, and end and edge rabbets.
9. What special attachments are available for the radial arm saws in the class shop? Practice using these accessories.

UNIT REVIEW

Multiple Choice

1. The radial arm saw is ideally suited for
   a. grooving.
   b. ripping.
   c. crosscutting.
   d. shaping.

2. The size of a radial arm saw is determined by the
   a. length of travel of the saw blade.
   b. diameter of the largest blade it can hold.
   c. length of the radial saw arm.
   d. height of radial arm saw column.

3. When crosscutting with most of the stock to the left of the blade,
   a. hold the stock with the right hand and pull the saw with the left hand.
   b. hold the stock with the left hand and pull the saw with the right hand.
   c. hold the saw with the right hand and pull the stock with the left hand.
   d. hold the saw with the left hand and pull the stock with the right hand.

4. To loosen the radial arm saw arbor nut,
   a. turn it clockwise.
   b. turn it against the rotation of the blade.
   c. turn it counterclockwise.
   d. turn it with the rotation of the blade.

5. A stop block must be attached to the radial arm saw to prevent the saw blade from hitting the column when bevel cutting if the angle is
   a. less than 45 degrees.
   b. more than 30 degrees.
   c. less than 50 degrees.
   d. more than 50 degrees.

6. To make miter cuts using the radial arm saw, the arm is usually
   a. moved to the left.
   b. moved to the right.
   c. locked in the 45-degree angle position.
   d. locked in the straight crosscutting position.

7. When ripping, the infeed end of the blade guard is lowered close to the work to
   a. prevent kickback.
   b. protect the operator from the saw blade.
   c. prevent sawdust from flying out.
   d. hide the saw blade from the operator.

8. When ripping, stock is fed
   a. to the left.
   b. with the rotation of the blade.
   c. to the right.
   d. against the rotation of the blade.

9. To make cove cuts with the radial arm saw,
   a. the saw blade is tilted.
   b. the arm is moved horizontally.
   c. the saw blade is rotated horizontally.
   d. the arm is moved up and down.

10. For a six-sided hopper box, the miter cut should be
    a. 45 degrees.
    b. 60 degrees.
    c. 67 1/2 degrees.
    d. 90 degrees.

Questions

1. What are the advantages of using the radial arm saw instead of the table saw for crosscutting operations?
2. Describe the method used for crosscutting pieces of identical length. What should you do if the lengths to be cut are longer than the saw table?
3. Explain how to find miter angles for hoppers without referring to tables or charts. Explain how to find the angles for making hoppers with butt joints.
4. What is the purpose of the power miter box?
5. List three safety precautions to take when using the radial arm saw for ripping operations.
6. Why is the infeed side of the saw blade guard lowered as far as possible for ripping operations?
7. Explain how cove cuts are made using the radial arm saw.
8. How is the travel of the saw limited when making a number of blind dadoes?
9. Explain how V grooves are made using the radial arm saw.
10. Name several other types of operations that may be performed with the radial arm saw using special equipment.

unit 9 Band Saw and Scroll Saw

OBJECTIVES

After studying this unit, the student will be able to:

- describe the parts of the band saw and scroll saw.
- select the proper blades for the band saw and scroll saw. Remove, coil, and replace blades and adjust them properly.
- use the band saw to cut curves and make straight and compound cuts.
- use a jig for circle cutting, use a pivot block for parallel curve cutting, and make duplicate parts on the band saw.
- make external and internal cuts using the scroll saw.

The band saw and scroll saw are designed mainly for making curved cuts. Although straight cutting can be done on these machines, it is better to use machines like the table saw or radial arm saw for these cuts.

BAND SAW

The band saw consists of two large wheels around which a continuous band of steel with cutting teeth is driven, figure 9-1. The lower wheel is power driven and does not tilt on its axis. The upper wheel is free running and is turned by the band saw blade. The upper wheel may also be tilted back or forth to center the blade on the wheels. To protect the teeth of the saw blade, a thin rubber tire covers the rim of both wheels. The upper wheel is also adjusted up or down to put tension on the saw blade. The band saw wheels and blades are covered with guards and are used during band saw operations.

The band saw table may be tilted. A scale indicates the degree of tilt. There is a slot in the table from the center to the edge for removing and replacing saw blades. A larger opening in the center is covered with a soft metal insert. Soft metal is used for the insert because the band saw blade occasionally comes in contact with it. It is replaced when worn.

At the bottom of a sliding post above the table is the saw guide assembly, figure 9-2. Adjustable side guide wheels on each side control the sideways movement of the saw blade. A rear guide wheel prevents the blade from being pushed off its wheels when stock is being cut. The back edge of the blade does not touch this rear guide wheel when the blade is running free. The only time the blade comes in contact with the rear wheel is when stock is being cut.

The guide assembly is moved up or down according to the thickness of the material being cut.

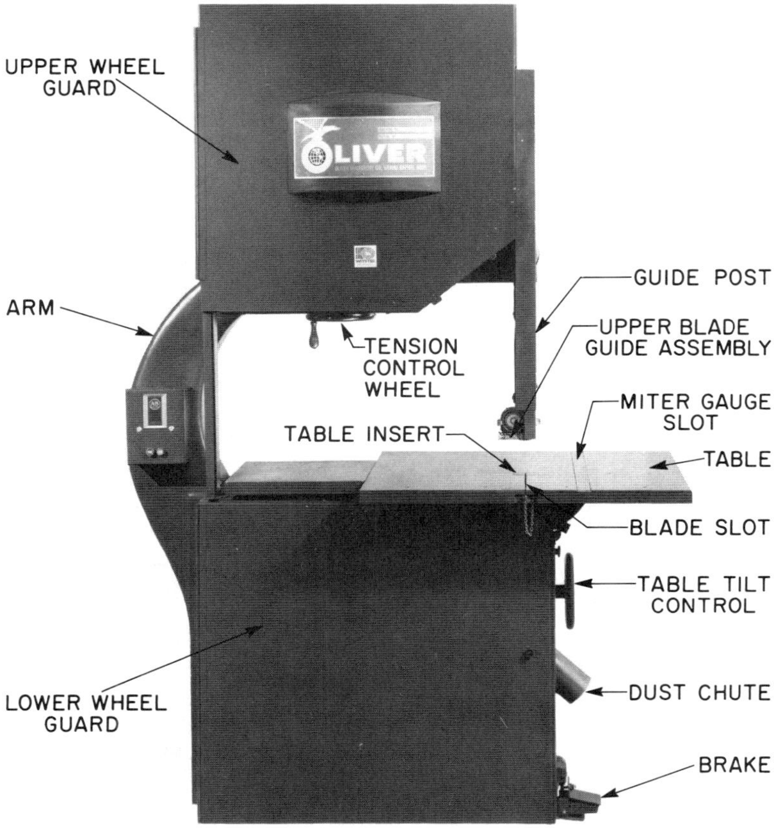

Fig. 9-1 Band saw *(Oliver Machinery Co.)*

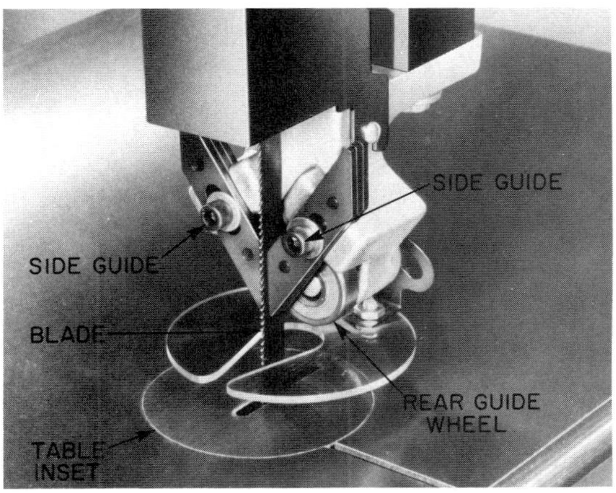

Fig. 9-2 Upper guide assembly *(Rockwell International)*

A similar guide assembly is held in a fixed position below the table.

The size of the band saw is indicated by the width of its throat. This is the distance from the blade to the arm. Usual sizes are from 14 to 35 inches.

BAND SAW BLADES

Blades are manufactured as narrow as 1/8 inch. Although they are made wider for production machines in large mills, widths rarely exceed 1 1/2 inches for small band saws. Lengths are cut from large rolls, and the cut ends are welded together. The teeth on band saw blades are alternately set and shaped to give a ripping action. Usually blades with 6 or 7 points per inch are suited for cutting wood.

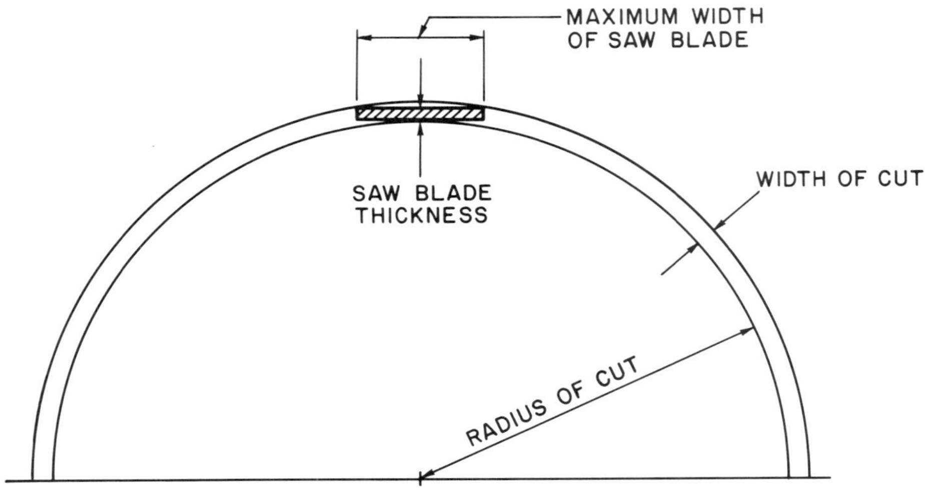

Fig. 9-3 Select the correct blade width.

Select a blade that is as wide as possible for the curve to be cut. Choose a narrow blade to cut sharp curves. Use the correct width blade for the cut, figure 9-3. The blade should not twist as the stock is guided through it.

Removing and Replacing Blades

To remove a band saw blade:

1. Open the guards that cover the wheels and those that cover the saw blade, figure 9-4.
2. Remove the table insert and lower the top wheel.
3. Remove the blade carefully through the slot in the tabletop.

> **Caution:** Do not let your hands or arms come in contact with the teeth of the blade. Sharp saw teeth can cause a severe cut.

Because some band saw blades are so large, they must be coiled to be stored. To coil a blade, grasp the blade at the top and hold the bottom with your foot. Twist the blade in either direction until the loops are formed, figure 9-5. Tie the loops together to prevent them from unwinding.

Fig. 9-4 Removing a band saw blade

Fig. 9-5 Coiling a band saw blade

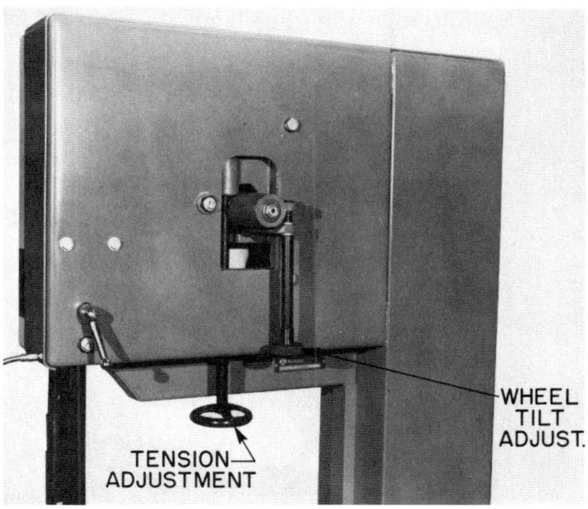

Fig. 9-6 Wheel tilt and tension adjustments

To replace a band saw blade:

1. Uncoil the blade, after untying it, by lifting one loop and tossing the blade on the floor while standing clear. The blade should open up. If it does not, repeat the procedure. If the blade is tossed on a wood surface, no substantial damage will be done to the teeth.
2. Open the side guides and move the rear guides back as far as possible.
3. Slip the new blade through the slot in the table and over the wheels with the teeth toward the operator.
4. The teeth should point downward on the right side. If the teeth point up, remove the blade and turn it inside out by twisting the blade along its length.
5. Remount the blade on the wheels and apply tension to the blade by adjusting the upper wheel upwards.

Adjusting the Tension

A common cause of blade breakage is applying too much tension. Narrow blades do not need as much tension as wider blades. Usually a scale graduated to the width of various blades shows the correct amount of tension. For instance, if a 1/2-inch blade is used, adjust the tension on the upper wheel until the pointer is even with the 1/2 mark. If the band saw has no tension scale, apply only enough tension to straighten the blade and keep it on the band saw wheels.

Centering the Blade

Rotate the upper wheel by hand to see if the blade is centered on the wheels. If necessary, tilt the upper wheel forward or backward until the blade is centered as the wheels are rotated, figure 9-6. The blade should be running free and clear without rubbing against any of the guides in the upper or lower assemblies.

Adjusting the Guide Assemblies

Move the rear guides forward until they almost touch the back edge of the blade. When the saw is running, the back edge of the saw blade should not contact the rear guide wheels unless stock is being cut. Lock the rear guide wheels in position.

Adjusting the Side Guides

Adjust each side guide in both assemblies toward the blade. The clearance between the blade and the guides should be the thickness of a piece of paper. Be careful not to push the blade sideways one way or the other. Move the guides forward or back until the faces of the guides are just in back of the teeth. The saw teeth should not run in between the side guides. Lock the side guides in position, see figure 9-2.

Rotate the wheels by hand to make sure the blade is running free, clear, and true. If the blade bends as it runs through the guide assemblies, it will break sooner. Replace the table insert and all guards.

Detecting Blade Breakage

If a clicking noise is heard when operating the band saw, it may indicate that the blade is about to break. Stop the saw, open the guards, and inspect the blade. Cracks develop from the gullet of the saw teeth to the back edge of the blade.

Another sign that a band saw blade is about to break is if the blade wobbles in and out from the rear guide. If this happens, stop the machine and inspect each gullet closely.

To keep blade breakage to a minimum, avoid excessive blade tension and twisting the blade when making cuts. Keep the wheel tires clean of impacted sawdust. Make sure the guides allow the saw blade to run in perfect aligment. Always use a sharp blade.

CUTTING ON THE BAND SAW

Planning the Cut

Before beginning any cut, make fine, accurate layout lines on the stock. Plan the path of the cut so the stock does not hit the arm of the band saw while cutting.

When more than one cut is to be made, plan to make the shorter cuts first to avoid backing out of long cuts. Waste will then fall away when the long cut is made. When possible, cut through the waste stock instead of backing out of the cut. At times, backing out of long cuts cannot be avoided.

> **Caution:** Back out of long cuts carefully. Twisting the blade in the saw cut while backing out may pull the blade off the wheels.

Freehand Cutting

To make freehand cuts:

1. Lower the upper guide assembly close to the top surface of the material. Stand facing the blade.
2. Feed the stock into the blade. Cut as close to the layout line as possible without removing it, figure 9-7. The stock will be smoothed down to the layout line later.
3. Feed the stock as fast as the saw will cut it in a continuous motion. Avoid stopping and

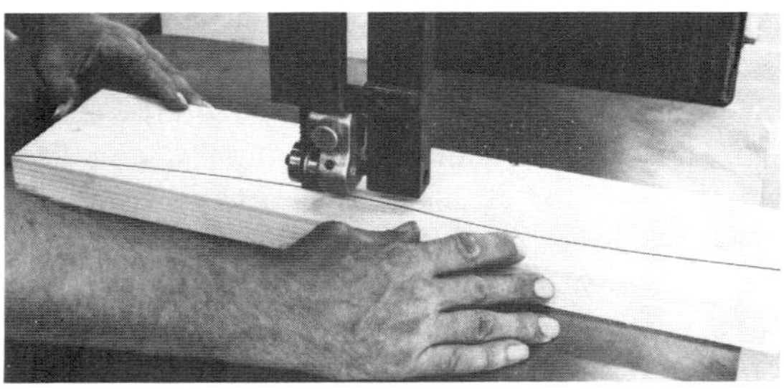

Fig. 9-7 Making a freehand curved cut

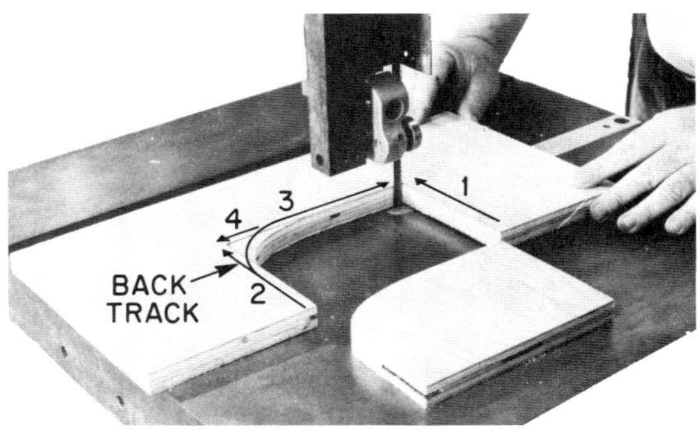

Fig. 9-8 Cutting wide rectangular openings

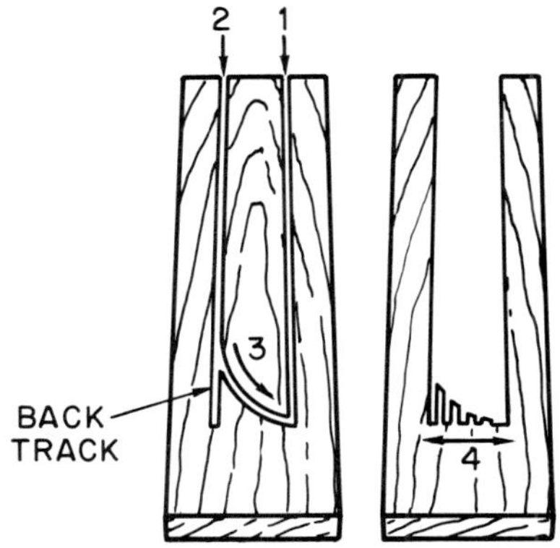

Fig. 9-9 Cutting narrow rectangular openings

Fig. 9-10 Ripping stock at a bevel

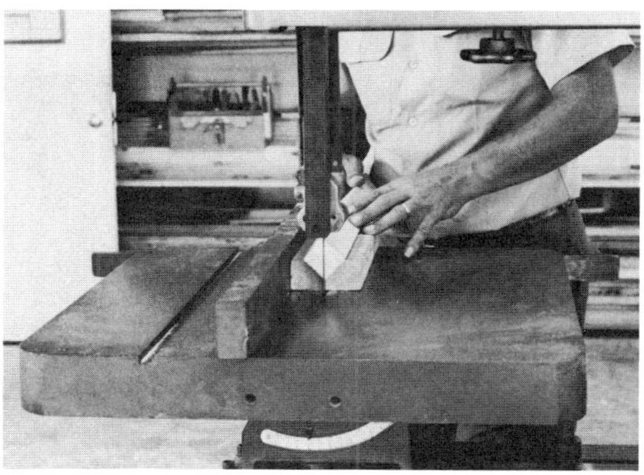

Fig. 9-11 Using a V-grooved block to bevel rip

backing up. A continuous forward motion results in a smoother cut. Too slow a feed will overheat the blade and burn the work.

4. Move the end of the stock sideways as necessary to follow the layout line as the work is fed into the blade.

Caution: Never place hands or arms in line with the saw blade. Keep fingers at least 2 inches away from the saw blade. Do not pick small scraps away from a moving blade.

Cutting Rectangular Openings

To cut large rectangular openings, cut a straight line into one corner and then back out. Cut into the second corner, back up slightly, and make a curved cut into the first corner. Turn the stock and cut the second corner square, figure 9-8.

Narrow rectangular openings are made in a similar manner except that the end is squared by making a series of cuts, figure 9-9.

Using a Ripping Fence

A ripping fence may be used either on the right or left side of the band saw blade to guide the work for straight cuts.

The ripping fence is used as a guide to rip stock at a bevel, figure 9-10. Note that the table is tilted. Another method of bevel ripping is by holding a V-grooved block against the fence and sliding the stock through the block, figure 9-11.

The ripping fence is also used to resaw material, figure 9-12. When resawing wide stock, use a sharp, wide blade. A straightedge clamped to the tabletop can be used in place of the ripping fence.

Fig. 9-12 The band saw is very useful for resawing operations.

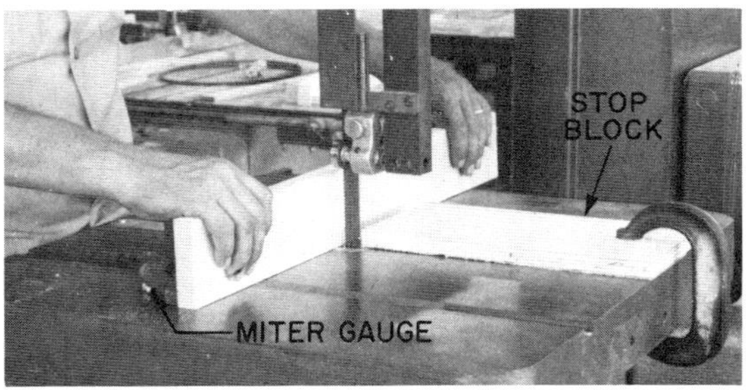

Fig. 9-13 Using the miter gauge to make a series of saw kerfs

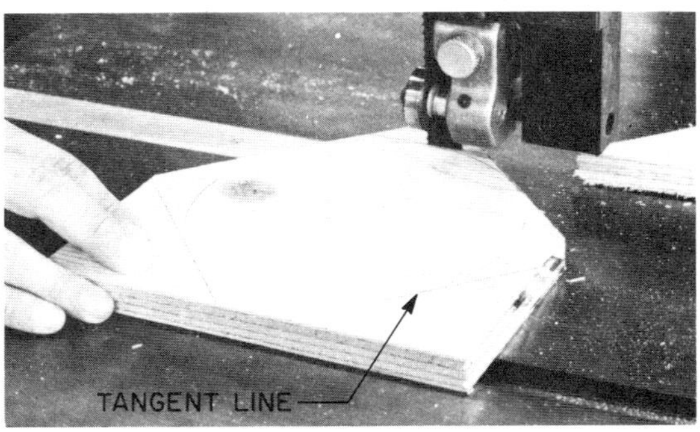

Fig. 9-14 Cutting a circle by making tangent cuts

Miter Gauge

A miter gauge is also used on the band saw, though not frequently. The miter gauge is sometimes useful when making a series of saw kerfs in the face of stock, figure 9-13.

Cutting Circles

Circles are made with a wide blade by making cuts tangent to the circumference. These are straight line cuts that touch the circle. After the tangent cuts have been made all around, the circle is cut without twisting the blade, figure 9-14.

A circle-cutting jig can also be used. A plywood extension is attached to the band saw table with a small dowel pin inserted in the center. The pin is at right angles to the side and in line with the blade. The distance from the pin to the saw blade is the radius of the circle to be cut, figure 9-15.

An adjustable circle-cutting attachment is also available for cutting circles on the band saw, figure 9-16.

Cutting Parallel Curved Edges

Edges with parallel curves may be cut by using a pivot block. A *pivot block* is simply a small strip of wood with one rounded end.

1. Lay out the curve to be cut on the stock and cut it freehand.
2. Clamp the pivot block to the table with its rounded end toward the saw blade. The

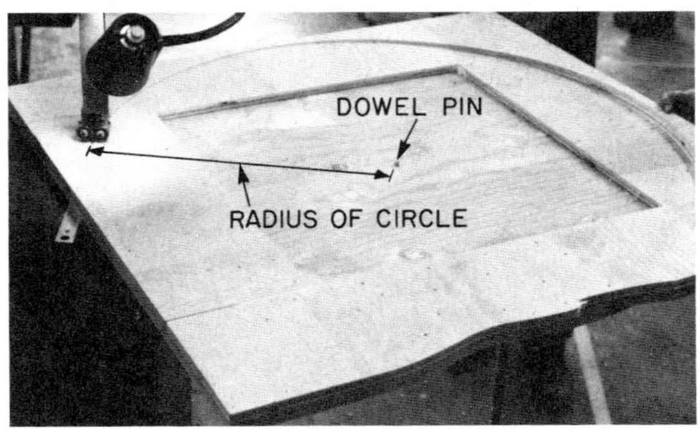

Fig. 9-15 Using a circle-cutting jig to cut a round tabletop

distance from the pivot block to the saw blade is the desired width of the piece.

3. Guide the precut curved edge against the pivot block while cutting the parallel curve, figure 9-17.

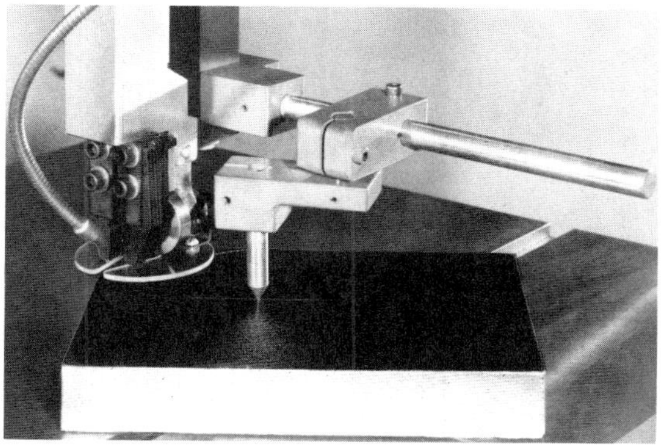

Fig. 9-16 An adjustable circle-cutting attachment may be mounted on the band saw. *(Rockwell)*

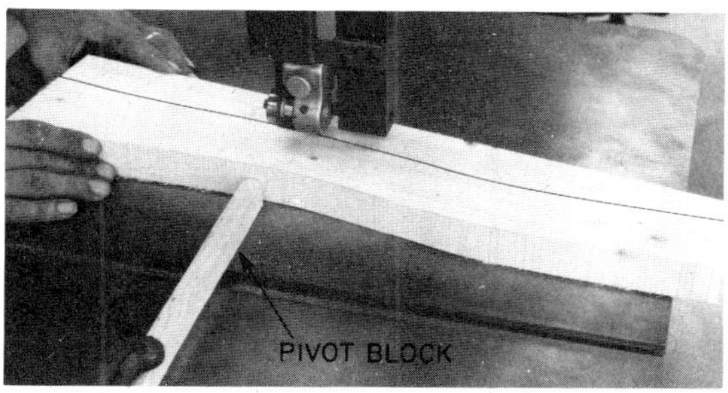

Fig. 9-17 Cutting parallel curves using a pivot block as a guide

Cutting Duplicate Parts

One method of making duplicate parts is by resawing. The desired shape is cut from thick stock and then resawed into thinner pieces, figure 9-18.

Another method is to fasten a number of pieces together and cut all at the same time. Fasteners should be in the waste stock, figure 9-19.

Compound Sawing

When two adjacent sides of stock are cut in similar curves, it is called *compound sawing.* To make a compound cut:

1. Lay out the curve on a pattern and trace it on the two adjacent surfaces.
2. Make the cuts on one side and fasten the waste stock back into position, figure 9-20.
3. Turn the stock and make the cuts on the other side.
4. Remove the waste stock.

SCROLL SAW

The scroll saw, figure 9-21, is designed mainly to make curved cuts on small work and for internal or external precision cutting of irregular shapes. The advantage of using the scroll saw, or jigsaw, is that internal cuts can be made without cutting through the stock from the outside edges.

Unlike the band saw, the blade of the scroll saw may be as short as 5 inches. It moves in a up and down motion similar to the portable electric

Fig. 9-18 Cutting duplicate pieces by resawing

Fig. 9-19 Make duplicate pieces by fastening the pieces together and then cutting the assembly.

Fig. 9-20 Making a cabriole table leg by compound sawing

Fig. 9-21 Scroll saw

saber saw. A motor drives a pulley by a belt. The crankshaft changes the circular motion of the pulley to an up-and-down motion. The belt-driven pulley usually is stepped with different diameters to vary the speed of the saw, figure 9-22. The blade is clamped in the V jaws of the lower chuck assemby, figure 9-23. The top end of the blade is held by the upper chuck at the end of tension sleeve.

The work rests on a tilting table. A scale indicates the degree of tilt. An adjustable hold-down keeps the work from vibrating as it is being cut. A blade guide controls the sideways sway of the blade, and a roller in back of the blade supports it as material is being cut.

The size of the scroll saw is determined by the width of its throat. The throat, as in the band saw, is the distance from the blade to the arm of the saw. Scroll saws are made in many sizes, from light bench models to heavy-duty production machines.

SCROLL SAW BLADES

With the proper blade, a variety of materials can be cut using the scroll saw. Besides wood, metals such as steel, lead, copper, and aluminum can be cut. Other materials such as paper, felt, leather, and plastics are also easily cut with the proper blade.

Use fine-tooth, thin blades for intricate cutting of thin material. Use wide blades for straight and large curve cutting operations. Select a blade that has at least three teeth in contact with the material thickness. Figure 9-24 shows a chart to aid in the selection of a correct blade.

Fig. 9-22 The stepped pulley of the scroll saw varies the speed.

Installing a Blade

To install a scroll saw blade:

1. Move the upper guide assembly back. Remove the table insert. Open the jaws of the lower chuck.
2. Insert the blade with the teeth pointing down at least 3/4 inch into the jaws. Tighten the lower chuck so the blade is at right angles with the table surface.
3. Bring the upper guide assembly forward. The blade should ride in the proper slot of the side guide. The rear roller should be just in back of the blade.
4. Rotate the motor by hand. As the blade moves up and down, it should not move away from or against the rear roller. Adjust the blade angle until it remains parallel to the rear roller as the motor is rotated by hand. Lock the guide assembly in position. The teeth of the blade should not ride in the slot of the side guide.

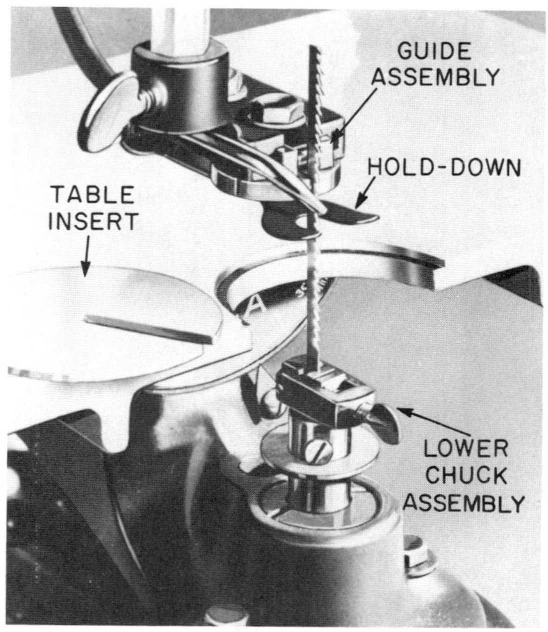

Fig. 9-23 The lower chuck assembly holds the blade.

Material	Gauge	Teeth Points Per Inch
Steel ■ Iron Lead ■ Copper Aluminum	.070	32
Pewter Asbestos Paper ■ Felt	.070	20
Steel ■ Iron Lead ■ Copper Brass	.070	15
Aluminum Pewter Asbestos	.085	15
Wood	.110	20
Asbestos ■ Brake Lining ■ Mica Steel ■ Iron Lead ■ Copper Brass Aluminum Pewter	.250	20
Wood Veneer Plus Plastics Celluloid Hard Rubber Bakelite Ivory Extremely Thin Materials	.035	20
Plastics Celluloid	.050	15
Bakelite	.070	7
Ivory ■ Wood	.110	7
Wall Board Pressed Wood Wood ■ Lead Bone ■ Felt Paper ■ Copper Ivory Aluminum	.110	15
Hard and Soft Wood	.110	10
	.187	10
	.250	7
Pearl ■ Pewter Mica	.054	30
Pressed Wood Sea Shells	.054	20
Hard Leather	.085	12

Fig. 9-24 Scroll saw blade selection chart *(Rockwell)*

Fig. 9-25 Making intricate external cuts using the scroll saw

5. Lower the tension sleeve to clamp the blade in the jaws of the upper chuck. Raise the tension sleeve to apply tension to the blade. Apply only enough tension to hold the blade straight and firm during cutting operations. A scale on the tension sleeve indicates the amount of tension on the blade.

6. Replace the table insert. Lower the hold-down to apply some pressure to the material to be cut. Rotate the motor by hand to make sure the saw runs free and clear.

CUTTING ON THE SCROLL SAW

External Cutting

Use the same techniques for external cutting with the scroll saw as in using the band saw. Lay out lines carefully. Cut on the waste side of the line. Avoid backing out of long cuts. Turn the stock slowly and avoid twisting the blade. Break up complicated curve cuts into a series of simpler cuts, figure 9-25.

Internal Cutting

To make internal cuts:

1. Bore holes in all cutouts to be made.
2. Release the blade from the upper chuck and raise the tension sleeve. For extensive internal

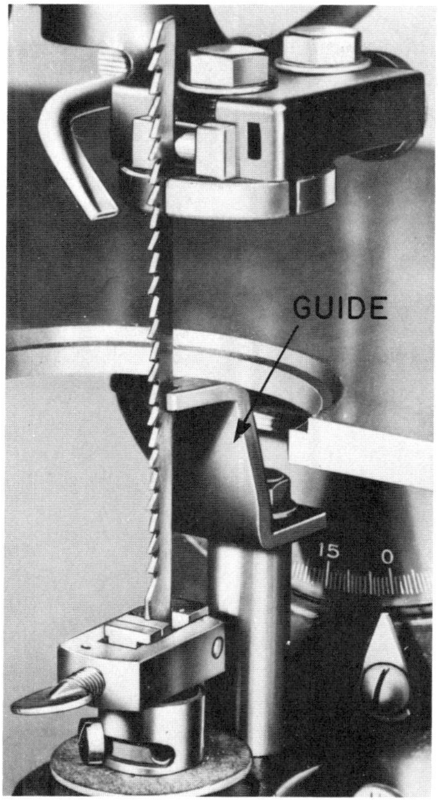

Fig. 9-26 A special guide is installed below the table when the upper chuck is not used. *(Rockwell)*

Fig. 9-27 Bore all holes before making internal cuts.

cutting, a special guard is used in place of the upper chuck to save time, figure 9-26.

3. Raise the guide assembly and hold-down. Slip the stock over the blade through the prebored holes. Lower the hold-down to hold the material firmly on the table surface.
4. Cut from the bored hole to the layout line until the cut is complete. Follow the layout lines carefully, figure 9-27.

Bevel Cuts

Bevel cuts are made by tilting the table to the desired angle. The hold-down is tilted to match the table tilt, figure 9-28.

Filing and Sanding

Files of different shapes may be inserted in the V jaws of the lower chuck for finishing metal or plastics. Sanding attachments are also available for sanding concave, convex, or flat surfaces.

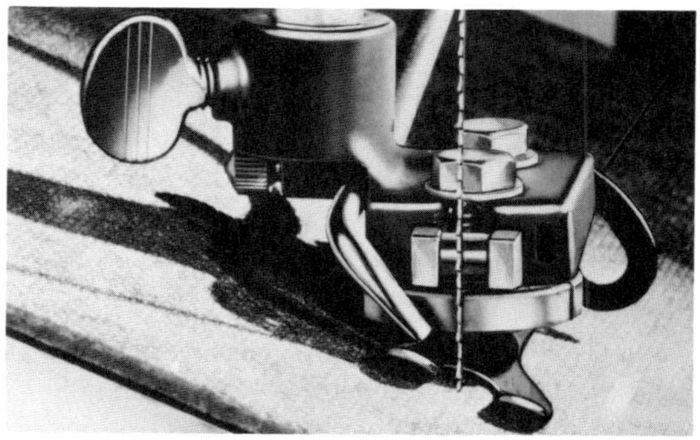

Fig. 9-28 Bevel cutting. The hold-down is tilted to the angle of the table. *(Rockwell)*

Fig. 9-29 Using the scroll saw to cut letters

ACTIVITIES

1. Lay out a design for a window valance and cut it out on the band saw. Follow the layout lines carefully and plan the cuts in advance to avoid backing out of long cuts. Select a band saw blade of the proper width and mount on the band saw making correct adjustments.
2. Make a circle-cutting jig. Cut a round tabletop of specified diameter on the band saw using the jig.
3. Design, lay out, and cut a cabriole leg for a table using the band saw. Refer to figure 9-20.
4. Lay out a design for a rectangular tissue box involving internal cuts. Set up the scroll saw with proper adjustments and make the cuts.
5. Design a raised nameplate using your initials. Cut out each letter on the scroll saw, figure 9-29. Fasten them to a wood block.

UNIT REVIEW

Multiple Choice

1. Only the upper wheel of the band saw
   a. can be tilted.
   b. has thin rubber tires.
   c. is power driven.
   d. drives the band saw blade.
2. Band saw guards are used
   a. at all times unless they are in the way.
   b. except when resawing.
   c. with all operations.
   d. except when the table is tilted.
3. A common cause of band saw blade breakage is
   a. cutting stock that is too thick.
   b. cutting green lumber.
   c. applying too much tension to the blade.
   d. cutting tight curves.
4. The amount of tension applied to a band saw blade depends on
   a. the thickness of the blade.
   b. the number of points per inch.
   c. the length of the blade.
   d. the width of the blade.
5. The clearance between the band saw blade and the side guides is
   a. none.
   b. 1/16 inch.
   c. the thickness of a piece of paper.
   d. dependent upon the type of guides.
6. Pushing the stock in a continuous forward motion when using the band saw
   a. overheats the blade.
   b. makes it difficult to follow layout lines.
   c. gives a smoother cut edge.
   d. binds the work in the blade.
7. If a clicking noise is heard when operating the band saw, it may indicate
   a. a worn wheel bearing.
   b. the blade is about to break.
   c. there are teeth missing in the saw blade.
   d. the saw blade has uneven set.

8. Small circles may be cut with the band saw without changing a wide blade by
   a. making cuts tangent to the circumference.
   b. making a series of saw cuts to the circumference.
   c. using a circle-cutting jig.
   d. using a pivot block.

9. To cut parallel curves on the band saw, use
   a. freehand cuts.
   b. a miter gauge.
   c. a pivot block.
   d. a ripping fence.

10. The scroll saw blade is clamped in
    a. the hold-down.
    b. the guide assembly.
    c. the pulley.
    d. the lower chuck assembly.

11. The hold-down on the scroll saw
    a. keeps the work from vibrating as it is cut.
    b. holds the scroll saw blade.
    c. adjusts the tension of the blade.
    d. drives the pulley.

12. The size of the scroll saw is determined
    a. by the length of the blade.
    b. by the number of steps in the pulley.
    c. by the diameter of the blade.
    d. by the width of its throat.

13. The throat of the scroll saw is the
    a. slot in the table insert.
    b. distance from the lower chuck to the upper chuck assemblies.
    c. maximum width between the guide pins.
    d. distance from the blade to the arm of the saw.

14. To save time when making internal cuts on the scroll saw,
    a. increase the speed of the saw.
    b. release the upper chuck.
    c. bore all holes in advance.
    d. raise the upper guide assembly.

15. Bevel cuts are made on the scroll saw by
    a. holding the work against the fence.
    b. tilting the table to the desired angle.
    c. twisting the blade to the desired angle.
    d. clamping a block under one end of the stock at the desired angle.

Questions

1. How is the band saw blade centered on the wheels of the band saw?
2. Describe the purpose of the rear guide of the band saw and its position in relation to the band saw blade.
3. Describe two conditions that may indicate a band saw blade is about to break. Give three reasons why blades break.
4. Describe how to apply the correct amount of tension to the blade.

5. Explain what must be done if, when replacing a band saw blade, the teeth are pointing upward.
6. Describe the procedure for cutting wide and narrow rectangular openings with the band saw.
7. Describe two methods of making duplicate pieces on the band saw.
8. How much tension should be applied to the scroll saw blade?
9. How are scroll saw blades selected?
10. Why are holes bored in the areas to be cut out internally on the scroll saw?

unit 10 Shaper and Overarm Router

OBJECTIVES

After studying this unit, the student will be able to:

- describe the parts of the shaper and the basic kinds of shaper cutters.
- install shaper cutters and adjust the shaper.
- shape straight work with a fence and shape curved work using collars and patterns on the shaper.
- describe the parts of the overarm router.
- set up and use the overarm router for routing operations.

SHAPER

The spindle shaper, figure 10-1, is designed for cutting shapes on the edges of stock. Most small shapers have a single spindle located in the center of the table, figure 10-2. Larger shapers sometimes have two spindles that rotate in opposite directions.

Shaper cutters and collars are placed on the spindle. The spindle can move up or down for different thicknesses of cuts. Looking from the top, the spindle rotates counterclockwise. Work is fed from right to left against the rotation. Some single-spindle shapers have a reversing switch to change the rotation of the spindle. For many jobs, reversing the rotation of the spindle is necessary to keep from cutting against the grain of the wood.

The size of a shaper is designated by the diameter of the spindle. Usual spindle sizes range from 1/2 inch to 1 inch with 3/4 inch being commonly used. Many shapers have interchangeable spindles so a variety of shaper cutters can be used.

Spindles rotate at speeds of 4000 to 10,000 rpm. *Because of its high speed and the difficulty of completely guarding the cutters, the shaper can be a dangerous machine to operate.*

The shaper table is grooved so that a miter gauge can be used. Threaded holes are located in the tabletop for fastening the fence and for installing starting pins.

SHAPER CUTTERS

There are various kinds of shaper cutters. Three-wing cutters are the safest to use because they cannot fly loose from the spindle, figure 10-3. They are available in high-speed steel or carbide-tipped types and come in many shapes.

The clamp-type cutterhead contains two individual cutters. Grooved collars above and below hold the cutters in place, figure 10-4.

Caution: The clamp-type cutterhead must be properly installed or the cutters will work loose and fly out. Make sure the grooves in the collars are clean, the beveled edge of the cutters match the V-groove in the collars, and the spindle nut is firmly tightened down. The cutters should be in at least half the length of the groove of the collar. Each of the two cutters should be of equal weight.

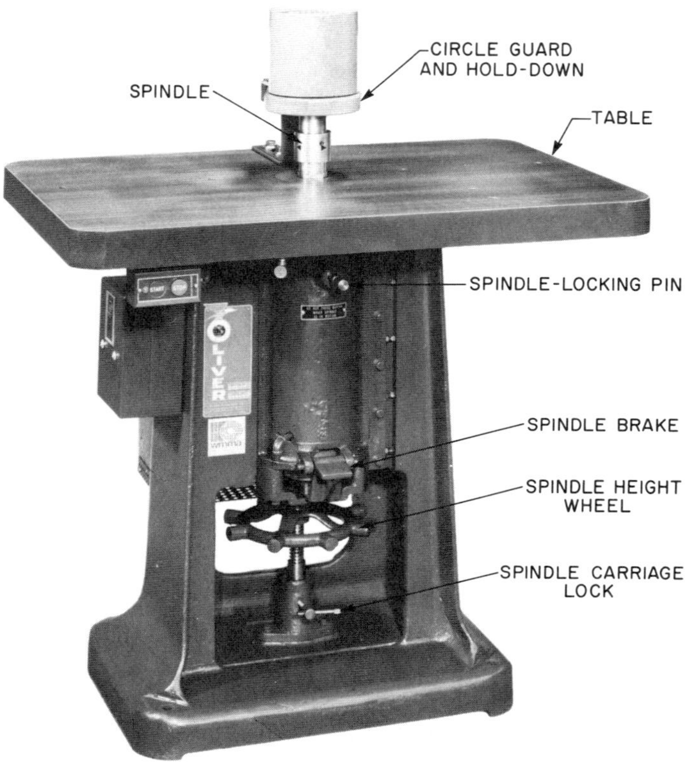

Fig. 10-1 The single-spindle shaper *(Oliver Machinery Co.)*

Fig. 10-2 The spindle is located in the middle of the table. *(Rockwell)*

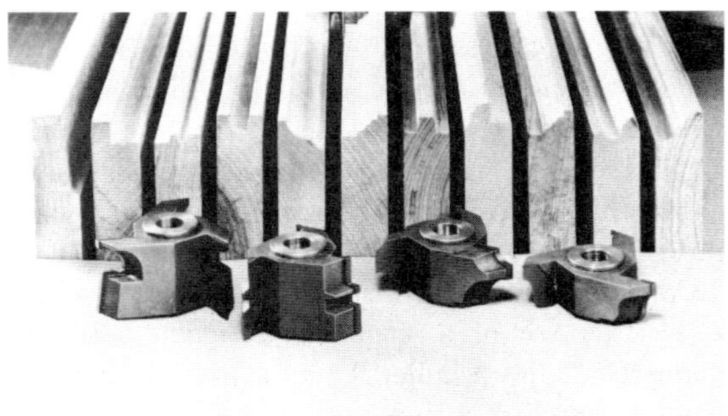

Fig. 10-3 Three-wing shaper cutters *(Rockwell)*

One advantage of using the clamp-type cutterhead with individual cutters is that the cutters are easily ground to special shapes. Three-wing cutters cannot be ground to shape on the job. The operator must use the standard shapes available. Three-winged cutters, however, may be sharpened by rubbing only the flat surface on an oilstone.

Installing Shaper Cutters

Shaper spindles are either held with a wrench or locked in place with a pin while the spindle nut is loosened. To install a shaper cutter:

1. Disconnect the machine. Remove the spindle nut, collars, and shaper cutter.
2. Replace the collars and the desired shaper cutter.
3. The cutting edge of individual cutters should project equally. To insure this, clamp a strip of wood on the table so the edge of one cutter touches the strip. Rotate the cutterhead and move the other cutter until it also touches the strip, figure 10-5.
4. Do not install cutters backwards. The cutting edge should be toward the operator. It should cut the bottom side of the stock if possible. The stock then covers the cutters as the cut is being made, figure 10-6.
5. Tighten the spindle nut securely.
6. Adjust the spindle up or down for depth of cut.

Fig. 10-5 The shaper cutters should project equally.

USING THE SHAPER

Shaping With the Fence

The two-part shaper fence, figure 10-7, may be adjusted forward or backward. The two halves may also be moved closer together or farther apart. The opening between the two fences is always as close as possible according to the cut to be made.

Each half of the fence can also be moved forward or backward. When all of the edge is removed during shaping, the outfeed side of the fence is moved forward to support the cut edge, figure 10-8. When only part of the edge is removed, the faces of each half of the fence remain in line.

To shape with the fence:

1. Adjust the fence as closely as possible.
2. Make sure the direction of feed and the rotation of the spindle is correct. Always rotate the spindle by hand to check if it is running clear.

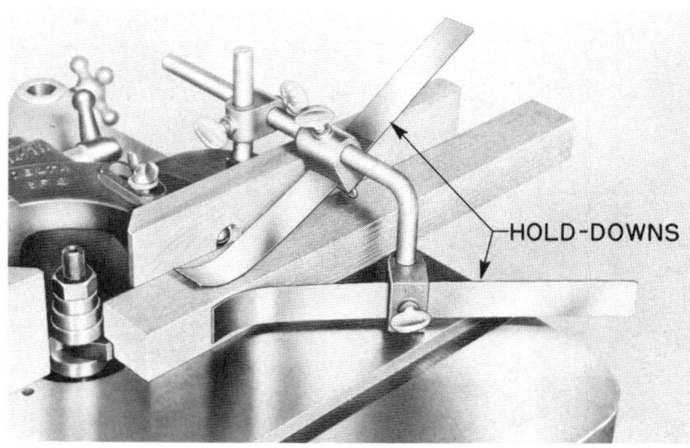

Fig. 10-6 Cuts should be made on the bottom side of the stock if possible. *(Rockwell)*

3. Make a small cut on the end of scrap stock. Examine the cut end and make adjustments to the fence or height of the spindle if necessary.
4. When all adjustments are made, cover the spindle with the guard, install hold-downs and feather boards as necessary, and make the cut, figure 10-9. Use an even rate of speed to obtain a smooth cut.

Caution: Always wear safety goggles when shaping.

End grain may be shaped by using a miter gauge, figure 10-10, or by backing the piece up with a wide board, figure 10-11. If ends and edges are to be shaped, cut the ends first. The splintering that results is then removed when shaping the edges.

Shaping With Collars

To shape curved pieces, collars are used as guides, figure 10-12. Other special work, such as shaping the inside of frames, also requires the use of collars. When using collars, only part of the edge can be shaped. The remaining part must ride against the collar. Collars cannot be used as guides when all of the edge is removed.

The guide collar is placed above the cutter, if possible, so the stock being cut covers the cutter, figure 10-13. When necessary, the guide collar is placed below the cutter. For other cuts, the guide collar is placed between two cutters, figure 10-14.

Solid collars rotate with the spindle. If the work is held too tightly against the collar or if the feed is too slow, the edge will burn. Ball-bearing

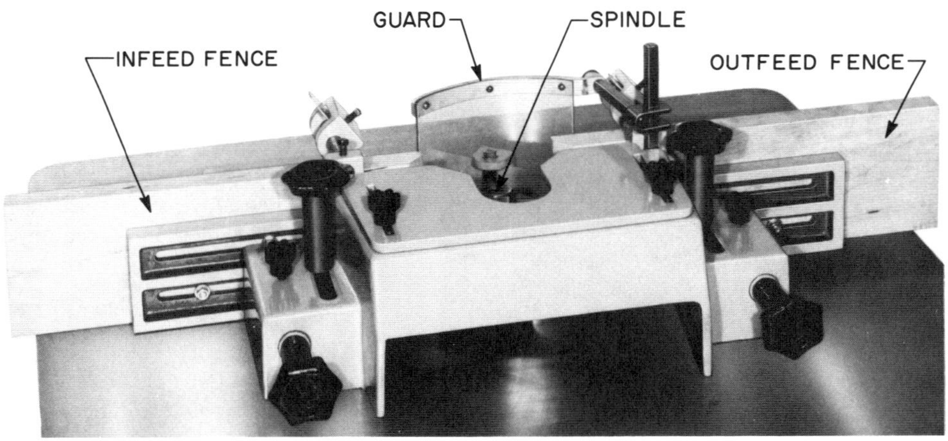

Fig. 10-7 The two-part shaper fence *(Rockwell)*

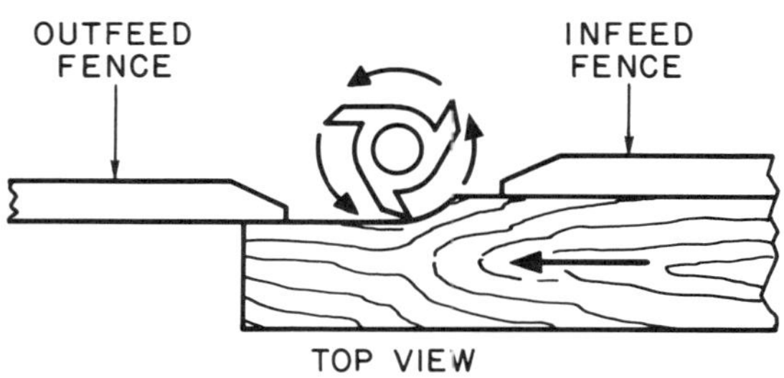

Fig. 10-8 The outfeed fence is moved forward to support the cut edge when all of the edge is removed during the shaping.

Fig. 10-9 Shaping with the fence *(Rockwell)*

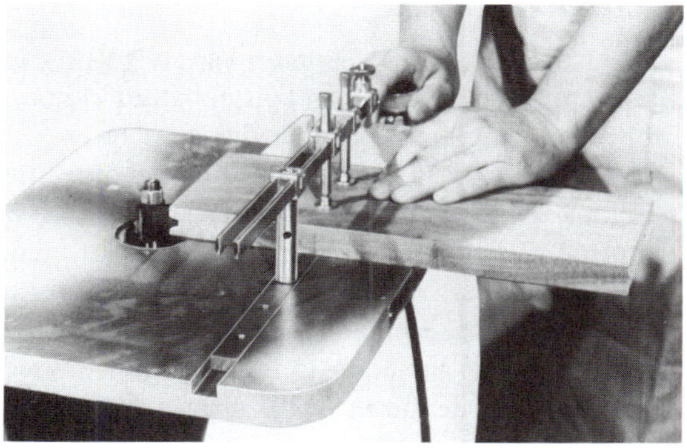

Fig. 10-10 Shaping end grain with a miter gauge *(Rockwell)*

Fig. 10-11 Shaping end grain by backing up the piece with another board (Hold down and guard removed for clarity.)

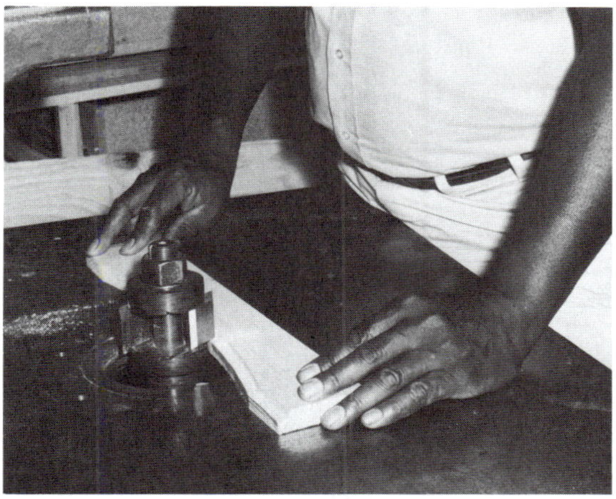

Fig 10-12 Shaping curved work using collars (Hold down and guard removed for clarity.)

collars should be used to eliminate burning or darkening of the edge.

When starting a cut using collars, a starting pin is used to steady the work and prevent kickback, figure 10-15. Hold the work against the starting pin and move the stock into the rotating cutter. When the edge of the piece comes in contact with the collar, move it away from the pin and continue the cut. A pivot block, similar to that used on the band saw, may be clamped to the shaper table and used in place of the starting pin, figure 10-16.

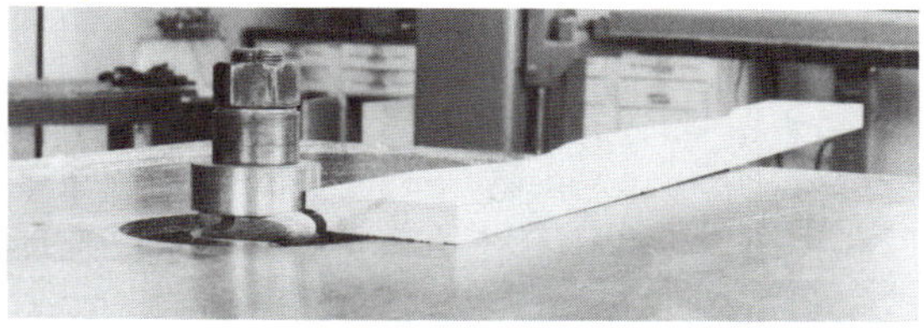

Fig. 10-13 Place the guide collar above the cut if possible. (Hold down and guard removed for clarity.)

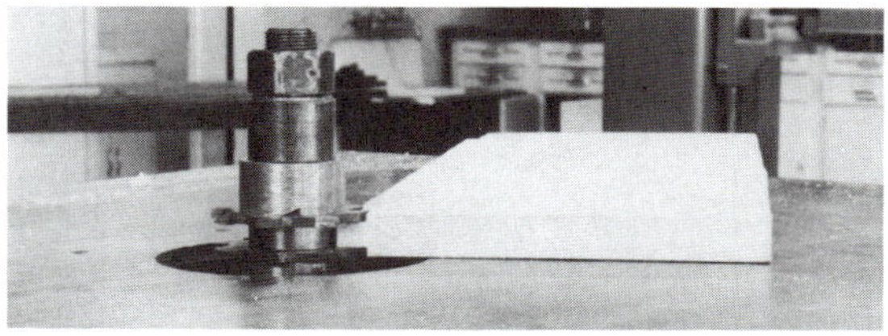

Fig. 10-14 For special cuts, the collar is placed between two cutter. (Hold down and guard removed for clarity.)

Fig. 10-15 A starting pin is used to prevent kickback when starting the cut. *(Rockwell)*

Shaping With Patterns

If all of the edge is removed when shaping curved edges, a pattern must be used. The pattern must be cut accurately to the desired shape and have smooth edges.

The pattern is fastened to the blank piece to be shaped by nails, screws, or clamps. The pattern rides against a collar allowing all of the edge of the blank to be shaped, figure 10-17. Shaping with patterns is similar to shaping with collars except that the pattern rides against the collar. The work may be placed either above or below the pattern. If possible, keep the work below the pattern.

Special jigs may be made for certain shaping operations such as fluting a round, tapered leg. In production work, many special forms, jigs, clamping devices, and fixtures are used for shaping operations.

Fig. 10-16 A pivot block may be used in place of a starting pin. (Hold down and guard removed for clarity.)

Fig. 10-17 If possible, keep the work below the pattern. *(Rockwell)* **(Hold down and guard removed for clarity.)**

OVERARM ROUTER

The overarm router has a variety of uses in the shop. It can make dovetails, mortises, tenons, rabbets, and hundreds of other shapes.

Clean, fast cuts are obtained by a motor that develops up to 20,000 rpm. The head, which is mounted above the table, can be raised or lowered, swiveled 360 degrees, and tilted to any angle. An adjustable fence, similar to the shaper fence, is mounted on the table. A miter gauge can be used in the slot provided in the table surface.

Some overarm routers double as a shaper and router by installing another motor below the table surface, figure 10-18. Router bits or three-lip shaper cutters are used in either motor.

One advantage of the overarm router is that the motor may be lowered by means of a foot pedal. The cutter then enters the wood like a drill to cut a recess in the surface. The motor may also be locked in position for depth of cut.

Another advantage of the overarm router is that cuts may be made on the side of fairly wide

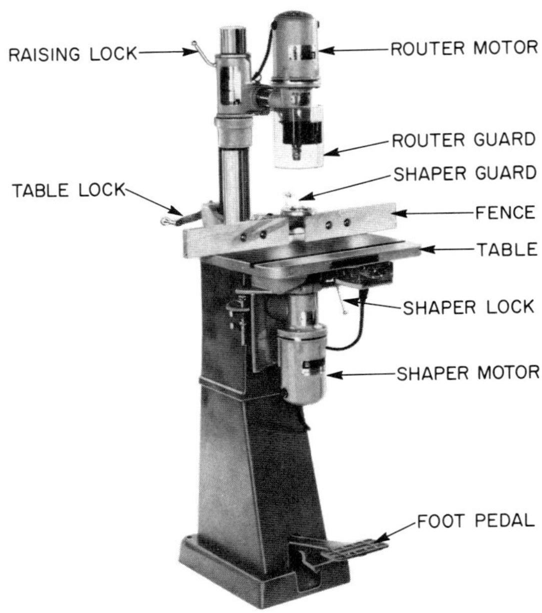

Fig. 10-18 The combination overarm router and shaper *(Rockwell)*

Fig. 10-19 Cutting grooves with the overarm router *(Rockwell)*

Fig. 10-20 Cutting a wide plough *(Rockwell)*

Fig. 10-21 Cutting a groove *(Rockwell)*

stock, an operation that cannot be done with the shaper. The cut is always clearly visible because the cutting tool is above the work.

USING THE OVERARM ROUTER

To use the overarm router, first install the router bit or shaper cutter of the desired shape. Whether using a fence, collars, or patterns, the operation is similar to the shaper except the motor is above the work.

The overarm router can be used to cut grooves or dadoes using the ripping fence or miter gauge, figure 10-19. With special bits, wide grooves or ploughs can be made, figure 10-20. Rabbets are cut in a similar manner by moving the fence to cut the edge of the piece. Designs can be cut in the faces of doors. Vee grooves can be made with the appropriate router bit, figure 10-21. Freehand routing is also easy because the work is clearly visible.

Collars may be installed to shape curved work. Pins of varying diameters may be installed in the center of the table and used as guides for shaping with a pattern. The edge of the pattern rides against this pin, figure 10-22.

Fig. 10-22 Routing using a guide pin and pattern. The pin and pattern are under the work. *(Rockwell)*

ACTIVITIES

1. Make a rectangular plaque. Shape the edge with a cove cut. Leave some of the original square edge.
2. Make a curved plaque shaping the edge by using a collar as a guide.
3. Make a curved plaque. Shape and form the edge using a pattern.
4. Make a rectangular design on a door face using the overarm router.
5. Make duplicate curved pieces by using a pattern against the guide pin of the overarm router.

UNIT REVIEW

Multiple Choice

1. Double-spindle shapers or single-spindle shapers with reversing switches are used
   a. to help cut with the grain.
   b. to feed from left or right.
   c. to speed up production.
   d. all of the above.
2. The size of a spindle shaper is designated by
   a. the diameter of the spindle.
   b. the length of the spindle.
   c. the size of the table.
   d. the amount of travel of the spindle.
3. The shaper is a relatively dangerous machine to operate because
   a. of the constant danger of kickback.
   b. the cutters are not completely guarded.
   c. the cutters often fly loose.
   d. of its counterclockwise rotation.
4. Although they are safer to use, three-wing cutters
   a. are difficult to sharpen.
   b. have limited shapes.
   c. cannot be ground to special shapes in the shop.
   d. are expensive.
5. In the clamp-type cutterhead, individual cutters should be installed
   a. with the cutting edges to the right.
   b. with the rotation of the spindle.
   c. only by an experienced machinist.
   d. with at least one half of the edge in the collar.
6. The opening between the two sections of the shaper fence should be
   a. as close together as possible.
   b. the width of the shaper collars.
   c. as far apart as possible.
   d. opened only enough to accommodate the guard.
7. Collars are limited in their use when shaping curved work because
   a. of the thickness of the stock being shaped.
   b. an edge cannot be completely shaped.
   c. of the shape to be cut.
   d. of the length of the shaper cutter.

8. If a rectangular piece is to be shaped all around,
   a. shape the ends first.
   b. shape the edges first.
   c. shape by moving it clockwise.
   d. shape by moving it counterclockwise.
9. When starting a cut while shaping a piece using collars, a starting pin or block is used to
   a. feed the work slowly into the cutters.
   b. make a test cut.
   c. prevent kickback.
   d. guide the work.
10. When all of the edge is to be removed when shaping curved edges,
   a. use a starting pin.
   b. use a pattern.
   c. adjust the outfeed side of the shaper fence to support the cut edge.
   d. use two guide collars instead of one.

Questions

1. What is the usual rotation of a single-spindle shaper?
2. Give the advantages and disadvantages of the three-wing and individual shaper cutters.
3. Explain the safety factors to follow when installing shaper cutters in the clamp-type cutterhead.
4. In what direction should stock be fed into the shaper?
5. If possible, on what side of the stock should cuts be made when operating the shaper?
6. Explain the position of the outfeed half of the shaper fence when shaping operations remove all of the edge.
7. Name two advantages of the overarm router over the shaper.
8. How is a pattern guided when shaping with an overarm router?
9. How is the pattern guided when using the shaper?
10. What is the disadvantages of using solid shaper collars?

unit 11 Mortiser and Tenoner

OBJECTIVES

After studying this unit, the student will be able to:

- name the parts of a mortiser and describe their functions.
- set up, adjust, and use the mortiser to make mortises of various sizes.
- name the parts of a tenoner and describe their functions.
- set up, adjust, and use the tenoner to make tenons of various shapes and sizes.

MORTISER

The mortiser, figure 11-1, is designed to make rectangular cuts in wood. These cuts are called *mortises.* Mortises form one-half of a mortise-and-tenon joint, figure 11-2.

The cutting tool consists of a bit that revolves in the center of a hollow square chisel, figure 11-3. As the tool is pressed down into the wood, the bit cuts a round hole. Then the chisel cuts and squares up the round hole.

Mortising Bit

The mortising bit is similar to an auger bit. It has spurs and cutting lips, but no feed screw. The cutting end is larger than the rest of the bit. This cuts a hole equal in size to the outside dimension of the chisel. The bit is made this way so that the chisel cuts away the least amount of stock to square up the hole.

Because of the flared end of the bit, there must be some clearance (about 1/16 inch) between the bit and chisel. If the bit rubs against the chisel, both will burn and probably be ruined, figure 11-4.

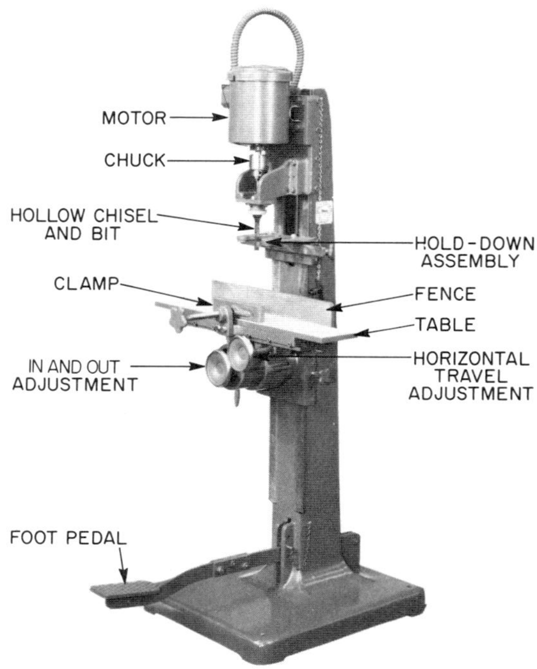

Fig. 11-1 Hollow chisel mortiser *(Oliver Machinery Co.)*

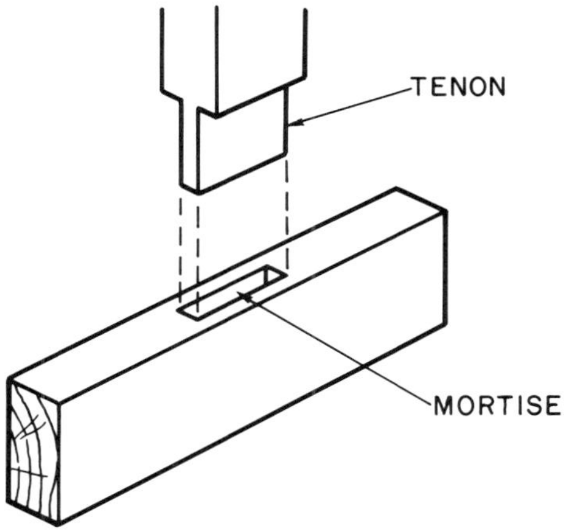

Fig. 11-2 A mortise-and-tenon joint

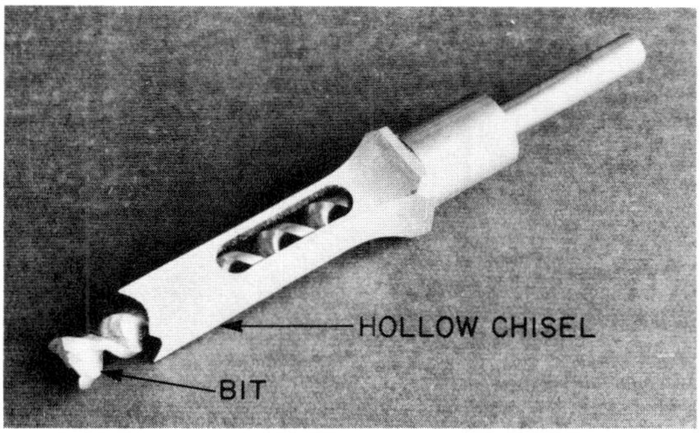

Fig. 11-3 The hollow chisel and mortising bit

Mortising bits are sharpened the same way as auger bits. Care must be taken to sharpen the inside of the spurs. Do not file the spurs excessively because this may destroy their height.

Hollow Chisel

The hollow chisel is square with an opening through its length in which the bit rides. It may have an opening in one or two sides to let wood chips escape. The bottom end of the chisel is ground on the inside to a sharp beveled edge. One method of maintaining this sharp edge is to sharpen it on a drill press, figure 11-5.

The top end of the chisel is securely held in the mortiser and does not rotate. Usual chisel sizes range from 1/4 inch to 1 inch, in increments of 1/8 inch. Other sizes are also available.

Parts of the Mortiser

The mortiser table moves sideways, up and down, or in and out as desired. Stops limit the sideway travel of the table. These are used when cutting a number of mortises of the same length. The table may also be tilted to cut mortises at an angle, figure 11-6.

The table has a fence against which the stock is placed. A clamp holds the stock against the fence. Hold-downs on both sides of the cutting tool hold the work firmly down against the tabletop.

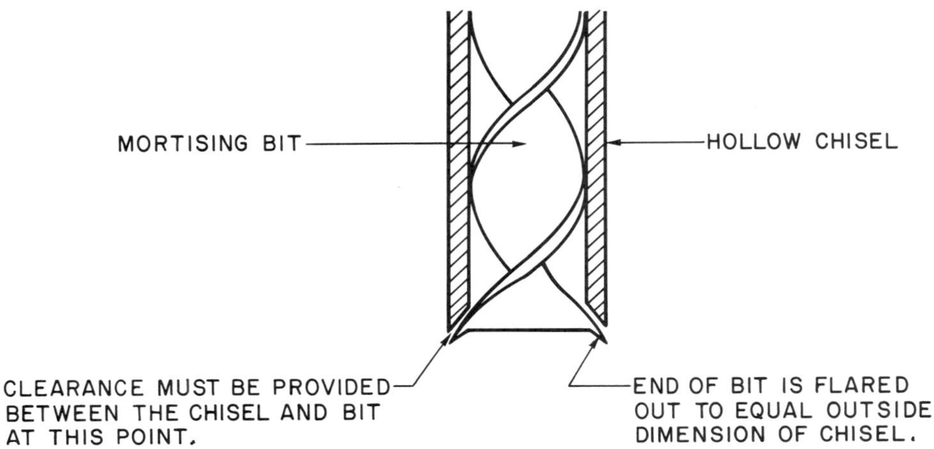

Fig. 11-4 Cross section of hollow chisel and bit

Above the table, the chisel is held in a framework. The bit is held in the chuck of the motor. The motor, chisel, and bit are moved downward toward the stock, usually by a foot pedal, to make the cut. The depth of cut is controlled by adjusting the table height or by adjusting the depth-of-cut stops.

Another type of mortiser is the chain-saw mortiser. This type cuts mortises using a continuous saw. The chain-saw mortiser leaves a rounded bottom to the mortise and is frequently used in production mills.

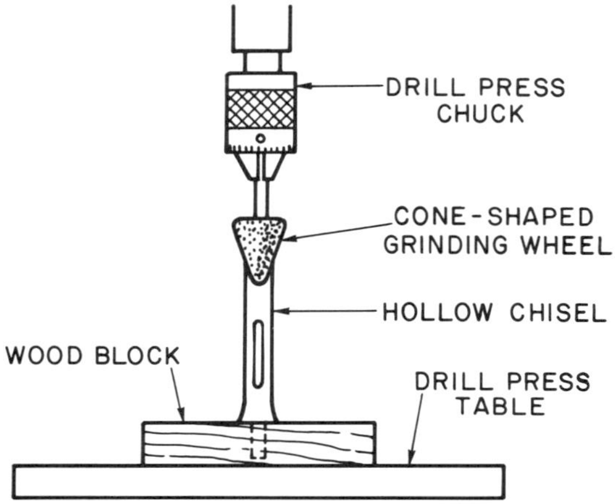

Fig. 11-5 A method of sharpening a hollow chisel

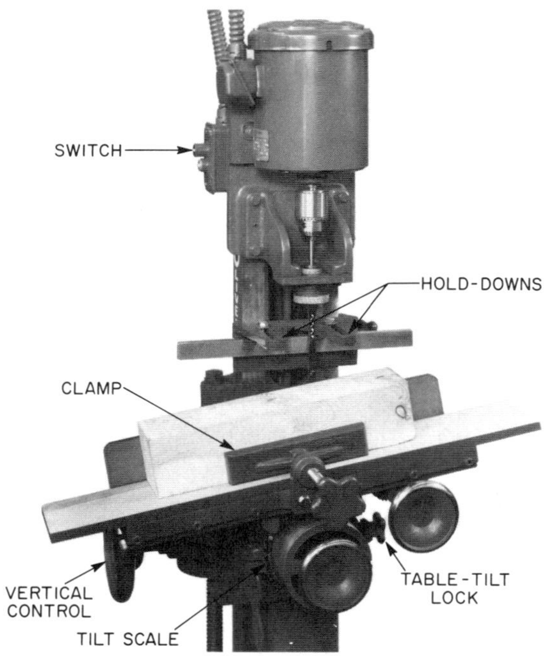

Fig. 11-6 The mortiser table may be tilted to cut mortises at an angle. *(Oliver Machinery Co.)*

USING THE MORTISER

Usually when making mortise-and-tenon joints, the mortises are cut first and the tenons then cut to fit the mortise.

Adjusting the Mortiser

1. Select the desired size chisel and install it with the proper size bushing, if required. Square the chisel with the fence and tighten it securely in place, figure 11-7.
2. Lower the table and insert the bit in the chisel. Tighten the bit in the motor chuck. Leave about 1/16-inch clearance between the bit and chisel.
3. Mark the end of the stock to be mortised for the depth of cut. Lay out the width and length of the mortise on the face of the stock.
4. Place the stock alongside the chisel and lower the table. Depress the foot pedal as far as possible. Raise the table until the end of the chisel is in line with the layout line, figure 11-8.

Fig. 11-7 Squaring the hollow chisel to the mortiser fence

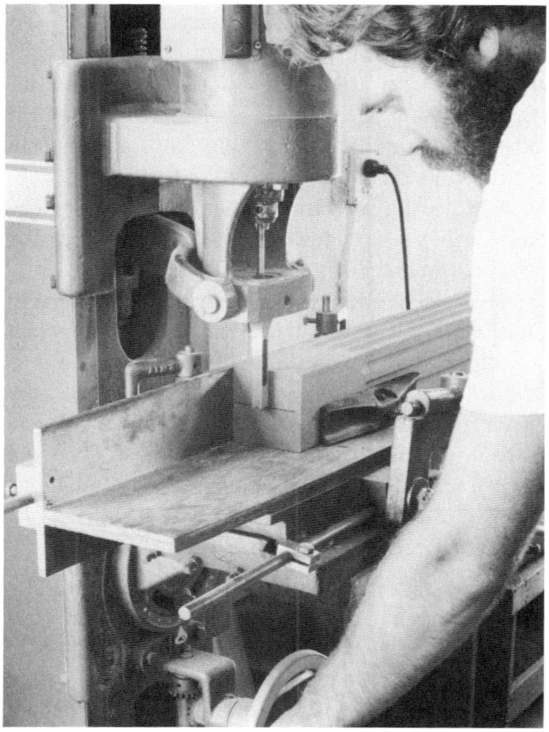

Fig. 11-8 Adjusting the mortiser for depth of cut

Fig. 11-9 Setting stops

5. Clamp the stock against the fence so the layout is about centered on the chisel. Move the chisel down until it is just above the work.
6. Move the table in or out until the front and back sides of the chisel line up with the layout lines. Move the table sideways until the chisel lines up with the end of the cut.
7. Clamp a stop against the end of the work. Set the table stop to limit travel beyond the end of the cut, figure 11-9.
8. Move the table sideways until the chisel is in line with the other end of the mortise. Set the table stop to limit travel beyond that end.
9. Adjust the hold-downs to hold the work firmly against the tabletop. The hold-downs should clear the motor assembly as it is pressed downward.

Fig. 11-10 Making mortises

Cutting the Mortise

1. Move the table to cut one end of the mortise. Start the machine.
2. Make the cut by depressing the pedal. If the mortise is deep, take small cuts, allowing the chips to escape, until the bottom of the mortise is reached, figure 11-10.
3. Move the table sideways and continue cutting to the other end of the mortise. Skip a little less than the width of the chisel with each cut, figure 11-11.

4. Cut the wood remaining between cuts. This method of cutting prevents any side strain on the chisel and bit.

Always cut with wood on four sides of the chisel or with wood on two opposite sides. Cutting mortises with wood on three sides, especially with small chisels, may cause the chisel to slip off the side of the mortise and bend or break the chisel and bit. Never attempt to lengthen a mortise by the width of the chisel or less. Cut all identical mortises required. Adjust the stops for making mortises in a different location.

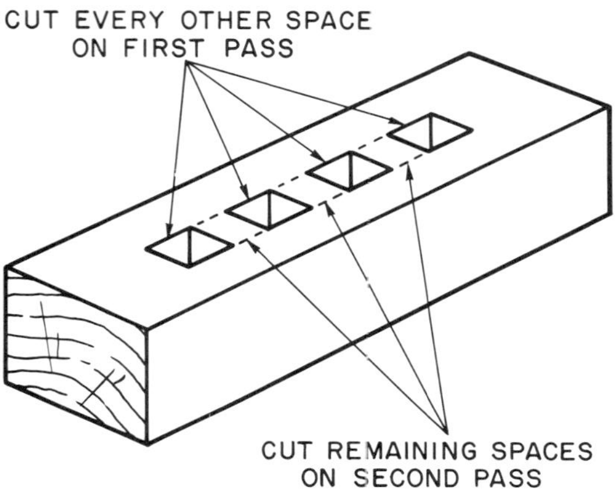

Fig. 11-11 Correct method of cutting a mortise

Mortising attachments are available for use on a drill press. However, use the mortiser for fast, accurate mortising and for ease of setup.

> **Caution:** When using the mortiser, observe all general safety precautions followed with other woodworking machinery. In addition, check the bit clearance at the end of the chisel, keep fingers away from the rotating bit, and make sure the work is securely clamped in position.

THE TENONER

The tenoner, figure 11-12, is used to make tenons for mortise-and-tenon joints. The table saw or radial arm saw is used to make tenons when only a few are to be made. The tenoner is used when many tenons of the same size and shape are needed.

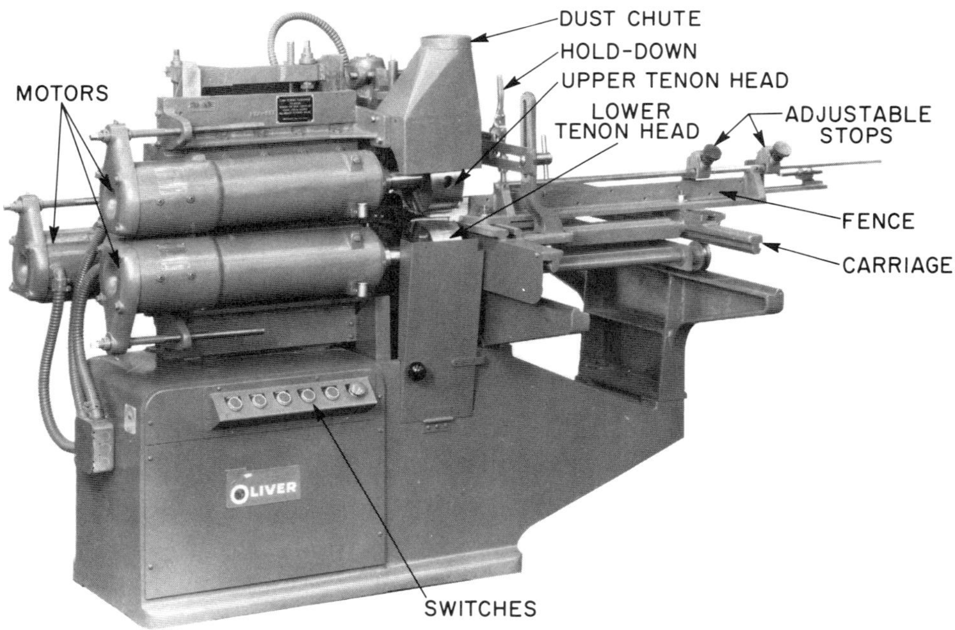

Fig. 11-12 Single-end tenoner *(Oliver Machinery Co.)*

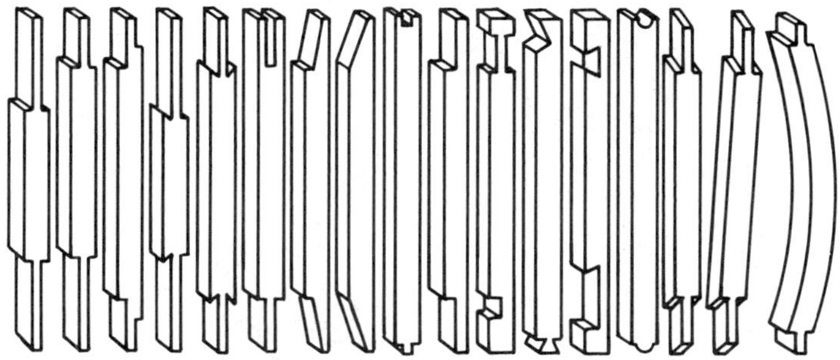

Fig. 11-13 Examples of tenons *(Oliver Machinery Co.)*

Double-end tenoners cut tenons on both ends of the stock with one pass. Single-end tenoners cut tenons on only one end of the stock with each pass, figure 11-13. Usually double-end tenoners are used only in high-production mills. In small shops, single-end tenoners are used.

CUTTING MECHANISM

The heart of the single-end tenoner is its cutting mechanism, figure 11-14. The stock first passes between two *tenon heads.* These tenon heads adjust vertically and closer together or farther apart for different tenon thicknesses. In addition, the top tenon head adjusts horizontally to cut tenons with offset shoulders. Separate motors operate the tenon heads. Knives in the tenon heads cut both sides of the tenon and score the shoulder cuts to prevent splintering.

The stock then passes between the *coping heads* which are mounted in a vertical position. Each coping head adjusts vertically or horizontally.

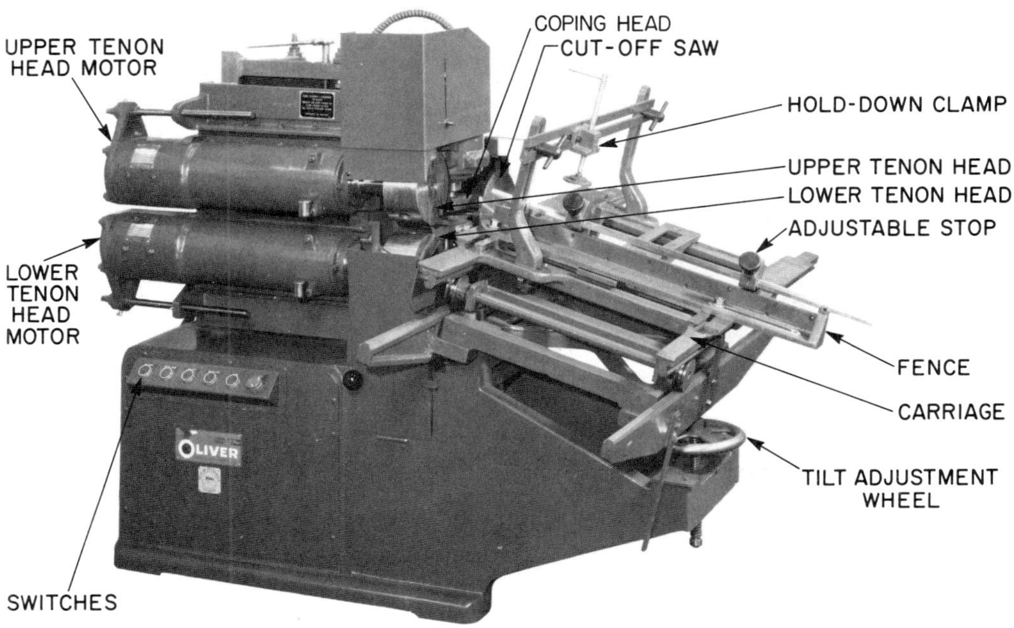

Fig. 11-14 Close up view of the tenoner's cutting mechanism *(Oliver Machinery Co.)*

Fig. 11-15 A tenon with coped shoulders

They are used to make the coped shoulders on the tenon to fit the shaped edge *(sticking)* of the mortised piece, figure 11-15. Because coping heads are not always used, a separate motor operates them. When tenons are made with square shoulders, for instance, the coping heads are not used.

The tenon is next cut to length by the *cut-off saw*. The cut-off saw also moves horizontally and vertically. Usually, the cut-off saw is operated by the same motor as the tenon head.

The stock is held by clamps on a rolling table, called the *carriage,* and moved back and forth through the cutting mechanism. In some models, the table may be tilted to cut tenons on an angle. In other models where the table is fixed, jigs hold the stock to cut tenons at the desired angle.

USING THE TENONER

Once the tenoner is set up and adjusted, the operation is fairly simple, figure 11-16. The stock to be tenoned is clamped to the table, the power turned on, and the carriage moved forward. The tenon heads cut both sides of the tenon. The coping heads, if used, shape the shoulders. Finally, the cut-off saw cuts the tenon to length.

Fig. 11-16 Using the tenoner

Once the tenoner is set up, any number of identical tenons may be made. Setup requires the most skill. All setups usually start at the lower tenon head because it is the only head that cannot be adjusted horizontally. After cuts are made in one end, the pieces are reversed and tenons are cut in the other end with the same setup.

Caution: When using the tenoner, make sure the setup is correct and that all saws and cutting knives are secure. Clamp the stock securely to the carriage. Keep fingers away from the revolving cutting tools.

ACTIVITIES

1. Explain the advantages and disadvantages of the hollow-chisel mortiser over the chain-saw mortiser.
2. Sharpen a mortising chisel and bit.
3. Make drawings of the many different kinds of tenons that can be made.
4. Devise a jig for cutting tenons at an angle on the single-end tenoner.
5. Cut tenons with square and coped shoulders.
6. Cut duplicate mortises using stops.

UNIT REVIEW

Multiple Choice

1. The mortising bit is similar to a
   a. auger bit.
   b. twist drill.
   c. circle cutter.
   d. multispur bit.

2. The mortising bit is flared out on its end to
   a. prevent undue strain on the bit.
   b. prevent the chisel from cutting excessive stock.
   c. prevent undue strain on the foot pedal.
   d. prevent overloading and burning out of the motor.

3. The amount of clearance between the bit and chisel of the mortiser should be about
   a. 1/32 inch.
   b. 1/16 inch.
   c. 1/8 inch.
   d. 1/4 inch.

4. If not enough clearance is maintained between the mortising bit and chisel,
   a. wood chips will not escape.
   b. the mortise cannot be made.
   c. the motor will burn out.
   d. the mortising bit and chisel will burn.

5. Mortises are cut with stock
   a. on one side of the chisel.
   b. on two sides of the chisel.
   c. on three sides of the chisel.
   d. on four sides of the chisel.

6. When making only a few tenons, use the
   a. double-end tenoner.
   b. single-end tenoner.
   c. shaper.
   d. table saw.

7. When tenons are made using the single-end tenoner, the stock is first passed through the
   a. cut-off saw.
   b. coping heads.
   c. tenon heads.
   d. shaper cutterhead.

8. The coping heads are used to
   a. cut the tenon to thickness.
   b. shape the shoulders of the tenon.
   c. cut the tenon to length.
   d. cope the end of the tenon.

9. The only cutting tool of the tenoner that cannot be adjusted horizontally is the
   a. cut-off saw.
   b. top tenon head.
   c. bottom tenon head.
   d. top cope head.

10. When the shoulders of a tenon are cut square, what part of the tenoner is not used?
   a. Tenon heads
   b. Coping heads
   c. Cut-off saw
   d. Top coping head

Questions

1. Why is clearance between the mortising bit and the chisel necessary?
2. Why are there openings in the sides of the mortising chisels?
3. How are the mortising chisel and bit sharpened?

4. Why is the end of the mortising bit flared out?
5. What are the usual sizes of mortising chisels and bits?
6. Describe the procedure for adjusting the mortiser for depth of cut.
7. Explain how to square up the mortising chisel with the mortiser fence.
8. Describe how to cut mortises. What precautions should be taken to avoid damaging the cutting tool?
9. How are duplicate mortises made?
10. Describe the order of cuts when making tenons using the single-end tenoner.

unit 12 Wood Lathe

OBJECTIVES

After studying this unit, the student will be able to:

- explain the purpose of the wood lathe and its parts.
- describe basic turning chisels and their uses.
- turn a cylinder to a specific diameter.
- cut shoulders from square to round and from round to round.
- cut V grooves, beads, and coves.
- make a faceplate turning of a specified design.

WOOD LATHE

The wood lathe is designed to make *turnings.* These are round pieces such as table and chair legs or parts of patterns. The wood lathe is not a production machine. It is used mainly by modelmakers and patternmakers or by cabinetmakers to turn an original piece.

The wood lathe, figure 12-1, consists of a bed on which the headstock and tailstock is mounted. The bed usually rests on a stand at a comfortable working height. The headstock is mounted in a fixed position on the left end of the bed. The tailstock can slide along the bed and lock in any position.

The size of the wood lathe is determined by the largest diameter that can be turned on it. The stock to be turned is either held between the head center and the tail center (spindle turning), figure 12-2, or secured to a faceplate (faceplate turning), figure 12-3.

Headstock

The headstock, figure 12-4, has a hollow spindle which is threaded on both ends to hold a faceplate. Work is screwed to faceplates when it is not turned between centers. Faceplates are mounted on the outer end of the spindle when large work is to be turned.

The head center is inserted in the hollow spindle facing the tailstock. Spurs on the end of the head center hold and drive the work that is held between centers. The head center is removed by tapping its end with a metal rod inserted in the hollow of the spindle.

The rotating speed of the spindle is controlled by a variable-speed motor or by a stepped pulley. Speeds range from 340 rpm to 3200 rpm. Slow speeds are used to rough down large cylinders. Higher speeds are used for finishing operations.

In some models, the headstock has an indexing head. This consists of one row of eight holes

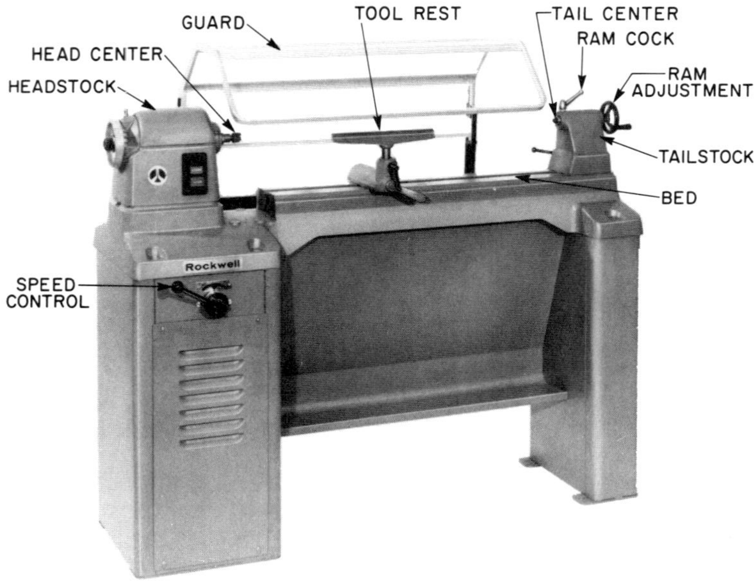

Fig. 12-1 Wood lathe *(Rockwell)*

Fig. 12-2 Spindle turning holds the work between centers. *(Rockwell International)*

Fig. 12-3 Work is fastened to faceplates when not held between centers. This is called faceplate turning. *(Rockwell International)*

and another row of 60 holes. The indexing pin is engaged to hold the work in a fixed position.

Tailstock

The tailstock also has a hollow spindle into which the tapered tail center is inserted. Some tail centers are called dead centers because they do not turn with the work. These types are coated with a light oil between the end of the work and the center to prevent burning the work or the center. Ball-bearing *live centers* revolve with the work and require no lubrication.

The tailstock spindle, called a *ram,* moves in or out by adjusting the ram wheel and locking it in position. The tail center is loosened and removed by bringing the spindle all the way back. Usually the ram travels about 2 1/4 inches.

Tool Rest

Turning chisels are supported by the tool rest. It can slide along the bed and be locked securely in any location. It may also be moved in or out, up or down, or rotated in a complete circle.

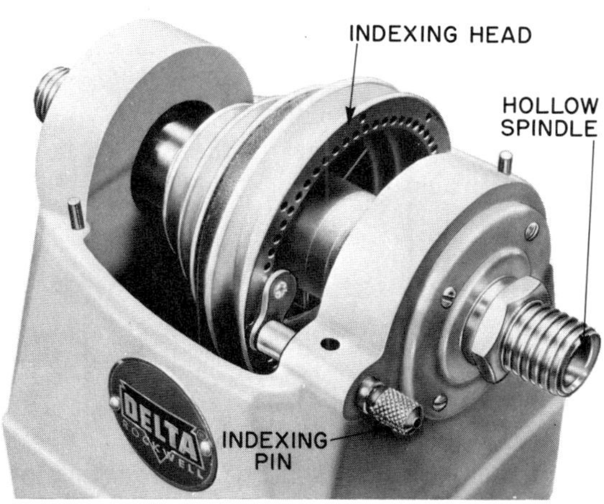

Fig. 12-4 Headstock *(Rockwell International)*

TURNING TOOLS

Tools used to cut and shape the wood as it turns in the lathe are called *turning chisels.* Most common shapes are the gouge, skew, round nose, and parting tool, figure 12-5. Other types are also available.

The *gouge* is used primarily to rough down the stock to its approximate diameter. It is also used to make cove cuts. The *skew* is used often on the wood lathe. It produces finishing cuts to smooth a cylinder and makes beads and V grooves. The skew is also used to scrape faceplate turnings. The *round nose* is usually used to scrape faceplate turnings for straight or curved cuts. The *parting tool* is a narrow chisel used in a scraping fashion to make narrow grooves for measuring diameters with calipers.

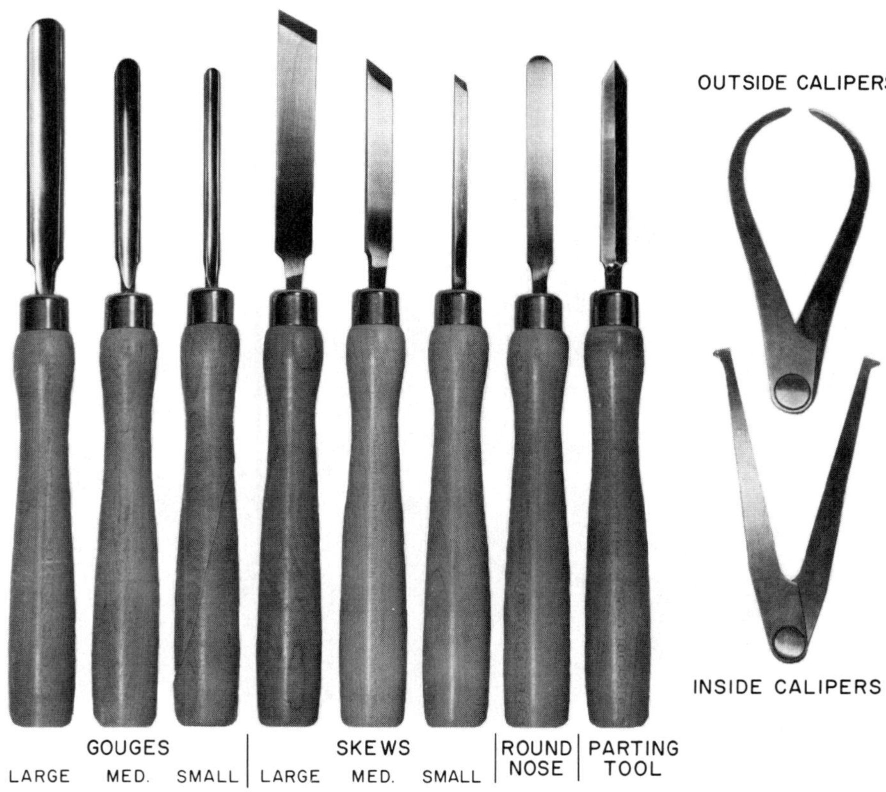

Fig. 12-5 Wood turning chisels and calipers *(Oliver Machinery Co.)*

Fig. 12-6 Spindle turning by the cutting method

Fig. 12-7 Spindle turning by the scraping method

Fig. 12-8 Centering the ends of a spindle turning

Turning chisels are ground and sharpened in a manner similar to wood chisels. Care must be taken to maintain their original shape. Never try to make a turning with a dull tool.

To test the diameter of the turning, inside and outside calipers are used. Dividers are used to mark the work.

SPINDLE TURNING

In spindle turning, the work is turned between centers and shaped by holding chisels against the work to cut or peel off wood, figure 12-6. This cutting method produces an extremely smooth surface that requires no sanding. However, unless the operator knows how to use the chisels properly, the chisel may slip and ruin the work when making some cuts.

Until this skill is acquired, a safer way to shape the work is by a scraping action. In scraping, the chisel is held straight into the work and the wood is scraped away, figure 12-7. The disadvantage of this method is that it leaves a rough surface that must be sanded.

A professional woodturner rarely uses the scraping method when spindle turning. However, because of the nature of the work, the scraping method is almost always used when faceplate turning.

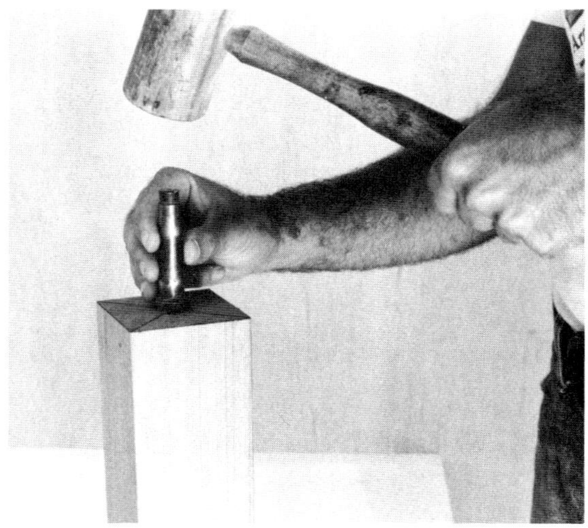

Fig. 12-9 Tapping the head center into the work

Turning a Straight Cylinder

Mount the spindle stock:

1. Cut the stock so it has the same thickness and width.
2. Find the center of each end by drawing lines from corner to corner, figure 12-8. Center punch the center. If the wood is hard, drill a 1/16-inch hole about 1/2 inch deep into each center. This insures that the work will not turn off center when mounted on the lathe.
3. Cut the stock so it has eight equal sides. Use the table saw or band saw. If part of the stock is to remain square, do not make the stock eight-sided.
4. Tap the head center into the center of one end with a mallet, figure 12-9. Move the tailstock back. Place the work and the head center into the headstock.
5. Slide the tailstock until the tail center is about 1 inch from the work. Lock the tailstock in position.
6. Bring the tail center into the work by moving the ram forward until the center is firmly into the end. Release the center slightly, apply a little oil, and seat the center again. Lock the ram into position.
7. Adjust the tool rest so that it is about 1/8 inch away from the work. Its top should be about even with the work's center.

Fig. 12-10 Use the gouge to rough down the stock.

8. Turn the work over by hand to make sure it is running free and clear.
9. Adjust the speed of the lathe. For roughing to size, speeds should be slow. For work over 8 inches in diameter, run the lathe at its slowest speed. For smaller work, the speed may be increased slightly.

Caution: Running rough stock that has not been trued up at a high speed may cause it to fly loose. Always wear eye protection during turning operations.

Rough down the cylinder:

1. Place the large gouge on the tool rest. Point it toward the turning work with the bevel down, figure 12-10. Twist the gouge slightly toward the end.
2. Start the cut about 3 inches from the end and work toward the end. Twisting the gouge makes the chips fly away from the operator.
3. Start another cut about 3 inches from the start of the first cut and work toward the same end.
4. Continue making cuts until about 2 or 3 inches from the other end. Make the last cut toward the opposite end.
5. Repeat this procedure until the stock is round and slightly larger than the desired diameter.

Caution: As turning progresses, stop the lathe and move the tool rest closer to the work. Always keep the tool rest as close to the work as possible or else the chisel may be flipped from the hands. Always stop the lathe when adjusting the tool rest.

Check the diameter:

1. Bring the cylinder close to size with the gouge.
2. Set the calipers for 1/8 inch more than the desired diameter of the cylinder.
3. Make cuts with the parting tool about every two or three inches while testing with the calipers, figure 12-11.

4. Continue cutting with the gouge just until the parting tool marks are removed.

Smooth the cylinder to size:

1. Adjust the lathe to a slightly higher speed to produce a smoother cut. Starting about 2 or 3 inches from one end, lay the skew on the tool rest so the cutting edge is above the work.
2. Gradually bring the cutting edge down and the handle up until a small amount of wood is peeled off, figure 12-12.
3. Move the skew toward the end. Hold it at the same angle and guide it with the fingers against the tool rest. Continue cutting until the end is reached.
4. Turn the skew in the opposite direction. Make a cut from the starting point to the opposite end. Continue making cuts in this manner until the desired diameter is reached.

Cutting a Shoulder From Square to Round

Many times a part of a turning must remain square. A part of a table leg may be left square to receive the table rail. Other parts of turned pieces, such as a baluster in a stair rail, are left square for appearance.

The object in turning a shoulder from square to round is to avoid splintering the corners of the square section.

Fig. 12-11 Use the parting tool to make narrow grooves close to the desired diameter. Measure with the calipers.

1. Lay out a line around the piece to be turned to indicate the shoulder.
2. Lay the skew on edge on the tool rest with the toe (longest point) down. Hold the skew at an angle so that the beveled edge is toward the square section and at right angles to the work. Bring the skew into the revolving work and nick the square corners about 1/8 inch deep, figure 12-13.
3. Use the parting tool to make a groove about 1/16 inch away from the nicked corners, figure 12-14. Space grooves with the parting tool along the section to be turned.
4. Bring the section to be turned down to a rough size with the gouge.
5. Holding the skew with the toe down and at the same angle, trim the shoulder by bringing the handle of the chisel down, figure 12-15.
6. Smooth the round section of the turning with the skew. Bring the skew into the shoulder to make a clean corner.

Cutting a Shoulder From Round to Round

Cutting a shoulder from round to round is similar to cutting a shoulder from square to round.

1. Since there are no square corners, the parting tool is used to cut a groove slightly away from the shoulder.
2. With a gouge, bring the work down to a size slightly more than the smaller diameter.
3. Cut the vertical part of the shoulder with the toe of the skew.
4. Trim and smooth both small and large diameters with the skew in the same manner as finishing a straight cylinder.

Cutting a Taper

To cut a taper:

1. Turn the piece to slightly more than its largest diameter.
2. With the parting tool, mark the small diameter.

Fig. 12-12 Smoothing the cylinder with the skew

Fig. 12-13 Nick the corners of a square shoulder with the toe of the skew.

Fig. 12-14 Make a groove with the parting tool.

3. With a gouge, rough cut the piece in a straight line from the largest to the smallest diameter. Use a straightedge to test the work. Always work from the large to small diameter to insure cutting with the grain.
4. Smooth the taper by finishing with a skew in the same direction.

Cutting V Grooves

To cut a V groove:

1. Outline the width of the grooves with a pencil.
2. Hold the skew on edge with the toe up and at a 45-degree angle with the work.
3. Raise the handle and cut on one side of the groove. Make a similar cut on the other side of the groove, figure 12-16.
4. Continue back and forth until the groove depth is obtained and a clean, clear cut is made.

Cutting Beads

Beads are convex or rounded-over cuts. They are perhaps the most difficult cuts of all to make without slipping.

1. Lay out the width of each bead.
2. Make a deep, vertical cut with the toe of the skew on the layout lines.

Fig. 12-15 The square shoulder is trimmed with a skew.

Fig. 12-16 Cutting a V groove

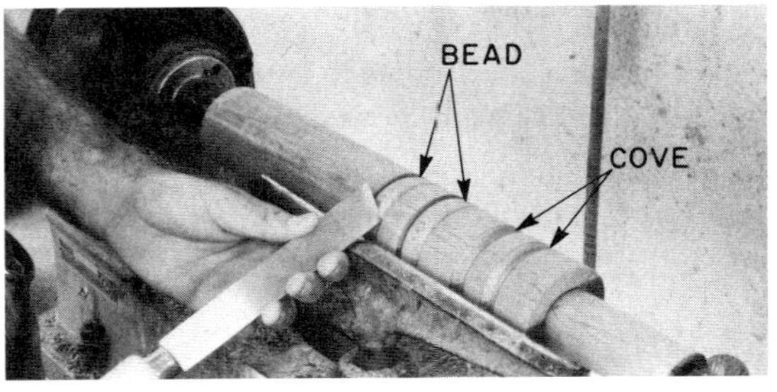

Fig. 12-17 Making a bead

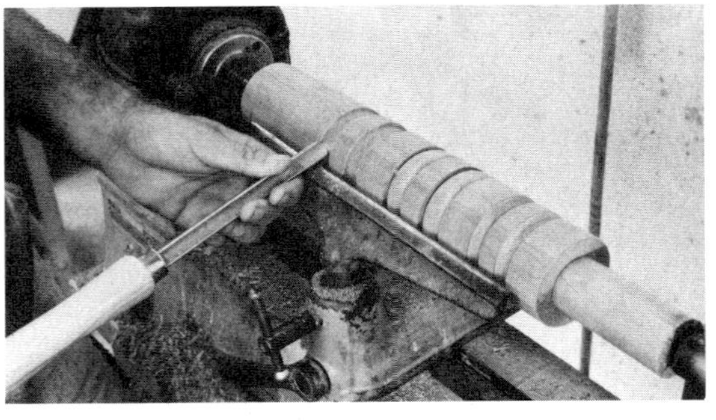

Fig. 12-18 Cove cuts are made with the gouge.

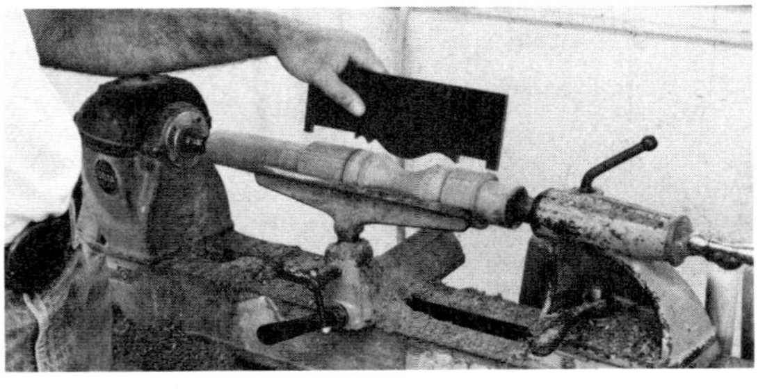

Fig. 12-19 Using a pattern speeds the work.

3. Lay the skew on the work as if smoothing a straight cylinder so a cut is started in the center of the bead.
4. Roll the skew over on its edge so it ends up on the layout line on one side, figure 12-17.
5. Turn the skew over and repeat the procedure to cut the other side of the bead.

Making Cove Cuts

Cove cuts are concave or hollowed-out cuts in a turned piece. Gouges are used to make cove cuts.

1. Lay out the width of the cove cut.
2. Use the gouge in a scraping fashion to make a concave cut to almost the layout line.
3. Lower the tool rest and hold the gouge on edge until its longest point is centered on the work. Hold the gouge so its bevel is at right angles to the work.
4. Rotate the gouge and bring it into the cut and over the work with the bevel down, figure 12-18. This makes one-half of the cut.
5. Make the other half in a similar manner.

Complicated Turnings

Complicated turnings are a combination of the cuts previously described. If the student can master these cuts in a professional manner, any design can be turned with speed, accuracy, and good workmanship. In order to speed up the work, a pattern of the turning is first made and then used to test the work as turning progresses, figure 12-19.

FACEPLATE TURNING

Faceplates are used to hold the work when turning small circular pieces, such as bowls and trays. The standard faceplate has screw holes near its outer edge and usually one in the center for holding the work. Faceplates are available in 3-inch and 6-inch diameters. For turning small delicate work, a screw center is used. The screw center has a tapered shank that fits into the head spindle. Its

Fig. 12-20 A rosette for a door casing is turned from a square piece.

face is 1 1/2 inches in diameter with a long replaceable screw in the center to hold the work.

If the work is to be round, it is band-sawed to a rough size about 1/4 inch larger than the finished turning. Sometimes the face of a square piece is turned, figure 12-20. If this is the case, find the center of the piece and carefully mount it on the faceplate using wood screws about 3/4 inch long. Do not use screws that are too long. These may come in contact with the turning chisels when shaping the inside of the piece.

To eliminate the screw holes, glue a piece of scrap stock to the turning with a piece of heavy brown paper between them, figure 12-21. After turning, separate the pieces by tapping a wide chisel with a mallet into the joint with the flat side against the turning.

Faceplate turning:

1. Screw the faceplate on the headstock spindle after removing the head center.
2. Adjust the tool rest close to the edge of the stock with its top centered on the work. Turn the work over by hand.
3. Set the lathe for a slow speed.
4. Using a round-nose chisel, make scraping cuts until the edge is round and true, figure 12-22.
5. Adjust the tool rest across the face and, with the round-nose chisel, true the face.

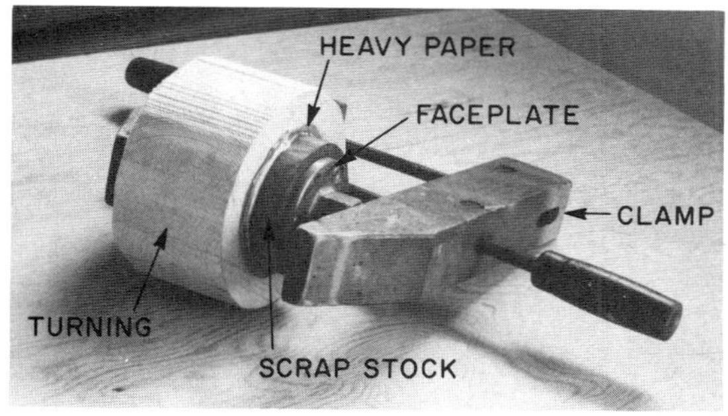

Fig. 12-21 A scrap block with a piece of heavy paper glued to the turning eliminates screw holes.

Fig. 12-22 Truing the edge of a faceplate turning with the round-nose chisel

Caution: Make cuts from the center outward so that the stock revolves down against the chisel. Cutting on the other side of center will push the tool upward and possibly out of the operator's hands.

6. Shape the inside of the work using a round-nose chisel or skew. Hold the chisels straight into the work and remove the stock in a scraping fashion, figure 12-23. Keep the tool rest as close to the work as possible.
7. Adjust the tool rest and shape the outside in a similar manner.

SANDING THE TURNING

After the turning is the desired shape, the tool rest may be moved out of the way. The piece is now ready to be sanded.

Sand the piece with a coarse abrasive first. Gradually use finer grit sandpaper to achieve a smooth surface.

Fig. 12-23 Shaping the inside of a faceplate turning *(Rockwell International)*

ACTIVITIES

1. Make a tapered table leg with a square shoulder.
2. Make a table leg with either bead or cove cuts.
3. Make a rosette for a door casing on a square piece (see figure 12-20). Use a screw center to hold the piece to the faceplate.
4. Make a small bowl on the wood lathe. Do not use screws to fasten the work to the faceplate.

UNIT REVIEW

Multiple Choice

1. The head center of the wood lathe is securely held in the headstock by
   a. a nut.
   b. its tapered shank.
   c. a clamp.
   d. its four spurs.
2. Slowest lathe speeds are used
   a. for finishing cuts on small work.
   b. for finishing cuts on large work.
   c. for roughing cuts on small work.
   d. for roughing cuts on large work.
3. The tool used to rough down cylinders is called a
   a. skew.
   b. parting tool.
   c. gouge.
   d. round-nose chisel.
4. Work turned between centers is called
   a. center-to-center turning.
   b. turning on center.
   c. spindle turning.
   d. faceplate turning.

5. To finish cuts when turning between centers, use a
   a. parting tool.
   b. skew.
   c. gouge.
   d. round-nose chisel.

6. For making V grooves and beads, use a
   a. parting tool.
   b. skew.
   c. gouge.
   d. round-nose chisel.

7. For making cove cuts, use a
   a. parting tool.
   b. skew.
   c. gouge.
   d. round-nose chisel.

8. When using a dead tail center,
   a. use a live head center.
   b. lubricate it.
   c. bring it lightly into the work.
   d. the lathe is run at high speed.

9. The tool rest should be adjusted often during turning operations
   a. to keep turning chisels from slipping.
   b. to keep the tool rest parallel to the work.
   c. as close to the work as possible.
   d. when making long cuts.

10. The turning chisel used to make cove cuts is the
   a. skew.
   b. gouge.
   c. parting tool.
   d. round-nose chisel.

Questions

1. What is the purpose of a wood lathe?
2. How are the head and tail centers of the wood lathe removed?
3. What is faceplate turning? Give some examples of faceplate turnings.
4. Name the common turning chisels and explain their use.
5. Describe two methods of spindle turning. Give the advantages and disadvantages of each method.
6. Which method of turning is used when work is mounted on the faceplate?
7. Why must the tool rest be kept as close to the work as possible?
8. Explain the danger of running stock that has not been trued up at high speed.
9. Describe the method used for roughing down a cylinder.
10. When making a faceplate turning, describe how to eliminate screw holes in the back of the turning.

unit 13 Drill Press

OBJECTIVES

After studying this unit, the student will be able to:

- explain the purpose of the drill press and its parts.
- describe common cutting tools and their uses.
- bore holes square to the work or at an angle.
- bore equally spaced holes, drill round work, bore close to an edge, and counterbore.
- bore holes for dowels in butt or miter joints and make pocket holes.

DRILL PRESS

The drill press consists of a metal column that is securely fastened to a base, figure 13-1. The table tilts 90 degrees, can be raised or lowered, and locks securely in any position on the column. A lock ring prevents the table from slipping downward and should be securely tightened when adjusting the table height. Side ledges and slots in the table also help to clamp the work.

The drill press head is mounted at the top of the column, figure 13-2. The motor is located here and drives the spindle by means of a stepped pulley and belt, figure 13-3. Speeds range from 680 to 4600 rpm.

The quill moves up and down by means of a handle. In most models the length of the quill stroke is about 4 inches. The quill may also be locked in any position. Adjustable stops are provided to limit the stroke of the quill to any desired depth. The quill returns to its original position upon release of the handle.

A chuck is inserted in the bottom of the quill. The chuck holds the cutting tool and is tightened by means of a chuck key. Some chucks eject the chuck key when it is released. When using models without this feature, the chuck key should be removed before turning on the power. Chucks on most drill presses usually have a capacity of 1/2 inch.

The size of a drill press is indicated by the distance from the column to the center of the chuck. A common size is 15 inches.

CUTTING TOOLS

High-speed twist drills are used primarily to cut holes in metal, figure 13-4. These drills are also commonly used to drill holes 1/4 inch or less in wood.

Twist drills are sharpened on a grinding wheel. A drill grinding attachment is used to grind the drills to the correct angles, figure 13-5. If a drill grinding attachment is not used, a drill point gauge

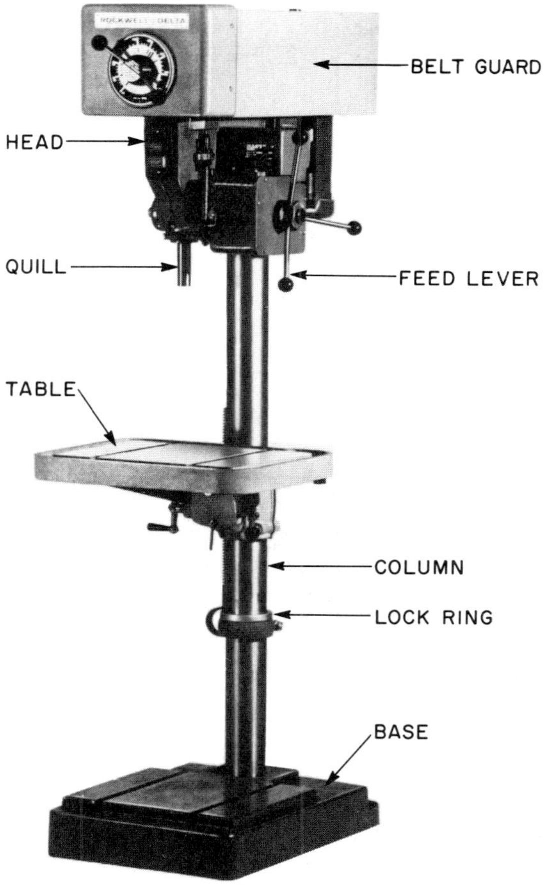

Fig. 13-1 Drill press *(Rockwell International Inc.)*

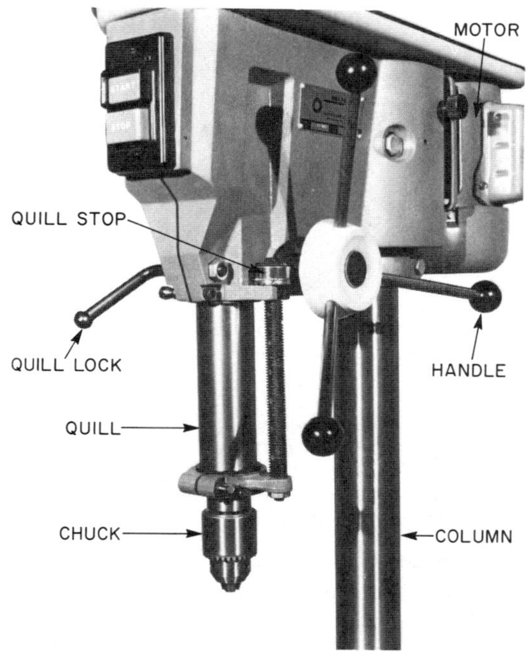

Fig. 13-2 Drill press head *(Rockwell)*

Fig. 13-3 The stepped pulley allows variable speeds. *(Rockwell)*

is used to test the angle, figure 13-6. Relief must be provided behind the cutting edge, figure 13-7.

The *machine spur bit* has a brad point and two cutting lips. It is commonly used for clean, fast cutting of dowel holes, figure 13-8. Machine spur bits range in size from 1/4 inch to 3/4 inch and are graduated in 16ths of an inch. It is sharpened on a narrow grinding wheel while maintaining its original shape. The bit must be cooled frequently while grinding to avoid burning it.

The *power-bore bit* is similar to an auger bit except it has no twist, figure 13-9. This bit makes clean holes in wood. Sizes range from 1/4 inch to 1 inch, graduated in 1/8ths of an inch. It must be installed with the three flats, ground on the end of the shank, lined up with the jaws of the chuck. It is sharpened with a bit file in the same manner as an auger bit. The spurs must not be filed excessively.

For clean cutting of large holes, the *multi-spur bit* is used, figure 13-10. Usual sizes range

Fig. 13-4 High-speed twist drills are used primarily to drill holes in metal.

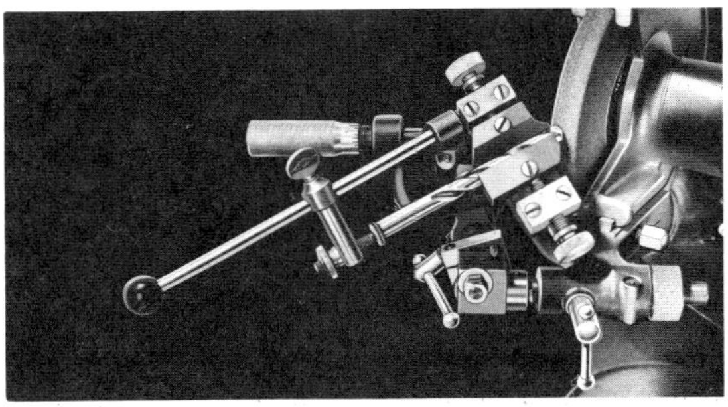

Fig. 13-5 A twist drill grinding attachment assures correct drill-cutting angles. *(Rockwell)*

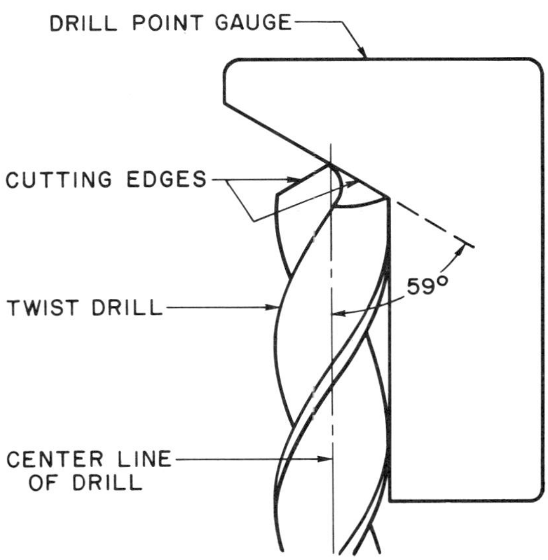

Fig. 13-6 Testing the angle of a twist drill with a drill point gauge

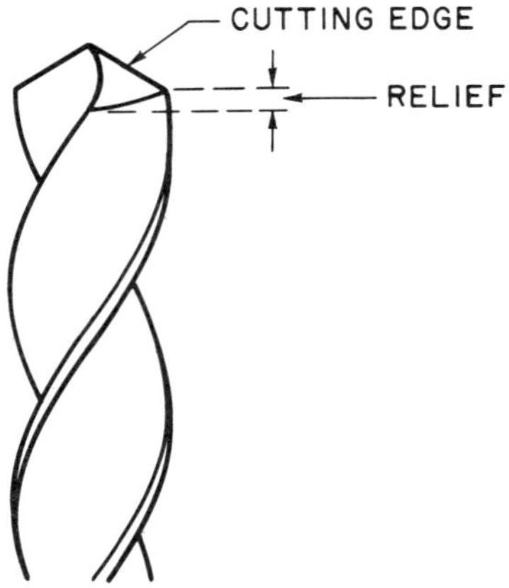

Fig. 13-7 Relief angle of a twist drill

Fig. 13-8 The machine spur bit is an excellent tool for drilling dowel holes.

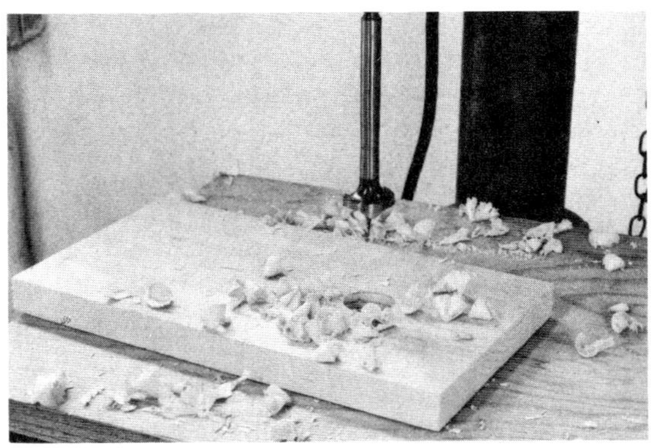

Fig. 13-9 The power-bore bit is similar to an auger bit except it has no twist.

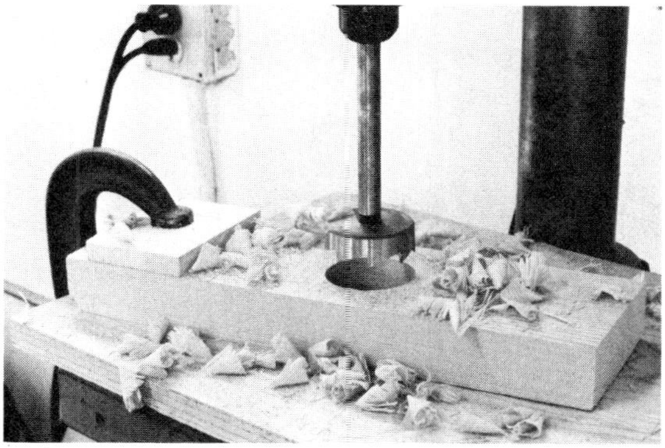

Fig. 13-10 The multispur bit is used to bore large holes in wood.

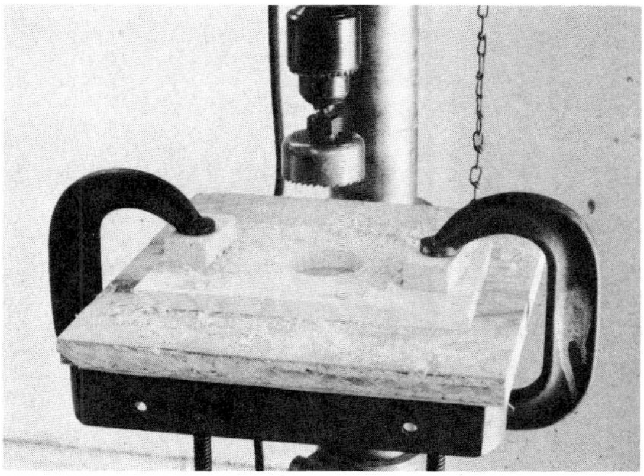

Fig. 13-11 The hole saw is used to cut large holes in thin material.

from 1 inch to 2 1/8 inches. Larger sizes are available. It has a cutting lip and a series of saw teeth around its rim. Sharpening is done with a file while maintaining the original shape.

The *hole saw* is also used to cut large holes, figure 13-11. However, because of its design, the depth of cut is limited. It is used to cut holes in thin wood, plywood, metal, or plastics. It ranges in diameter from 5/8 inch to 3 1/2 inches. The hole saw is sharpened with a tapered file similar to the handsaw.

Another tool used to cut large holes is the *circle cutter,* figure 13-12. Common types adjust to cut holes from 13/16 inch to 5 inches. Larger sizes are available. When using this tool the work must be securely clamped to the drill press table. The drill press must also operate at its slowest speed. The cutting end of the circle cutter is brought to a sharp edge on a grinding wheel while maintaining its original shape. Care must be taken to avoid burning the tool while grinding.

The *forstner bit* has no point and is guided by its outer rim. This tool is used primarily for boring holes close to the bottom surface or close to an edge, figure 13-13. It is available in sizes of 1/4 inch to 2 inches. The two cutting lips of the forstner bit are sharpened with a bit file like the auger bit. Its outer rim may be sharpened with a curved oilstone, but this is rather difficult. The forstner bit is therefore only used for the special purposes it is intended.

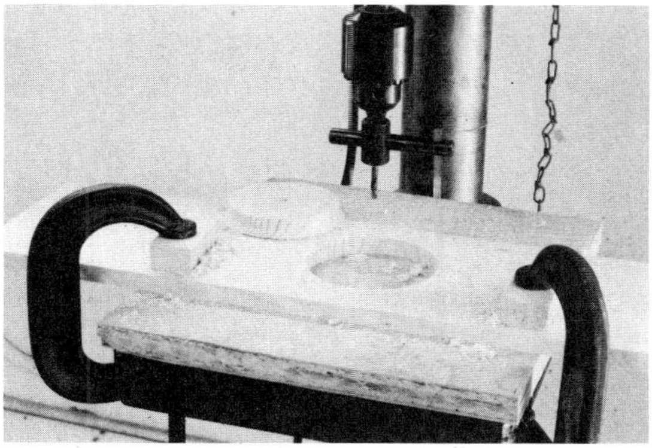

Fig. 13-12 This circle cutter is adjustable to cut holes up to 5 inches in diameter.

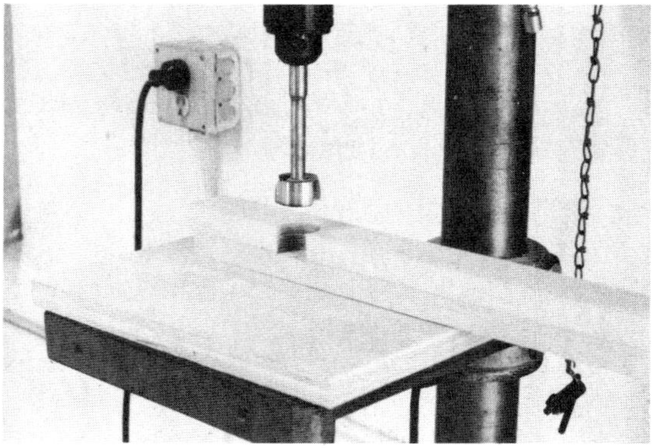

Fig. 13-13 The forstner bit is guided by its outer rim and can bore holes close to an edge.

Fig. 13-14 The countersink makes recesses for flathead screws.

Fig. 13-15 Combination drill and counterbore

The *countersink* is used to make recesses for flathead screws so that they can be driven flush or below the surface of the work, figure 13-14. When screwheads are to be sunk below the surface and covered with wood plugs, a combination drill and counterbore is sometimes used, figure 13-15.

Holes may be made and counterbored for screws and other fasteners by using two cutting tools. When counterboring in this fashion, bore the larger hole first, figure 13-16. If the smaller hole is drilled first, the center is harder to locate with the larger boring tool.

To make plugs for counterbored holes, *plug cutters* are used, figure 13-17. Plug cutters range in size usually from 1/4 inch to 1 inch and come in many different types. Plugs are made from scrap stock of the same material that is to be plugged or in wood of contrasting color. For instance, oak is sometimes plugged with walnut for a contrasting effect.

Fig. 13-16 Counterboring with two tools. The larger hole must be bored first. (Stock is cut away to show both holes.)

BORING AND DRILLING

The term *drilling* usually refers to holes cut in metal. Drilling also refers to small holes made in wood. *Boring* refers to larger holes that are cut in wood and similar materials. Boring is never used in reference to metal.

Use low speeds for cutting large holes and high speeds for cutting small holes or for cutting plugs. The reason for this is that the larger the

Fig. 13-17 Making plugs with a plug cutter

cutting tool, the greater its rim speed. Also, high speeds give smoother cuts when using tools of small diameter.

Caution: When boring or drilling, always clamp the work securely to the table. If the work is not securely clamped in position, the operator may not be able to hold it. The rotating work could then cause severe injury to the operator or others nearby.

When possible, place a scrap block of wood on the drill press table and under the work. This prevents splintering the work as the tool comes through. This also prevents damage to the cutting tool by not allowing it to come in contact with the metal table.

When making deep cuts, back out often to clear the wood chips from the cutting tool.

Never wear loose clothing or hold rags near the revolving bit.

Never use a bit with a threaded feed screw. This type of bit feeds itself into the work. The operator will not be able to control it. If it is necessary to use a bit of this type, first grind or file the threads off the point.

Boring Holes At An Angle

To bore holes at an angle:

1. Tilt the table to the desired angle. Test the angle with a sliding T bevel or a wood block cut to the desired angle, figure 13-18.
2. Adjust the table to the desired height.
3. Clamp the stock to be bored securely on the drill press table. Use scrap pieces under the clamps to avoid marring the work.
4. Insert the desired cutting tool.
5. Set the quill stop for the desired depth of cut.
6. Bore the hole slowly to avoid bending the cutting tool, figure 13-19.

Boring Spaced Holes

To bore holes at a specified spacing, lay out the location of the individual holes and then bore each one, figure 13-20.

A jig is frequently used to drill equally spaced holes.

1. Using a wood table with a fence, clamp a block with a stop pin in it to the fence.
2. Locate the stop pin according to the spacing desired.
3. Drill the first hole.
4. Drill the second and succeeding holes by putting the stop pin in the previously drilled hole, figure 13-21.

Fig. 13-18 Testing the tilt of the drill press table

Fig. 13-19 Boring holes at an angle with a multi-spur bit

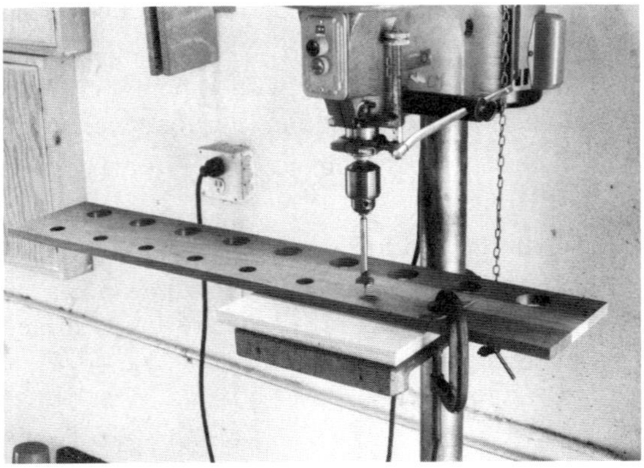

Fig. 13-20 Boring spaced holes with a power-bore bit

Fig. 13-21 Boring equally spaced holes using a stop pin in the previously drilled hole

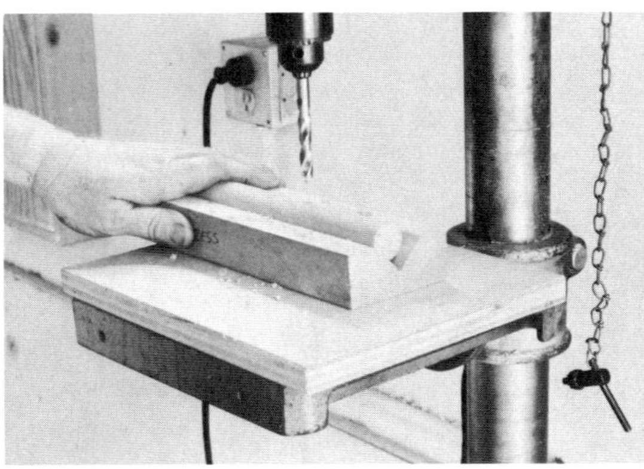

Fig. 13-22 Drilling holes in round stock using a V block

Drilling Holes in Round Stock

To drill holes in round stock:

1. Mark and center-punch the location of the holes.
2. Hold the stock in a V-grooved wood block, figure 13-22. Make sure the center of the vee is centered on the drill.
3. Drill the holes in the usual manner. Never attempt to drill holes in round stock without a V block.

Drilling Dowel Holes in the End of Stock

1. Lay out the location of the dowel holes carefully and center-punch their centers.
2. Turn the drill press table parallel to the cutting tool.
3. Clamp a guide for the edge of the stock to the table surface. The center line of the hole should line up with the drill center. Clamp the stock in position against the guide.
4. Drill the hole to the desired depth, figure 13-23.

Fig. 13-23 Drilling dowel holes for butt joints

Fig. 13-24 The first dowel hole is drilled for a mitered joint by holding the stock in a jig.

Drilling Dowel Holes in Miters

To drill holes in miters, a jig is often used to hold the work. The jig is constructed so that the miter cut is held at right angles to the cutting tool.

1. Place the work in the jig with the mitered end flush with the top of the jig.
2. Clamp the jig and the work in position to drill the first hole, figure 13-24.
3. Without moving the jig, raise the work, then clamp it again to drill the second hole. A spacing block is used to raise the work the desired distance, figure 13-25.
4. Reset the depth stop to drill the second hole. It should be the same depth as the first hole. A piece of masking tape can be wrapped around the drill to mark the depth.

Making Pocket Holes

Pocket holes are used to recess screws when tabletops are fastened to the rails and in other similar construction. They are cut on the drill press using a forstner bit. The stock is held at an angle in a jig which has a fence to guide the bit. The jig is clamped in the correct position on the drill press table. The depth stop is set for the depth of cut.

Fig. 13-25 Use a spacing block to raise the work to drill the second hole.

Fig. 13-26 Making pocket holes. The forstner bit is guided by the fence.

Pocket holes are then bored by sliding the stock in the jig to the desired location of the holes, figure 13-26.

Making Plugs

Insert the desired size plug cutter in the chuck. Set the drill press at a high speed to obtain a clean cut. Make the cut the depth of the plug cutter. This will give a beveled end to the plug for easy insertion in the counterbored hole. After the desired number of plugs are cut, pop them from the scrap block with a screwdriver.

Other Operations

The drill press is an extremely useful tool. Jigs can be made to help do faster and more accurate work. Many attachments are available for use on the drill press, such as routing, mortising, and sanding attachments, figure 13-27.

Fig. 13-27 Using a sanding attachment to sand a part for a gun cabinet

ACTIVITIES

1. Sharpen a twist drill with and without a grinding attachment.
2. Sharpen a machine spur bit, a power-bore bit, a multispur bit, a hole saw, and a circle cutter.
3. Bore dowel holes for a rectangular frame with butt joints.
4. Make a jig for boring dowel holes in a mitered rectangular frame.
5. Make a jig for boring pocket holes.
6. Bore five, 2-inch diameter holes at a 60-degree angle. Space the holes 3 inches apart.
7. Bore three, 1/2-inch diameter holes in a round piece of stock using a V-grooved wood block. Space the holes 2 inches apart.
8. Make a pocket cut using a forstner bit.

UNIT REVIEW

Multiple Choice

1. A safety feature that prevents the drill press table from dropping is the
   a. double-locking nut. c. lock ring.
   b. threaded-raising mechanism. d. stop pin.
2. That part of the drill press head that moves up and down is called the
   a. column. c. feed lever.
   b. quill. d. depth stop.
3. The term boring refers to cutting
   a. small holes in metal. c. small holes in wood.
   b. large holes in metal. d. large holes in wood.
4. Slow drill press speeds are used when
   a. drilling small holes. c. making plugs.
   b. boring large holes. d. drilling metal.

5. When boring deep holes in wood,
   a. run the drill press at a high speed.
   b. plunge the cutting tool into the work.
   c. back out often to clear wood chips.
   d. use twist drills only.

6. On the drill press, never use a bit
   a. with no feed screw.
   b. with spurs.
   c. with no twist.
   d. with a threaded feed screw.

7. High-speed twist drills are used primarily
   a. to drill large holes in wood.
   b. to drill small or large holes in metal.
   c. on hard woods and metal.
   d. to produce a clean, smooth hole.

8. Machine spur bits are commonly used
   a. to make dowel holes.
   b. to drill metal.
   c. to bore large holes in wood.
   d. to bore holes with a flat bottom.

9. An excellent choice to bore clean, large holes in wood is the
   a. twist drill.
   b. multispur bit.
   c. machine spur bit.
   d. countersink.

10. A disadvantage of the hole saw is its
   a. limited depth of cut.
   b. difficulty to be sharpened.
   c. fragility.
   d. inability to cut hardwood.

Matching

1. Twist drill
2. Machine spur bit
3. Power-bore bit
4. Multispur bit
5. Hole saw
6. Forstner bit
7. Countersink
8. Plug cutter
9. Circle cutter
10. Pocket hole

a. Similar to an auger bit except it has no twist
b. Used to cut large holes in thin wood
c. Best to run drill press at high speed when using this cutting tool
d. Has a brad point and two cutting lips
e. Used to cut holes in metal
f. Used to attach rails to tabletops
g. Makes recesses for flathead screws
h. Has a cutting lip and saw teeth around a rim
i. Bores holes with a flat bottom
j. Best to run drill press at slowest speed when using this cutting tool

unit 14 Sanding Machines

OBJECTIVES

After studying this unit, the student will be able to:

- describe, adjust, and use the long-belt stroke sander, the abrasive-belt finishing machine, the disc sander, and the spindle sander.

SANDING MACHINES

Sanding machines smooth wood surfaces by using various types of abrasives. There are many types available to do a variety of finishing work. They can be classified as belt, disc, drum, and spindle sanders. Each is built in a number of different styles and sizes according to the work to be done.

Sanding machines use *coated abrasives.* These abrasives are glued to paper or cloth. They come in the form of sheets, discs, drums, or belts. The quality of the abrasive depends on the kind of abrasive, backing, and the type of adhesive used. Abrasives are discussed in more detail in *Unit 27 Finishing.*

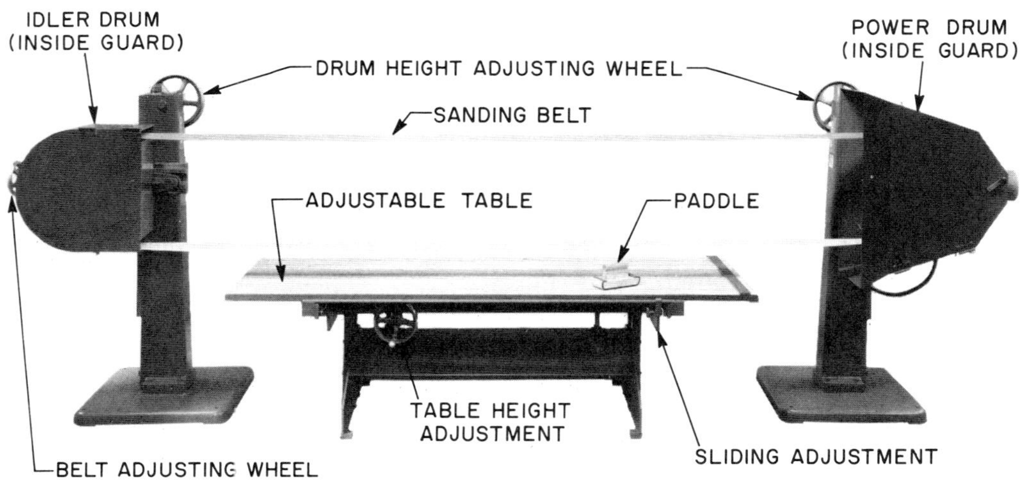

Fig. 14-1 Long-belt stroke sander *(Oliver Machinery Co.)*

Although most sanding can be done by machines, there are some places machines cannot reach. In these instances, sanding is best done by hand.

LONG-BELT STROKE SANDER

The *stroke sander,* figure 14-1, has a long belt that moves around two large drums. The usual width of the belt is 6 inches. The table under the belt can be raised or lowered and slides back and forth. The size of the sander is determined by the length of the table or by the length of the piece that can be sanded.

The work is laid on the table. The table is raised so that the belt is slightly above the work to be sanded. As the belt moves across the work, it is pressed to the work by means of a sanding block. The operator moves the block back and forth along the length of the work and slides the table in or out to sand the piece.

One drum of the sander is driven by the motor and remains in a fixed position. The other drum, called the *idler drum,* moves back or forth to apply enough tension to stretch the belt. The belt should not be overtightened. The idler drum may also be tilted to center the belt on the drum, figure 14-2. The drums are turned over by hand to center the belt. After centering the belt on the idler drum, the belt is checked for tension.

The *sanding block* has a felt pad glued to its bottom and covered with canvas. Wax is rubbed thoroughly into the canvas covering to reduce friction between the sanding block and the belt. Special canvas is available that is made with a frictionless material. Sanding blocks may be shaped for sanding irregular or curved surfaces.

The stroke sander is the last machine used to sand before finishing. For most woods, the first sanding is usually done with a 120-grit belt. For a higher quality surface, finish sanding is done with a 180-grit belt.

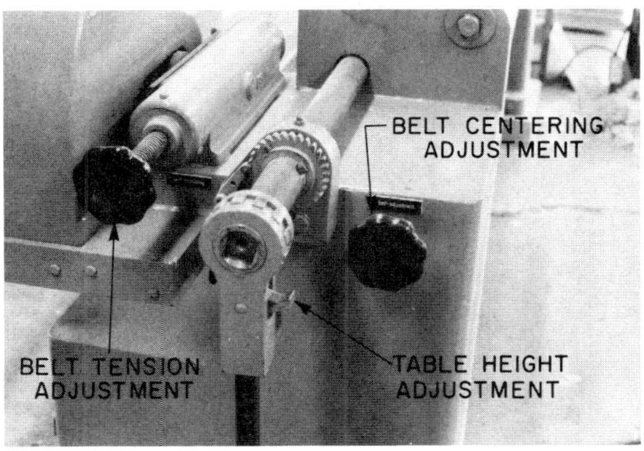

Fig. 14-2 Long-belt sander idler drum adjustments

Using the Stroke Sander

When installing a new belt, note that an arrow stamped on the inside of the belt shows the direction the belt should move, figure 14-3. Installing a belt backwards may cause the joint to tear.

To operate the stroke sander:

1. Make sure all guards are in place.
2. Place the work on the table and against the stop. The stop keeps the work from sliding as it is being sanded.
3. Raise the table so the work is about 1/2 inch below the belt. Start the machine.
4. Place the sanding block on the belt and press downward. Use light pressure. Move the sanding block back and forth along the total length of the work, figure 14-4.
5. Move the table to a new position and repeat the procedure. Be careful not to tilt the sanding block over the edges or ends of the work as this will bevel those edges.

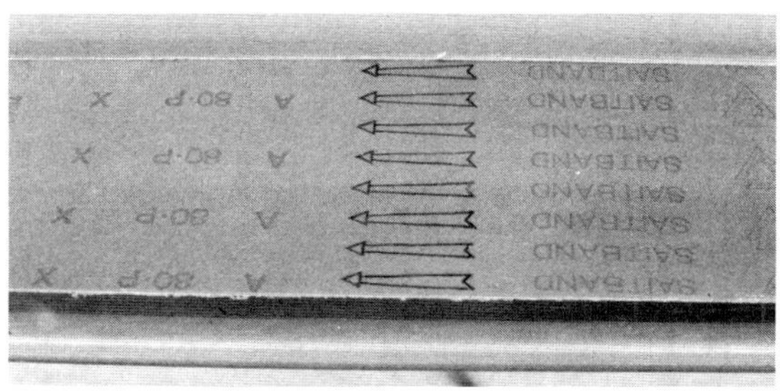

Fig. 14-3 Install long sanding belts in the correct direction.

Another method is to sand all the way across the work by moving the table back and forth. Then the sanding block is moved to a new position. This procedure is repeated until the end of the work is reached.

If the guard of the idler drum can be removed, this drum can be used for sanding curved parts, figure 14-5. The top of the belt may also be used for sanding assembled parts, such as drawer sides, if an auxiliary table is installed under the belt.

One of the most common causes of a streaked or unevenly sanded surface is a worn belt. Belts should be replaced when they become worn. Woods containing much pitch or resin will clog a sanding belt. The stroke sander should not be used to sand woods such as yellow and Ponderosa pine.

Fig. 14-4 Using the long-belt stroke sander

Fig. 14-5 Sanding a curved piece over the drum of the long-belt sander

ABRASIVE-BELT FINISHING MACHINE

The most common abrasive-belt finishing machine holds a belt 6 inches wide by 48 inches in length, figure 14-6. In its upright position, the lower drum is power driven. The upper drum is the idler drum. The idler drum is raised to apply tension to the belt and can also be tilted to center the belt on the drums.

The abrasive-belt sander may be used in the vertical, horizontal, or slant position. The table tilts 20 degrees towards the belt and 40 degrees away from the belt. A miter gauge can be attached to the tilting table. Fences can also be attached and used as stops when sanding small, flat surfaces.

The 6-inch abrasive-belt sander is widely used in school and cabinet shops for sanding small parts. Abrasive belts range from 80 to 120 grits. Worn belts are indicated by burned surfaces and should be replaced. Woods that contain much pitch or resin should not be sanded as they will clog the belt.

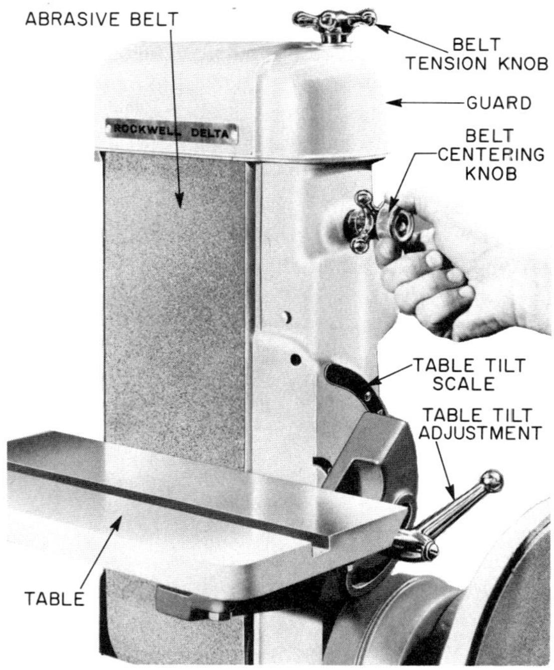

Fig. 14-6 Abrasive-belt finishing machine *(Rockwell International)*

Fig. 14-7 Sanding end grain on the abrasive-belt sander.

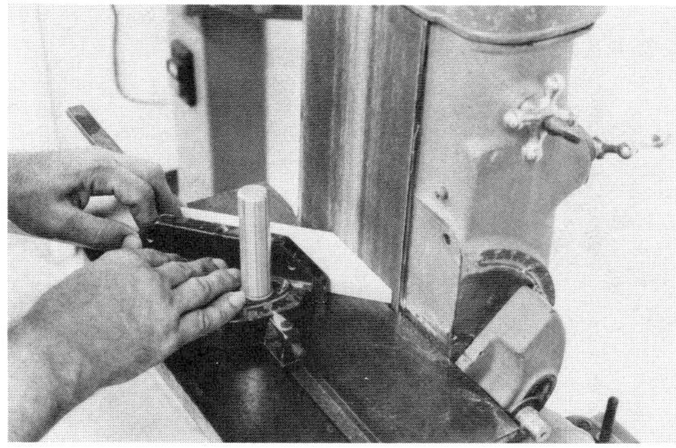

Fig. 14-8 To sand a miter, guide the work with a miter gauge.

Replacing the Belt

To replace a belt on the abrasive-belt sander:

1. Place the belt on the drums so that it will move in the direction indicated by the arrow stamped on the back.
2. Apply a small amount of tension.
3. Turn the belt over by hand and center the belt on the drums.
4. Increase the tension. Check the belt for centering by again turning the belt over by hand.
5. When the belt is centered, lock it in position. Replace all guards before turning on the power.

End-Grain Sanding

The abrasive-belt sander is used in the vertical position for end-grain sanding, figure 14-7. The table supports the stock and is adjusted as close to the sanding belt as possible. Light pressure is used when sanding as excessive pressure will burn the material and may possibly damage the motor.

By using the miter gauge, square or mitered ends may be sanded smooth, figure 14-8. Bevels and chamfers are sanded by tilting the table and holding the work against the slanted belt, figure 14-9.

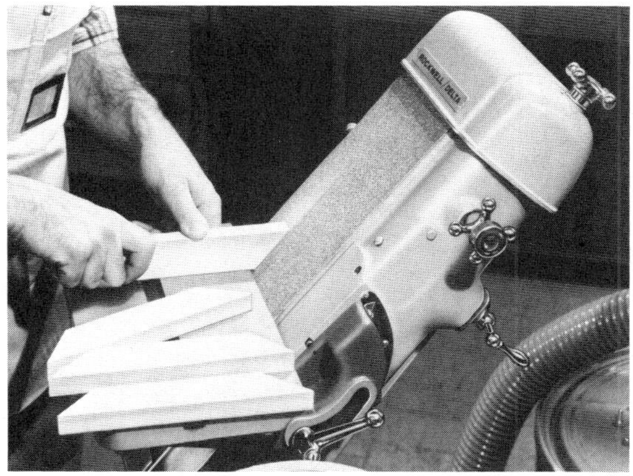

Fig. 14-9 Sanding a bevel or chamfer *(Rockwell)*

Edge and Surface Sanding

The abrasive-belt machine is usually locked in the horizontal position for edge and surface sanding. For surface sanding, install a stop to hold the work, figure 14-10. Place the work against the stop and apply light pressure against the moving belt.

Caution: When surface sanding, keep the tips of the fingers away from the belt. When edge sanding, a stop is not required. A fence is used to guide the work at the required angle. Curved edges can be sanded freehand over the open idler drum, figure 14-11.

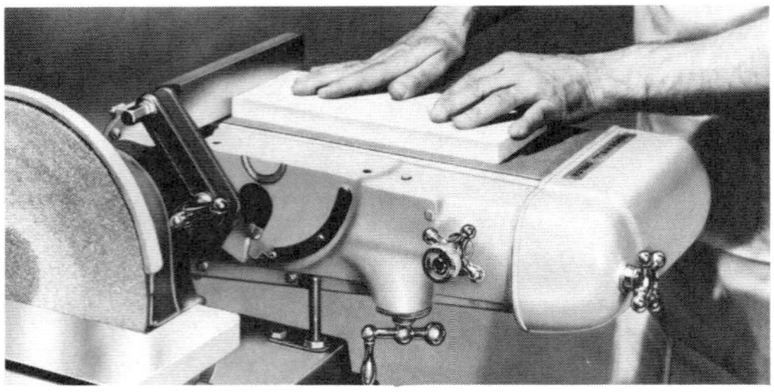

Fig. 14-10 Use a stop when sanding flat surfaces. Keep finger tips clear of the belt. *(Rockwell)*

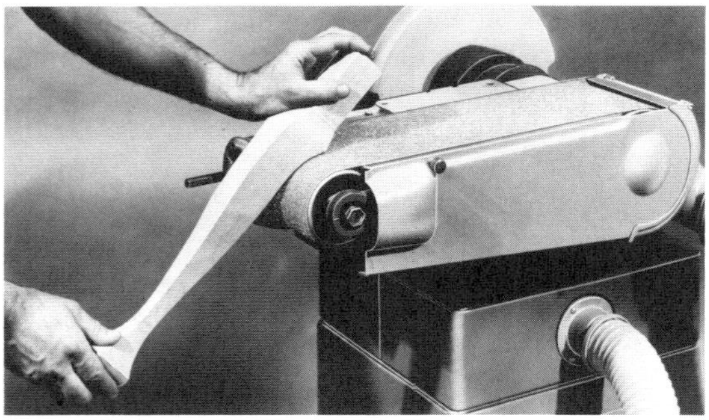

Fig. 14-11 Sanding a cabriole table leg over the idler drum of the abrasive-belt sander. *(Rockwell)*

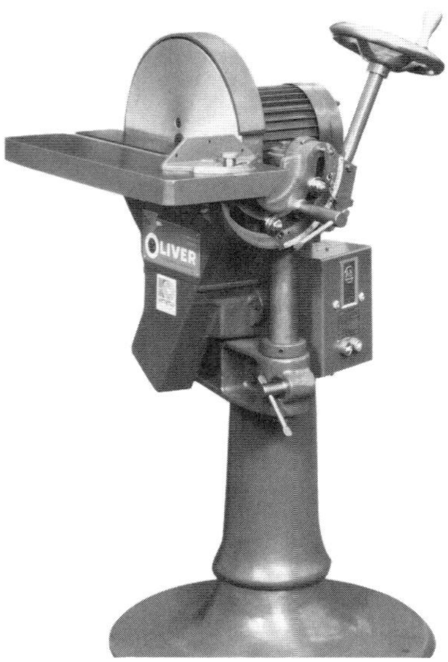

Fig. 14-12 The disc sander *(Oliver Machinery Co.)*

DISC SANDER

The *disc sander,* figure 14-12, consists of a stand supporting a motor that drives a metal disc. Sanding discs are cemented to the metal disc. A tilting table supports the work, figure 14-13. A miter gauge can also be attached to guide the work.

The size of a disc sander is determined by the diameter of the disc. Common disc sanders range in size from 12 inches to 36 inches.

The disc sander is a simple machine to operate. The work is held on the table and brought in contact with the revolving disc. Sanding is done only on the side of the disc that revolves downward, figure 14-14. Sanding on the upward side will lift the work off the table and out of the operator's hands.

Like all machine sanding controlled by hand, only light pressure should be applied to the sanding disc. Burned work is usually caused by worn sanding discs.

The disc sander is usually restricted to end-grain sanding. Rounded corners or pieces are disc-sanded, but the circular motion of the disc leaves

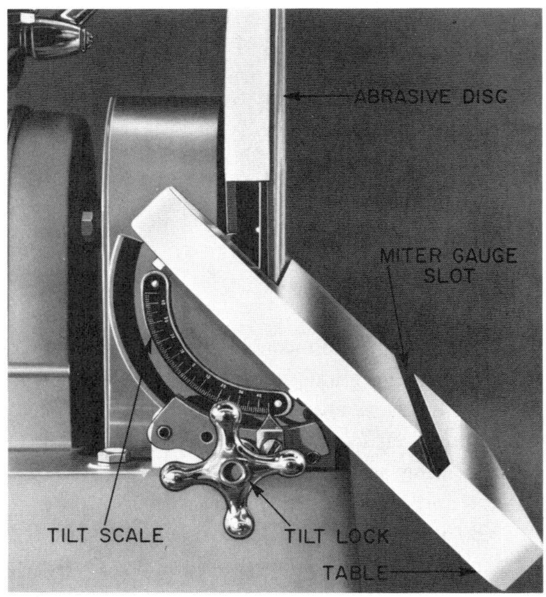

Fig. 14-13 The disc sander table tilts 45 degrees in either direction. *(Rockwell)*

scratches across the grain. Surfaces sanded in this manner are usually not exposed, except when the appearance of the surface is not important.

With the miter gauge, square or mitered ends can be sanded. Bevels and chamfers are sanded by tilting the table.

A combination finishing machine with belt and disc sanding equipment is also available, figure 14-15.

Replacing Abrasives

To replace the abrasive on a disc sander, the disc must be removed. This is done by removing the bolts that hold the disc to the motor shaft. The sandpaper is peeled from the metal disc. Then the old adhesive is scraped from the metal with a putty knife. After the residue is removed, the disc is polished with a solvent and sandpaper.

Solvents may be water, denatured alcohol, or mineral spirits according to the adhesive used. A commonly used adhesive is specially prepared for disc sanders and comes in a stick form. Use the solvent recommended by the manufacturer. Other adhesives are shellac, rubber cement, and water glass. Shellac is dissolved with denatured alcohol. Rubber cement and water glass may be removed with hot water.

Fig. 14-14 Disc sand on the side that revolves downward.

Some sanding discs have a pressure-sensitive adhesive applied to the back. When applying this type of disc, the backing is simply peeled off and pressed to the clean metal. Other types coat both the metal disc and the back of the sandpaper with a uniform amount of adhesive. The abrasive is then clamped to the disc with a flat piece of thick plywood.

It is better to have an extra disc available. This will avoid time lost in replacing abrasives.

SPINDLE SANDER

The *spindle sander,* figure 14-16, is constructed similar to a shaper. A motor, mounted below the table, drives a revolving and oscillating (up-and-down movement) spindle. The spindle is located in the center of the table. It holds a drum to which sandpaper is fastened. The oscillating movement of the spindle spreads the wear of the abrasive.

Fig. 14-15 Combination belt and disc sander *(Oliver Machinery Co.)*

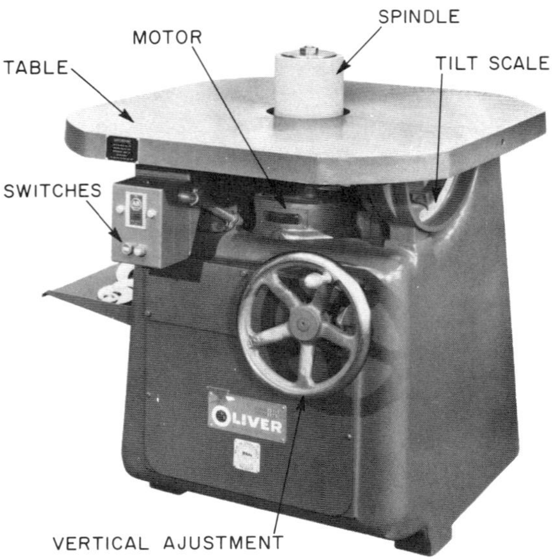

Fig. 14-16 Spindle sander *(Oliver Machinery Co.)*

The spindle adjusts vertically, usually by means of a handwheel. The entire sanding surface of the drum can therefore be used. In some models, the table can be tilted. In other models, the spindle can be tilted.

The spindles are detachable and come in many different sizes. Sanding drums may be made of metal, wood, or rubber. Drums are available from 1/2 inch to 6 inches in diameter and lengths up to 9 inches. The largest size drum should be used according to the work to be sanded.

The spindle sander is designed for sanding irregular curved edges. The piece is supported by the table and brought in contact with the revolving drum. Any grit abrasive may be installed on the drum according to the degree of finish desired. Vertical adjustments to the spindle are needed occasionally to make full use of the abrasive. The work is hand-fed against the revolving drum. Only enough pressure to smooth the edge is applied.

Replacing Abrasives

To remove the drum from the spindle, the nut on the top of the spindle is loosened. The spindle is usually held in place by a stop pin. Then

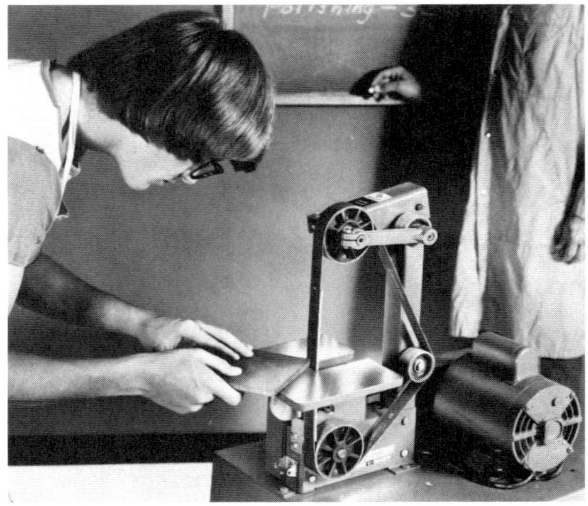

Fig. 14-17 Variety sander *(Rockwell)*

the worn paper is removed and a new piece cut to size.

For certain work, steel drums are used. In this case, the abrasive paper or cloth is cemented to the drum with an adhesive designed for that purpose. Other drums contain a groove. The sandpaper is inserted in the groove and locked into place with a key.

Sanding sleeves are cardboard tubes around which sandpaper is cemented. These sleeves are used on rubber drums. They are held tight by the expansion of the rubber under pressure.

Wood drums are also used. On these the paper is held by a wedge-shaped piece fastened into a groove on the drum.

Whatever type, the drum with the new abrasive is mounted back on the spindle. The spindle nut is then tightened securely.

OTHER SANDING MACHINES

The sanders described in this unit are the basic sanding machines used in many shops. Other types of sanders, such as the variety sander, may also be used to do special work, figure 14-17.

Industrial sanding machines, such as the drum sander, are discussed in *Unit 35 Wood Mills and Machines.*

ACTIVITIES

1. List the advantages and disadvantages of each of the sanding machines available in your shop.
2. Practice installing a new belt on the long-belt stroke sander and the abrasive-belt finishing machine.
3. List the abrasives available for use on each of the sanding machines in your shop. Read the information on abrasives in *Unit 27 Finishing.*
4. Remove the abrasive from the disc sander. What type of adhesive is used? Use the proper solvent to clean it. Install a new abrasive.
5. Note the type of drum your spindle sander has. Practice replacing the abrasive.
6. Make a form for holding a molded piece and shape a sanding block for use on the stroke sander to sand it.
7. Sand a beveled end grain on the abrasive-belt sander. Sand the surface of a piece of stock using a stop fence to hold the work.
8. Use a miter gauge to sand a mitered end on the disc sander.
9. Make a list of the spindle sizes available for your shop spindle sander.
10. Cut out an irregular curved piece. Sand the edges on the spindle sander.

UNIT REVIEW

Multiple Choice

1. Irregular curved edges are best sanded on the
   a. belt sander.
   b. disc sander.
   c. spindle sander.
   d. stroke sander.

2. A sanding machine that is similar to a shaper is called a
   a. belt sander.
   b. disc sander.
   c. spindle sander.
   d. stroke sander.

3. If shellac is used to cement the abrasive to the disc sander, it is dissolved with
   a. hot water.
   b. denatured alcohol.
   c. mineral spirits.
   d. cold water.

4. The last machine sanding of a flat wood panel is usually done with a
   a. belt sander.
   b. disc sander.
   c. spindle sander.
   d. stroke sander.

5. Installing sanding belts backwards
   a. makes sanding difficult.
   b. may cause the joint to tear.
   c. makes it difficult to center the belt on the drums.
   d. produces a burned or polished surface.

6. One of the most common causes of a streaked or burned surface when machine sanding is
   a. a worn abrasive.
   b. applying too much pressure.
   c. uneven sanding.
   d. caused by running the machine at too high a speed.

7. Besides burning the work, applying excessive pressure by hand when machine sanding
   a. will cause the operator to lose control of the work.
   b. will tear the sanding belt.
   c. will possibly damage the motor.
   d. will cause the belt to run off the center of its drums.

8. Interior curves may be sanded by using the
   a. disc sander.
   b. belt sander.
   c. stroke sander.
   d. any of the above.

9. The disadvantage of using a disc sander for exposed sanded surfaces is
   a. only stock of limited thickness can be sanded.
   b. sanding can only be done on one side of center.
   c. only coarse grit sandpaper can be used.
   d. it leaves cross grain scratches on the surface.

10. The reason for the vertical adjustment and the oscillating movement of the spindle sander is to
    a. sand wood of different thicknesses.
    b. sand a smoother surface.
    c. spread the wear over all the drum.
    d. sand a greater amount in a shorter time.

Questions

1. Why must a sanding belt be installed in the direction of the arrow stamped on the back of the belt?
2. Explain why a machine-planed surface must be sanded.
3. What determines the quality of a coated abrasive?
4. If a sanding machine cannot reach a certain part, how is it sanded?
5. Explain two methods of stroke sanding.
6. How is friction between the sanding block and the belt reduced when using the stroke sander?
7. Explain the consequences of machine sanding woods that contain too much pitch or resin.
8. What is the most common cause of streaked or burned surfaces when machine sanding?
9. What materials are used to cement sanding discs?
10. Name three methods of attaching coated abrasives to spindle sander drums.

section 3

Joinery

unit 15 Making Joints

OBJECTIVES

After studying this unit, the student will be able to:

- describe basic woodworking joints.
- list the advantages and disadvantages of each joint.
- select the best joint for the job.
- make the joint selected.

SELECTING A JOINT

There are many different types of woodworking joints. Selecting the right one will greatly affect the finished product. There are a number of points to consider when selecting a joint.

- Time. The simplest joint to make is the one to use if it will do the job. Complicated joints take time to make.
- Appearance of the finished product. End grains should not show in quality cabinetwork.
- Strength of the joint. A joint that is subjected to much use requires a stronger joint than one that is not.
- Kind and location of fasteners. Nails and screws should not show in quality cabinetwork.
- Use of the work.
- Kind of material.

BUTT JOINTS

The *butt joint* is the most common and easiest joint to make, figure 15-1. It is usually fastened with nails or screws in addition to glue. Disadvantages of the butt joint is that it shows the end grain and fasteners. However, it is quick and easy to join.

Butt joints can be strengthened by using glue blocks or dowels, figure 15-2. *Glue blocks* are triangular-shaped pieces glued in the corner of the joint. The block is fastened with hot glue and pressed into place.

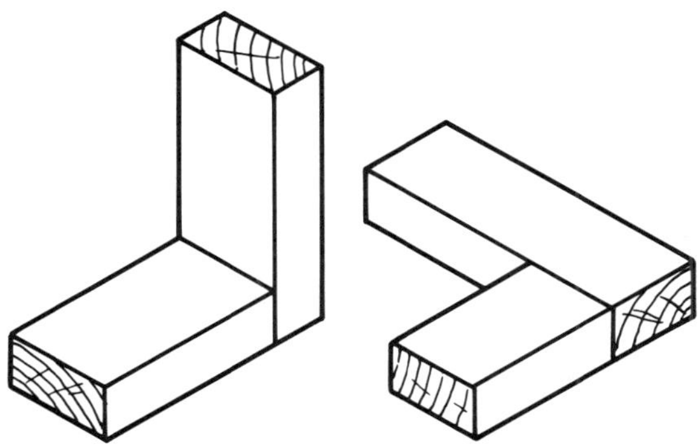

Fig. 15-1 Butt joints

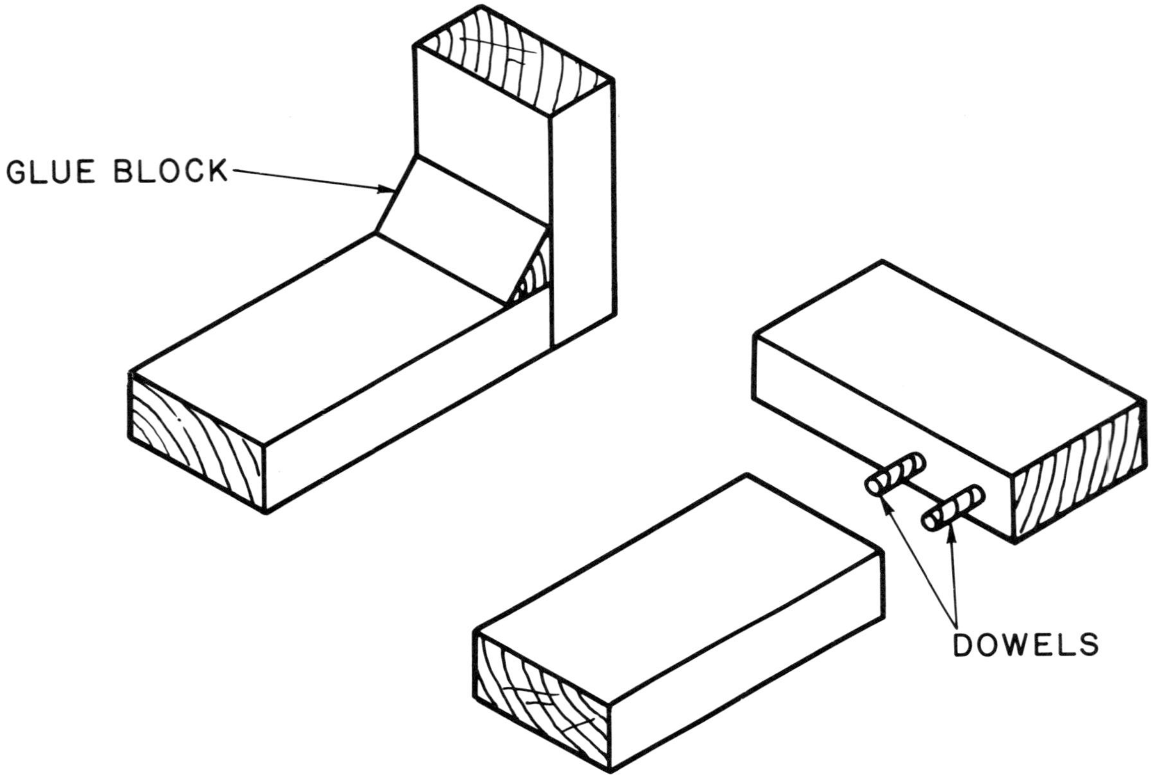

Fig. 15-2 Strengthening butt joints

Dowels are small round rods of hardwood, usually birch. They are often made with spiral grooves. The grooves let excess glue escape from the dowel hole and give the dowel more gluing surface. Dowels keep the joint from twisting, align the joint when assembling, and make the joint stronger. At least two dowels are used in each joint.

Making a Doweled Butt Joint

To make a doweled butt joint, matching holes are bored in the pieces to be joined. The most important part in making this joint is laying out the centerline of the holes accurately, figure 15-3.

1. Hold the pieces together and mark the location of the dowels with a combination square. Square these lines across the edges.
2. Set the blade of the combination square for half the thickness of the stock. Hold the body of the square on the face of the pieces. Mark across the edge lines to find the centerline of the dowel hole.
3. Center punch each centerline accurately.
4. Bore the holes slightly deeper than half the length of the dowel. This prevents the dowel from bottoming and allows the joint to come up tight.
5. To make the joint, glue the dowels in one piece. Apply glue to the edges and in the holes of the other pieces. Clamp the joint in place.

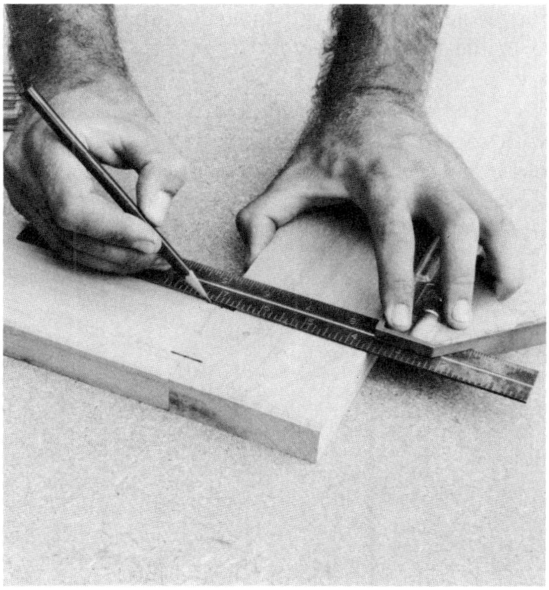

A. Lay out the dowels.

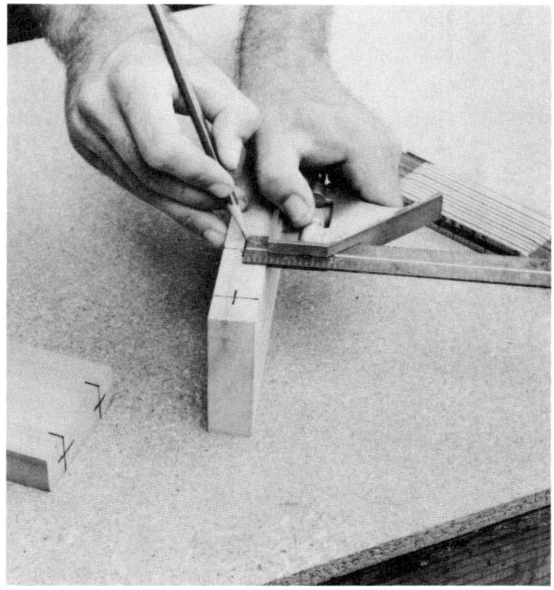

B. Locate the centerline.

C. Center punch and drill the dowel holes.

D. Apply glue, insert dowels, and clamp the joint.

Fig. 15-3 Making a doweled butt joint

EDGE JOINTS

Pieces joined edge to edge may be plain butted and fastened with glue. They may also be doweled, tongue and grooved, splined, or shaped, figure 15-4.

The splined and tongue-and-groove joints are seen from the ends of the stock. They are frequently used for watertight joints. The doweled, tongue-and-grooved, and splined joint are easier to assemble than the plain butt joint. However, they take more time to make.

The shaped edge joint is made on a shaper with special glue-joint cutters. It is easy to make once the setup is done, easy to assemble, and provides a large gluing surface. Its disadvantage is that the milled edges show at the end of the pieces.

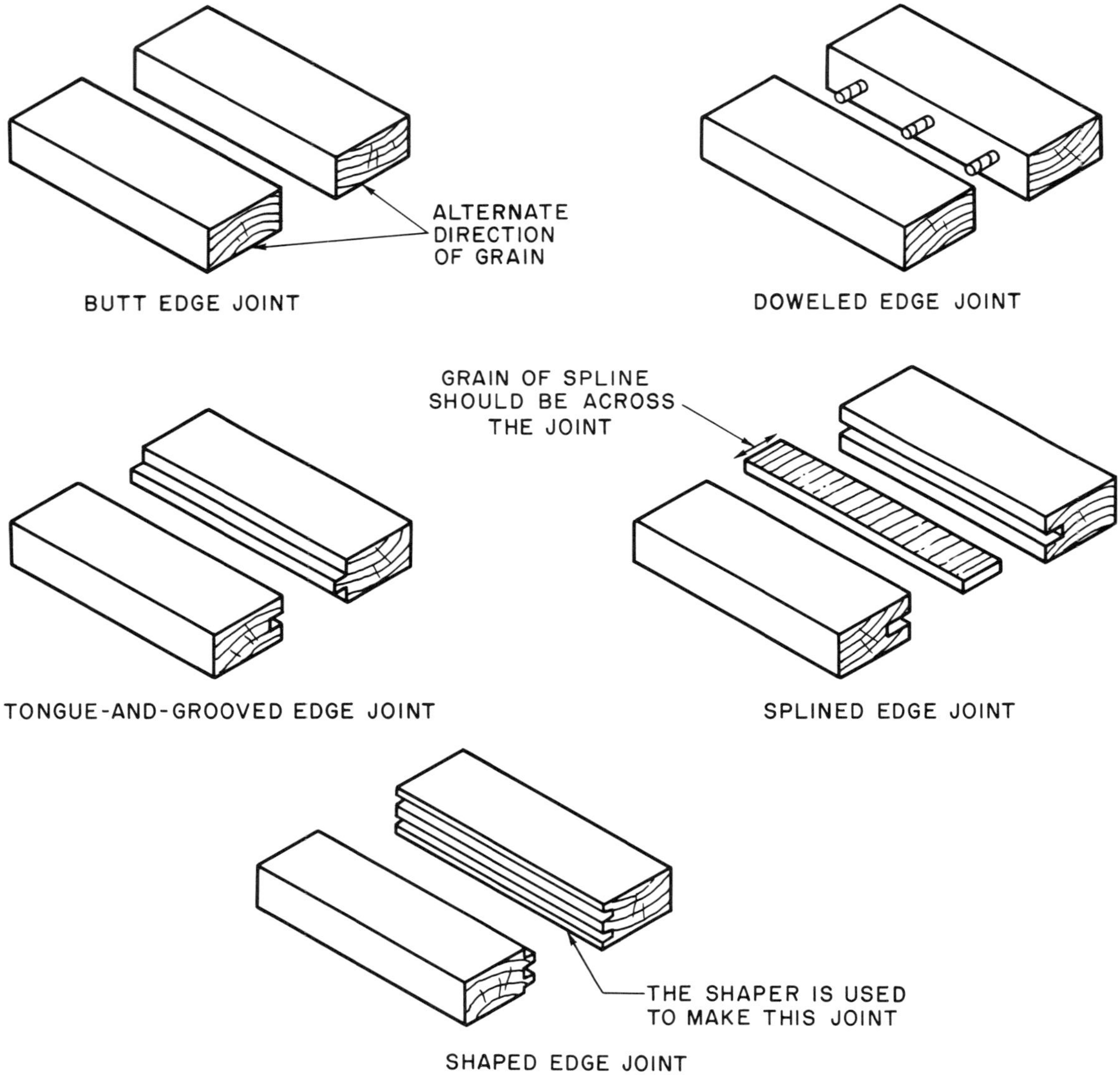

Fig. 15-4 Edge joints

Making Splined Joints

Splined joints are made by cutting grooves in the center of the edges to be joined and inserting a spline. The width and depth of the groove is generally one-third the thickness of the stock. The spline is cut slightly narrower than the depth of the two grooves. This keeps the spline from bottoming in the groove and allows the joint to come up tight.

The spline fits snugly in the grooves. If the spline is made of solid wood, it is cut so the grain runs at right angles to the joint. If the spline is made of plywood or hardboard, the direction of the grain does not matter.

Grooves are made with a dado head. The dado head is adjusted in height for the depth of the groove. One side of each piece is marked with an X. The X side is held against or away from the rip fence when cutting, figure 15-5. Then the X faces are placed on the same side when assembled. In this manner, the sides of the pieces will still be flush even if the grooves are slightly off center.

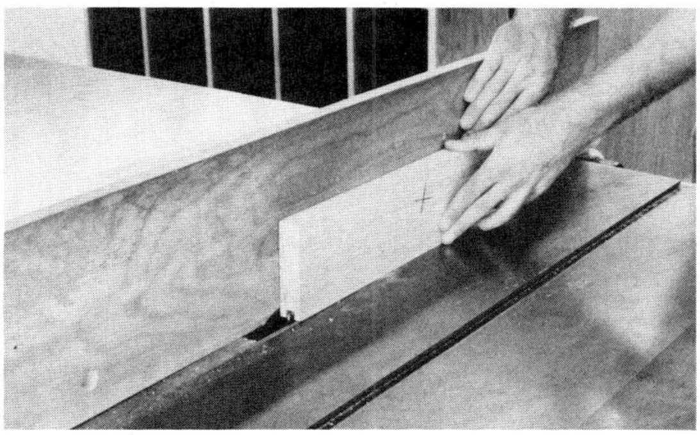

Fig. 15-5 Grooves are cut for a splined joint with a dado head (guard removed for clarity).

The joint is made by applying glue to the groove and edges of the pieces. The spline is inserted and the pieces are clamped together.

RABBETED JOINTS

End rabbeted joints are used to reduce the amount of exposed end grain. *Edge rabbeted joints* are used to set the backs of cabinets to hide their edge grain, figure 15-6.

The rabbet joint is easy to assemble, but needs to be strengthened with fasteners or other means. It is a simple joint to make on a table saw. It can also be made on a radial arm saw with a dado head, or by using the shaper with a straight blade. (See Units 7, 8, and 10.)

MITER JOINTS

Miter joints are made by cutting the two pieces at equal angles and joining together, figure 15-7. They are hard to assemble but give a pleasing appearance. They have the advantage of hiding the end grain of both pieces. Machines used to make these joints must be set up accurately. Being a fraction of a degree off will cause the joint to open on the inside or outside.

Miter joints are weak and must be strengthened with fasteners, dowels, or splines, figure 15-8.

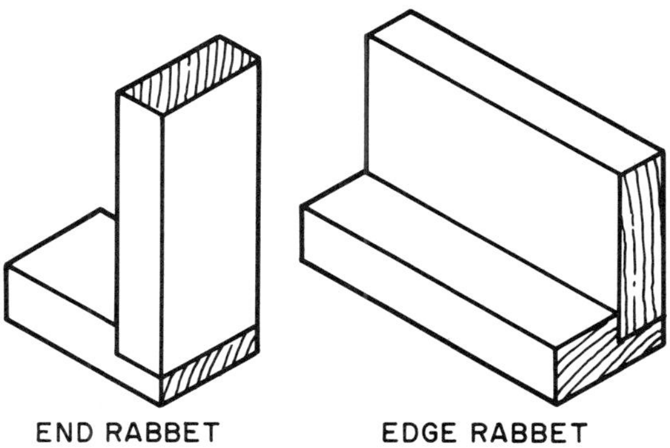

Fig. 15-6 Rabbeted joints

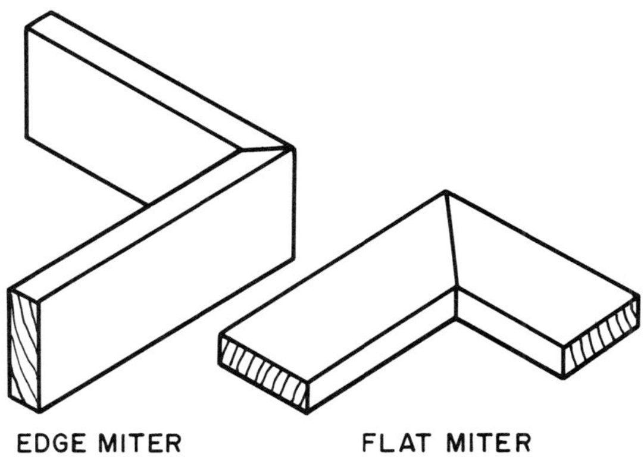

Fig. 15-7 Miter joints

The ends of through-splined miters can be seen. If the spline is hidden, it goes only part way through and is called a *blind spline*. A blind-splined miter joint is difficult to make and takes time. Nails are not used on quality work.

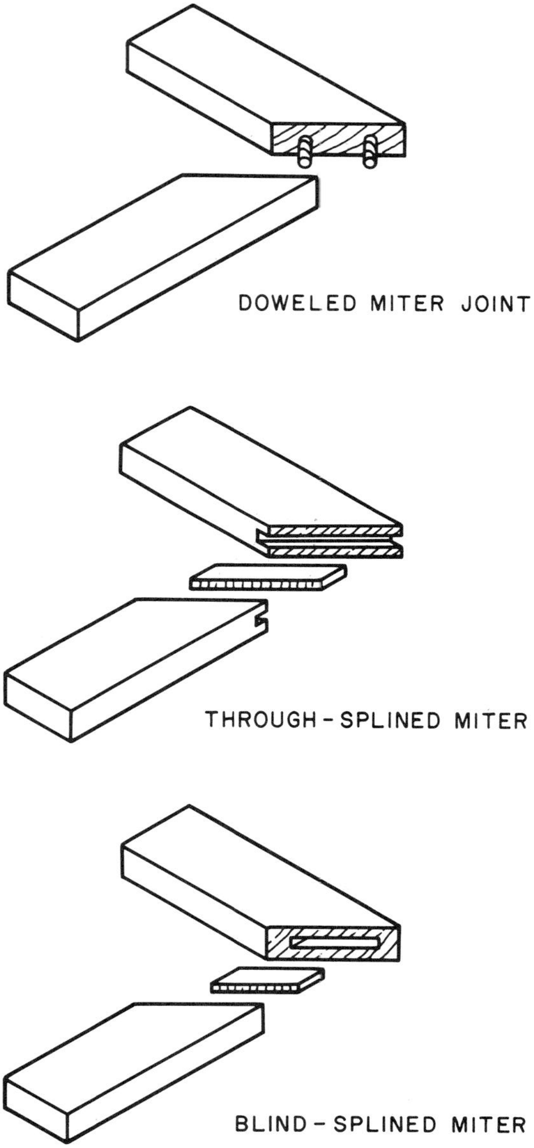

Fig. 15-8 Splined and doweled miter joints

DADO JOINTS

Dado joints are used in bookcases and cabinets that have shelves. They provide support and are easy to assemble. Common dado joints are the *through dado, blind dado, dovetail dado,* and the *rabbet and dado,* figure 15-9.

The blind dado is harder to make, but hides the dado. The dovetail dado joint locks one piece into the other and is used when pulling stress may be put on the pieces. This joint is used sometimes between the front and side of a drawer.

A *rabbet and dado,* although a little harder to make, is preferred over a plain dado joint. The rabbets can be machined to fit snugly into the dado. If dadoes are made for the full thickness of the stock, variations in the thickness of the shelf stock may cause problems in fitting.

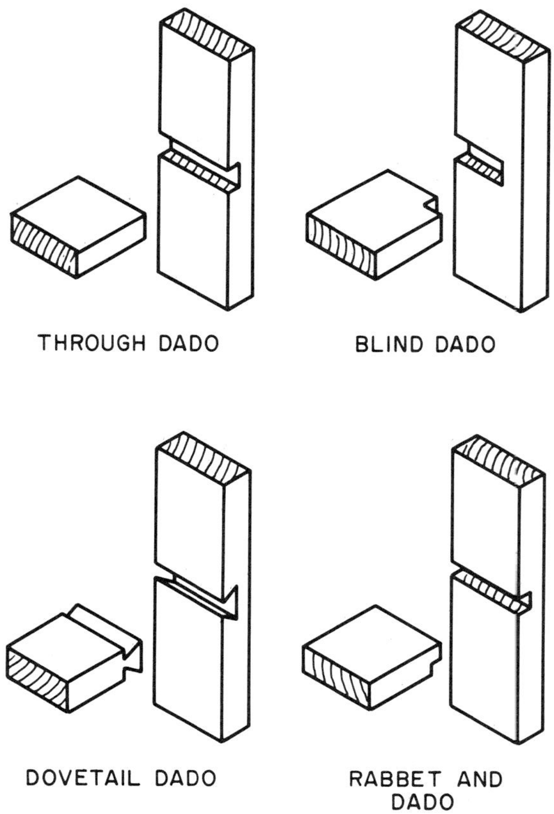

Fig. 15-9 Dado joints

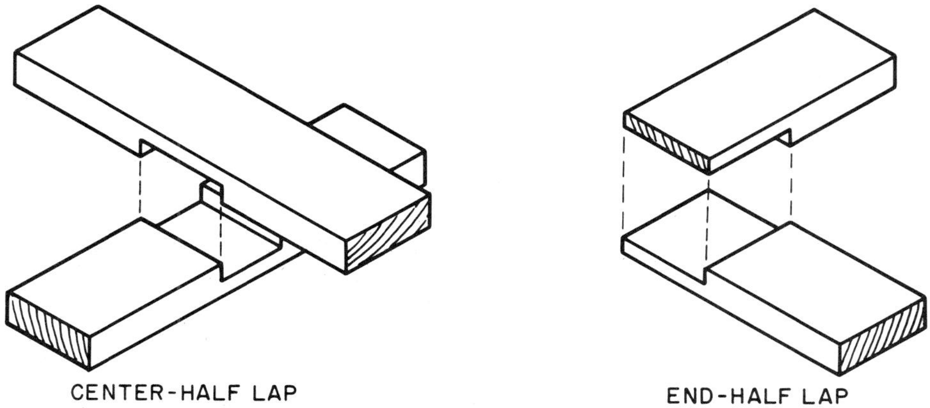

Fig. 15-10 Half-lap joints

HALF-LAP JOINTS

Half-lap joints are made when pieces cross over each other. Half the thickness of the face is removed from one piece, and half the thickness of the back is removed from the other piece. When assembled, the faces are flush with each other, figure 15-10.

Half-lap joints are used on framework, like the face frame of kitchen cabinets. They are easy to make and are held securely when glued and clamped. They do not need additional fasteners. Care must be taken when cutting the width and depth of the lap joint to assure a snug and flush fit.

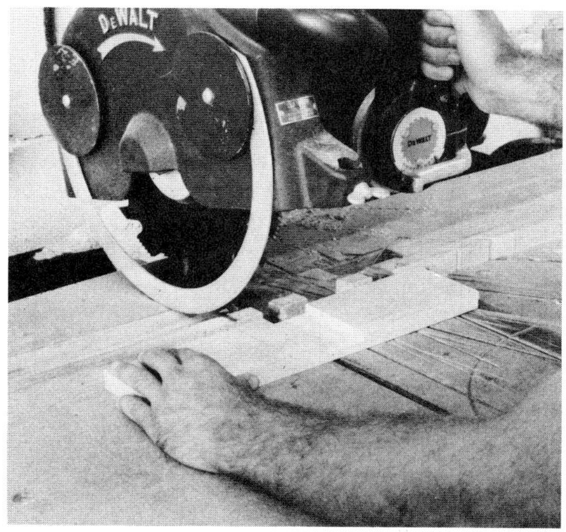

Fig. 15-11 Making half-lap joints with a dado head

Making A Half-Lap Joint

Half-lap joints are made in a number of ways. One method uses a dado head mounted on the radial arm saw.

1. Adjust the height of the dado head.
2. Mark off half the thickness of a scrap piece of stock of the same thickness. Adjust the dado head to the layout line.
3. Make cuts on two pieces of scrap stock and hold together by turning one piece over. The sides of the pieces should be flush. If not, make adjustments for height. Continue making cuts on the scrap material until the sides come flush.
4. When the dado head is adjusted, lay out the width of the cut on the good stock.
5. Cut to the layout lines. Cut out one piece on the face side and one piece on the back side, figure 15-11. If a number of pieces are to be half-lapped, use stop blocks on both ends.
6. Glue and clamp the half-lap joints together, or fasten together with glue and screws.

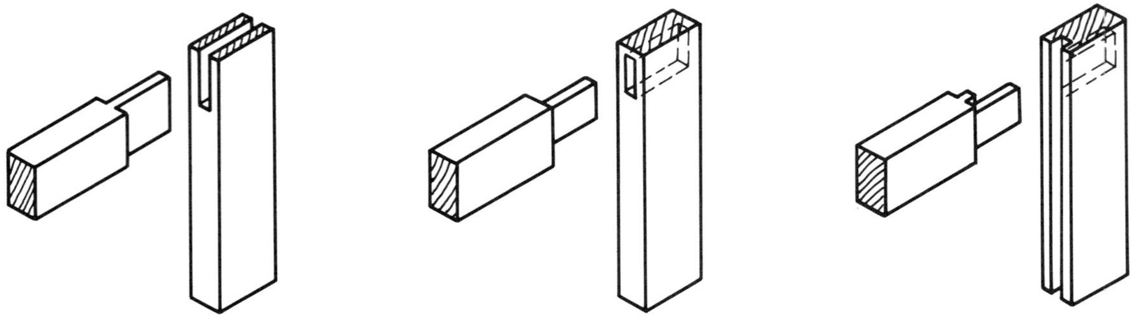

Fig. 15-12 Mortise-and-tenon joints

MORTISE-AND-TENON JOINTS

Mortise-and-tenon joints have been used for hundreds of years. The mortise is the opening into which the tenon fits. Although they require a little time to make, they are strong, easy to assemble, and present a good appearance. They are widely used for doors and frames.

Mortise-and-tenon joints can be classified as the through mortise and tenon, the blind mortise and tenon, and the haunched mortise and tenon, figure 15-12. The *through-mortise-and-tenon joint* is used when it is not objectionable to show the joint. This is a good joint to use on wood window screens.

The *blind-mortise-and-tenon joint,* although a little more difficult to make, is just as strong. It hides the joint completely and looks like a butt joint when assembled. The tenon is held securely with pegs, pins, screws, or glue.

The *haunched-mortise-and-tenon joint* is used with grooved pieces. Usually pieces of a frame are grooved to hold a panel such as doors. If the opening must not show where the groove comes through, part of the tenon is left on to fit in the groove. This is called *haunching the tenon.*

Like all joints, mortise-and-tenon joints depend on their snugness of fit for their strength. Directions for making mortises and tenons are given in *Unit 11 Mortiser and Tenoner.* They can also be made on the table saw. Doweled joints are used in place of the mortise-and-tenon joint to save time.

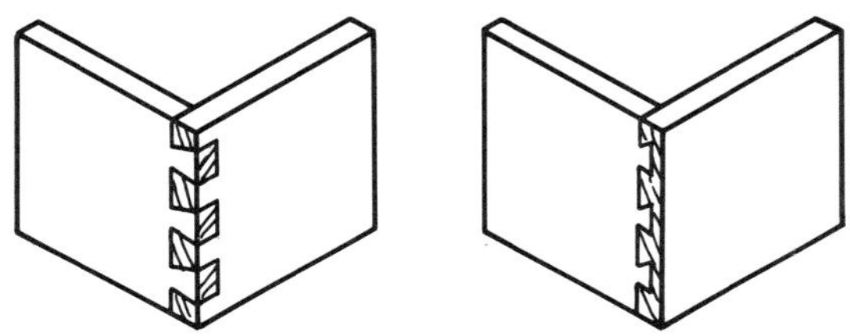

Fig. 15-13 Dovetail joints

DOVETAIL JOINTS

One of the strongest joints found in quality cabinetwork is the dovetail joint. Dovetail joints are made *through* or *blind,* figure 15-13. They are often used as joints in drawers where a lot of pulling stress is put on the pieces.

The dovetail joint is difficult to make without special equipment and takes a lot of time to cut by hand. Production shops use dovetail machines to make these joints in a minimum of time. However, these machines are expensive and take experience to set up.

Dovetail joints can also be made using a router and dovetail template, figure 15-14. To make dovetail joints using a router and dovetail template:

1. Select the correct router bit, template, and template guide for the stock to be dovetailed.
2. Adjust the router bit for the specified depth of cut.

Fig. 15-14 Dovetail joints can be made with a router and dovetail template. *(Rockwell)*

3. Place two scrap pieces of the same size in the dovetailing jig according to the manufacturer's directions.
4. Make cuts on the two pieces, following the template carefully.
5. Remove the pieces from the jig and make a trial fit. Adjust the depth of cut accordingly if the pieces do not fit properly.
6. Continue making cuts on the scrap pieces until the correct depth of cut is obtained.
7. When fully adjusted, make the necessary cuts on the pieces to be dovetailed.

LOCK JOINT

The lock joint can be used in place of the dovetail joint and is easier to make, figure 15-15. It is an extremely strong joint, easy to assemble, and can be held securely with glue. It is used frequently for drawer and box construction. The lock joint can be made with the shaper or a dado head.

BOX JOINT

The box joint, as its name implies, is used for making the end joints of boxes, figure 15-16. It is used when showing a quantity of end grain is not objectionable. It is an extremely strong joint and is sometimes used on Danish modern furniture. This joint provides tremendous amount of gluing surface for an extremely strong joint.

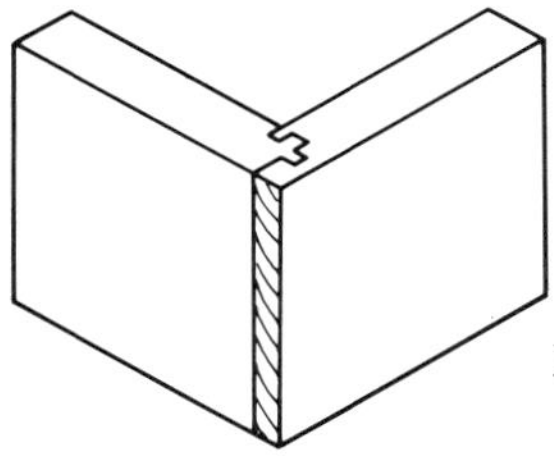

Fig. 15-15 Lock joint

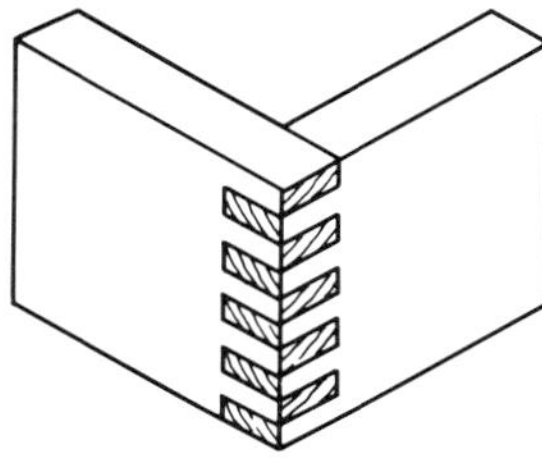

Fig. 15-16 Box joint

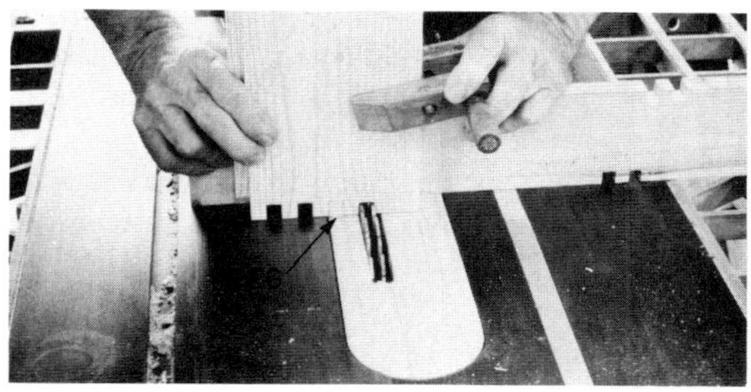

Fig. 15-17 Making a box joint (guard removed for clarity)

Making A Box Joint

1. Use a wood extension on the miter gauge and a dado head that cuts the widths of the joint.
2. Adjust the dado head to the correct height.
3. Clamp two pieces together and offset them the width of the joints.
4. Install a peg in the fence of the miter gauge the width of the joint and the same distance from the dado head.
5. Make the first cut with the pieces against the peg.
6. Make succeeding cuts by placing each cut made over the peg and continuing until all cuts are made, figure 15-17.

ACTIVITIES

1. Examine pieces of cabinetwork and list the different kinds of joints used in their construction.
2. Study and diagram different ways fasteners are concealed.
3. Strengthen a butt end joint with a glue block. With dowels.
4. Construct a wood box using box joints.
5. Select three of the following joints and construct them in class: rabbeted joints, splined joint, miter joint, dado joint, half-lap joint, mortise-and-tenon joint, dovetail joint, lock joint.

UNIT REVIEW

Questions

1. Name six points to consider when selecting a joint.
2. Which joint should be used to join the parts of a picture frame?
3. What is a dado that does not go all the way through a joint called?
4. Why would a rabbet and dado joint be preferred over a plain dado?
5. Why might doweled joints be used in place of mortise-and-tenon joints?
6. What would be a good joint to use on wood window screens?
7. Which joint is frequently used in place of the dovetail joint when making drawers and boxes? Why?
8. How are miter joints strengthened?
9. Which is the best joint to use between the front and side of a drawer?
10. What does *haunching the tenon* mean?

Identification

Identify each joint. Be specific as to its type.

1.
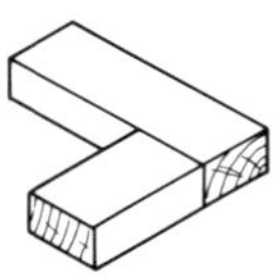

2.
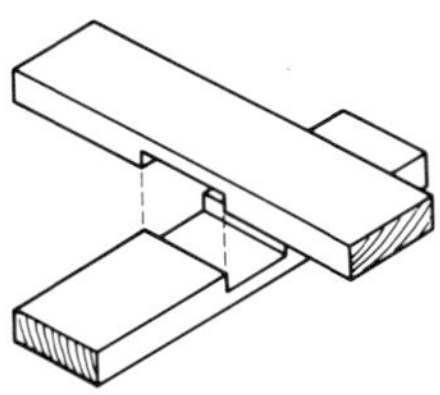

3.
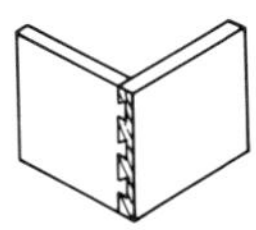

4.
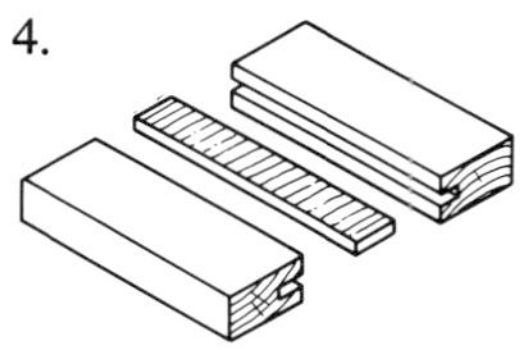

5.
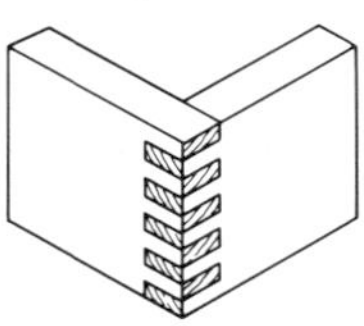

6.
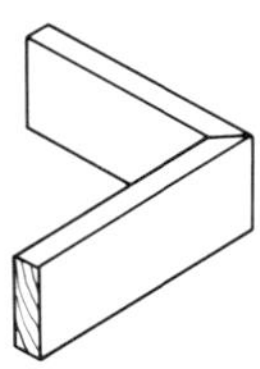

7.
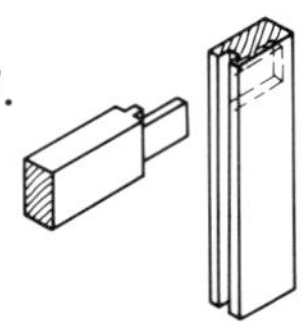

8.
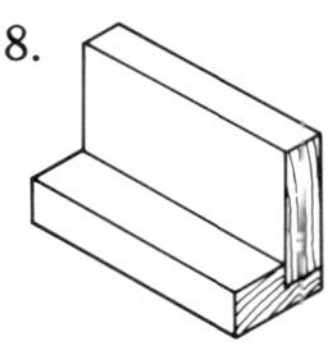

9.
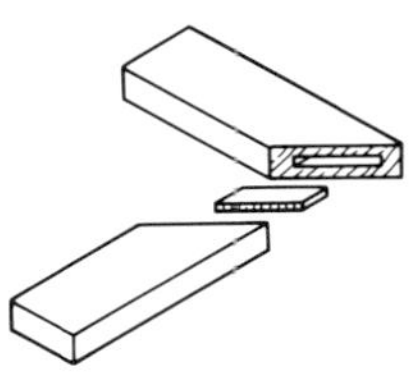

10.
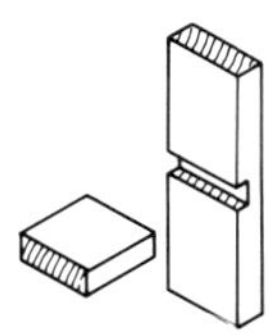

unit 16 Gluing and Clamping

OBJECTIVES

After studying this unit, the student will be able to:

- name the commonly used adhesives and describe their characteristics.
- describe the basic kinds of clamps and give examples of their use.
- explain the general procedure for gluing and clamping.

ADHESIVES

An *adhesive* is a glue, cement, or other substance that causes bodies to stick together. There are many different kinds of adhesives, figure 16-1. Some glues are made especially to *bond* or stick to a certain material, such as wood, metal, plastics, leather, or rubber. However, only those glues that are commonly used in small woodworking operations are discussed in this unit.

NATURAL GLUES

Animal Glue

Animal glue is usually made from animal hides. This type of glue has been used for a long time. It generally comes in a flake form which is mixed with water. It is soaked until the proper amount of water is absorbed. After soaking, it is heated and melted, usually in an electric glue pot.

Animal glue is a quick-setting glue. It requires little time under pressure to stick. It has good gap-filling properties and will strengthen even poorly fitted joints. However, because it sets quickly, it must be clamped or pressed by hand immediately after application. It is an ideal glue for small pieces that are difficult to clamp, such as reinforcing corner blocks. The glue can be applied and the pieces pressed and held by hand in position for a short while.

Caution: Hot glue will blister the skin. Avoid contact with the glue.

Joints bonded with animal glue are not waterproof. Animal glue can only be used on interior work.

Animal glue also comes in liquid form and is applied cold. It is not as quick setting as the hot glue. Some other kinds of cold liquid natural glues are made from fish skins. These cold glues are often used for small repair work in the home.

Casein Glue

Casein glue is another natural glue. Its principal ingredient is the curd of sour milk. It usually comes in a powdered form which is mixed with water. The manufacturer's directions should be followed carefully when mixing the glue.

TYPE	ADVANTAGES	DISADVANTAGES	USES	PREPARATION	APPLICATION	CLAMPING TIME
Animal Glue	Strong bond Quick setting Good filling quality	No water resistance	Interior woodwork only	Ready to use	Apply thin coat to both surfaces; let it get tacky before joining; clamp immediately	70°F Press for 5-10 min. Then do not disturb for 2 hours
Casein Glue	Can be used down to 32°F Pot life – 6 to 8 hrs Good filling quality Bonds oily wood Long assembly time	Stains certain wood Only slight water resistance	Gluing at low temperatures Gluing oily woods General woodwork	Mix with water according to manufacturer's directions	Apply thin coat to both surfaces; let glue thickens before bonding; clamp	32°-70°F Bonds – 4 hours Cures – 24 hours
Polyvinyl Acetate (white glue)	Non-staining, odorless Transparent when dry Strong bond Sets quickly Good filling quality	No water resistance Not as strong as animal glue	All-purpose glue Interior work only	Ready to use	Apply thin coat to both surfaces; bond; clamp immediately	70°F Bonds – 1 hour Cures – 24 hours
Aliphatic Resin (yellow glue)	Stronger than white glue and sets faster Easy to apply	No water resistance	Interior work only	Ready to use May need stirring	Apply thin coat to both surfaces; bond; clamp immediately	50°-110°F Bonds – Under 1 hour Cures – 24 hours
Urea Resin (plastic resin)	Highly water-resistant Nonstaining Pot life – 6 to 8 hrs 30-min. assembly time	Slow setting Low bond strength with oily woods	General gluing where moisture resistance is needed	Mix 2 parts powder with 1/2 to 1 part water to cream-like consistency	Apply thin coat to both surfaces; bond; clamp. Use within 8 hours after mixing	70° or warmer 70° – 16 hours 80° – 12 hours 90° – 5 hours
Resorcinol Resin	Completely waterproof Long pot life Long assembly time	High price Long setting time under pressure	Exterior work, boat work, etc. where waterproof glue is needed	Mix powder with liquid catalyst according to manufacturer's directions	Apply thin coat to boat surfaces; bond; clamp. Use within 8 hours after mixing	70° or warmer Clamp for 12 to 24 hours
Contact Cement	Instant bonding Waterproof	Poor heat resistance Low strength	Applying plastic laminates to countertops and cabinets	Ready to use	Let first coat dry for 20 min., then apply second coat to both surfaces; position surfaces exactly before contact is made	No clamping necessary Bonds instantly

Fig. 16-1 Types of adhesives

Casein glue can be mixed and used at any temperature above freezing. It has a long pot life, once mixed, and can be used for a long time. It has good gap-filling properties and will glue oily woods, such as teak, with a tough, durable bond.

Casein glue has a fairly long assembly time. After applying the glue, clamping can be delayed for up to 20 minutes. In fact, a delay in clamping the wood will greatly improve the strength of the bond. This lets the glue thicken somewhat before applying pressure. However, too long a delay will allow the glue to dry.

Casein glue is widely used in the woodworking field to glue fine veneers and to laminate timberwork. It is not waterproof, however, and will stain certain woods like oak and mahogany. Also, stock must be kept under pressure after gluing for a minimum of four hours, or preferably longer.

SYNTHETIC GLUES

Polyvinyl Acetate Glue

Probably the most popular synthetic glue is *polyvinyl acetate,* commonly known as *white glue.* It comes ready-mixed, is nonstaining, colorless, and odorless. It dries transparent and sets quickly. It is strong and has good gap-filling properties. Because it sets quickly, it must be clamped as soon as possible, usually no longer than 5 minutes. It is not resistant to moisture and should only be used for interior work.

Aliphatic Resin Glue

Aliphatic resin glue, sometimes called *yellow glue,* is similar to white glue. However, it is stronger than white glue and sets faster. It also comes ready-mixed and is easy to apply. Pieces glued with yellow glue must be clamped immediately. Like white glue, this type of glue is not resistant to moisture and is used only for inside work.

Urea Resin Glue

Urea resin glue usually comes in powdered form and is commonly called *plastic resin glue.* The powder is mixed with water to a consistency like heavy cream. It produces a highly water-resistant bond. It is nonstaining, has a fairly long pot life of about 6 to 8 hours, and has an assembly time of up to 30 minutes. This is important if a lot of time must be taken before the glued pieces can be clamped together.

Pieces bonded with plastic resin glue must be kept in clamps for a long period of time. The minimum pressure time is about 5 hours, depending on conditions. Longer periods of time in clamps insure stronger joints. Wood should not be worked until the glue has set firmly.

Resorcinol Resin Glue

Resorcinol resin glue is a completely waterproof glue. It is used extensively on boat work and other exterior work that is exposed to a lot of moisture. It comes in a powdered form together with a liquid that must be mixed with it.

Resorcinol resin has a long pot life. Speed is not so important in working with this type of glue. There is plenty of time to get the work in clamps before the glue starts to set. However, the pieces must be clamped for a long time under pressure, usually overnight.

Contact Cement

Contact cement is so named because it needs no clamps for bonding. Pieces glued with contact cement bond on contact with each other. It is extremely important, therefore, that the pieces are positioned accurately before contact is made.

Contact cement is widely used to apply plastic laminates such as those found on kitchen cabinet countertops. It is also used to bond any thin or flexible material that otherwise requires elaborate clamping devices.

Contact cement comes ready-mixed. It is made to be brushed, rolled, or sprayed. Both surfaces to be bonded must be coated with the cement. Usually two coats are required on porous surfaces like wood and particleboard. Follow the manufacturer's directions when using contact cement.

The use of contact cement is fully described in *Unit 25 Plastic Laminating.*

HOT GLUE GUNS

Hot, melted glue is usually applied with hot glue guns, figure 16-2. The gun contains an electrical heating element that melts cakes of glue that are inserted in the gun. After a warm-up time of a few minutes, the glue melts on demand by pressing a trigger, figure 16-3.

Hot glue requires no clamps. Just apply it to one surface and press the parts together.

Caution: Both the tip of the gun and the glue are very hot. Avoid contact with the skin.

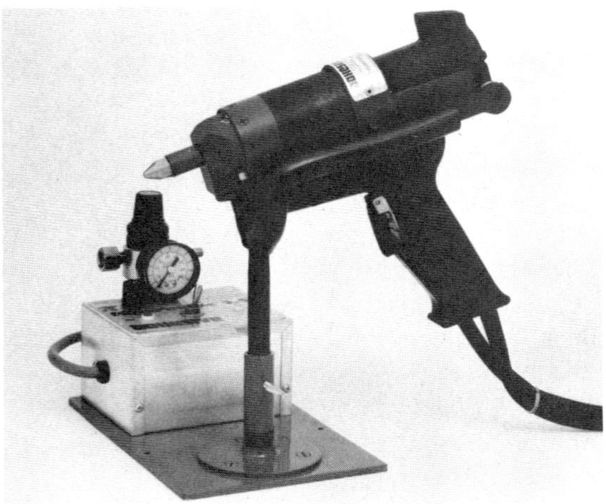

Fig. 16-2 Hot glue gun *(Adhesive Machinery Corporation)*

Fig. 16-3 Apply hot glue

Hot glue guns are made in many styles. A variety of glue cakes are available for bonding many different kinds of material. These guns are used extensively in industry for rapid gluing of small pieces.

CLAMPS

Most glues need to be applied under pressure. This forces the glue into the fibers of the wood and makes a strong bond. *Clamps* are used for this purpose.

Parallel Clamps

One of the oldest type of clamp is the *wood parallel clamp,* figure 16-4, also called *handscrews.* It can be used directly on the pieces to be clamped without damaging the surface. It comes in a variety of sizes and is extremely useful for all kinds of clamping work.

To use parallel clamps:

1. Close up the clamp and line up the jaws.

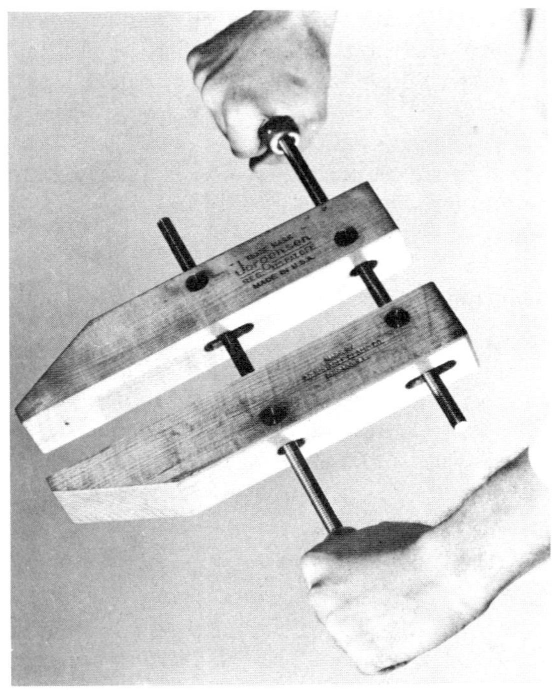

Fig. 16-4 Parallel clamps *(Adjustable Clamp Co.)*

Fig. 16-5 Tighten the center screw first and then tighten the end screw. *(Adjustable Clamp Company)*

Fig. 16-6 Jaws should rest firmly and equally on the work. *(Adjustable Clamp Co.)*

2. With one hand on each screw, rotate the clamp until the jaws open just enough to go over the work to be clamped.
3. Place the clamp on the work and tighten the center screw first, figure 16-5.
4. Tighten the end screw. The clamp is now in the correct position with the jaws resting firmly and equally on the work, figure 16-6.

Bar Clamps

Bar clamps are usually used to clamp pieces edge to edge, figure 16-7. They come in lengths of 2 to 10 feet. One end has a short screw and the other end has an adjustable stop. Pieces of scrap wood should be placed between the clamp and the work to protect the work.

To use the bar clamp, back off the screw as far as it will go. Adjust the other end until it is slightly larger than the stock to be clamped. Turn the screw in until the desired pressure is reached.

Fig. 16-7 Bar clamps

C Clamps

C clamps, figure 16-8, are used like parallel clamps. C clamps come in a wide variety of sizes. Jaw openings vary from 2 inches to 12 inches and can be classified as light or heavy duty.

Fig. 16-8 C clamps *(Adjustable Clamp Co.)*

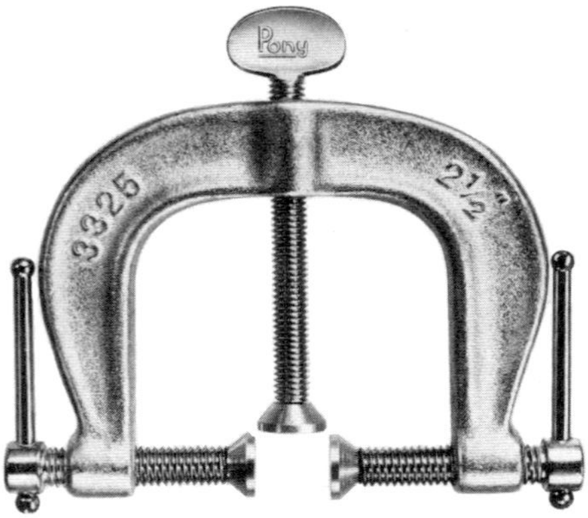

Fig. 16-9 Three-way edging clamp

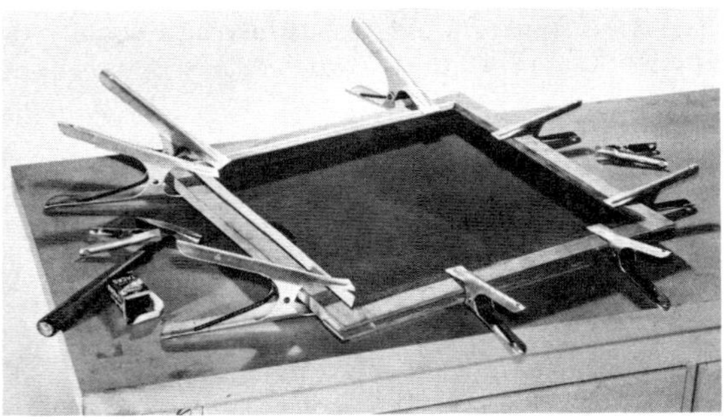

Fig. 16-10 Spring clamps hold light material quickly. *(Adjustable Clamp Co.)*

Fig. 16-11 Band clamps are used on irregular surfaces. *(Adjustable Clamp Co.)*

Use only hand pressure to tighten the C clamp. Using a wrench to tighten the clamp may bend the screw and make the clamp practically useless. Use scrap pieces of wood to protect the work when using C clamps.

Edging Clamps

The *three-way edging clamp* is a type of C clamp, figure 16-9. It comes in a variety of sizes and is very useful for clamping banding strips to the edge of the stock. A scrap piece of wood protects the stock when the clamp is screwed on. The center screw clamps the edge band to the stock.

Spring Clamps

Spring clamps, figure 16-10, are very useful for light-duty clamping where heavy pressure is not needed. They come in different sizes and their jaws are usually covered with plastic to protect the work.

Band Clamps

Band clamps, figure 16-11, consist of a canvas strap and some sort of take-up device. These clamps are used on irregular shaped or round pieces, such as chairs.

To use this type of clamp, adjust the band by hand around the piece to be clamped. Using a wrench or screwdriver, take up on the clamp as necessary.

Miter Clamps

Miter clamps, figure 16-12, are specially made to clamp pieces at right angles. These clamps are

Fig. 16-12 Miter clamps *(Adjustable Clamp Co.)*

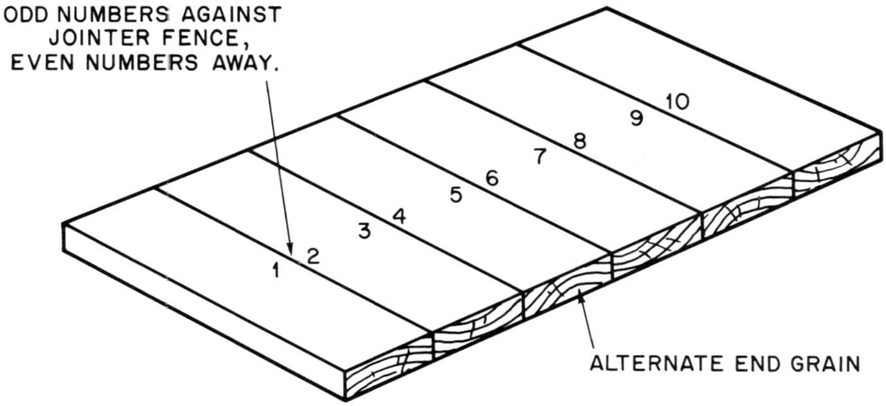

Fig. 16-13 Arranging edge jointed pieces

very useful when clamping picture frames and similar objects.

A recess is provided in the miter clamp to run a saw cut through the joint if the pieces do not fit perfectly. The saw cut is made by holding both pieces in the clamp. After the saw cut is made, the pieces are repositioned and clamped again until a perfect joint is obtained. The outside corner of the clamp is open to allow for fastening the joint.

GLUING AND CLAMPING PROCEDURE

Factors to Consider

A number of factors contribute to a successfully glued joint. A strong and durable joint depends on the moisture content of the wood, width of the pieces, direction of the grain, and the fit of the joint.

The wood to be bonded must have the proper moisture content. Wood that has a moisture content of over 10 percent will shrink. This will cause the joint to break or the wood to split. A moisture meter can measure the moisture content.

When gluing edge to edge, narrow pieces are used. Wide pieces will result in a warped surface. Pieces no wider than 6 inches should be used.

The pieces should be arranged to present the most pleasing grain appearance. Alternating the direction of the end grain prevents cupping.

The strength of a glued joint depends on the fit of the joint. To join pieces edge to edge, run the edges over the jointer. Keep the odd numbers against the fence and the even numbers away from the fence. Then if the edges have a slight bevel, the pieces will still lie flat and straight, figure 16-13. For all kinds of gluing, make sure the joints fit well before attempting to bond.

Advance Planning

All clamps and blocks should be adjusted and ready. The pieces should be clamped together without glue to make sure all joints come up tight. Some glues are quick setting. Having all clamps and blocks ready reduces the assembly time and prevents the glue from setting before pressure is applied. Amount of assembly time should be determined beforehand.

Applying Glue

Too little glue results in a weak joint. Too much glue is a waste and makes a mess. Usually both surfaces to be joined are coated with glue. When the pieces are clamped, small beads of glue should appear all along the joint. If glue drips out of the joint, too much has been applied. If little or no glue is squeezed out, too little has been used. Follow the directions carefully when using any type of glue. Use appropriate glue for the job to be done.

Amount of Pressure

When using a variety of clamps, each clamp is tightened a little at a time until all are tight. This allows all pieces to be drawn up equally. Only enough pressure to bring the joint up tight and to squeeze out the glue is used. Too much pressure may crush the fibers of the wood. It also may squeeze out too much glue resulting in a *starved joint.* Wax paper placed between blocks of wood and the glued pieces prevents them from sticking to the work. Clamps should be alternated on the top and bottom of the stock.

Excess glue is scrapped off after glue is dry. It is important to remove all excess glue if pieces are to be stained later on. Glue left on the surface will seal it and not let the stain penetrate. This produces blotches on the stained surface.

Curing and Surfacing

Pieces should remain in clamps for the recommended time. Glue should set firmly before the piece is worked. If the piece is surfaced before the glue is thoroughly dry, the bond may not hold. Also, because glue tends to swell the wood at the joint, a sunken joint may be produced if the piece is surfaced prematurely.

Summary

Important points to remember when gluing and clamping are:

1. Select the correct type of glue.
2. Use dry lumber.
3. Have all clamps ready.
4. Clamp up dry and check for fit of joints.
5. Apply proper amount of glue and pressure.
6. Let glue dry thoroughly before removing from clamps.

ACTIVITIES

1. Study the chart in figure 16-1 of the different kinds of adhesives and their characteristics.
2. What types of adhesives are available in the class shop? Read the manufacturer's directions before using them.
3. Test the holding strength of different types of glues.
4. Practice using the various clamping devices available in the class shop.
5. Glue and clamp pieces edge to edge with a tight joint using bar clamps, C clamps, and parallel clamps.
6. Make a picture frame with splined miter joints by fitting, gluing, and clamping with miter clamps.
7. Band the edge of stock with tongue-and-grooved joints and mitered corners using three-way clamps.

UNIT REVIEW

Multiple Choice

1. Too much pressure when clamping glued pieces together may cause
   a. the pieces to buckle.
   b. a starved joint.
   c. a sunken joint.
   d. a swollen joint.
2. The correct amount of glue is indicated by
   a. glue dripping out of the joint.
   b. no glue visible along the joint.
   c. small beads of glue along the joint.
   d. a solid bead of glue along the joint.

3. Pieces to be glued should first be clamped together without glue to
   a. check the fit of the joints.
   b. have all clamps and blocks ready.
   c. to reduce the assembly time.
   d. all of the above.
4. When jointing edges in preparation for gluing,
   a. all numbers are kept against the fence.
   b. odd numbers are kept against the fence and even numbers away.
   c. all numbers are kept away from the fence.
   d. disregard the numbers and cut with the grain.
5. When gluing pieces edge to edge, no piece should be wider than
   a. 4 inches.
   b. 6 inches.
   c. 8 inches.
   d. 10 inches.
6. A clamp used to hold pieces at right angles is called a
   a. C clamp.
   b. band clamp.
   c. parallel clamp.
   d. miter clamp.
7. Parallel clamps are made primarily of
   a. steel.
   b. wood.
   c. canvas.
   d. aluminum.
8. A type of glue that requires no clamps is
   a. hot melt glue.
   b. white glue.
   c. yellow glue.
   d. plastic resin glue.
9. A type of glue that is used to bond thin, flexible material without elaborate clamping is
   a. plastic resin.
   b. aliphatic resin.
   c. contact cement.
   d. polyvinyl acetate.
10. The most important characteristic of resorcinol resin glue is that
    a. it sets up quickly.
    b. it is waterproof.
    c. it is highly water-resistant.
    d. no clamps are needed.

Questions

1. What are two kinds of natural glue?
2. What type of glue should be used to obtain tough, durable bonds on oily woods such as teak?
3. What is the common name for polyvinyl acetate glue?
4. What is the common name for aliphatic resin glue?
5. Which glue is completely waterproof?
6. Which type of glue is used to bond plastic laminates?
7. Why is casein glue not used on oak or mahogany?
8. What percentage moisture content should lumber have if a strong, durable, glued joint is desired?
9. What are hot glue guns used for?
10. How is contact cement applied?

unit 17
Fasteners

OBJECTIVES

After studying this unit, the student will be able to:

- describe the kinds and sizes of nails, staples, screws, bolts, lag screws, anchors, and miscellaneous fasteners.
- use correct techniques to install fasteners.

Many kinds of fasteners are used to assemble furniture and cabinets and to anchor the completed product to walls, floors, and other parts of a building. It is important that the woodworker know the best fastener to use for each job.

NAILS

There are many different kinds of nails. They differ according to their shape, material, coating, and in other ways. Nails are made of aluminum, brass, copper, steel, and other metals. Different coatings are applied to reduce corrosion, increase holding power, and for appearance. Some nails are also hardened so that they can be driven into masonry.

Sizes

Nails are sized according to the *penny system*, figure 17-1. The symbol for penny is a *d*. For instance, a six-penny nail is written as 6d and is two inches long. In the penny system, the shortest nail is a 2d which in one inch long. The longest nail is a 60d which is six inches long. The thickness

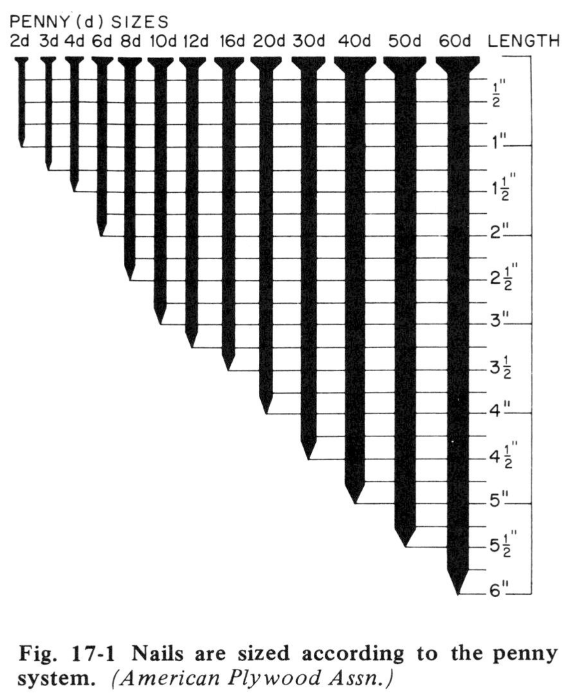

Fig. 17-1 Nails are sized according to the penny system. *(American Plywood Assn.)*

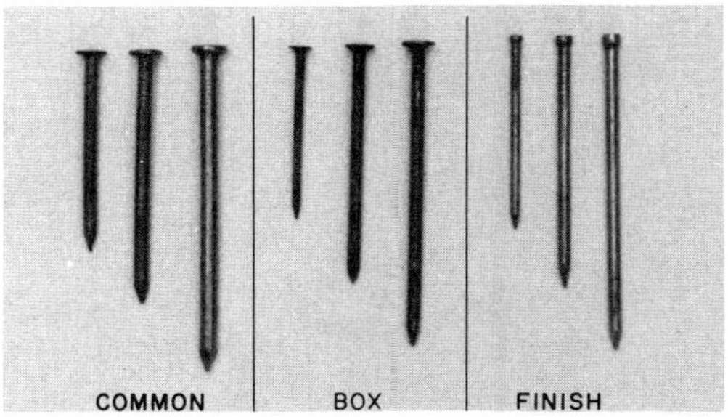

Fig. 17-2 Left to right: common, box, and finish nails

of the nail is called its *gauge.* The gauge of a nail depends on its kind and length. Long nails, 16d and over, are usually called *spikes.* Nails longer than six inches and shorter than one inch are listed by inches and fractions of an inch.

Kinds of Nails

Common nails are widely used in construction and seldom used in cabinetwork, figure 17-2. They are a fairly thick nail with a medium-size head.

Box nails are similar to common nails but are not as thick, see figure 17-2. Because of their small gauge, they can be used on thin pieces without splitting the wood. Commonly used on boxes, they also come cement-coated to increase their holding power.

Finish nails are thin with a very small head, see figure 17-2. The small head is sunk into the wood with a nail set and covered with a filler. The small head of the finish nail does not detract from the appearance of a job as much as a nail with a larger head.

Brads are small finishing nails, figure 17-3. They are sized according to their length in inches and their gauge. Usual lengths range from 1/2 inch to 1 1/2 inches. Gauges range from 14 to 20. The higher the gauge number, the thinner the brad. Brads are used for fastening thin material like small molding or thin cabinet backs.

Escutcheon pins are small nails, usually brass, with round heads, figure 17-4. They are available in lengths of 3/16 inch to 2 inches and gauges

Fig. 17-3 Brads are small finishing nails.

Fig. 17-4 Escutcheon pins are small nails with round heads.

from 10 to 24. They are used to install some types of hardware when the head of the pin adds to the appearance.

Tacks are small, pointed nails with a head, figure 17-5. Sometimes used to fasten lightweight material, such as cloth or wire screen, they are available in sizes 1 to 24. The larger the number of the tack, the longer it is.

Masonry nails, figure 17-6, are hardened to prevent bending when being driven into concrete or other masonry.

Fig. 17-5 Tacks are often used to fasten lightweight material.

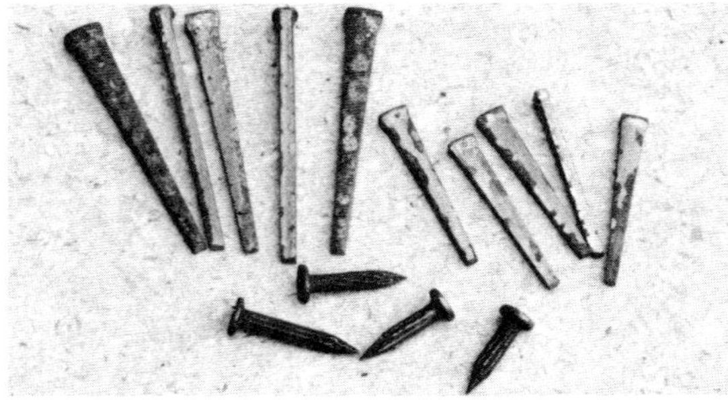

Fig. 17-6 Masonry nails are hardened so that they can be driven into concrete and similar materials.

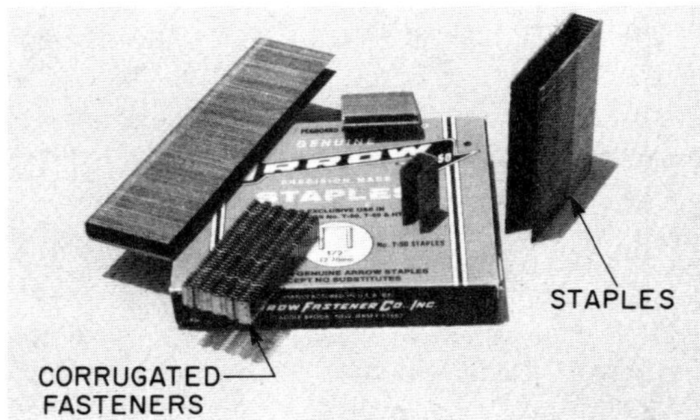

Fig. 17-7 Staples and corrugated fasteners come glued together and are driven by hand and power-operated tools.

Caution: Safety goggles should be worn when driving masonry nails. The hardened metal of the nail could chip off masonry which may fly into the eye.

Staples are U-shaped fasteners, figure 17-7. They may be purchased to be driven individually with a hammer. For industry use, staples also come glued together in rows. These are driven by air-operated and hand-operated staplers. Widths range from 3/8 inch to 1 inch. Lengths generally are from 1/2 inch to 2 inches. Common gauge sizes are from 14 to 16. There are many other sizes for special work. Make sure the staples fit the machine being used. Follow the manufacturer's instructions for the operation of the stapler being used.

The *T nail,* figure 17-8, is a specially designed nail also used in nailing machines. These nails are widely used for rapid assembly of cabinet parts in places where the exposed head does not detract from the appearance of the product.

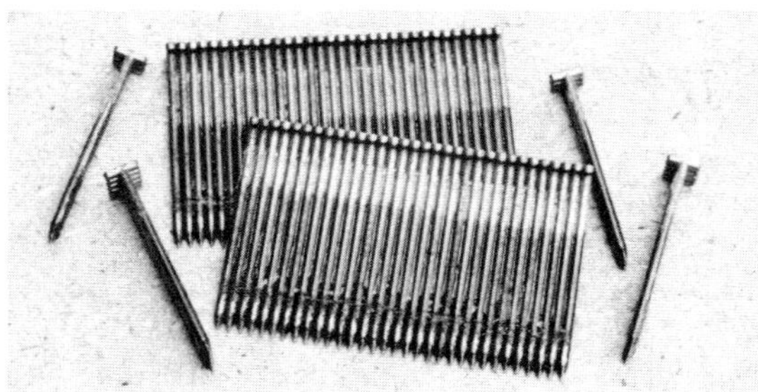

Fig. 17-8 The T nail is designed to be used in power nailers.

Nailing Techniques

Select nails three times longer than the thickness of the material to be fastened. Hold the hammer handle firmly. Hit the head of the nail squarely. Rub the face of the hammer head occasionally with sandpaper to help prevent glancing off the nailhead.

Drive finish nails almost home, then set the nail below the surface with a nail set. Finish nails are set at least 1/8 inch deep so that the filler used to cover the nailhead will not drop out.

Fig. 17-9 When nailing close to an edge, predrill holes for nails. *(American Plywood Assn.)*

If nailing in hardwood or close to an edge, drill a hole slightly smaller than the diameter of the nail, figure 17-9. This prevents the wood from splintering or the nail from bending. A little wax applied to the shank of the nail makes driving the nail easier.

When possible, stagger nails from edge to edge. This technique provides greater strength and is less likely to split the wood when nailing in a straight line, figure 17-10.

SCREWS

Because of their great holding power, wood screws are more widely used in quality cabinetwork than nails or staples. Screws can also be removed without damaging the work. They present a neat appearance when fastening all kinds of hardware. However, screws cost more than nails and staples and require more time to drive.

Kinds of Screws

The shape of the screwhead and screwdriver slot determines the kind of screw used. Three of

Fig. 17-10 Stagger nails from edge to edge for greater strength and to avoid splitting the wood.

Fig. 17-11 Common kinds of screw heads

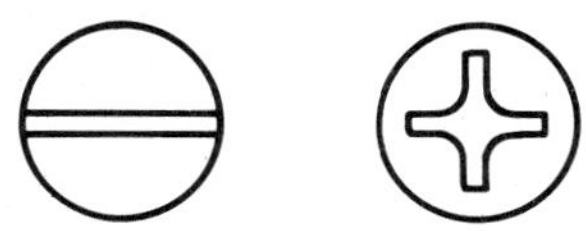

Fig. 17-12 Common kinds of screw slots

the most common shapes are the *flathead, roundhead,* and *ovalhead* screw, figure 17-11. Other shapes include the *panhead* and *truss head.*

Screwheads containing a straight, single slot are called *common screws. Phillips screws* are those containing a cross-slot, figure 17-12.

Wood screws are threaded only part way to the head. *Sheet metal screws* have a deeper thread all the way up to the head. Sheet metal screws are recommended for fastening hardboard and particleboard because of their deeper thread.

Another type of screw used with power screwdrivers is the *self-drilling wood screw.* This screw has a cutting edge on its point to eliminate predrilling a hole.

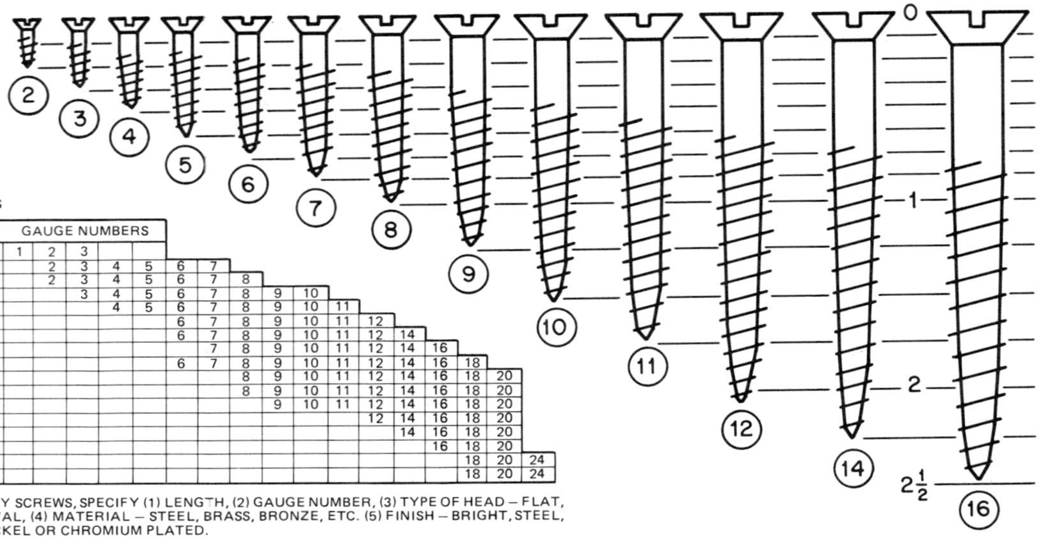

WOOD SCREWS

LENGTH	GAUGE NUMBERS																
1/4 INCH	0	1	2	3													
3/8 INCH			2	3	4	5	6	7									
1/2 INCH			2	3	4	5	6	7	8								
5/8 INCH				3	4	5	6	7	8	9	10						
3/4 INCH					4	5	6	7	8	9	10	11					
7/8 INCH							6	7	8	9	10	11	12				
1 INCH							6	7	8	9	10	11	12	14			
1 1/4 INCH								7	8	9	10	11	12	14	16		
1 1/2 INCH							6	7	8	9	10	11	12	14	16	18	
1 3/4 INCH									8	9	10	11	12	14	16	18	20
2 INCH									8	9	10	11	12	14	16	18	20
2 1/4 INCH										9	10	11	12	14	16	18	20
2 1/2 INCH													12	14	16	18	20
2 3/4 INCH														14	16	18	20
3 INCH															16	18	20
3 1/2 INCH																18	20	24
4 INCH																18	20	24

WHEN YOU BUY SCREWS, SPECIFY (1) LENGTH, (2) GAUGE NUMBER, (3) TYPE OF HEAD – FLAT, ROUND OR OVAL, (4) MATERIAL – STEEL, BRASS, BRONZE, ETC. (5) FINISH – BRIGHT, STEEL, CADMIUM, NICKEL OR CHROMIUM PLATED.

Fig. 17-13 Screw sizes *(American Plywood Assn.)*

Many other kinds of screws are available that are designed for special work. Like nails, screws come in a variety of metals and coatings.

Screw Sizes

Wood screws are made in many different sizes, figure 17-13. Usual lengths range from 1/4 inch to 4 inches. Gauges range from 0 to 24. The higher the gauge number, the greater the diameter of the screw. Screws are not available in every gauge. The lower gauge numbers are for shorter screws, and higher gauge number are for longer screws.

Screwdriving Techniques

If possible, select screws so that two-thirds of their length penetrate into the piece in which the screws are gripping.

To drive screws, two holes must be drilled. One hole, called the *shank hole,* is for the threadless part of the screw. The other hole, called the *pilot hole,* is for the threaded part of the screw.

To select a drill for the shank hole, hold the drill bit against the shank and determine by eye if it is the same size as the shank. To select a drill for the pilot hole, hold the drill against the threaded portion of the screw. The drill should cover the solid center section and leave the threads visible.

Select drills carefully. Smaller drills may be used for a pilot hole in softwoods. However, in hardwoods, if the pilot hole is too small, it may be difficult to drive the screw. Also, if too much pressure is applied when driving the screw, the head may twist off. This is particularly true when driving screws of soft metal like aluminum or brass. Rubbing some wax on the threads will make driving easier. Of course, if the pilot hole is too large, the screw will not grip.

Always drill holes for screws, whether in softwood or hardwood. Drill the pilot hole deep enough so that the screw will not bottom. When there is danger of drilling through the material, use a stop on the drill.

Countersinking and Counterboring

In addition to drilling shank and pilot holes, ovalhead and flathead screws must be countersunk. Countersink flathead screws flush with or slightly below the surface. Countersink ovalhead screws with its head flush with the surface.

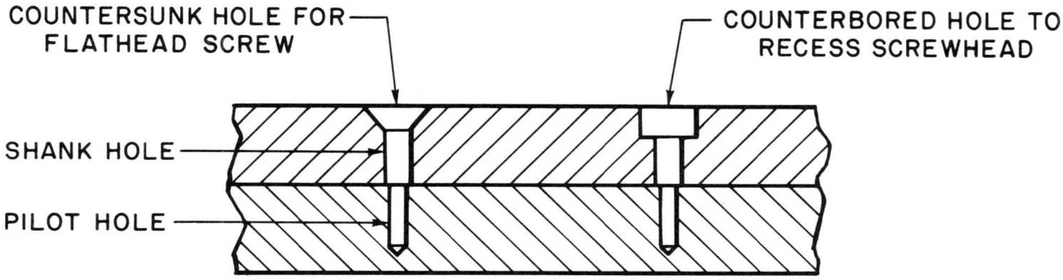

Fig. 17-14 Detail of shank, pilot, countersunk, and counterbored screw holes

If the screwhead is to be covered with a wood plug (sometimes called *bungs*), the head is sunk into a counterbored hole. Make the counterbored hole first, drill the shank hole second, and the pilot hole last. The center will be lost for the larger holes if the smaller holes are drilled first, figure 17-14.

It is good practice to make the counterbored hole in a scrap piece of stock first. Then test the plug to be used in the hole. The plug should fit tightly. If it does not, the bit is cutting oversize. Test other bits until the plug fits tightly in the bored hole.

Counterbored holes should be at least 3/8 inch deep so the plug will not be too thin and eventually fall out of the hole. When driving screws into counterbored holes, do not scrape the sides of the hole with the screwdriver because the plug will not fit tightly.

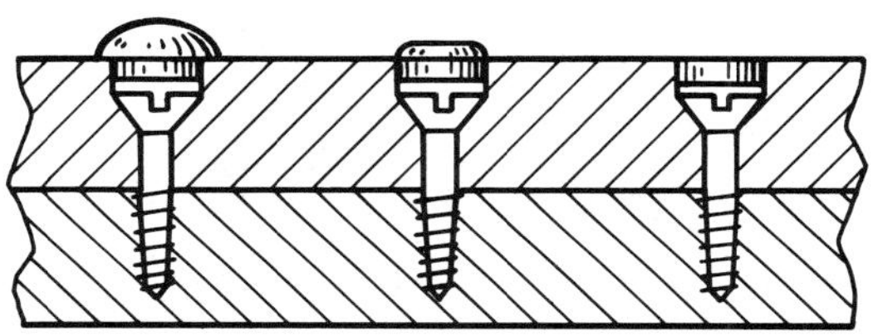

Fig. 17-15 Various types of plugs are used to conceal screwheads.

Fig. 17-16 Trimming a plug with a wood chisel

Plugs

Some plugs are ready-made with tapered edges to be driven into counterbored holes, figure 17-15. These plugs only need to be driven and require no trimming or sanding.

Other plugs are made in the shop using plug cutters. These plugs are tapped into the counterbored hole and trimmed flush to the surface with a wood chisel. The chisel is held with the bevel down, and cuts are made at least 1/8 inch above the surface, figure 17-16. Notice the direction of the grain of the plug. If the grain runs downward toward the counterbored hole, make cuts from the other side. If the grain runs uphill, continue cutting from the same side. Trimming plugs in this manner results in a plug flush with and not below the surface in which the counterbored hole is made.

Driving Screws

Screwdrivers should fit snugly, without play, into the slot of the screw being driven. The screwdriver should not be wider than the screwhead, nor should it be too narrow.

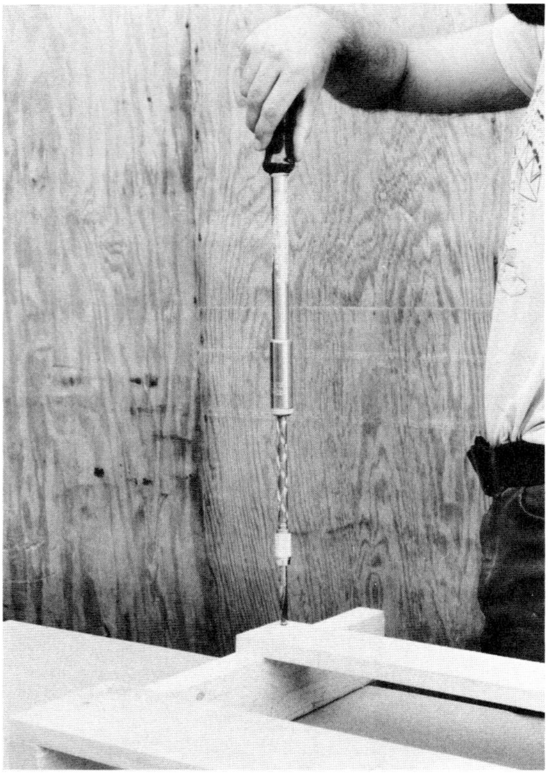

Fig. 17-17 Using a spiral screwdriver makes driving screws easier.

Exert firm pressure downward while turning the screwdriver. Seat the head firmly. When seated, the screw slot should not have burred edges.

Spiral screwdrivers save much energy when many screws need to be driven, figure 17-17. Screwdriver bits are selected with the same consideration as hand screwdrivers. When using spiral screwdrivers, care must be taken not to slip off the screw head. This can be avoided by twisting the handle while pushing downward when the screw nears its seat.

Slipping is also avoided when using a bit brace and screwdriver bit by using the bit brace ratchet. When the screw is almost driven home, the ratchet is engaged. The handle is pulled a quarter turn or less each time until the screw is seated, figure 17-18.

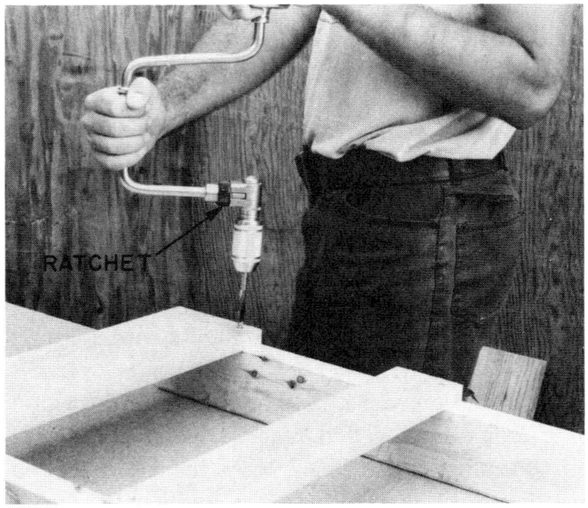

Fig. 17-18 Ratchet the bit brace to avoid slipping when driving screws.

BOLTS

Commonly used bolts are the *carriage, machine,* and *stove bolts,* figure 17-19.

The *carriage bolt* has a square section under its ovalhead. The square section is embedded in the wood and prevents the bolt from turning as the nut is tightened.

The *machine bolt* has a square or hexagonal (six-sided) head. It is held with a wrench to keep the bolt from turning as the nut is tightened.

The *stove bolt* has either a round or flathead with a screwdriver slot. It is usually threaded all the way up to the head.

Bolt Sizes

Bolt sizes are specified by the diameter and length of the bolt. Carriage and machine bolts range from 3/4 inch to 20 inches in length and from 3/16 inch to 3/4 inch in diameter. Stove bolts are smaller. They come in 3/8-inch to 6-inch lengths and in 1/8-inch to 3/8-inch diameters.

Fastening Bolts

The hole for the bolt should be the same diameter as the bolt. Washers are used under the head (except for carriage bolts) to distribute the

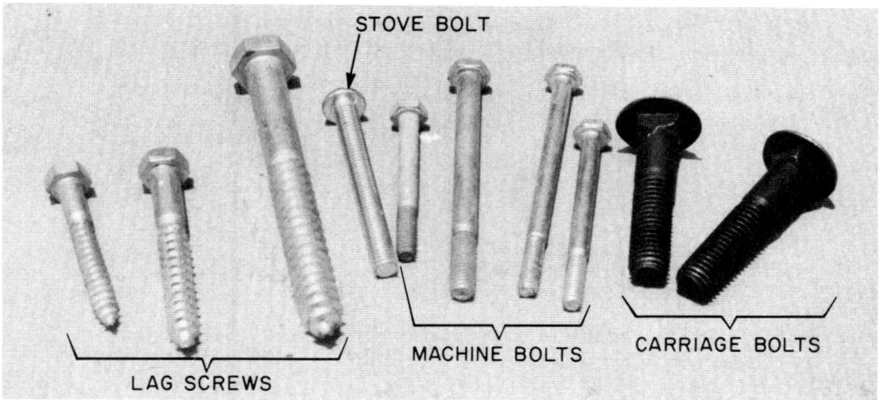

Fig. 17-19 Kinds of lags and bolts

pressure over a wider area. A little light oil is applied to the threads before the nut is turned on. Use a washer under the nut.

LAG SCREWS

Lag screws are similar to wood screws except they are larger and have a square head, see figure 17-19. They are designed to be turned with a wrench instead of a screwdriver.

Lag screws are sized by their diameter and length. Diameters range from 1/4 inch to 1 inch. Lengths range from 1 inch to 12 inches and up.

Lag screws are used when great holding power is needed to join heavy parts and where a bolt cannot be used. Shank and pilot holes are drilled to receive lag screws in the same manner as wood screws. A flat washer is placed under the head to prevent the head from digging into the wood as the lag screw is tightened down. A little wax is applied to the threads to turn the screw easier and to prevent twisting the head off.

ANCHORS

Anchors are used to fasten parts to solid masonry or to hollow walls of various materials. There are hundreds of different types available.

Solid Wall Anchors

Expansion shields take either lag screws or machine bolts, figure 17-20. The shield is a split sleeve of soft metal which is inserted in a hole drilled in the wall. As the fastener is threaded in, the shield expands tightly and securely in the drilled hole.

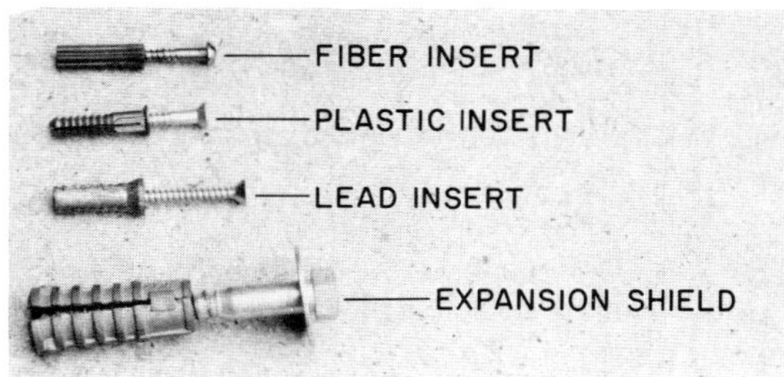

Fig. 17-20 Solid wall anchors

Expansion sheilds are usually used to fasten heavy objects. To fasten lighter objects to solid masonry, *lead, plastic,* and *fiber inserts (rawl plugs)* are commonly used, see figure 17-20. These inserts have an unthreaded hole into which a wood screw is driven. The threads of the screw cut into the soft material of the insert and cause it to expand and tighten in the drilled hole.

The *lead expansion screw anchor,* commonly called a *tamp-in,* takes a stove bolt and consists of two parts, figure 17-21. A lead sleeve slides over a cone-shaped piece containing threads for the bolt. The lead sleeve is driven over the cone-shaped piece with a setting tool. This expands the sleeve and holds it securely in the hole.

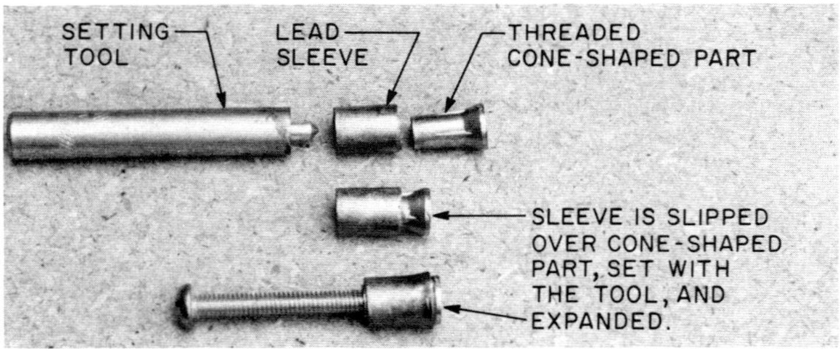

Fig. 17-21 The expansion screw anchor is commonly called a tamp-in.

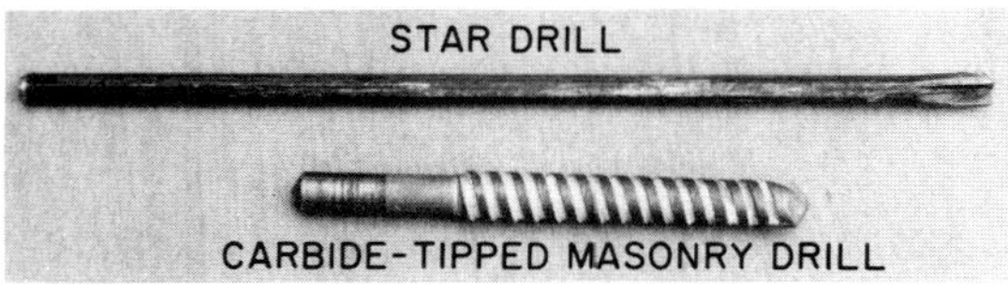

Fig. 17-22 Cutting tools used to drill holes in masonry.

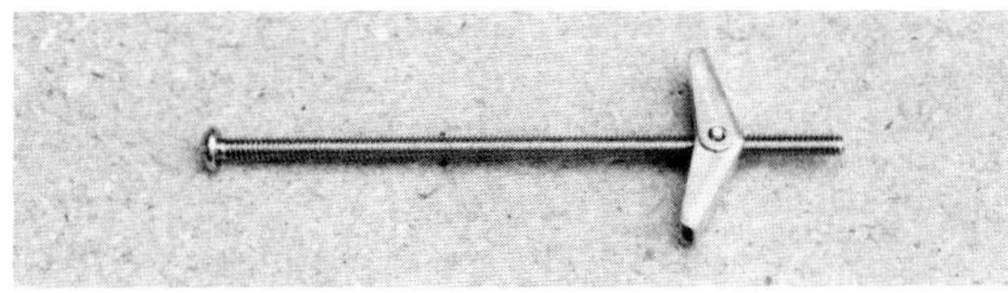

Fig. 17-23 Toggle bolt

To drill holes in solid masonry, use a carbide-tipped bit in an electric drill or use a star drill and hammer, figure 17-22.

Hollow-Wall Fasteners

Hollow-wall fasteners are commonly called toggle bolts, figure 17-23. They have either a wing head or a tumble head. The wing head is fitted with springs which cause the toggle to open as it passes through the hole. The tumble head falls into a vertical position when passed through the hole.

A hole must be drilled large enough for the toggle of the bolt to slip through. A disadvantage of using toggle bolts is that if removed, the toggle falls off inside the wall.

Star Expansion Anchors

Star expansion anchors are commonly called *molly screws,* figure 17-24. The anchor consists of a shield and a bolt. The unit is inserted in the drilled hole, and prongs on the shield are tapped into the surface of the wall. These prongs prevent the shield from turning while the anchor bolt is tightened. Tightening the bolt expands the shield against the inside of the wall. The bolt is then removed, inserted through the part to be attached, and screwed back into the shield. The advantage of using this type of anchor is that the fastened part can be removed and replaced without losing the shield inside the wall.

Star expansion anchors are available in many different sizes and are designed for different wall thicknesses. The wall thickness must be known in order to use the correct size anchor. The anchor will specify the hole size to drill.

MISCELLANEOUS FASTENERS

Hanger bolts have a wood screw thread on one end and a machine screw thread for a nut on the other end, figure 17-25. They are frequently used to secure table legs or other parts that must be removed occasionally. Since only the nut is unscrewed to remove the part, the wood threaded portion remains secured and does not loose its holding power.

To drive a hanger bolt, a hole of the correct size is predrilled. Two nuts are locked together on the bolt. The top nut is turned with a wrench. When the bolt is driven, the nuts are removed by holding

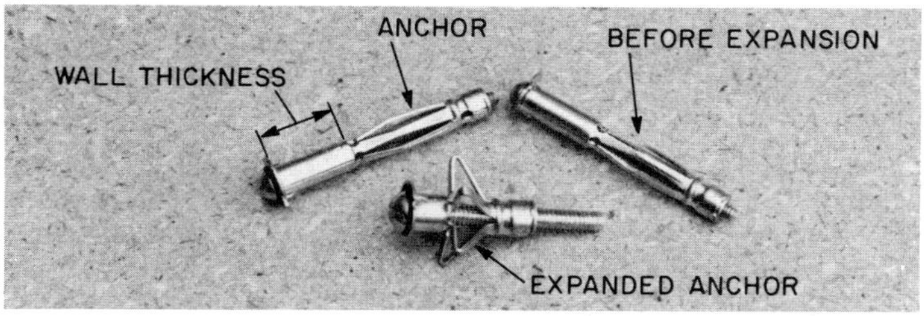

Fig. 17-24 The star expansion anchor is commonly called a molly screw.

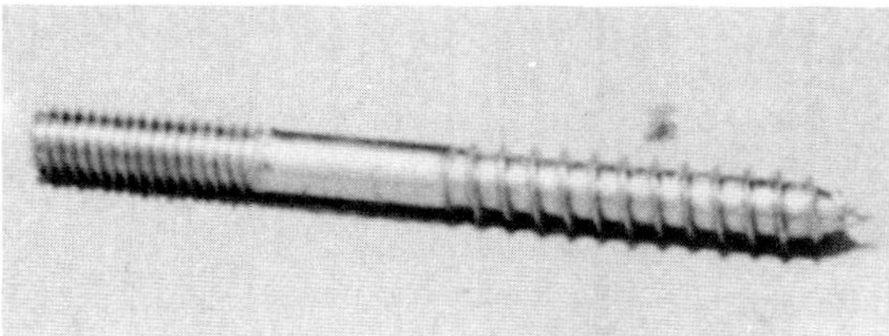

Fig. 17-25 Hanger bolt

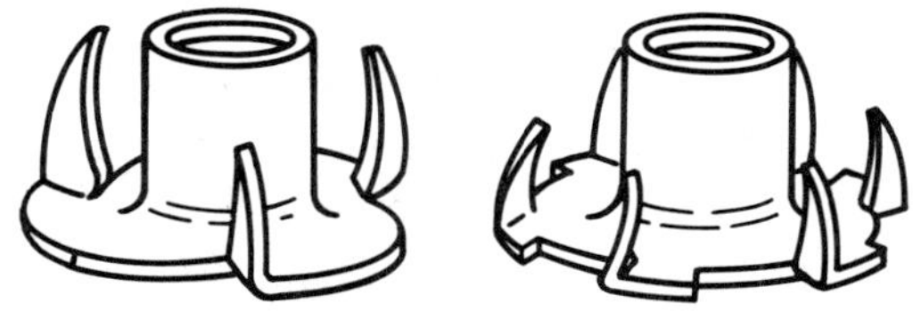

Fig. 17-26 Tee nuts *(Mohawk Finishing Products, Inc.)*

the bottom nut with a wrench while loosening the top nut.

Tee nuts are used when the head must be set flush with the surface, figure 17-26. A hole is bored into the wood just large enough for the center part of the nut. The nut is inserted into the hole and the prongs are driven into the wood to keep the nut from turning.

Corrugated fasteners are used in rough work to hold butt or miter joints together, figure 17-27. They are available in various widths. Other types of joint fasteners are used in a manner similar to corrugated fasteners. Designs differ according to the manufacturer. They are ordered by their trademark name such as Chevrons® or Skotch® fasteners.

Dowels are frequently used to strengthen joints, figure 17-28. Usual diameters range from 1/4 inch to 1 inch. Dowels are made from hardwood, usually birch. Grooves cut into the dowel provide a means for air to escape when the dowel is inserted in the hole. Grooves also provide greater holding power because they distribute the glue better. When selecting dowels, the diameter should be 1/3 to 1/2 the thickness of the material to be joined.

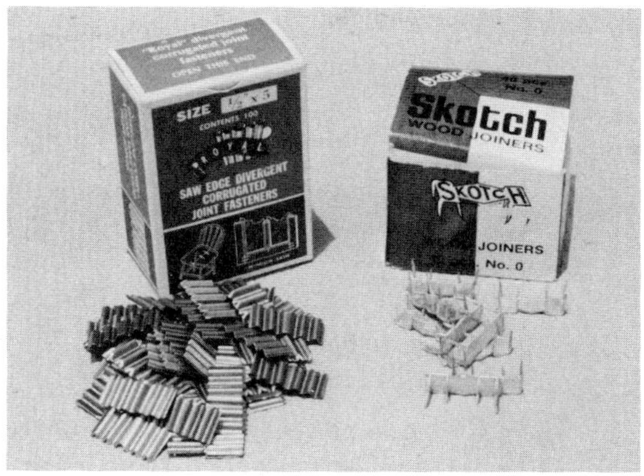

Fig. 17-27 Corrugated joint fasteners and wood joiners

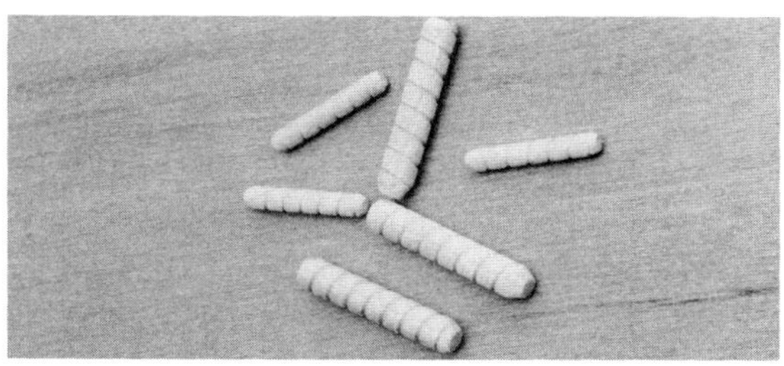

Fig. 17-28 Dowels

ACTIVITIES

1. Make display boards of the following, labeling each item displayed.
   a. nails and similar fasteners
   b. staples
   c. wood and sheet metal screws
   d. bolts and lag screws
   e. solid and hollow wall anchors
   f. miscellaneous fasteners such as hanger bolts, tee nuts, corrugated fasteners and other joint fasteners, and dowels
2. Visit a hardware store and study its inventory of fasteners.
3. Study the fasteners listed in manufacturers' catalogs.
4. Countersink and counterbore for wood screws. Make and trim plugs to cover counterbored screw holes.
5. Select a screwdriver and demonstrate how to use it for driving screws.
6. Show how to drive a nail properly. Set a finish nail.

UNIT REVIEW

Multiple Choice

1. A six-penny nail is
   a. 1 inch long.
   b. 1 1/2 inches long.
   c. 2 inches long.
   d. 2 1/2 inches long.
2. The diameter of a nail or screw is called its
   a. thickness.
   b. shank.
   c. gauge.
   d. round.
3. Brads are
   a. large box nails.
   b. small finishing nails.
   c. large finishing nails.
   d. a type of screw.
4. To help prevent the hammer from glancing off a nail,
   a. rub the hammer face with sandpaper.
   b. start the nail square with the work.
   c. choke up on the hammer handle.
   d. drill holes for the nail.
5. If a pilot hole is drilled too small in hardwood,
   a. the screw cannot be driven.
   b. the head of the screw may be twisted off.
   c. the slot will be ruined.
   d. the screw will not grip.
6. The first hole to be drilled when counterboring for screws is the
   a. shank hole.
   b. pilot hole.
   c. counterbored hole.
   d. any of the above.

7. When driving screws in counterbored holes that are to be plugged, care must be taken to
   a. use the proper size screwdriver.
   b. avoid scraping the sides of the counterbored hole with the screwdriver.
   c. use a screw with a head smaller than the counterbored hole.
   d. prevent the screwdriver from slipping off the screwhead.

8. The bolt with a square section just under its ovalhead is called a
   a. carriage bolt.
   b. machine bolt.
   c. hanger bolt.
   d. stove bolt.

9. Hollow-wall expansion anchors are commonly called
   a. rawl plugs.
   b. toggles.
   c. tamp-ins.
   d. molly screws.

10. When selecting dowels to strengthen joints, use those with a diameter of
    a. 1/4 the thickness of the material.
    b. twice the thickness of the material.
    c. the same thickness as the material.
    d. 1/3 to 1/2 the thickness of the material.

Identification

1. 

2. 

3. 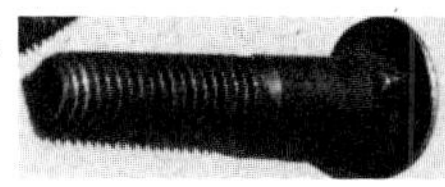

4. 

5. 

6. 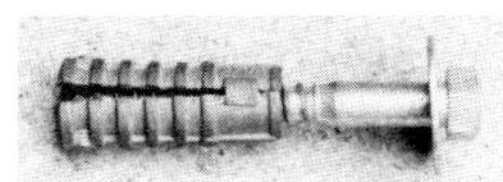

7. 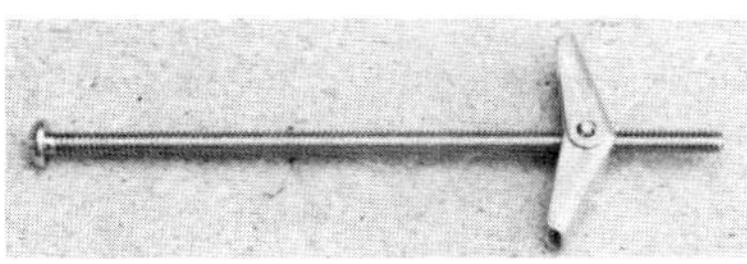

8. 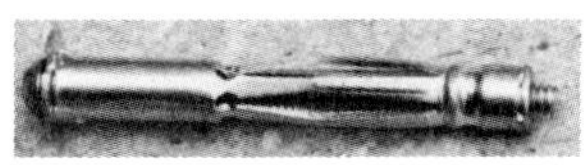

9. 

10. 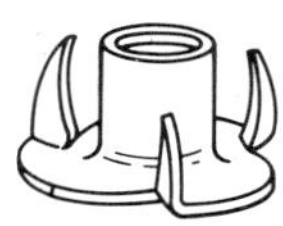

unit 18 Making Curved Pieces

OBJECTIVES

After studying this unit, the student will be able to:

- describe the methods of making curved pieces and list the advantages and disadvantages of each method.
- make a curved piece by band sawing from solid stock.
- make a curved piece by cutting saw kerfs.
- make a curved piece by assembling and gluing shaped parts.
- make a curved piece by steaming and clamping in a form.
- make a curved piece by gluing and clamping thin strips in a form (laminating).

MAKING CURVED PIECES

Often it is necessary to make curved parts for furniture and cabinets. Drawer fronts, doors, table rails, and chair backs and legs use curved pieces. Boat building uses curved pieces in the door and window frames, hatch covers, and other parts. Circular windows, like the rose windows in churches, require curved parts. Arches use curved parts for their frames, casings, doors, and windows. Bow windows which are popular in house construction, also require curved parts.

It is important for the woodworker to know how to produce curved pieces. Curved pieces can still be made even without special, expensive equipment and machinery.

CUTTING FROM SOLID STOCK

The quickest and easiest method of making curved pieces without special equipment is by cutting the curve out of solid stock with a band saw. Only curves with large radii can be cut with this method. Tight curves will produce weak, cross-grained pieces. Also, this method produces a lot of waste and the rough-sawn surface must be sanded smooth.

To make a curved piece on the band saw:

1. Trace the desired curve on the edge of the stock.
2. Cut to the line using a band saw, figure 18-1.
3. If the curve is on two sides of the stock, cut one side first.
4. Fasten the waste back onto the stock to provide a working surface to cut the curve on the other side. Keep fasteners for the waste clear of the cutting line.
5. Cut the second side the same as the first side, figure 18-2.

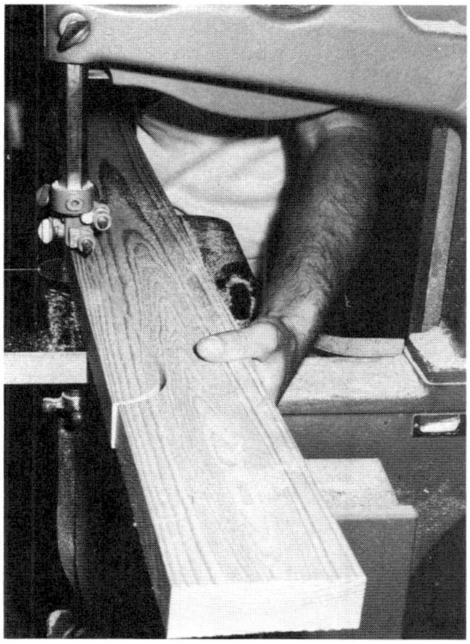

Fig. 18-1 Making a curved piece by band sawing from solid stock.

MAKING SAW KERFS

A curved piece can be made by cutting closely spaced saw kerfs on the back side of the piece, figure 18-3. Tight curves can be obtained by this method quickly and easily. However, kerfing the piece produces a weaker curve. These curves can only be used where the outside face of the curved piece shows. If the inside face or edges are exposed, they must be covered with some other material.

To make a curved piece by cutting kerfs, the correct spacing of the saw kerfs must be determined. This depends on the width of the saw cut and the curve to be made.

1. Measure off a distance equal to the radius of the curve on a piece of stock.
2. Make a saw cut at one mark to within 1/8 inch of the face.
3. Clamp the stock down on a bench and raise the end with the other mark until the saw cut closes, figure 18-4.
4. Measure the vertical distance at the mark. This is the spacing of the saw cuts.

Fig. 18-2 Curved lamp bases made on the band saw *(Woolums Mfg. Co.)*

Fig. 18-3 The rails of the round table are curved by saw kerfing. *(Woolums Mfg. Co.)*

Fig. 18-4 Determining the spacing of the saw kerf

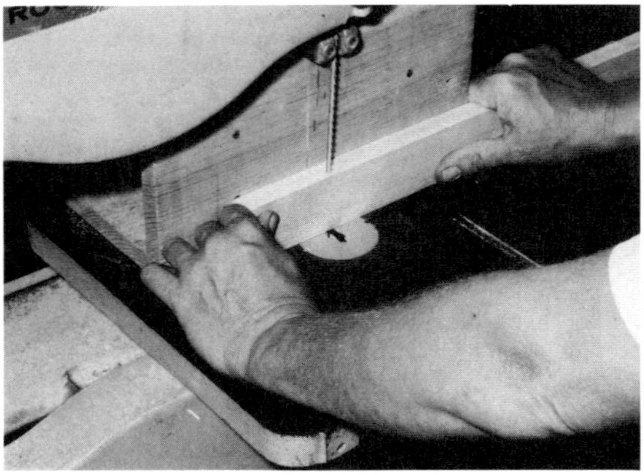

Fig. 18-5 Making a curved piece by cutting closely spaced saw kerfs on the back side of the stock

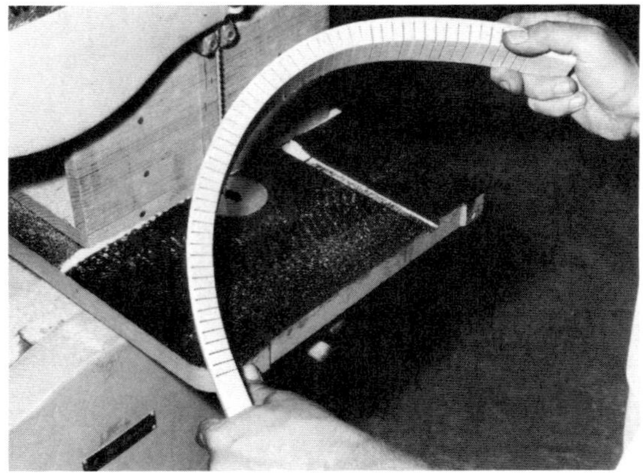

Fig. 18-6 Wet the board before bending.

5. Mark off the determined spacing of the cuts for the length of the curve.
6. Set up a block behind the band-saw blade so the cuts will come to within 1/8 inch from the face of the piece, figure 18-5.
7. Make the cuts on the band saw until the piece hits the block. Continue until all the cuts are made.
8. The piece is now ready for bending. Wet the board before attempting the bend. This will keep the piece from splitting or breaking, figure 18-6.

Spiral-type bends can also be made in this manner by tilting the band saw table. The saw cuts may also be made on a table saw or radial arm saw.

THE BUILT-UP METHOD

Curved pieces are also made by shaping small parts and building them up like bricks in a curved wall, figure 18-7. This makes a very strong piece, but exposes a great number of joints. For a good appearance, the individual pieces must be trued up and covered with a thin veneer. This method also takes quite a bit of time.

To make curved pieces by the built-up method:

1. Lay out the inside and outside radii on pieces of equal length.

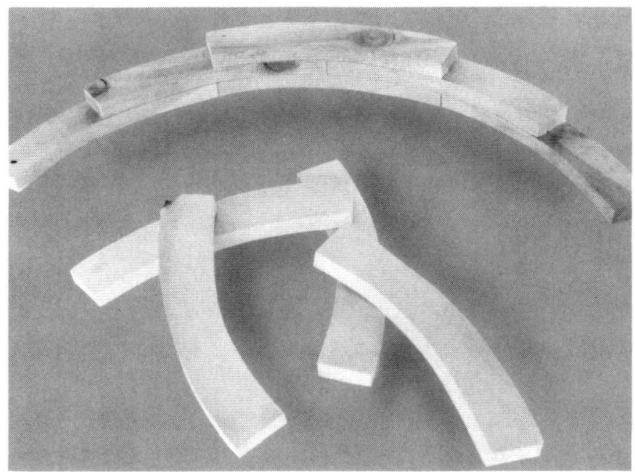

Fig. 18-7 Making a curved piece by the built-up method

2. Mark and cut equal and matching angles on each end of the pieces. This is done by laying out a full-size curve and dividing it into equal parts.
3. Cut the angles on both ends of the pieces. Then cut the inside and outside curves.
4. Build up the pieces by staggering the joints.
5. Glue and clamp the pieces together until the desired curve is obtained. Three or more layers are needed to provide strength.
6. After the glue has set, true up the edges of the pieces to the desired shape.

Fig. 18-8 Making a curved piece by steaming and bending with a shop-made rig

Fig. 18-9 Removing stock from the steam box. Notice the heavy gloves. *(Woolums Mfg. Co.)*

STEAMING SOLID WOOD

One of the most popular ways of making curved pieces is by steaming the wood and then clamping it in a curved form of the desired shape until it dries. This method needs steaming equipment, curved forms, and a variety of clamps. It produces strong, excellent-looking, curved pieces. In order to bend a piece by steaming and clamping, the form and clamps must be ready. Once the piece has been taken from the steaming chamber, it must be clamped in the form as soon as possible.

Oak, hickory, and ash can be bent by this method extremely well. Woods such as pine and spruce and other softwoods do not have good bending qualities. The wood selected for bending by this method must not be dried below a moisture content of 16 to 18 percent.

To steam and bend curves:

1. Put the pieces to be bent into a box and cover the opening with canvas, figure 18-8.
2. Run steam into the box. Steam the pieces for one hour for each one inch of thickness.
3. Remove the pieces from the steam box, figure 18-9, and clamp them immediately into the prepared form, figure 18-10.

> **Caution:** Since the pieces will be extremely hot, handle them carefully with thick, heavy gloves. Also, keep clear of the steam when opening the box.

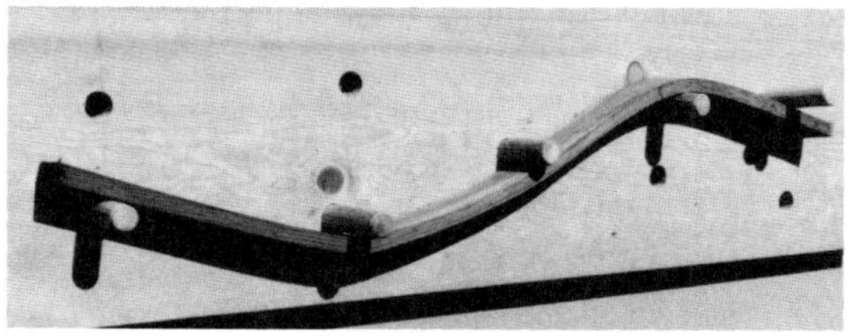

Fig. 18-10 Clamp the pieces immediately after steaming.

4. Let the pieces dry thoroughly before removing them from the form. When dry, they will retain the approximate shape of the form, figure 18-11.

LAMINATING THIN STRIPS

Another method of making curved pieces is by building up layers of thin strips of wood. They are glued and clamped together in a form until the glue sets. This is called *laminating.* When removed from the form, the piece keeps its shape.

Laminating produces strong, neat-appearing curved pieces. Very small parts for bows, skis, chairs, and tables are easily made in this manner. Also, extremely large pieces are laminated to make curved roof beams for churches, schools, and commercial buildings.

Pound for pound, laminated beams are said to be stronger and more resistant to fire than steel

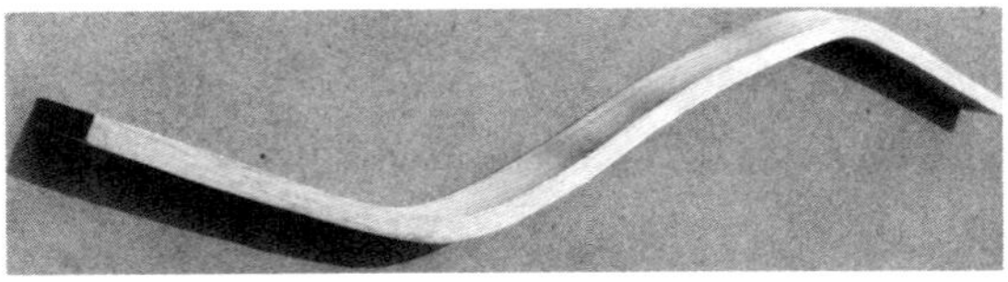

Fig. 18-11 When dry, the steamed and bent piece will retain its shape.

beams. However, when removed from the form, these pieces need to be trued and smoothed. Also the glue joints between the strips are visible, but in many cases this is not objectionable.

To make curved pieces by laminating:

1. Have the form and clamps ready.
2. Select wood with good bending quality such as oak, hickory, ash, teak, birch, maple, mahogany, walnut, or cherry.
3. Cut the wood into strips of 1/28 to 1/16 inch thick. Allow extra length on the strips for final trimming.
4. Apply glue to both surfaces of each strip. Do not use a quick-setting glue if time is needed to clamp the strips into the form. The glue may set before the pieces are clamped into position.

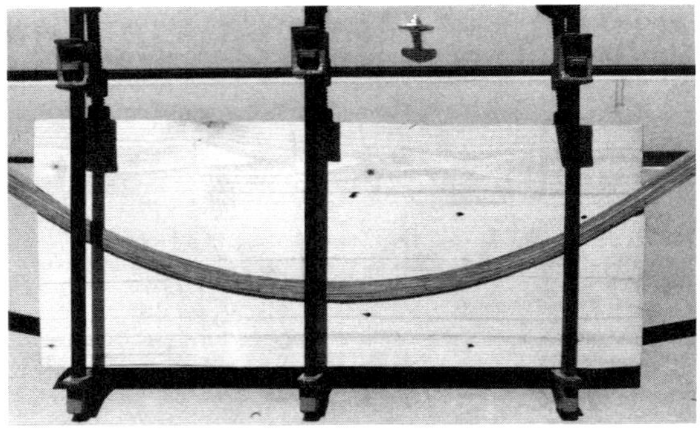

Fig. 18-12 Making a curved piece by laminating.

5. Clamp the pieces together in the form. Apply wax paper between the form and the strips to keep them from sticking to the form, figure 18-12.
6. Let the glue set for the recommended time.
7. Remove the piece from the form. True up and smooth the edges. Cut the piece to the desired length.

ACTIVITIES

1. Make a chart listing woods in order of their bending quality.
2. Using the band saw, make a curved base for a table lamp.
3. Make a 2-inch wide rail for a round table using the saw kerfing method.
4. Make a curved casing for an arched doorway using the built-up method.
5. Make a form to shape a curved piece by steaming and bending. Make two duplicate pieces.
6. Make a 7-ply, curved veneer strip by laminating. Make your own form or use one that is available in the classroom.

UNIT REVIEW

Multiple Choice

1. A disadvantage of making curved pieces by cutting from solid stock is that
   a. it takes too much time.
   b. many joints are exposed.
   c. there is too much waste.
   d. it produces a weak piece.

2. An advantage of making curved pieces by cutting from solid stock is that
   a. it is a quick method.
   b. there is little waste.
   c. no smoothing is required.
   d. tight curves can be made.

3. When making a curved piece by kerfing the back of the stock, it is necessary to determine the
   a. width of the saw cut.
   b. spacing of the saw cuts.
   c. curve to be made.
   d. length of the saw cut.

4. An advantage of using the saw kerfing method to produce curved pieces is that
   a. tight curves can be made.
   b. no joints are exposed.
   c. the piece is not weak.
   d. all four sides can be exposed.

5. Spiral-type bends can be made on the band saw by the saw kerfing method by
   a. spacing the saw cuts closer.
   b. clamping the stock in a form.
   c. tilting the saw table.
   d. twisting and clamping the stock.

6. A kind of wood that has excellent bending qualities is
   a. pine.
   b. hickory.
   c. spruce.
   d. redwood.

7. In order to bend wood, it must be steamed for
   a. one hour per one inch of thickness.
   b. one hour per half inch of thickness.
   c. one half hour per one inch of thickness.
   d. two hours per one inch of thickness.

8. A disadvantage of bending wood by steaming is that
   a. the piece becomes weaker.
   b. many joints are exposed.
   c. a lot of equipment is needed.
   d. there is a lot of waste.

9. Bending by gluing and clamping layers of thin strips of wood in a form is called
   a. building up.
   b. steaming.
   c. saturating.
   d. laminating.

10. A kind of wood that is not suitable for bending is
   a. ash.
   b. hickory.
   c. oak.
   d. pine.

QUESTIONS

1. Why may only large curves be cut on a band saw?
2. What is a disadvantage of making curved pieces by saw kerfing?
3. If the inside, saw-kerfed face of a curved piece must be shown, what should be done?
4. How deep are saw kerfs made?
5. What effect does wetting the board before bending have when making curves by the saw kerfing method?

6. In the built-up method, how many layers are needed to provide strength to the piece?
7. In the built-up method, what must be done to the curved piece after the glue has set?
8. What should be the moisture content of wood that is to be steamed and bent?
9. Why is a quick-setting glue not suitable when laminating?
10. After a laminated strip has been removed from the form, what must be done to it?

section 4

Cabinetmaking

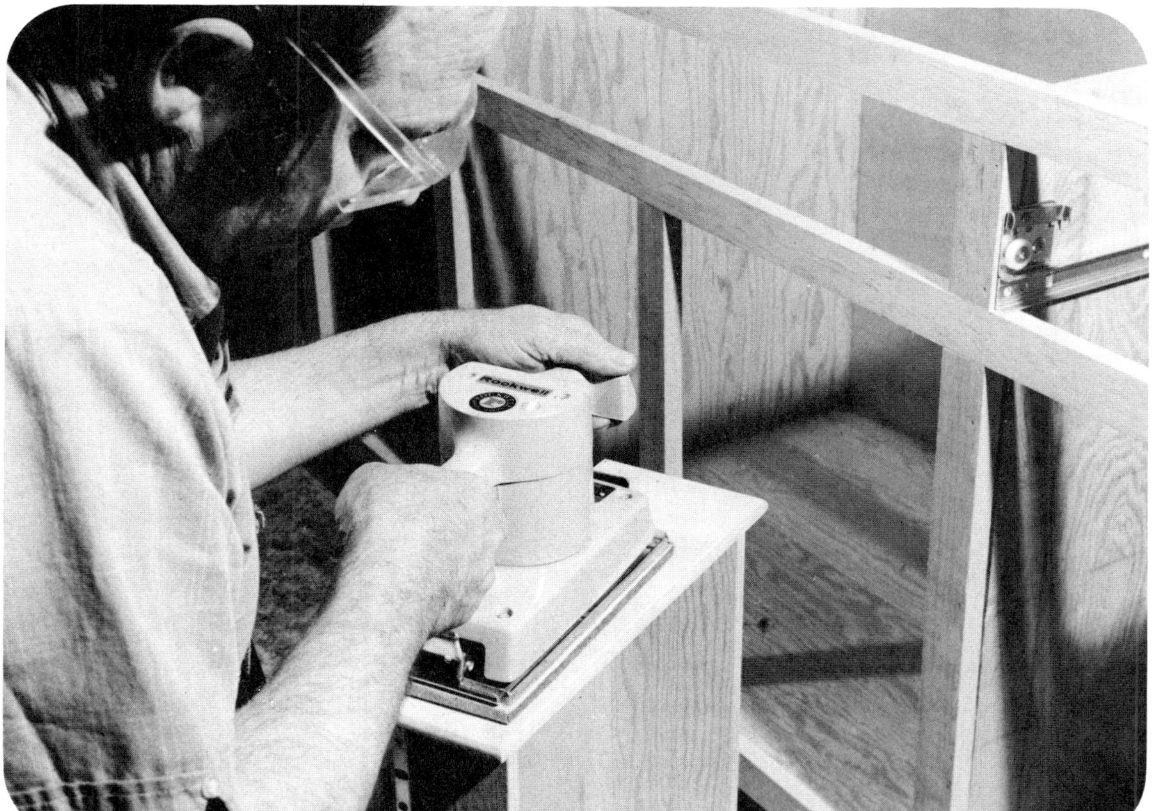

unit 19
Design

OBJECTIVES

After studying this unit, the student will be able to:

- explain how to develop a good cabinet design.
- describe several popular cabinet and furniture styles.
- make sketches to develop a product that meets specifications and conforms to the principles of good design.
- make a working drawing from the developed sketch.

DESIGNING A PRODUCT

Cabinetmakers are often provided with complete working drawings of the product to be constructed. These are drawn by professional architects, designers, and drafters. The drawings usually contain information to build the product, such as size, style, material, construction, and finish.

Sometimes the cabinetmaker must design, sketch, and make a working drawing from the customer's verbal specifications or from simple line drawings. This is especially true in small custom shops. In order to make a working drawing from the incomplete instructions supplied by the customer, the cabinetmaker must:

- know the principles of good cabinet design.
- be familiar with popular cabinet styles.
- develop a sketch that meets the customer's specifications and conforms to good design principles.
- make a working drawing from the sketch in order to build the product.

The cabinetmaker should obtain the customer's approval on completion of the developed sketch, and again when the working drawings are finished and before any work is started on the product.

DESIGN CONSIDERATIONS

There are many things to consider when designing a product. The end result must, of course, satisfy the customer. To do this the designer must consider the purpose, strength, size, shape, proportion, appearance, time, and cost of the product.

The time spent in designing is well worth the effort. This helps avoid mistakes and saves time in the long run.

Purpose

One of the most important considerations in designing a product is its purpose. A product's purpose may be the deciding factor in determining the design. For instance, bookcases must be the proper size and strength to hold the desired quantity and kind of books. A game table does not have to be as strong as a dining table. A door must be constructed with stronger joints than a picture frame. Outdoor furniture must be made of wood that resists decay.

Strength

Usually cabinets and furniture are made only strong enough to fulfill their purpose. The strength required of an object may determine such things as the type of joint, the size, and the kind of wood. Table legs, for instance, should be attached to the table rails with strong joints. In many cases these legs are detachable. A chair must not collapse when a person sits on it.

It is often better to use strong woods like oak, ash, and maple to give the strength required by a product. Oversized pieces of soft wood like pine, spruce, and poplar may also be used. However, using oversize parts gives a massive, awkward appearance besides using unnecessary materials.

Size, Shape, Proportion

Some furniture and cabinets must be built to standard sizes in order to serve their purpose. A dining table that is too low will not serve its purpose. A kitchen cabinet countertop that is too narrow will not accommodate a sink. Figure 19-1 shows some standard sizes for furniture and cabinets.

Cabinets are made of different shapes in order to harmonize or balance with existing cabinets. Some of the more popular shapes are square, rectangular, round, elliptical (oval), octagonal, and hexagonal.

In addition to size and shape, the designer must also consider proportion. The *proportion* of a cabinet is the relationship between its dimensions. This is its width, height, and length.

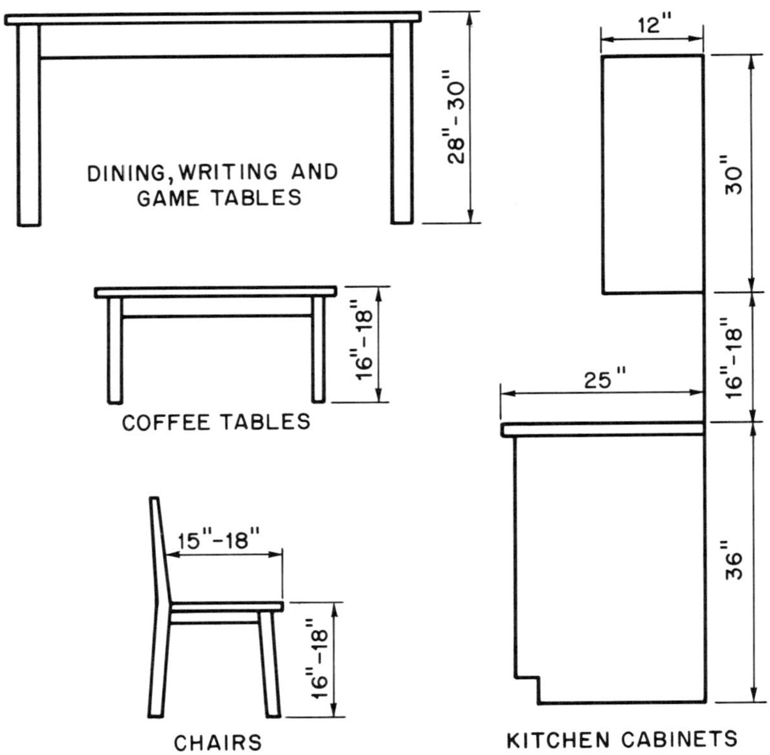

Fig. 19-1 Average sizes of furniture and cabinets

Some proportions are more pleasing to the eye than others. A rectangle considered to have the most pleasing proportion of length to width is called the *golden mean rectangle.* The sides are in a proportion of approximately 5 to 8. Figure 19-2 shows how to lay out the golden mean rectangle.

Appearance

The appearance of a cabinet may be largely due to its purpose, location, and finish. If the object is to be painted, a less expensive material can be used. If it is stained with a clear finish, a better quality material should be used. A rustic or primitive appearance calls for woods like pecky cypress or knotty pine. Exotic woods like mahogany or walnut would not be used.

Kitchen cabinets are easier to clean and look better if a plastic laminate is used over an inexpensive core material. However, the customer may prefer the warmth and beauty of natural wood. The cabinetmaker then has to search for the most readily available and least expensive hardwood plywood and matching solid wood, perhaps selecting birch for the customer's approval.

The appearance of kitchen cabinet doors may be changed by cutting shapes in them, by applying molding to the surfaces, or by using paneled doors instead of solid doors, figure 19-3. The edges of doors may be lipped or cut square-edge according to the appearance desired.

Color also plays an important part in appearance. Green and blue shades are cool colors, while browns and yellows suggest warmth. Dark woods and stains are often associated with certain cabinet styles. Light stains and natural finishes are usually applied to more modern pieces.

Time and Cost

One of the important considerations in designing cabinetwork is the time required to construct it. Time may affect the type of joint, kind of material and fasteners, method of construction, and kind of finish.

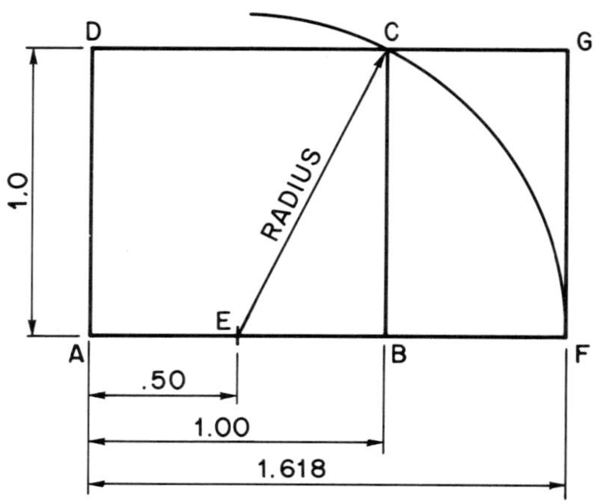

1. LAY OUT SQUARE ABCD ANY CONVENIENT SIZE.
2. FIND THE MIDPOINT OF ONE SIDE AB AT E.
3. WITH EC AS RADIUS AND E AS CENTER, STRIKE AN ARC AS SHOWN.
4. EXTEND SIDE AB TO F.
5. CONSTRUCT FG AT RIGHT ANGLES TO AF.
6. EXTEND DC TO G.
7. RECTANGLE AFGD HAS SIDES IN RATIO OF 1.00 TO 1.618 OR APPROXIMATELY 5 TO 8.

Fig. 19-2 How to lay out the golden mean rectangle

Fig. 19-3 Different designs change the appearance of cabinet doors.

Fig. 19-4 Early American furniture has simple, curved lines.

Fig. 19-5 This traditional-style table has cabriole legs.

Fig. 19-6 The graceful curves and elaborate carvings of this dresser are characteristics of French provincial style.

In order to save time, a butt joint reinforced by power-driven corrugated fasteners may be used instead of the more complicated and time-consuming mortise-and-tenon joint. For ease in cutting and sanding, a wood with better workability, like poplar, may be used instead of maple. To save time, nails may be used instead of screws, or a quick-drying sprayed finish may be selected.

The quality of the finished product is also a factor in designing. High-quality products take more time and cost more to construct. The cabinetmaker must decide the minimum quality level the customer will accept and produce it at minimum cost.

CABINET STYLES

A knowledge of the basic styles of cabinetwork and furniture helps the cabinetmaker design a product that satisfies the customer. The six basic styles that are popular today are early American, traditional, French provincial, Italian provincial, Spanish Mediterranean, and modern. Modern styling includes Danish modern, oriental, and contemporary designs.

Early American

Early American styling is characterized by its sturdy, bulky appearance, figure 19-4. It is a simple style that uses many turnings and very little ornate moldings or carved work. It also displays black wrought iron or brass hardware. Fasteners are usually plugged for a decorative appearance. Colors range from light to dark brown, and the finish has a hand-rubbed, satin appearance. The woods used are those that were readily available during the colonial period, namely pine, maple, birch, and oak.

Traditional

Traditional styling, figure 19-5, includes designs made during the eighteenth century in England and the United States by Chippendale, Hepplewhite, Sheraton, Phyffe, the Adams brothers, and others. Traditional cabinetry may sometimes be difficult to identify, but it has a few distinctive features.

Exquisitely matched veneers is one distinctive feature. Leg shapes, especially the cabriole and the Queen Anne, are tapered and slender and almost never turned. Many carvings are used around mirrors, on doors, and on drawer fronts. Most chairs have open backs. Traditional furniture is usually made of mahogany which is stained brown to dark red and finished to a high gloss.

French Provincial

French provincial is easily identified because of the graceful use of curves, figure 19-6. It appears dainty and fragile. Most pieces have slender cabriole legs. There are few turnings, but elaborate and delicate carvings are used abundantly. A popular finish is white paint trimmed in gold and distressed with black specks. Curved moldings are also used extensively on doors and drawer fronts.

Italian Provincial

Italian-styled furniture's distinctive characteristic is the use of straight, fluted lines, figure 19-7. Many chairs, tables, and cabinets have straight, tapered legs that may be fluted or inlaid. Almost all furniture in this style contains fluting which may run in any direction. Veneers are matched in a number of ways and used extensively. This style has a balanced proportion based on the golden mean rectangle. It is heavier and seems more bulky than

Fig. 19-7 This Italian provincial coffee table is distinguished by its straight, fluted lines.

Fig. 19-8 The massive, geometric shapes of Spanish Mediterranean styling usually have deep carvings.

French provincial. Hardware of brass or bronze is usually quite prominent. Most pieces are finished in light tan to brown.

Spanish Mediterranean

Spanish Mediterranean styling uses massive, geometric shapes, primarily the horseshoe and the crescent. Deep carvings are used on doors and drawer fronts, but no carvings of plants or animals are found in this style, figure 19-8. The color is usually dark and hardware is ornate. Ceramics and glass are used on tabletops, and cane webbing is frequently used on chair backs.

Danish Modern

Danish modern furniture is easily identified by its free-flowing, slender, graceful look. Danish modern pieces are usually made of teak. It has simple, straight lines with rounded edges and corners, figure 19-9. Usually no finish is applied to the wood. It is hand-rubbed to bring out the natural oils to give a satin finish. The pieces seem to be very fragile, but actually are sturdy and well constructed with doweled joints.

Oriental

Oriental-style furniture is characterized by a high-gloss black background with painted accents of gold or red and oriental figures and scenes, figure 19-10. Balance and proportion is an important feature. Painted borders of intricate design and shape are also common.

Fig. 19-9 Danish modern style has a fragile appearance but is actually quite sturdy.

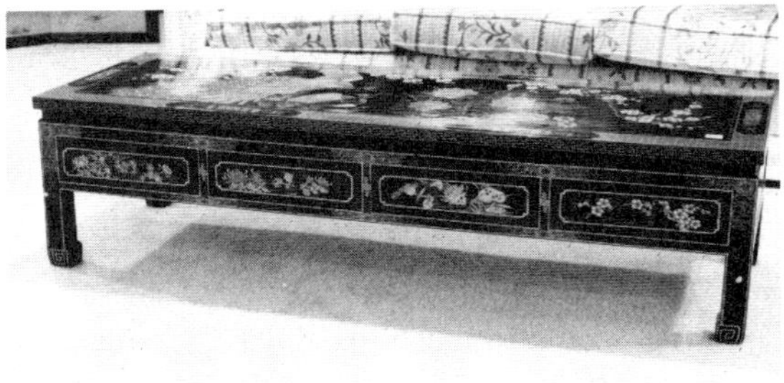

Fig. 19-10 This oriental-style coffee table has a high-gloss black background with painted scenes in red and gold.

Fig. 19-11 The straight lines and box-line appearance identifies this cabinet as contemporary.

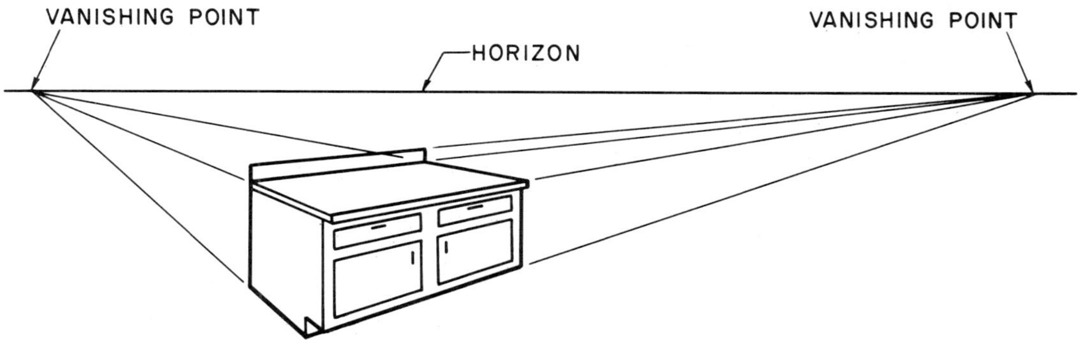

Fig. 19-12 Two-point perspective drawing

Contemporary

Contemporary furniture combines the characteristics of many styles. Its appearance is often box-like with straight lines, hidden handles and door hinges, and a built-in look, figure 19-11. Laminated plastics are often used on doors, cabinet faces, and countertops. Contemporary cabinetwork is made from a wide variety of woods, plastics, and metals. Finishes range from blonde to walnut.

KINDS OF DRAWINGS

There are several kinds of drawings used to describe an object in sufficient detail so that it can be built. *Pictorial drawings* show the general appearance of the object. Pictorial drawings include perspective, isometric, and exploded cabinet drawings. Orthographic drawings usually show the object from three views.

Perspective Drawings

A *perspective drawing,* like a photograph, shows the object like the eye sees it, figure 19-12. The lines representing the sides of an object diminish in size as they converge toward vanishing points on a line called the *horizon.*

Isometric Drawings

In an *isometric drawing,* the vertical lines of an object are drawn vertically. The horizontal lines are drawn at a 30-degree angle from the horizontal, figure 19-13. All lines are drawn to actual scale and do not diminish in size or converge as in perspective drawings.

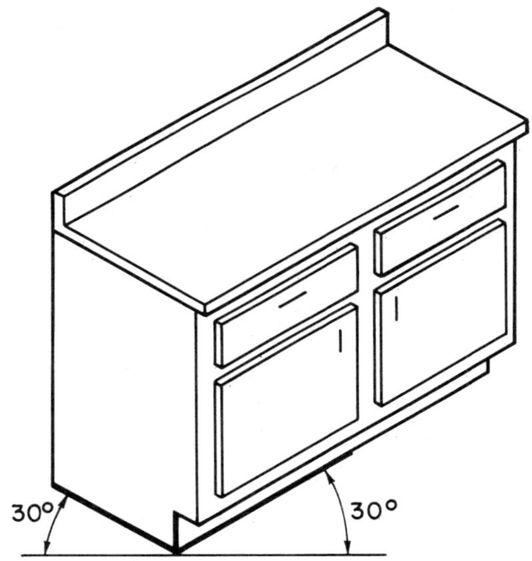

Fig. 19-13 Isometric drawing

Cabinet Drawings

In a *cabinet drawing,* the front view is drawn in exact scale. The horizontal lines of the object are drawn horizontally, and the vertical lines are drawn vertically. The lines representing the sides of the object are usually drawn at a 45-degree angle from the horizontal with dimensions reduced to one-third or one-half, figure 19-14. Cabinet drawings, therefore, appear more like the object than isometric drawings.

Exploded Drawings

Pictorial drawings of cabinets are often made as exploded drawings. *Exploded drawings* separate the individual parts of the cabinet. The parts are placed in their relative positions with lines showing the direction in which they are joined. This type of drawing is used to show the construction and joinery of the cabinet more clearly, figure 19-15.

Orthographic Projection

An *orthographic projection* or *multiview drawing* is usually a three-view drawing, although more or less views are drawn as necessary. Usually the object is drawn as viewed from the top, front,

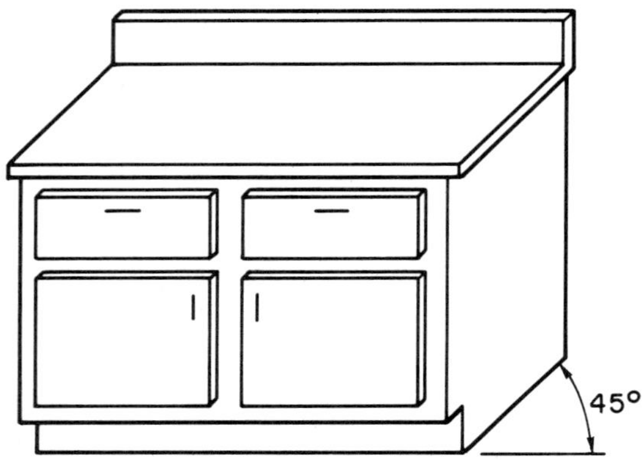

Fig. 19-14 Cabinet drawing

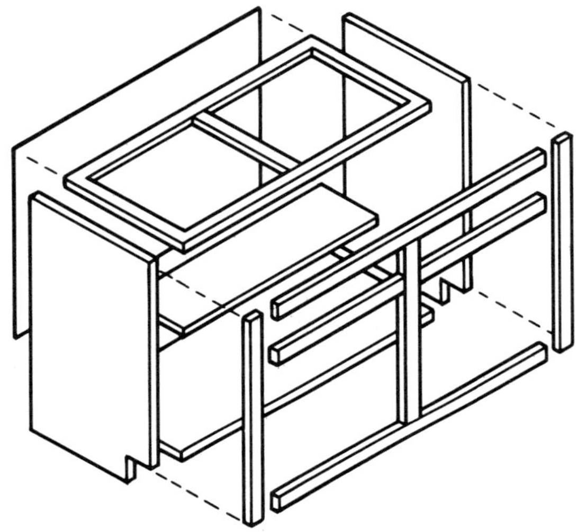

Fig. 19-15 Exploded drawing

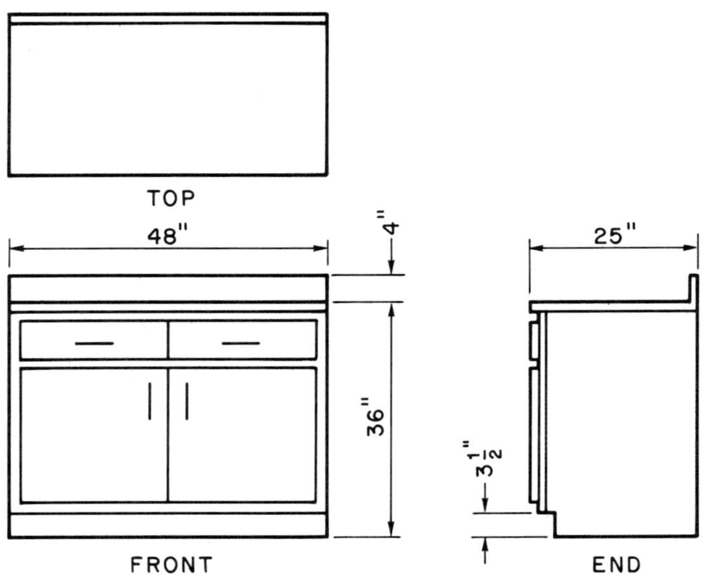

Fig. 19-16 Orthographic projection (multiview drawing)

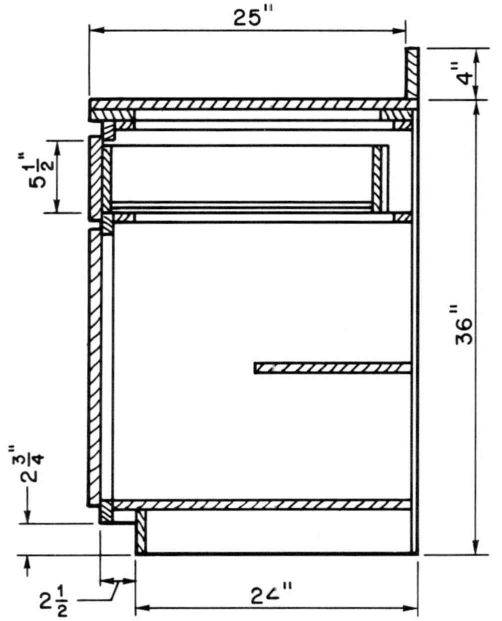

Fig. 19-17 Typical section of a kitchen base cabinet

and right end, figure 19-16. The lines are drawn exactly to scale and dimensioned as necessary.

Section Drawings

A *section drawing* is obtained by making an imaginary cut through the object and viewing the exposed parts, figure 19-17. A section view is used when the interior contains a number of details that cannot otherwise be shown.

Detail Drawings

Details are drawings that show parts of an object at a larger scale, figure 19-18. They present more information about those parts. Details may be drawn to show the construction of doors and drawers in a kitchen cabinet.

USING THE SCALE RULE

Large objects cannot be drawn full size on paper. They must be *scaled down* or drawn to scale using a smaller unit of measure. Usually an

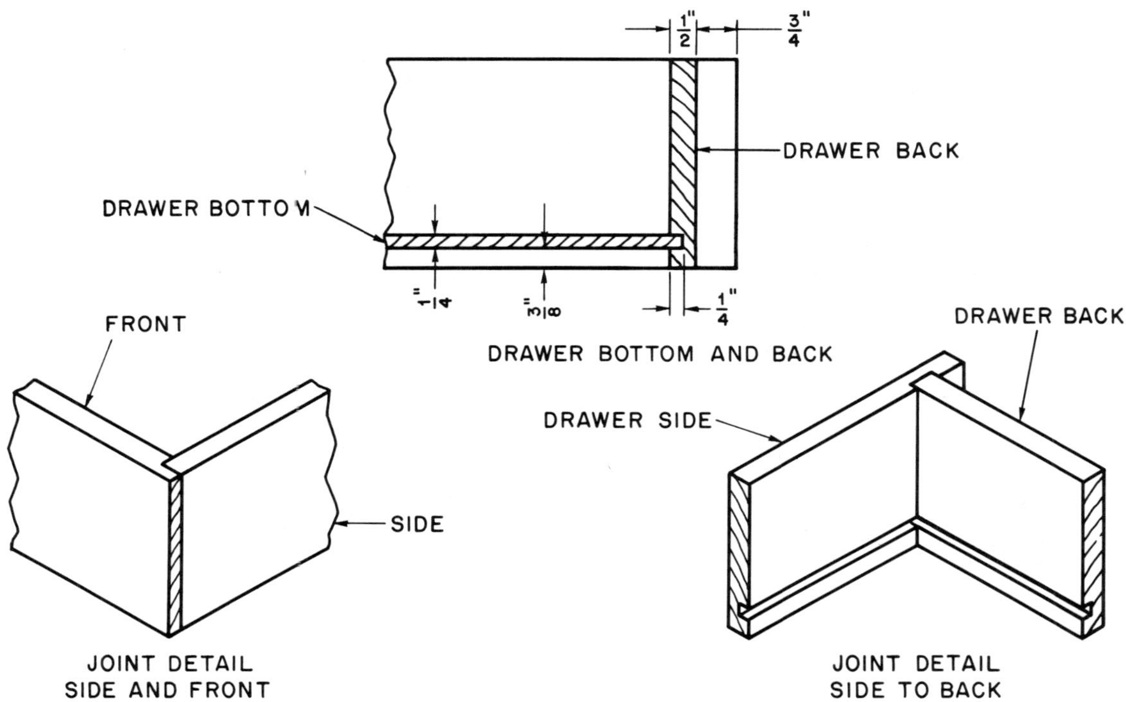

Fig. 19-18 Drawer details

architect's scale is used which contains a number of smaller units of measure, figure 19-19.

It first must be decided what scale to use. Use as large a unit of measure as possible that will still fit the number of drawings required on the paper. More than one sheet of paper is often required to present all the information necessary to build the object. Usually 3/4" = 1'-0" or 1 1/2" = 1'-0" are good scales to use for cabinet drawings. Details are drawn using a scale of 3" = 1'-0" or larger, even full size, if possible.

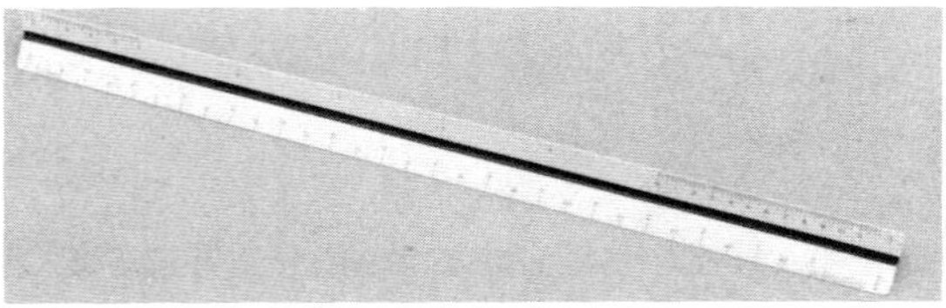

Fig. 19-19 Triangular architect's scale

SKETCHING

A *sketch* is a freehand drawing usually made with only a pencil and a scale rule. Sketching to scale keeps the drawing in proportion to better visualize the shape of the finished product.

Sketching lets the designer experiment with the elements of design. It is the preliminary step from which a working drawing is made.

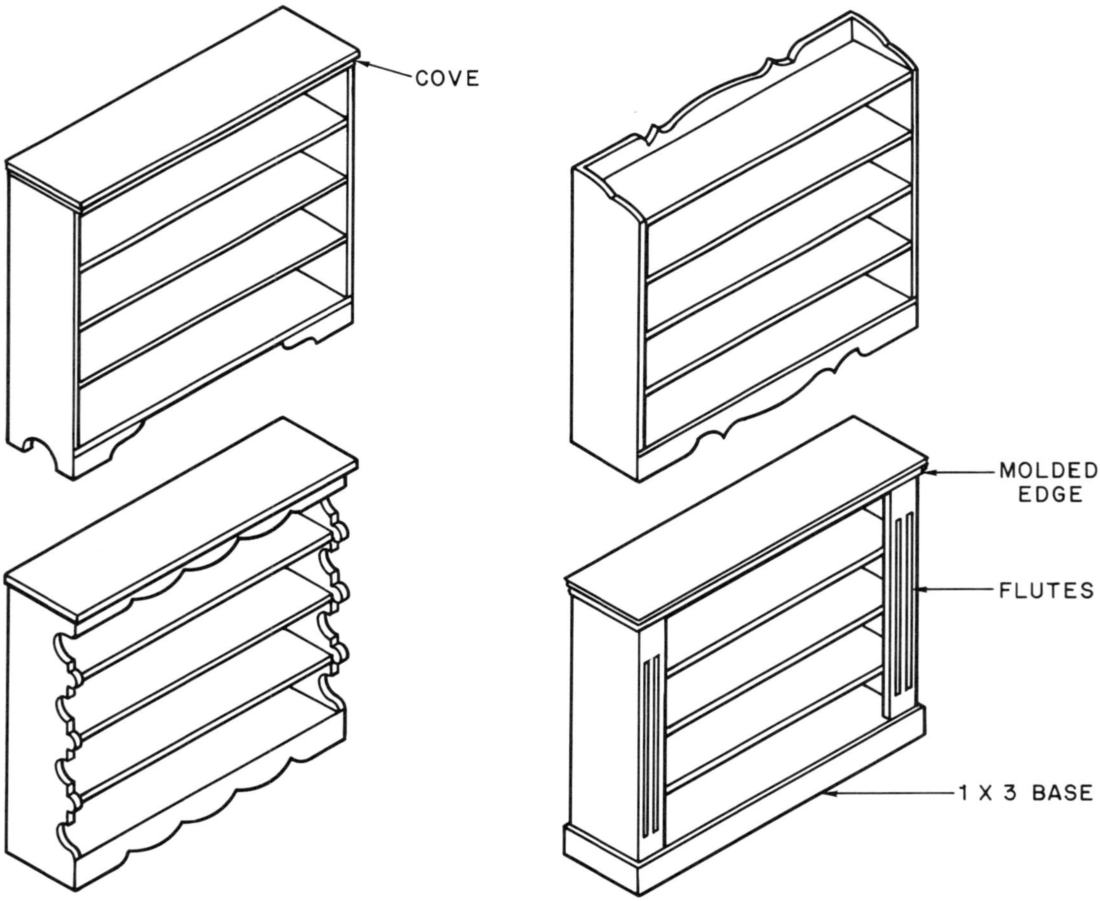

Fig. 19-20 Making several drawings of a bookcase to experiment with design

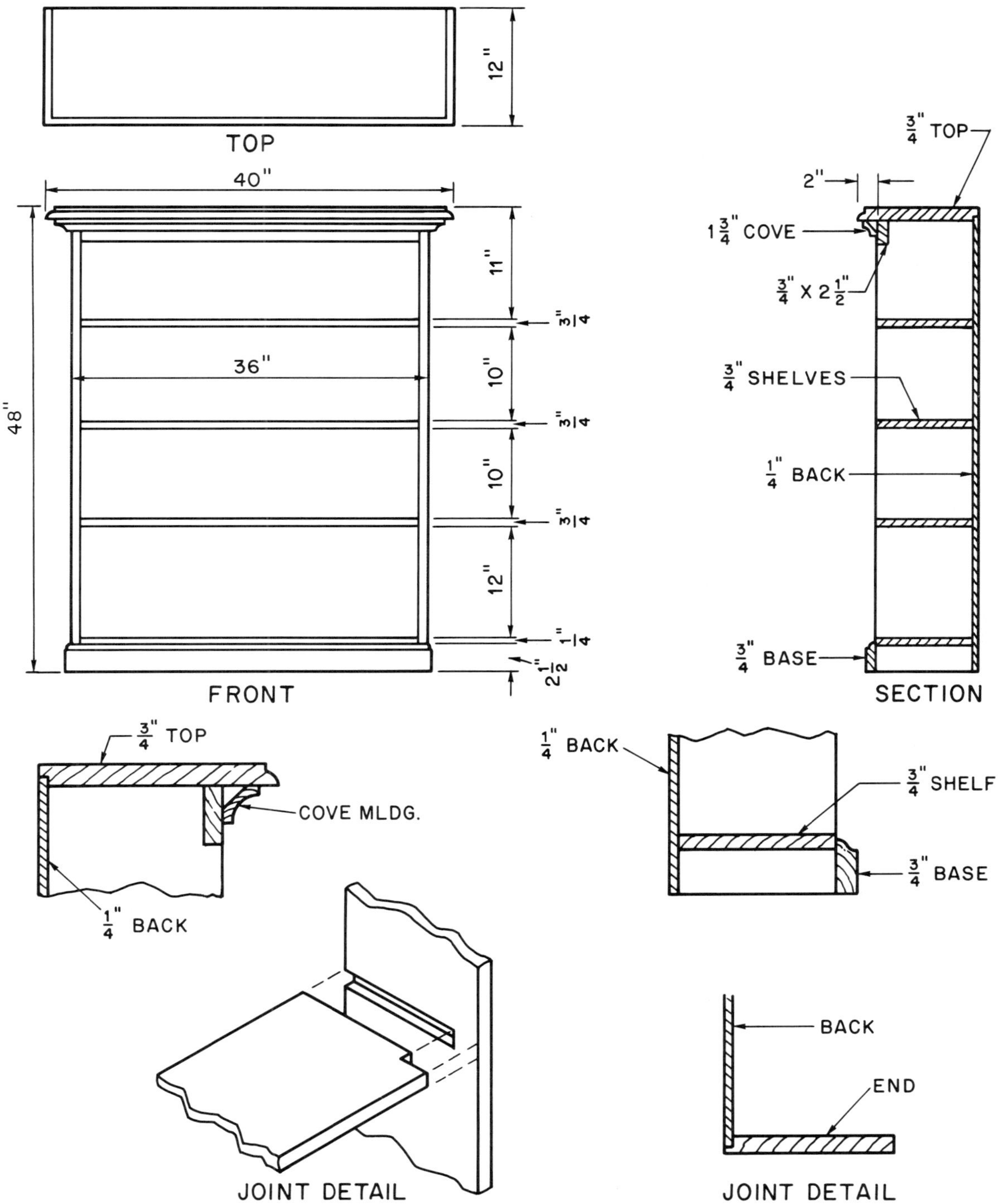

Fig. 19-21 A working drawing and details of the bookcase

Making a Sketch

Usually the first step in designing and building an object is to make several drawings to experiment with design, size, and proportion. For example, a simple bookcase may require several sketches before its design and size are finally determined, figure 19-20.

After the design and size have been determined, the kind of wood, finish, and sizes of the pieces are selected. Details of construction, such as joints, style of curved pieces, location and type of fasteners, are then added to the sketch. Different kinds of drawings may be used according to which best illustrates the information.

WORKING DRAWINGS

When enough sketches have been made, the ideas developed are put into the form of working drawings. A *working drawing* is one made with drawing tools, such as the T square, triangles, and compass. It is drawn to exact scale, figure 19-21. It provides most of the information required to build the object. Some features, like the type of joint, glue, or fasteners, are left to the discretion of the cabinetmaker however.

When developing working drawings, drafting standards are followed closely. The drawing should be centered on the page. Lines should be standard weights. The drawing should be adequately dimensioned and include all necessary notes. Lettering should be neat and legible.

SUMMARY

Use the following steps in designing and building cabinets:

1. Determine the use and purpose of the piece to be made.
2. Consider the desired style and appearance of the object and select the materials and finish.
3. Consider the strength, size, shape, and proportion of the object.
4. Study the time required and the cost of the finished product.
5. Make several sketches and present the final sketch to the customer for approval.
6. Make a working drawing and present it to the customer for approval before starting the actual construction.

ACTIVITIES

1. Select a cabinet or piece of furniture from a magazine. Point out the features of good or poor design.
2. Select a cabinet or piece of furniture in the classroom. Point out if the piece was designed properly for use, strength, appearance, shape, size, proportion, time, and cost.
3. Make a notebook of furniture and cabinet styles. Describe the distinguishing features of each style.
4. Make a perspective drawing of an existing cabinet.
5. From the simple specifications provided by the instructor, draw several sketches to create a design of a cabinet that meets the specifications.
6. Make a working drawing from a sketch selected from #5.

UNIT REVIEW

Multiple Choice

1. One of the most important considerations to be taken when designing a cabinet is
   a. the size of the cabinet.
   b. the color of the cabinet.
   c. the purpose of the cabinet.
   d. the shape of the cabinet.

2. The average height of dining tables is
   a. 26 inches.
   b. 28 inches.
   c. 30 inches.
   d. 32 inches.
3. The average height of a chair is
   a. 12 to 14 inches.
   b. 14 to 16 inches.
   c. 16 to 18 inches.
   d. 18 to 20 inches.
4. What is the approximate ratio of the sides of the golden mean rectangle?
   a. 2 to 4
   b. 4 to 6
   c. 5 to 8
   d. 8 to 10
5. The most distinguishing feature of a French provincial cabinet is its
   a. carved geometric shapes.
   b. use of graceful curves.
   c. bulky, massive look.
   d. use of turned legs.
6. What style cabinet makes use of horseshoe and crescent-shaped carved pieces?
   a. Early American
   b. Danish modern
   c. Italian provincial
   d. Spanish Meditteranean
7. An abundance of fluted pieces appears in what style cabinetwork?
   a. Danish modern
   b. Contemporary
   c. Oriental
   d. Italian provincial
8. The drawing in which the lines converge to a vanishing point is called
   a. perspective drawing.
   b. orthographic drawing.
   c. isometric drawing.
   d. cabinet drawing.
9. Usually the first step in developing a design is to sketch
   a. exploded drawings.
   b. details.
   c. sections.
   d. several types of drawings.
10. When the sketches are completed and approved, the developed ideas are
    a. used to build the cabinet.
    b. transformed into a working drawing.
    c. kept for future reference.
    d. blueprinted.

Questions

1. Give several examples of how a cabinet's use will affect its design.
2. What happens to the design if unnecessarily oversized parts are used?
3. Give an example of where the strength of a piece of cabinetwork may determine the kind of joint to be used.
4. Demonstrate how to lay out a golden mean rectangle. What is the approximate ratio between the width and length?
5. What are some of the distinguishing features of early American cabinets?
6. Explain why Danish modern furniture and cabinets are so easily identifiable.
7. What is the distinguishing feature of French provincial styling?
8. What is usually the first step in creating a design for a cabinet?
9. What is a section drawing?
10. Summarize the steps to follow in designing and building cabinets.

unit 20 Layout and Planning

OBJECTIVES

After studying this unit, the student will be able to:

- make a layout rod for a simple cabinet.
- write a cutting list from the layout rod.
- develop a plan of procedure for building the cabinet.

LAYOUT AND PLANNING

After the working drawings are completed and approved, a layout and plan of procedure are done. These save time and eliminate mistakes. A good cabinetmaker will always lay out and plan the work before starting to build. A number of construction problems are solved during this planning period.

Layout and planning is even more important when the supervisor lays out the work for others. With a good layout and plan, those who do the work have few questions.

THE LAYOUT ROD

One of the most common ways to lay out work is to use a rod. A rod is usually a strip of 1" x 2" lumber that is a little longer than the longest dimension of the piece to be built. Full-size marks are made on the rod that indicate the actual location of all the parts of the job.

One side of the rod is used for marking the width. Another side is used for marking the height. The third side shows the depth. The fourth side is used if other cabinets are to be built that differ in only one dimension. This dimension is marked here.

A small hole is drilled in one end of the rod so that it can be hung on a wall. The rod is kept in case similar cabinets are built at a later date. Not only does the rod answer questions about construction, it is also used to make a cutting list. A *cutting list* is simply a list of the exact sizes of each piece of the cabinet.

The kitchen cabinet wall unit in figure 20-1 is 36 inches wide, 30 inches high, and 12 inches deep. The stiles and rails are made of 3/4" x 1 1/2" stock. The back is rabbeted into the end panels with a 1/4-inch projection for scribing. (*Scribing* fits a unit to an irregular surface.) The top, bottom, and shelf are made of 3/4-inch thick material.

To lay out a rod for this cabinet, select a smooth, straight piece of 1" x 2" stock, 42 inches long. This is 6 inches longer than the longest dimension of the cabinet.

Laying Out the Width

1. On the 2-inch face of the rod starting at the left end, mark off a full 36 inches for the

width of the cabinet. Square a line across the face, figure 20-2.

2. Mark the left end panel 3/4 inch from the left end.
3. Measure 1 1/2 inches in from the left end for the width of the left stile and square a line across the face.
4. From the 36-inch mark, measure in toward the left 3/4 inch for the right end panel and 1 1/2 inches for the right stile. Square lines across the face of the rod.
5. Find the center of the 36-inch width and measure 3/4 inch in each direction to lay out 1 1/2 inches for the center stile.
6. Mark *WIDTH* on the right end of the rod. The layout for the width of the cabinet is complete.

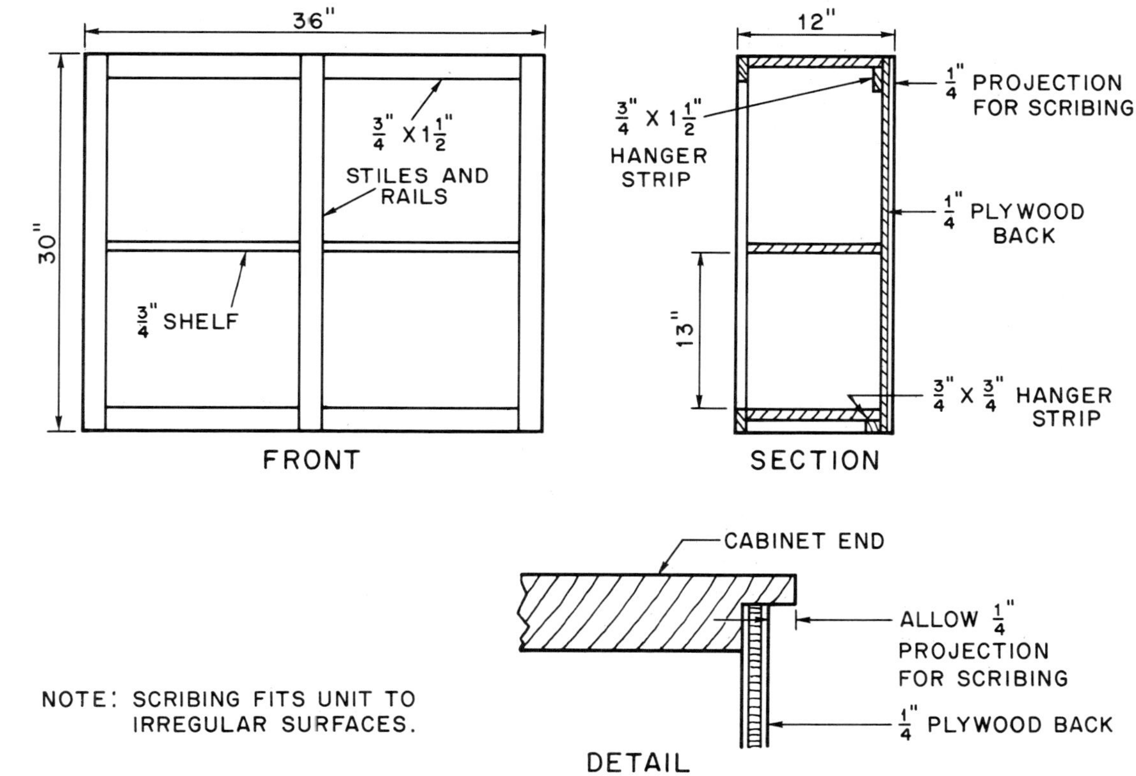

Fig. 20-1 Kitchen cabinet wall unit

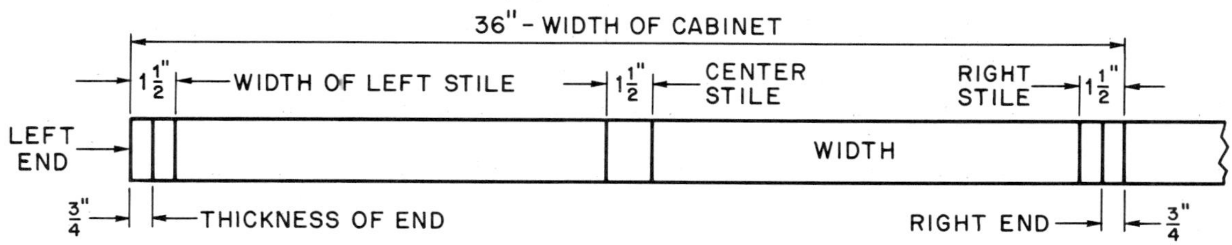

Fig. 20-2 Laying out the width on a rod

Laying Out the Depth

1. To lay out the depth of the cabinet, turn the rod over to its opposite face, figure 20-3. Starting at the left, mark off 12 inches for the overall depth.
2. From the left end, mark off 3/4 inch for the thickness of the face frame.
3. From the 12-inch mark (or back of cabinet), measure and mark off 1/4 inch for the projection allowed for scribing.
4. From that mark, measure another 1/4 inch for the plywood back.
5. Mark off 3/4 inch for the thickness of the mounting strips.
6. On the right end of the rod, mark *DEPTH* and this layout is complete.

Laying Out the Height

1. Turn the rod to one of its edges. Measure and mark from the left end 30 inches for the overall height of the cabinet, figure 20-4.
2. From the left end, mark off 3/4 inch for the thickness of the top and 1 1/2 inches for the width of the top rail.
3. From the 3/4-inch mark, measure 1 1/2 inches and mark the rod for the width of the top mounting strip.
4. From the 30-inch mark, which is the bottom of the cabinet, measure up 1 1/2 inches for the width of the bottom rail. This mark is also the top of the bottom shelf.
5. To the right, measure and mark 3/4 inch. This is the top of the bottom mounting strip and also the bottom of the lower shelf.
6. To locate the middle shelf, measure up 13 inches from the top of the lower shelf. Mark off 3/4 inch more for the thickness of the middle shelf.

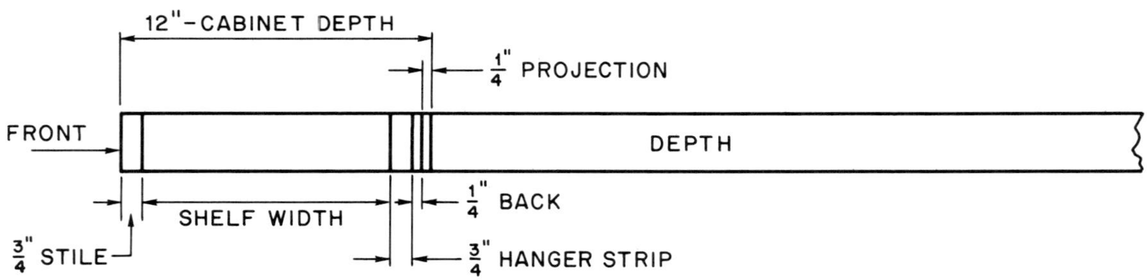

Fig. 20-3 Laying out the depth on a rod

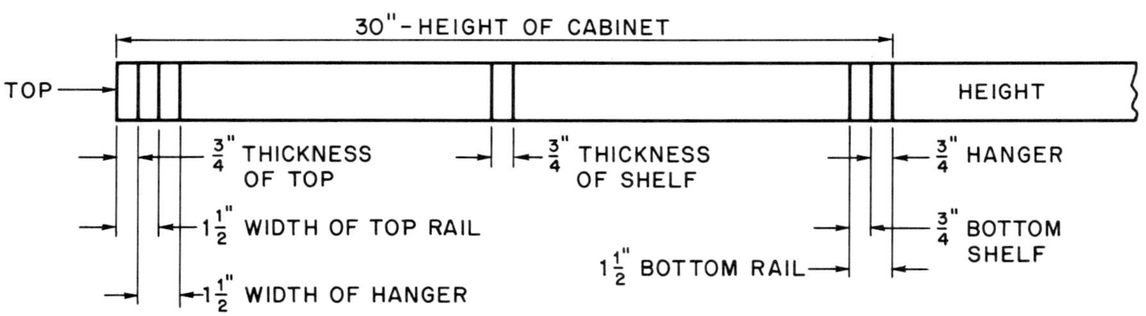

Fig. 20-4 Laying out the height on a rod

7. Mark *HEIGHT* on the right end and the rod layout is complete.

Other Rod Layouts

Laying out a rod becomes more difficult with more complicated work. Cabinetmakers use different methods to make reading the rod easier, figure 20-5. Some use different colored pencils for certain marks, others use dotted lines to show joints, etc. With each rod layout, a system is gradually developed so that the rod is read with little trouble.

MAKING A CUTTING LIST

Once the rod layout is complete, all measurements for cutting the stock can be taken from it. A cutting list can then be made listing all the parts and their sizes. A cutting list for the cabinet shown in figure 20-1 and taken from the rod layout is shown in figure 20-6.

DEVELOPING A PLAN OF PROCEDURE

A plan of procedure should be developed before making a piece of cabinetwork. This involves writing down all the steps of construction.

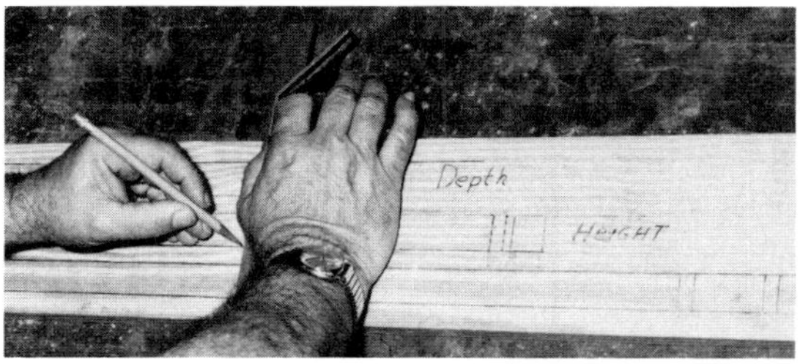

Fig. 20-5 Different methods are used to read the rod easier.

The complexity of the work may determine the order of the steps to be taken to complete a job. In most cases, the following order can be used:

1. Make a layout rod from the sketch or drawing.
2. Make a cutting list using the measurements obtained from the layout rod.
3. Cut the stock to rough lengths. Rough length is two or three inches longer than actually required. Cutting to rough lengths makes handling the stock easier and facilitates machining.
4. Face one side of the stock. Facing produces a straight surface and eliminates any cup, bow, or twist.
5. Plane the stock to thickness. This is the first step to bring the stock to size. Make sure all parts are planed at the final setting of the planer to insure equal thicknesses.
6. Joint one straight edge on each piece. This straight edge will be held against the fence of the table saw for ripping to width.
7. Rip the stock to the required width. Use the correct saw blade for the smoothness of the edge desired. Rip all pieces of the same width without changing the setting of the rip fence.
8. Cut the stock to the overall length. This is the last step in cutting the pieces to their overall finished size. Use a stop block to cut equal lengths.
9. Make rabbets, dadoes, mortises, and tenons; bore holes; and perform other machining as

QUANTITY	PART NAME	THICKNESS	WIDTH	LENGTH
2	ENDS	3/4″	11 1/4″	30″
3	SHELVES	3/4″	10 3/4″	34 1/2″
1	TOP MOUNTING STRIP	3/4″	1 1/2″	34 1/2″
1	BOTTOM MOUNTING STRIP	3/4″	3/4″	34 1/2″
3	STILES	3/4″	1 1/2″	30″
4	RAILS	3/4″	1 1/2″	15 3/4″
1	BACK	1/4″	30″	35 1/2″

Fig. 20-6 Cutting list for wall unit in figure 20-1

necessary. Set up machinery accurately and make all similar cuts without changing the setup.

10. Sand inside faces before assembling. After assembly the inside of a cabinet is difficult to sand. These surfaces are smoothed prior to assembling.

11. Assemble the parts. If possible, assemble the parts using only clamps (no glue or fasteners) to check the quality of fit. Then assemble the piece permanently as required. After assembly, wipe off any excess glue that may make finishing difficult.

12. Prepare exterior surfaces for finishing by sanding, if the exterior surfaces were not sanded prior to assembly. Handle the pieces carefully to avoid marring the finished surfaces.

13. Apply the finish. Finish may consist of filling, staining, and applying clear or pigmented coatings.

14. Install the necessary hardware. Hardware is often installed before finishing, then removed and replaced after finishing. If there is no danger of marring the finish, the hardware is installed after finishing. Finish is not usually applied to the hardware.

LAYING OUT ANGLES

When machining stock, it is sometimes necessary to cut angles. It is difficult to read the degree scale on certain power machines because the scale is so small. The pointer is often fragile and is easily bent. An accurate reading is not always assured. Stops on machines for 45-degree and 90-degree angles are sometimes moved by vibration. These stops should not be completely trusted for accuracy.

In some shops, blocks of wood with their ends cut at different angles are used to test the tilt of blades or tables. The block is laid on the table and the blade is tilted until it is flush with the block angle.

To make these blocks, it is necessary to know how to lay out different angles. To lay out the angles to make these blocks requires a square, dividers, and straightedge.

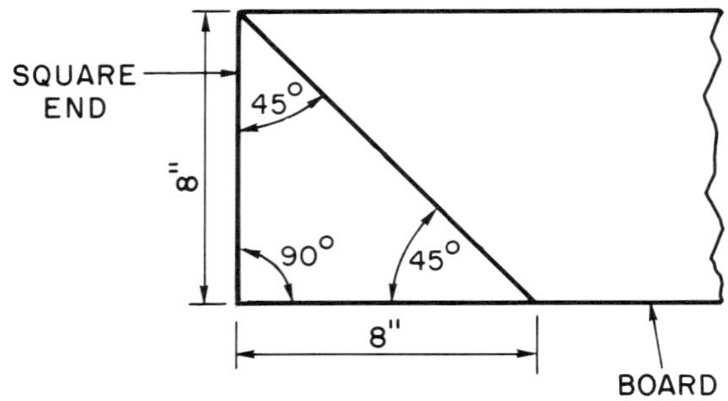

Fig. 20-7 Laying out 45-degree angles

Laying Out a 45-Degree Angle

A line connecting points an equal distance from a square corner produces 45-degree angles, figure 20-7.

1. Square a line across a board.
2. Mark off a distance equal to the width of the board from the squared line.
3. Draw a line to connect each end point. This produces a 45-degree angle at each end.

Laying Out a 60-Degree Angle

A triangle with three equal sides forms equal angles of 60 degrees, figure 20-8.

1. Along the edge of a board mark off a distance (AB) of 8 inches.
2. From point A, swing an 8-inch arc with the dividers. Do the same from point B.
3. Mark the point where both arcs meet (point C). Draw lines from point C to points A and B. This forms a triangle with three equal sides.

Laying Out 30, 15, and 7 1/2-Degree Angles

1. Lay out the same 60-degree angle triangle as in figure 20-8.
2. Measure 4 inches along BC and mark this point D. This cuts the line in half and produces a 30-degree angle at AD, figure 20-9.

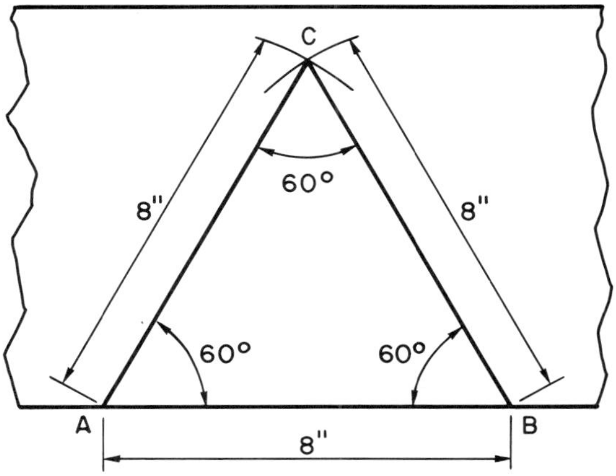

Fig. 20-8 Laying out a 60-degree angle

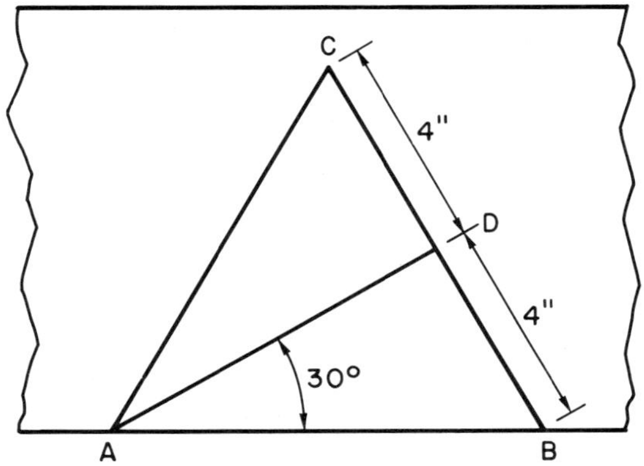

Fig. 20-9 Laying out a 30-degree angle

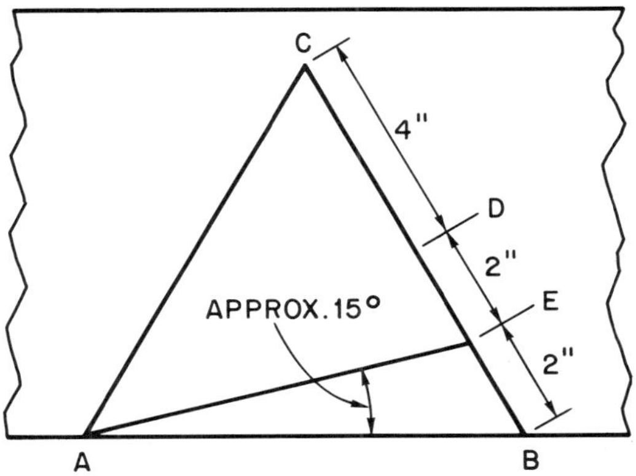

Fig. 20-10 Laying out an approximate 15-degree angle

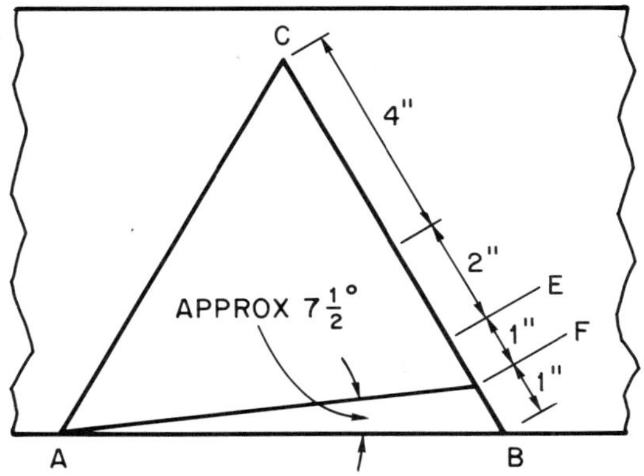

Fig. 20-11 Laying out an approximate 7 1/2 degree angle

3. Repeat steps 1 and 2. Cut BD in half and mark this point (E). Draw AE to produce an approximate 15-degree angle, figure 20-10.
4. Repeat steps 1, 2, and 3. Cut BE in half and mark this point (F). Draw AF to produce an approximate angle of 7 1/2 degrees, figure 20-11.

Laying Out Other Angles

Combine any of these triangles to make 67 1/2, 37 1/2, 22 1/2-degree angles and others. Mark these angles and cut the ends of blocks to use for testing the setting of machines for cutting bevels and miters. Drill holes in the end of these blocks and hang on the wall for use when needed.

ACTIVITIES

1. Make a layout rod for the bookcase in figure 19-21.
2. Using the layout rod constructed in #1, prepare a cutting list for the bookcase.
3. Write a plan of procedure for the bookcase in figure 19-21. Be as specific as possible.

4. As a class project, make a series of wood blocks to test blade angles. Make blocks to test 90°, 60°, 45°, 30°, 15°, and 7 1/2° angles, as well as any other angles that are frequently used in the classroom.

UNIT REVIEW

Multiple Choice

1. A layout rod is usually a
   a. piece of iron rod.
   b. strip of wood.
   c. length of wood dowel.
   d. metal angle.

2. Marks made on the rod
   a. are scaled down.
   b. show depth.
   c. are half size.
   d. are actual size.

3. The layout rod shows
   a. the style of the piece.
   b. the kind of finish to use.
   c. the exact size of the parts.
   d. the kinds of joints to use.

4. The height, width, and depth of a cabinet are shown on the
   a. plan of procedure.
   b. layout rod.
   c. steel square.
   d. wood layout blocks.

5. The layout rod should be longer than the
   a. depth.
   b. longest dimension.
   c. height.
   d. width.

6. A cutting list includes
   a. the layout for cutting.
   b. the plan of procedure.
   c. all parts and their sizes.
   d. the tools to use for cutting.

7. A piece of stock cut to rough length is actually
   a. 1/2 to 3/4 inch longer than required.
   b. 1 inch longer than required.
   c. 2 to 3 inches longer than required.
   d. 5 inches longer than required.

8. One side of the stock is faced to
   a. eliminate bows, cups, and twists.
   b. bring the stock to thickness.
   c. cut the stock to the overall length.
   d. prepare the surface for finishing.

9. Stock is planed to
   a. eliminate bows, cups, and twists.
   b. bring the stock to thickness.
   c. cut the stock to the overall length.
   d. prepare the surface for finishing.

10. The inside faces of a cabinet are sanded
    a. before the stock is cut to rough length.
    b. before the stock is cut to finish size.
    c. before assembling the cabinet.
    d. after assembling the cabinet.

Questions

1. What are the advantages of laying out and planning a project?
2. What is a layout rod used for?
3. What is marked on the fourth side of a layout rod?
4. What is scribing?
5. What is included in the cutting list?
6. What is a plan of procedure?
7. Why is stock cut to rough length?
8. When are the exterior surfaces of a cabinet sanded?
9. When is hardware installed on a cabinet?
10. What may affect the accuracy of a degree scale on a miter gauge?

unit 21 Casework Construction

OBJECTIVES

After studying this unit, the student will be able to:

- describe the basic parts of casework and their construction.
- lay out and build a simple bookcase and a hinged chest.
- lay out, cut, and install a butt hinge.

CASEWORK CONSTRUCTION

Casework is defined as box-like articles of cabinetwork. Usually rectangular, they may contain shelves or drawers for storage. Doors or covers are sometimes fitted to enclose the storage space. Examples of casework are bookcases, chests, desks, display cases, and kitchen cabinets, figure 21-1.

Casework consists of a skeleton frame, face frame, two ends, legs, and a bottom, back, and top, figure 21-2.

Ends

The case ends are made of solid edge-glued lumber or plywood. They may also be a paneled frame with stiles and rails and plywood or hardboard panels. Paneled ends are made similar to paneled doors using either doweled or mortise-and-tenon joints (see *Unit 24 Cabinet Doors*).

The back edge is usually rabbeted to receive the cabinet back, figure 21-3. If the case is to be fitted to the wall, the rabbet is cut deep to recess the back and allow the projecting material to be scribed to the wall.

The front edge is joined to the face frame with a butt, rabbeted, or mitered joint. If a butt joint is used, the front stile of the case end is made narrower than the back stile because of the thickness of the face frame.

Case ends may also be dadoed to receive the top, bottom, fixed shelves, skeleton frames, and dust panels of the case.

Skeleton Frames

The *skeleton frame* is made to fit in the interior of the case. It consists of stiles (vertical members) and rails (horizontal members) only. Panels fitted into the frame are called *dust panels.* The skeleton frame serves a number of purposes:

- It provides a means of fastening the case top to the case and holding the ends together at the top.
- It fastens and holds the ends together at the bottom.
- It separates and supports drawers.
- It is used vertically as divisions when solid partitions are not required.

Fig. 21-1 Examples of casework *(Hardwood Plywood Manufacturer's Assn.)*

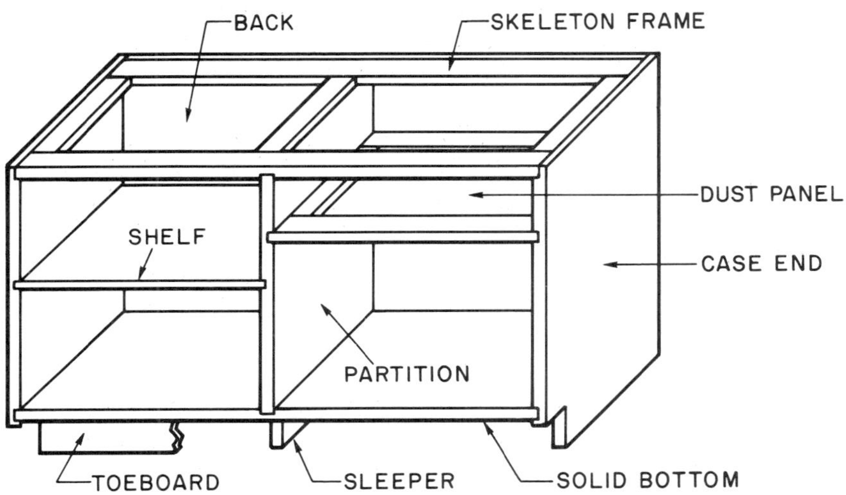

Fig. 21-2 Typical case construction. The face frame is later fitted to the front of the case.

Skeleton frames are assembled prior to being installed in the case. As with paneled ends, doweled or mortise-and-tenon joints are used to make skeleton frames.

Legs

Sometimes the stiles of the case ends extend below the bottom and act as legs. The front stile of the ends also acts as a stile for the front frame. In this type of construction, it is usual for the skeleton frames to be notched around the leg. It then extends to the front and becomes the face frame and dividing rails for the drawers, figure 21-4.

Partitions and Sleepers

Partitions are vertical members dividing the interior of the case into sections. They tie the top and the bottom of the case together and are usually dadoed into the top and bottom. The skeleton frame, dust panels, and shelves are cut in between the partitions and are usually dadoed into the partitions. Partitions are also known as *divisions* or *standards.*

Sleepers extend from the bottom of the case to the floor and are located directly under the partitions. They provide support of the case to the floor and keep the bottom from sagging.

Shelves

Shelves must be strong enough to support the weight to be placed on them. They must also be wide enough and correctly spaced for their purpose. Shelves may be made of solid wood, plywood, particleboard, or glass.

Bookcase shelves should be from 8 to 10 inches wide and spaced 10 to 14 inches apart. The

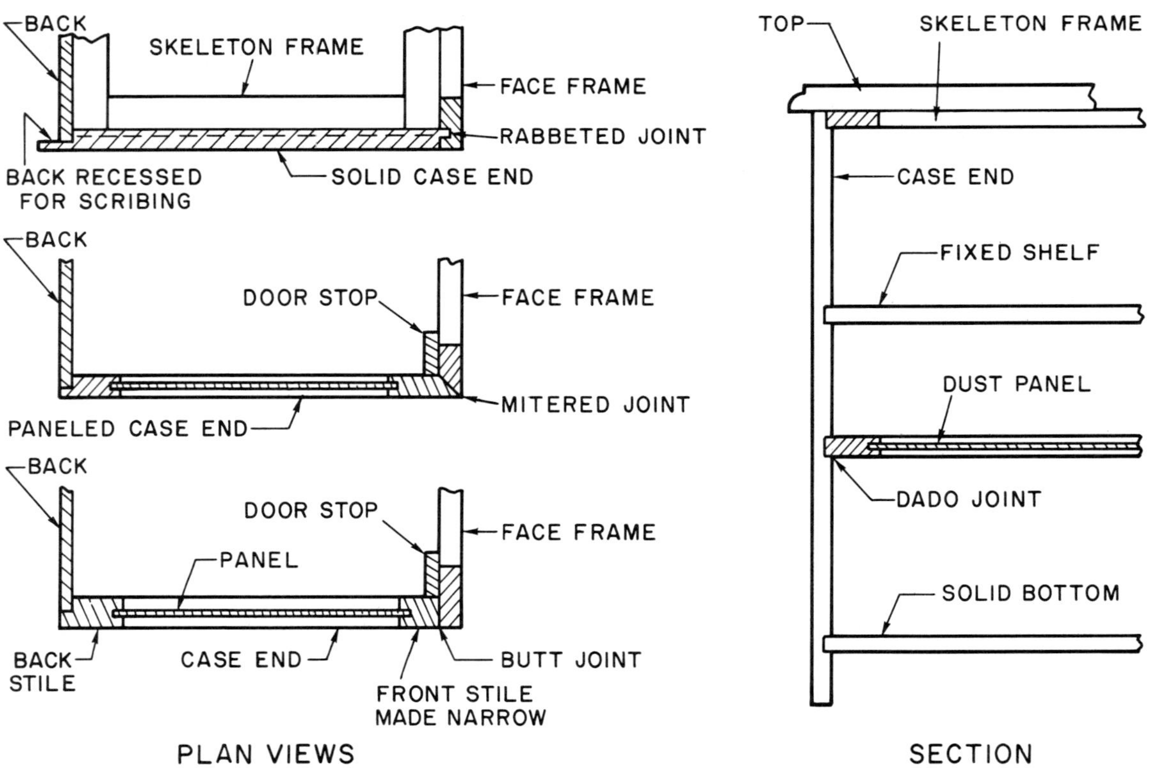

Fig. 21-3 Typical case end construction

length of a 3/4-inch thick shelf should be no more than 36 to 42 inches without intermediate supports. Supports should be spaced close enough to keep shelves from sagging under the weight placed upon them.

One way of increasing the strength of a shelf is by installing strongbacks. A *strongback* is a strip of wood screwed on edge to the underside of the shelf. It is placed either on or near the front or back edge of the shelf or both edges of the shelf.

Fixed shelves are usually dadoed in or supported on wood cleats, figure 21-5. A *cleat* is a small strip of wood screwed to the inside of the case to support the shelf. A through dado or dovetail dado may be used to support a shelf. A better method is to use a blind dado to conceal the joint.

Adjustable shelves may be supported with metal shelf standards and clips that are either surface mounted or set flush in grooves, figure 21-6. A pair of notched and numbered standards supports the shelves at both sides of the case. They are fastened 1 to 2 inches in from the back and front edges. When installed, the same number appears right-side up at the bottom of all four standards. The clips can then be inserted in the correct notch so the shelf lies flat.

Another method of supporting adjustable shelving is by inserting wood *dowel pins* or commercial *shelf pins* into four holes at each shelf location. Two vertical rows of equally spaced, 1/4-inch holes are drilled on either side of the case about 1 to 2 inches in from the front and back edges. The

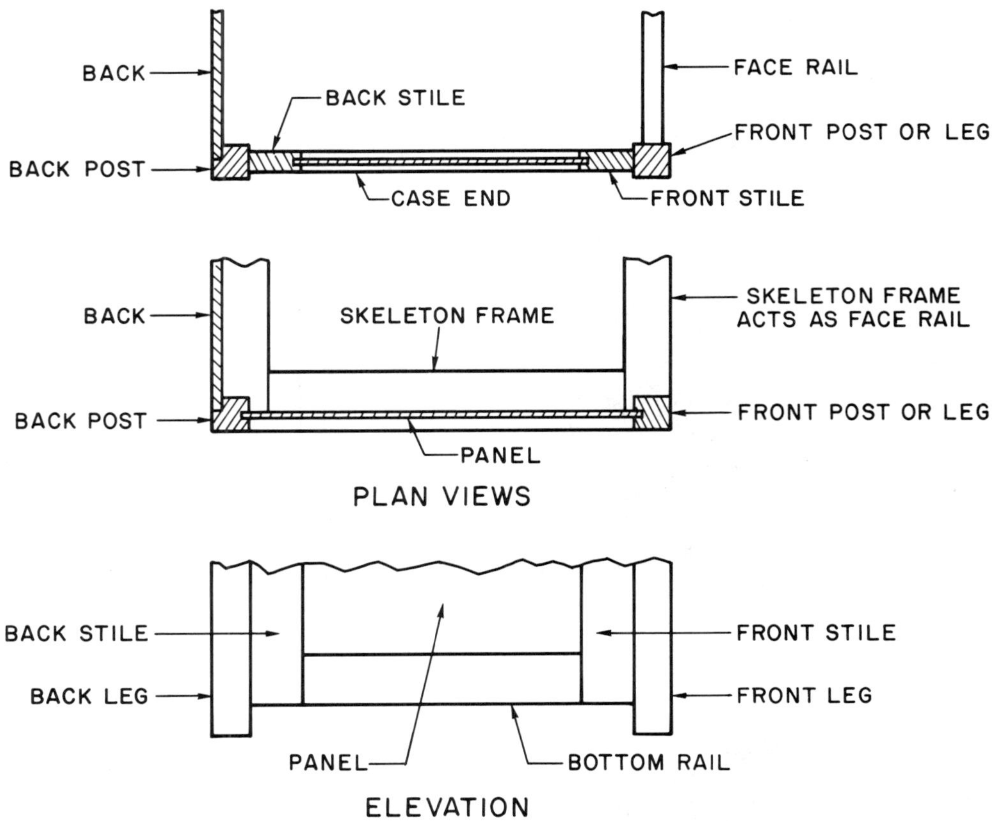

Fig. 21-4 Typical leg construction

holes are spaced approximately 2 inches apart for ordinary work. The holes should be drilled deep enough so the pins will not fall out when the shelf is placed upon them.

Adjustable shelves are sometimes installed by using ratchet strips, figure 21-7. *Ratchet strips* are strips of wood with notches cut at equal intervals on one edge. These strips are fastened to the

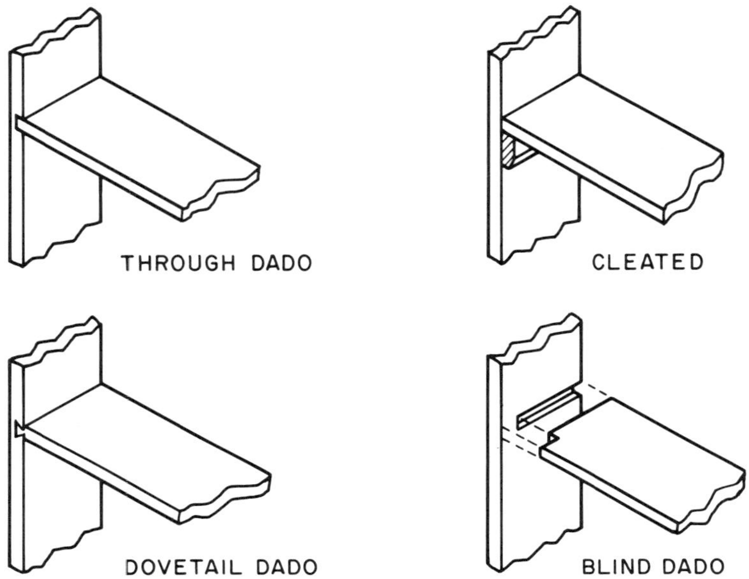

Fig. 21-5 Fixed shelf construction

Fig. 21-6 Adjustable shelf standard and clips

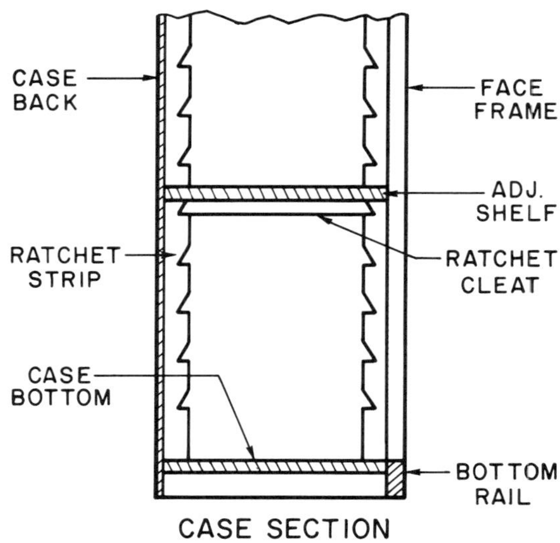

Fig. 21-7 Adjustable shelves using ratchet strips and cleats

front and back edges of the case on the inside. A ratchet cleat is cut to length with ends matching the notches to fit in between the ratchet strips. The ratchet cleat may be moved to any notch to support the shelf.

Another method of making ratchet strips is by boring a series of equally spaced, 3/4-inch holes along strips of 1" x 4" lumber. The strips are cut in half along the centerline of the holes. Ratchet cleats with rounded ends are then cut to match the ratchet strips.

Bottoms and Toeboards

The bottom of a case is usually made of solid lumber, particleboard, or plywood, unless a dust panel is used when a drawer is supported by the bottom, figure 21-8. Case bottoms are sometimes raised above the bottom rail of the face frame to act as a stop for doors. Another design eliminates the bottom rail of the face frame. The door or drawer then covers all of the bottom edge which also acts as a stop.

To cover the space between the bottom and the floor and to provide toe clearance, a *toeboard* is installed. The toeboard is usually set back from the face of the case 2 1/2 to 3 inches.

Face Frames

Face frames are preassembled units, usually joined with doweled or mortise-and-tenon joints, into which drawers and doors are fitted, figure 21-9. Face frames are joined to cabinet ends with a butt, rabbeted, or mitered joint. The face frame must fit the case accurately so doors and drawers may be installed easily at a later stage.

If flush doors are to be hung on the face frame, the frame is made about 1/16 inch thicker than the doors to be hung. This prevents the doors from binding against the door stops. If the end of the case is to be fitted against a wall, approximately 1/2 inch is added to the width of the stile on that end for scribing.

Door stops are installed on the back side of the face all around the door openings if flush doors

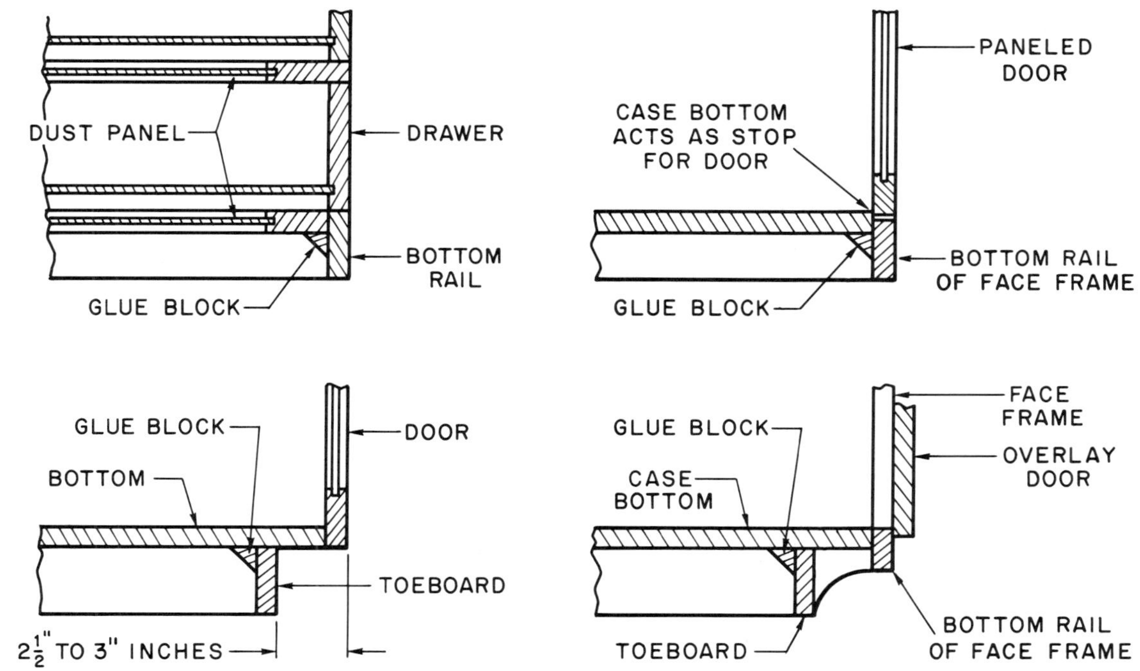

Fig. 21-8 Typical case bottom and toeboard sections

are to be hung. Door stops project about 5/16 inch inside the opening and usually are made of thinner material than the face frame. They are applied with screws and glue because they take much abuse and a strong joint is needed.

Tops

The case top is installed according to its location. If it is above the line of vision, the top is cut in between the case ends so the ends of the top are not visible. The top may also be lowered between the case ends. This provides clearance between the ceiling and the top of the case and also acts as a stop for the top ends of doors. If the top is below the line of vision, it is placed above the case ends.

In most cases, tops are fastened to the cabinet with screws driven up through the top skeleton frame. More complete information on constructing cabinet tops is found in *Unit 26 Tables.*

MAKING A BOOKCASE

Layout and Cutting List

Make a layout rod for the bookcase in figure 21-10. Lay out the width, height, and depth as described in *Unit 20 Layout and Planning.*

From the layout rod and drawing, make a cutting list for all the parts, figure 21-11. Cut the parts out to rough size. Cut the stock to size by reducing to the desired thickness, width, and length.

Machining

1. Rabbet the two ends and top at the inside back edge 1/4″ x 1/2″, see detail A.
2. Rabbet the two ends of the top 1/2″ x 3/4″. The top needs no additional milling and is set aside.
3. Rabbet both ends of each shelf and the toeboard 3/8″ x 1/4″, see detail B.

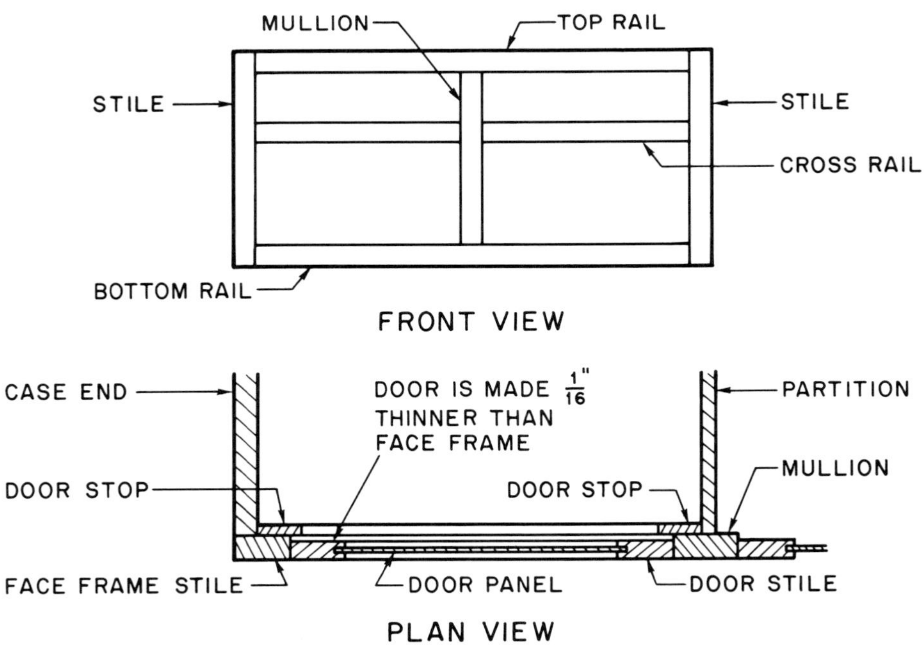

Fig. 21-9 Typical face frame construction

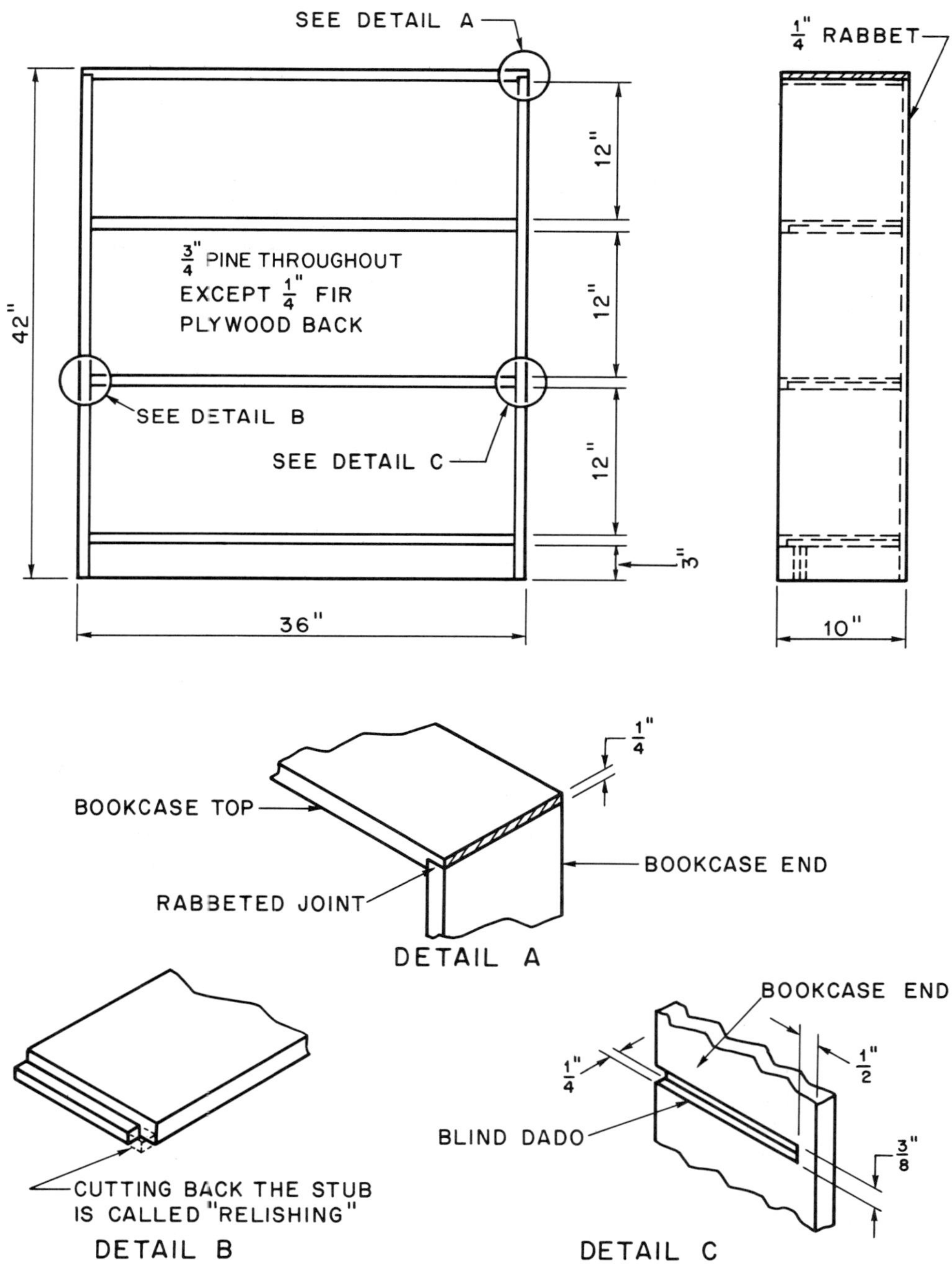

Fig. 21-10 Bookcase

NO. OF PIECES	PART NAME	THICKNESS	WIDTH	LENGTH
2	ENDS	3/4″	10″	41 3/4″
1	TOP	3/4″	10″	36″
3	SHELVES	3/4″	9 3/4″	35″
1	TOEBOARD	3/4″	3″	35″
1	BACK	1/4″	35 1/2″	41 3/4″

Fig. 21-11 Cutting list for bookcase in figure 21-10.

4. *Relish* (cut back) the stubs of the rabbet on the ends of the shelves. Set aside the shelves and toeboard.
5. Locate the position of the shelves on the ends of the bookcase.
6. Make blind dadoes 3/8-inch wide by 1/4-inch deep, see detail C. Stop the cut 1/2 inch from the outside edges. Finish the cut with a mallet and chisel.
7. Make a groove for the end rabbets of the toeboard.
8. Sand all inside surfaces.

Assembling

1. Gather all clamps and adjust them. Clamp the pieces without glue as explained in *Unit 16 Gluing and Clamping.*
2. Use bar clamps across the front and back at each shelf and top. Clamp also from top and bottom. Do not attach the back at this point.
3. Check the assembly for squareness by measuring diagonally from corner to corner, figure 21-12. If the measurements are equal, the case is square. Adjust if necessary.
4. Remove the clamps. Apply glue and reclamp. Check again for squareness.
5. When the glue has set, remove the clamps and fasten the back in position. The best face should be toward the inside of the bookcase.
6. Sand the outside surfaces in preparation for finishing. Make sure all sharp edges are rounded off. A finish will be applied to the bookcase in *Unit 27 Finishing.*

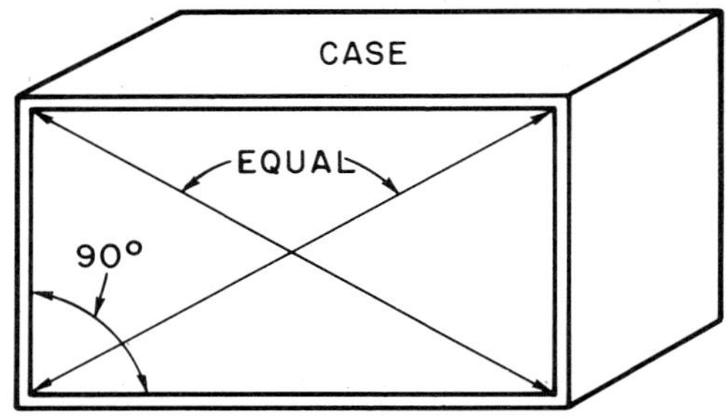

Fig. 21-12 Casework is square when the diagonal measurements are equal.

MAKING A CHEST

The chest in figure 21-13 is made as a closed box. After assembling, the lid is cut off by running the box through the table saw. Each vertical corner of the chest has a rabbeted joint. The top and bottom of the chest has rabbeted edges and ends. These fit into grooves made in the front, back, and ends of the chest.

1. Make a layout rod and cutting list, figure 21-14.
2. Cut out all the parts to their overall dimensions. Be sure to allow enough stock to make the required joints.
3. Groove the inside faces of the front, back, and ends 3/16 inch deep by 1/4 inch wide to receive the rabbeted top and bottom. The back needs no further machining.
4. Rabbet both ends of the front 5/16″ x 1/2″ to receive the ends. Rabbet the ends in the same manner in the back.

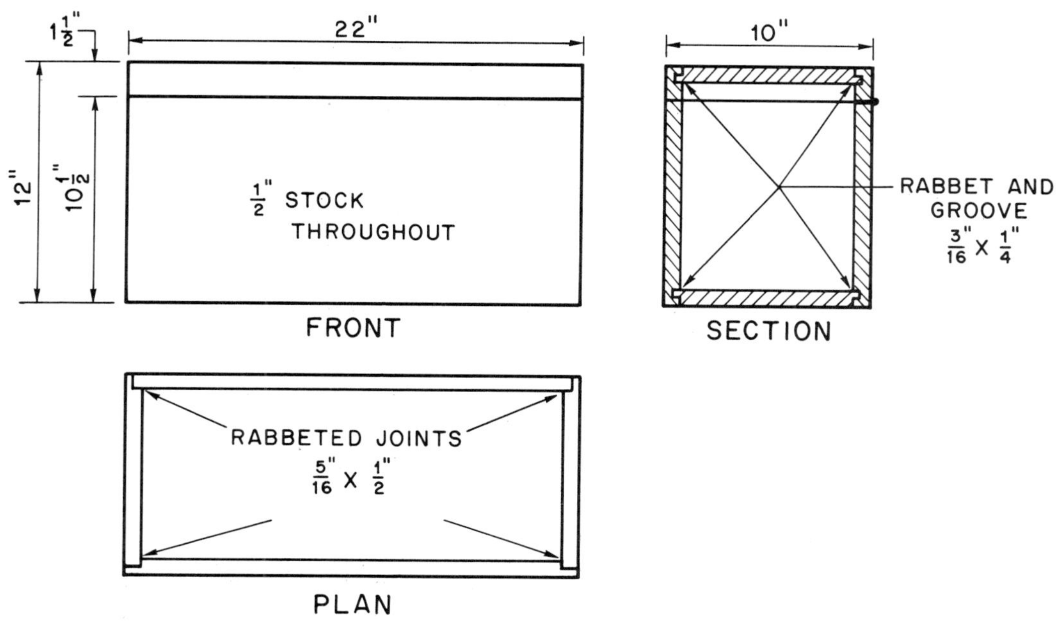

Fig. 21-13 Small chest with a hinged cover

NO. OF PIECES	PART NAME	THICKNESS	WIDTH	LENGTH
2	END	1/2"	12"	9 13/16"
1	FRONT	1/2"	12"	22"
1	BACK	1/2"	12"	21 5/8"
1	TOP	1/2"	9 3/8"	21 3/8"
1	BOTTOM	1/2"	9 3/8"	21 3/8"

Fig. 21-14 Cutting list for small chest in figure 21-13

5. Rabbet both edges and ends of the top and bottom to fit in the grooves made in the sides and ends of the chest.
6. Sand all inside surfaces.
7. Clamp the pieces together without glue and adjust until square. Remove the clamps.
8. Apply glue and reclamp. Do not apply too much glue because it is impossible, at this time, to remove excess glue from the inside of the chest.
9. After the glue has set, remove the clamps. Sand the outside surfaces. Make sure all edges and ends are flush and smooth.
10. Saw off the top of the chest by running the chest through the table saw on all four sides, figure 21-15. The chest is now ready for hinging.

HINGING THE CHEST

A butt hinge is used to hinge the cover to the chest. (Other hinges are discussed in *Unit 24 Cabinet Doors.*) A *butt hinge* has both leaves *gained* (cut) into the wood, figure 21-16. One half is gained into the cover and the other half into the chest. The *butt gauge,* figure 21-17, is used to lay out the measurements for setting hinges.

Fig. 21-15 Using the table saw to cut off the chest cover
Caution: guard must be removed for this operation.

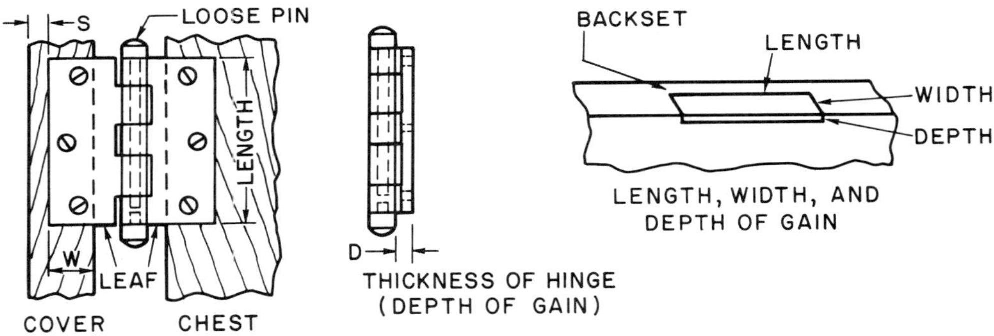

Fig. 21-16 Layout of a butt hinge

Fig. 21-17 Butt gauge

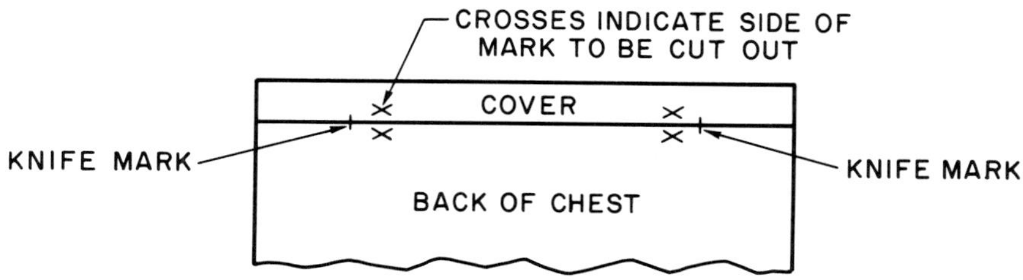

Fig. 21-18 Marking the location of the hinges

Locating The Hinges

Place the cover in position on the chest. Lightly mark the location of one end of each hinge across both cover and chest with a knife, figure 21-18. Do not make the knife marks longer than the depth of the hinge or they will show after the hinges are installed. Pencil an x on the side of the knife marks to be cut out. This will prevent cutting out on the wrong side.

Laying Out the Length of the Gain

Lay one leaf of the hinge on the cut edge of the chest back so that one end of the leaf lines up with the knife mark. The leaf should cover the x. With a sharp knife, mark that end of the hinge, figure 21-19. Hold the hinge firmly so it does not move. Do not mark across the whole edge.

Tap the hinge, while holding it, toward the knife mark until the end of the hinge covers the knife mark. Mark across the other end of the hinge with the knife. Lay out the length of the gain in the same manner on the cover.

Laying Out the Depth of the Gain

To lay out the depth of the gain, set the butt gauge for the thickness of the hinge leaf. Use the rod with the single marker. Set the gauge accurately. Too deep a gain will cause the cover to bind. Too shallow a gain will show too much joint between the cover and the chest.

Using the butt gauge, score a line from one end of the gain to the other. Do not mark beyond the end of the gain because these marks cannot be removed and will show on the finished product, figure 21-20. Mark all gains in the same manner.

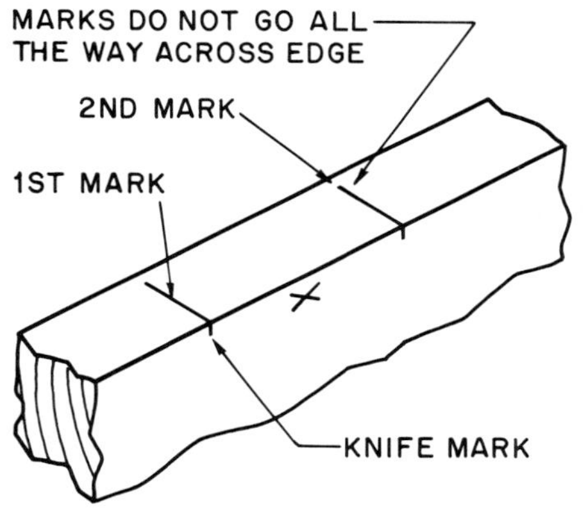

Fig. 21-19 Laying out the length of the gain

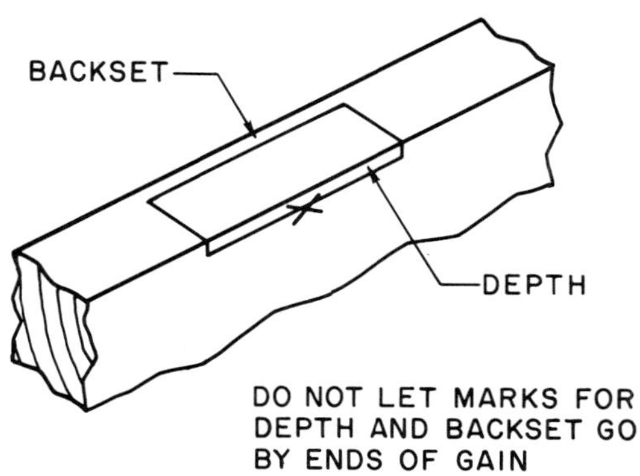

Fig. 21-20 Laying out the depth of the backset of the gain.

Laying Out the Backset

A gain for a hinge should never be cut across the whole edge. Part of the edge should be left on. This hides the edge of the hinge, and when the cover is opened, the hinge edge is not exposed. The part of the edge that remains is called the *backset* of the hinge.

To lay out the backset, adjust the other rod of the butt gauge to mark in about 1/16 inch. Ride the butt gauge against the inside frame of the cover and the bottom of the chest. Mark the backset for the length of the gains. Again be careful not to mark any more than the length of the gain.

Cutting Out the Gain

Score the backset with a knife. Using a chisel with the bevel side down, take a small cut at each end of the gain. The knife mark made at these ends will cause the waste to break off at the desired location. This also gives a shoulder against which to place the chisel and deepen the cut at each end of the gain.

Make small cuts along the gain with the bevel of the chisel down. Remove the excess wood. With the flat of the chisel down, work in from the depth mark and pare the wood to the correct depth. Continue in this manner until all gains are cut out.

Applying the Hinges

Each leaf of the hinge should press fit into the gains and be flush with the edge. Drill holes of the proper size and drive screws to apply the hinges. Hold the cover in position and insert the hinge pins.

The chest is now complete except for the application of additional hardware such as locks and handles. Other trim, such as molding, may be added if desired. All sharp, exposed edges are rounded off and the surface sanded in preparation for finishing.

ACTIVITIES

1. Select a piece of casework in your classroom or home. Describe its construction.
2. Design and make a jewelry box. Write a detailed plan of procedure specifically naming the equipment to use to cut and machine each part and the type of construction required.
3. Install adjustable shelves using metal shelf standards and cleats.

UNIT REVIEW

Questions

1. If the back of a case is to be fitted against a wall, how are the ends milled to receive the case back?
2. Why is the front stile of a case end made narrower if the face frame is to be butted against it?
3. Name three purposes of the skeleton frame.
4. What is the location and purpose of a sleeper?
5. Why are metal adjustable shelf standards numbered?
6. Describe the location of the case bottom in relation to the face frame when the bottom acts as a door stop.
7. What is the thickness of the face frame in relation to flush doors?
8. How can casework be checked for squareness?
9. When laying out hinge gains, where and why is an X placed on the chest?
10. What instrument is used to lay out the depth and backset for a butt hinge gain?

Multiple Choice

1. To receive the case back, the ends are usually
   a. dadoed.
   b. grooved.
   c. rabbeted.
   d. splined.

2. Partitions are usually fitted to tops and bottoms of a case with a
   a. rabbeted joint.
   b. doweled joint.
   c. dadoed joint.
   d. butt joint.

3. Without intermediate supports, the length of a 3/4-inch shelf should not exceed
   a. 18 to 24 inches.
   b. 24 to 30 inches.
   c. 30 to 36 inches.
   d. 36 to 42 inches.

4. A good joint to use for fixed shelves is a
   a. butt joint.
   b. reinforced dowel joint.
   c. blind dado joint.
   d. rabbet and dado joint.

5. Shelf standards should be installed in from the front and back edges of the case approximately
   a. 1/2 to 1 inch.
   b. 1 to 2 inches.
   c. 2 to 3 inches.
   d. 3 to 4 inches.

6. A toeboard is usually set back from the front of the case about
   a. 1 to 2 inches.
   b. 2 to 2 1/2 inches.
   c. 2 1/2 to 3 inches.
   d. 3 to 3 1/2 inches.

7. Cutting back the stubs of the rabbet on the ends of the shelf is called
   a. tenoning.
   b. dadoing.
   c. relishing.
   d. rabbeting.

8. When laying out the length of the gain, the end of the hinge is tapped to cover the first mark before making the second mark because
   a. the hinge will move when a mark is made against it.
   b. allowance must be made for the wedge shape of the chisel.
   c. this helps assure a press fit.
   d. this helps prevent too deep a cut.

9. Caution must be taken when marking the depth and backset of the gain
   a. to set the gauge accurately.
   b. not to mark beyond the length of the gain.
   c. not to score the gain too deeply.
   d. not to splinter the edge of the gain.

10. A hinge gain should never be cut across the whole edge. The part that remains is called a
    a. strongback.
    b. backset.
    c. cleat.
    d. ratchet strip.

unit 22 Kitchen Cabinets

OBJECTIVES

After studying this unit, the student will be able to:

- lay out a design for typical base and wall kitchen cabinet units.
- construct a typical base and wall kitchen cabinet unit of specified size.
- make a countertop with a backsplash.
- install kitchen cabinets.

DESIGNING KITCHEN CABINETS

The cabinetmaker must know how to build kitchen cabinets. This makes up a large part of the cabinetmaking industry. The average home probably has more cabinetwork in the kitchen than in any other room, figure 22-1.

The two basic kinds of kitchen cabinets are the base unit and the wall unit. The *base unit* sets on the floor; the *wall unit* hangs on the wall. Standard measurements for kitchen cabinets are shown in figure 22-2.

The distance between the wall and base units is usually 16 inches. This distance is enough to accommodate most articles that are placed on countertops, like coffee pots, toasters, blenders, and mixers. The top shelf in the wall unit should not be over 6 feet from the floor if it is to be in easy reach.

Countertop

The standard kitchen countertop is 36 inches high, 25 inches deep, and 1 1/2 inches thick. This provides enough room for an average-size sink and

Fig. 22-1 Making kitchen cabinets is a large part of the cabinetmaking industry.

ample working space on the surface. The countertop is held in place by driving screws up through the top frame of the base unit.

The countertop usually has a 3/4-inch overhang made of plywood or particleboard. It is doubled up by fastening a 2 1/2-inch wide strip flush with the edges and ends. This gives the appearance of a heavier countertop.

If a backsplash is used, it is usually 4 inches high. It has a 1/4-inch projection on the side that goes against the wall, figure 22-3. This projection allows the installer to scribe the countertop to uneven wall surfaces.

Preformed countertops may be purchased, figure 22-4, and cut to any desired length. Special fasteners hold the lengths together.

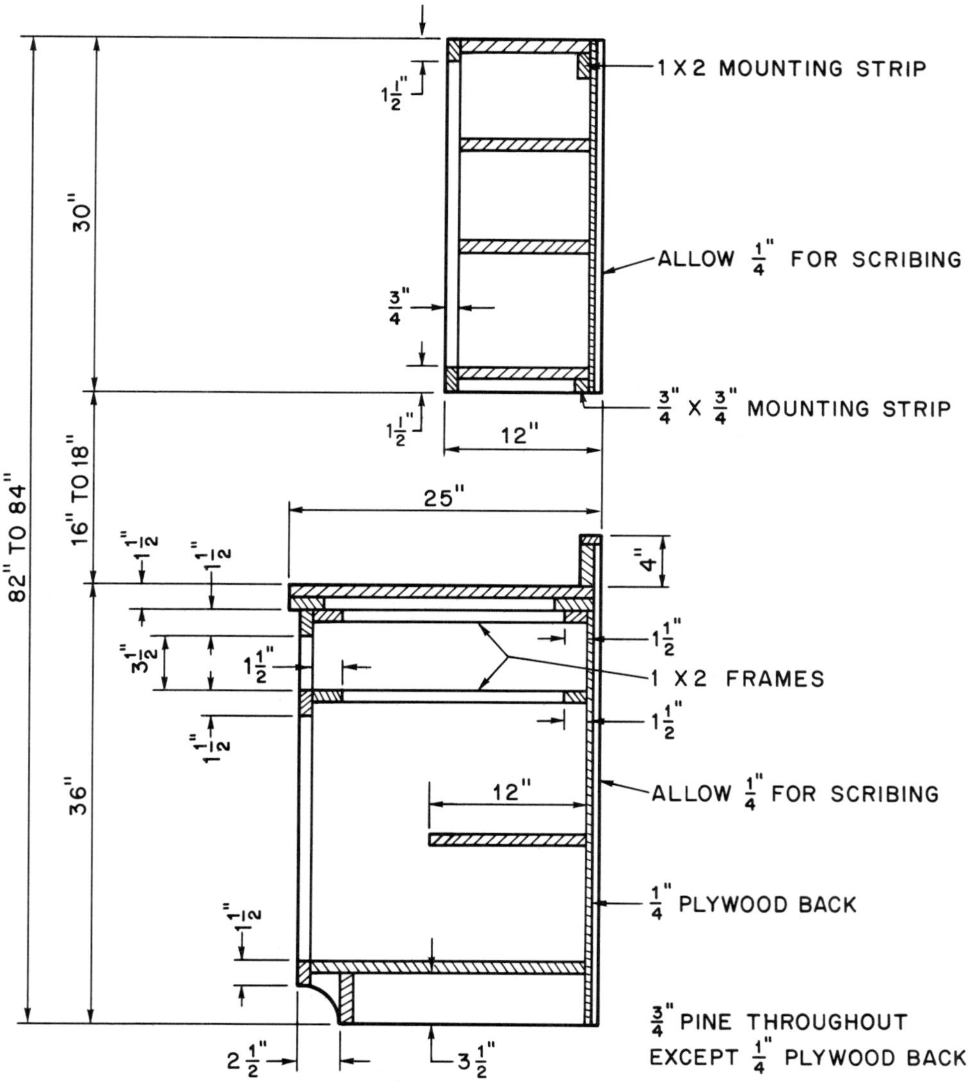

Fig. 22-2 Typical section of kitchen cabinets

Base Unit

The height of the base unit must be 36 inches to the surface of the countertop. The width must allow for the 3/4-inch overhang of the countertop. Therefore, a standard base unit is usually 34 1/2 inches high by 24 1/4 inches deep, figure 22-5. Toe space is provided beneath the base unit. The back edges of the end pieces project 1/4 inch beyond the cabinet back to allow for scribing, figure 22-6.

Usually the base unit is constructed with drawers just below the countertop. The drawer opening height is 5 1/2 inches. Some base units contain all drawers or all doors.

The base unit has a face frame and skeleton frames. A number of joints can be used to fasten the stiles and rails together. The blind mortise-and-tenon joint is widely used for this purpose. The stiles are always mortised and the rails are always tenoned, figure 22-7.

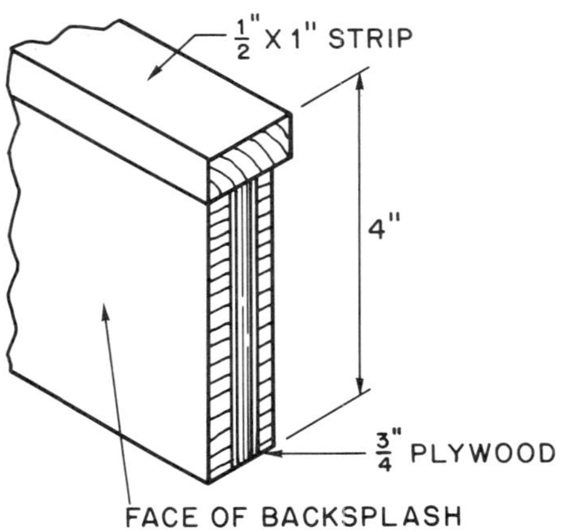

Fig. 22-3 Detail of backsplash showing construction to allow for scribing to wall

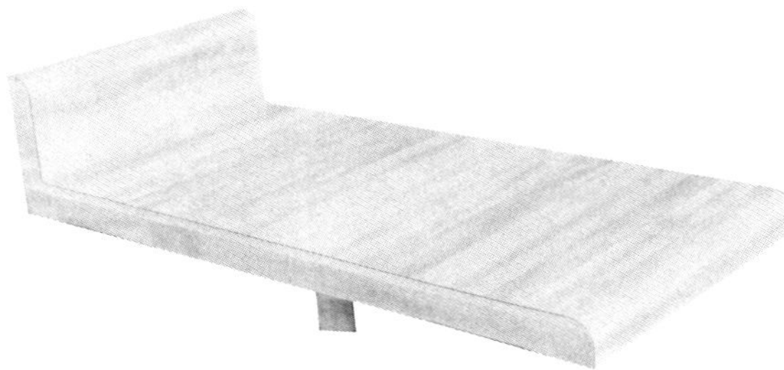

Fig. 22-4 Preformed countertop

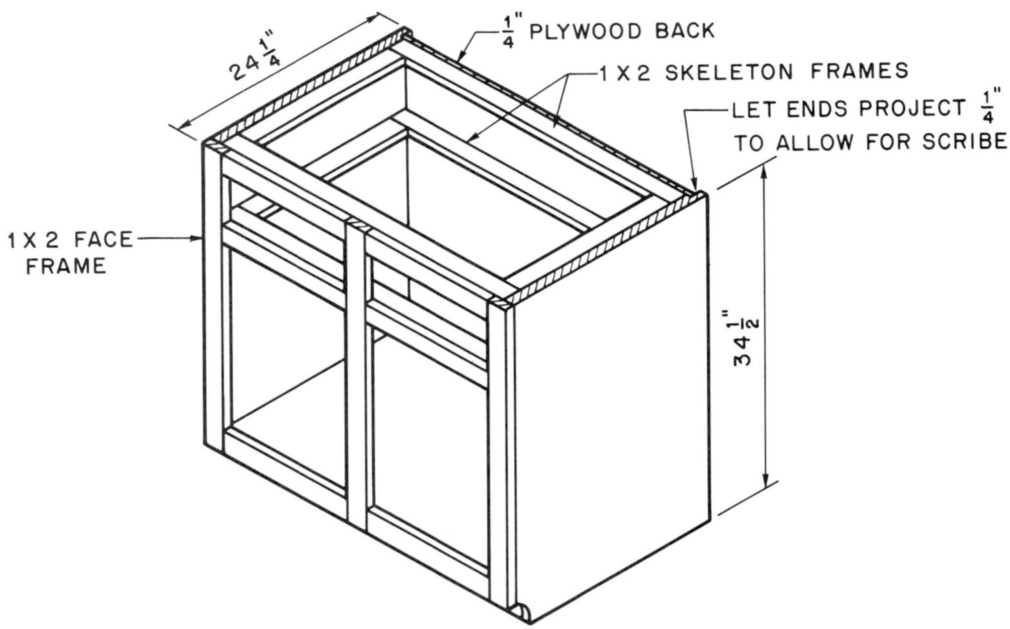

Fig. 22-5 Isometric view of typical base unit without countertop

Wall Unit

A full-size wall unit is usually 12 inches deep and 30 inches high, figure 22-8. A wall unit above a range is 18 inches high. Above a sink, it is 22 inches high, while over a refrigerator it is only 15 inches high.

The number and spacing of shelves depends on the purpose of the cabinet. Shelves may be fixed or adjustable. Shelves are usually spaced from 3 to 12 inches according to the customer's wishes.

The wall unit, like the base unit, has a face frame on which to fit and hang doors. These face frames are usually made of 1" x 2" solid lumber. The actual size is 3/4" x 1 1/2".

Note how the top edge of the bottom rail and the top of the bottom shelf are flush in figure

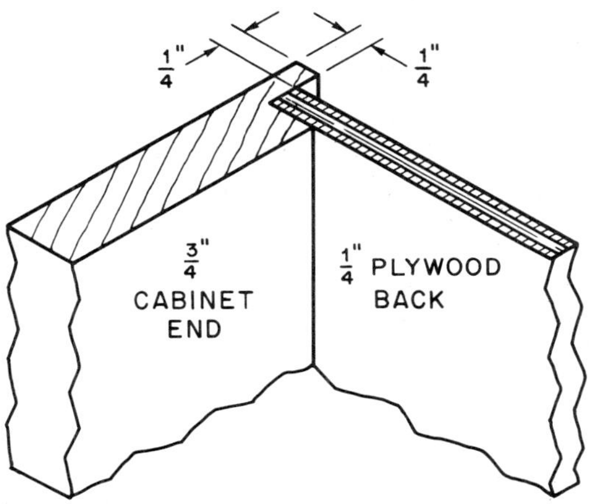

Fig. 22-6 Detail of back edge of cabinet showing allowance made for scribing cabinet to wall

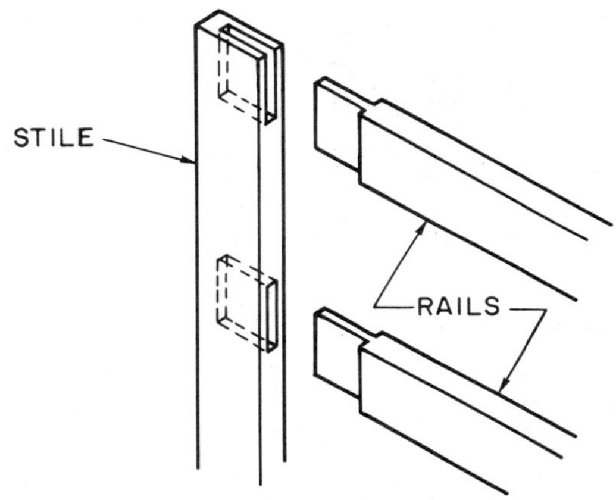

Fig. 22-7 Blind mortise-and-tenon joints join the stiles and rails.

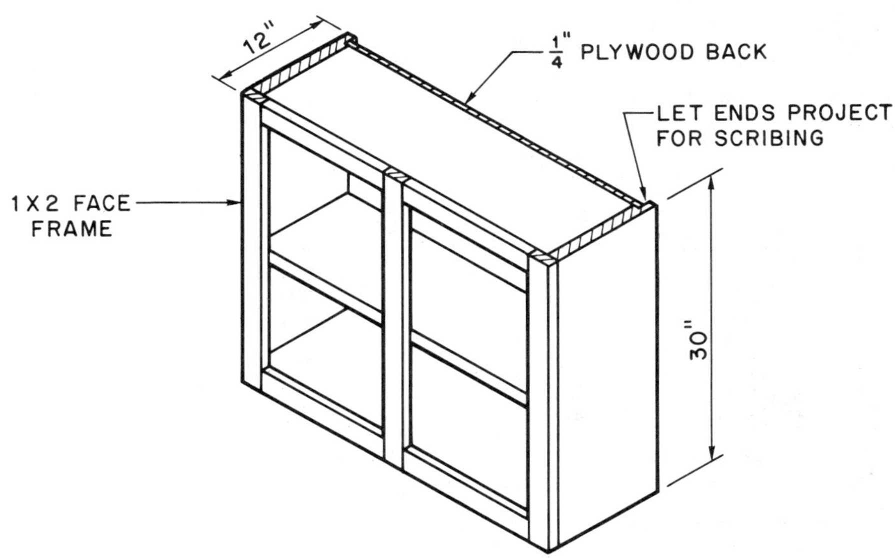

Fig. 22-8 Isometric view of typical wall unit

22-8. This makes it easier to clean the bottom shelf.

An allowance is made on the back edges of the end pieces for scribing. Mounting strips must be included in the wall units. Screws are driven through these strips to hold the cabinet on the wall.

MAKING A BASE UNIT

Layout and Cutting

1. Make a layout rod for a 3-foot wide base. Use figure 22-1 as a guide.
2. Develop a cutting list from the layout rod and drawing, figure 22-9.
3. Cut all the parts to their overall size.

Machining

1. Lay out and mortise the three face frame stiles. Make the mortises on the end stiles 7/8 inch deep so that the rail tenons will not bottom. Mortise the center stile from both edges so that the mortise goes all the way through.
2. Make the tenons on the four face frame rails 3/4 inch long. Relish the tenons as necessary to fit the mortises in the stiles.
3. Assemble the face frame by gluing and clamping. Remove from clamps after the glue has set, sand both faces, and set aside.
4. For this base unit, two skeleton frames are required. Cut a groove 1/4 inch wide by 1/2 inch deep on one edge of the four skeleton frame stiles.
5. Cut a stub tenon 1/4 inch thick by 1/2 inch long on all the skeleton frame rails, figure 22-10.
6. Lay out the location of the rails on the stiles and assemble the skeleton frames by clamping and gluing. When dry, remove the clamps and set aside until needed.
7. Rabbet the back edges of the ends 1/2" x 1/2". This rabbet will allow the end panels to project out beyond the plywood back for scribing the cabinet to the wall. Make sure the rabbet is cut from the inside face of the plywood so that the good face will show on the outside of the cabinet.
8. Lay out and cut the toe space on each end panel. The top of the cut is flush with the bottom rail of the face frame. The bottom of the cut is flush with the toe board.
9. Cut out the toe spaces so that a right hand and left hand end panel is obtained. Cut from the inside face to avoid splintering the face side.
10. Dado the inside of the end panels 3/16 inch deep and 3/4 inch wide for the bottom, shelf, and two skeleton frames. Make a blind dado for the cabinet shelf.

NAME OF PART	NO. OF POS.	THICKNESS	WIDTH	LENGTH
ENDS	2	3/4"	23 1/2"	34 1/2"
BOTTOM	1	3/4"	23"	34 7/8"
SHELF	1	3/4"	12"	34 7/8"
TOE BOARD	1	3/4"	3 1/2"	34 1/2"
SKELETON FRAME STILES	4	3/4"	1 1/2"	34 7/8"
SKELETON FRAME RAILS	4	3/4"	1 1/2"	21"
SKELETON FRAME CENTER RAIL	2	3/4"	3 1/2"	21"
FACE FRAME STILES	3	3/4"	1 1/2"	31 3/4"
FACE FRAME RAILS	6	3/4"	1 1/2"	17 1/4"
BACK	1	1/4"	34 1/2"	35 1/2"

Fig. 22-9 Cutting list for 36-inch base unit

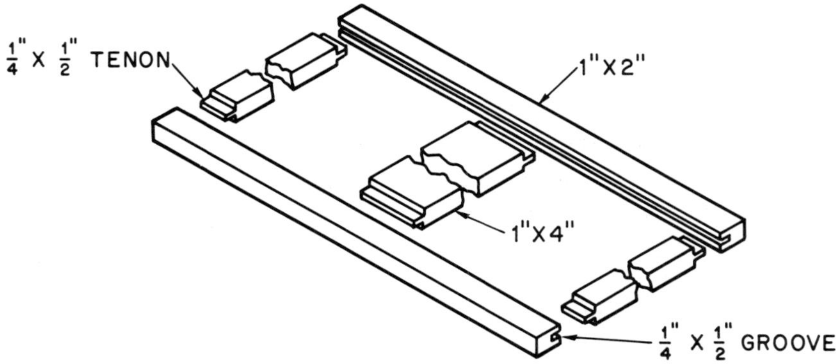

Fig. 22-10 Skeleton frame for base unit

Assembling

1. Assemble the end panels, bottom, shelf, and skeleton frames with glue and clamps, finish nails, or screws. Make sure all edges line up.
2. Install the toe board between the end panels by fastening with glue and finish nails through the end panels and down through the case bottom. Remove any bow by keeping the toe board the same distance from the front edge of the bottom all along its length.
3. The back helps hold the case rigid. The case must be absolutely square before fastening the back.
4. Fasten the back to the end panels, bottom, shelf, and skeleton frames. Straighten any bow when fastening.
5. Fasten the face frame to the front of the case, figure 22-11. Keep the top edge of the bottom rails flush with the top surface of the case bottom, and the outside stiles flush with the face of the case ends.

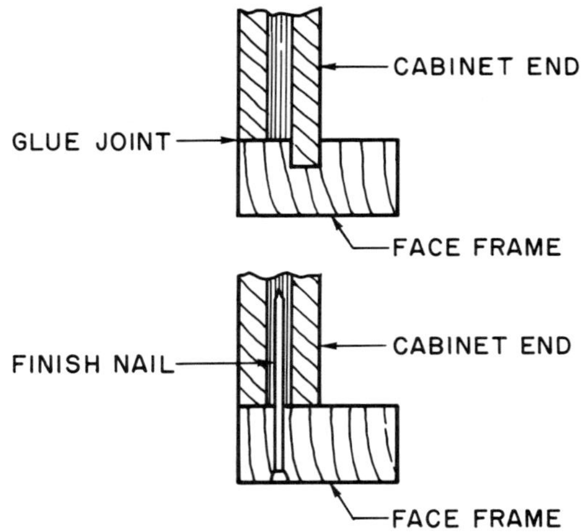

Fig. 22-11 Methods of fastening the face frame to the cabinet end

MAKING A WALL UNIT

First lay out a rod and make a cutting list, figure 22-12. Cut the parts to their overall size.

Machining

1. Lay out and make mortises on the three face frame stiles in a manner similar to that of the base unit.
2. Tenon the face frame rails 3/4 inch long and relish the tenons to fit the mortises.
3. Assemble the face frame by clamping and gluing. When dry, remove from the clamps and sand both faces. Set the face frame aside until needed.
4. Rabbet the back edges 1/2" x 1/2" on the end panels for the case back and to allow for scribing.
5. Lay out and dado the end panels for the top, bottom, and shelves 3/16 inch deep by 3/4 inch wide.

Assembling

1. Assemble the end panels, top, bottom, and shelves. Use glue and clamps or glue and finish nails according to the quality of the work.
2. Fasten the top and bottom mounting strips to the underside of the top and bottom of the case. Keep the strips flush with the back edges of the top and bottom.
3. Fasten the plywood back to the end panels, top, bottom, and shelves. Make sure the assembly is square. Take out any bow in the shelves when fastening the back.
4. Fasten the assembled face frame to the case. Keep the top edge of the bottom rails flush with the top surface of the case bottom, and the outside edges of the stiles flush with the case ends.

MAKING THE COUNTERTOP

1. From the layout rod of the base unit, make a cutting list for the countertop, figure 22-12. Cut out the parts.
2. Fasten the doubling strips to the underside of the countertop. Keep the edges flush with the countertop edges and ends.
3. To the top edges and ends of the backsplash, fasten the strips of 1/2" x 1" pieces with finish nails and glue.
4. After the backsplash and countertop are laminated, they are fastened together by driving screws of sufficient length up through the bottom of the countertop into the backsplash. Laminating is discussed in *Unit 25 Plastic Laminating.*

OTHER CONSTRUCTION METHODS

Kitchen cabinets are constructed in a number of different ways than those described in this unit, figures 22-14 and 22-15. Construction depends on the quality desired, the time required, and the materials used.

Kitchen cabinets may be made of hardwood plywood and solid hardwood or a combination of softwood plywood and solid softwood lumber. Often particleboard is used, sometimes with a vinyl coating on one side to eliminate finishing the inside

NAME OF PART	NO. OF POS.	THICKNESS	WIDTH	LENGTH
ENDS	2	3/4"	11 1/4"	30"
SHELVES	3	3/4"	10 3/4"	34 7/8"
BACK	1	1/4"	30"	35 1/2"
FACE FRAME STILES	3	3/4"	1 1/2"	30"
FACE FRAME RAILS	4	3/4"	1 1/2"	17 1/4"
TOP MOUNTING STRIP	1	3/4"	1 1/2"	34 1/2"
BOTTOM MOUNTING STRIP	1	3/4"	3/4"	34 1/2"

NOTE: Face frame rails are tenoned 3/4" long.

Fig. 22-12 Cutting list for 36-inch wall unit

NAME OF PART	NO. OF POS.	THICKNESS	WIDTH	LENGTH
TOP	1	3/4"	25"	36"
DOUBLING STRIPS	2	3/4"	2 1/2"	36"
	2	3/4"	2 1/2"	20"
BACKSPLASH	1	3/4"	3 1/2"	35"
BACKSPLASH SCRIBING	2	1/2"	1"	3 1/2"
STRIPS	1	1/2"	1"	36"

Fig. 22-13 Cutting list for 36-inch countertop

of the cabinet. Cabinets are also covered with plastic laminate (see *Unit 25 Plastic Laminating*).

In order to save time in many cases, the end panels are not dadoed to receive the interior piece. Skeleton frames are eliminated by some manufacturers. The end panels are then held together at the top by the back and the face frame. Sometimes the back is not installed and a 1″ x 3″ or 1″ x 4″ strip is used between the ends at the top flush with the back edge.

In the case of a sink unit, often only the face frame and toe board is provided and installed

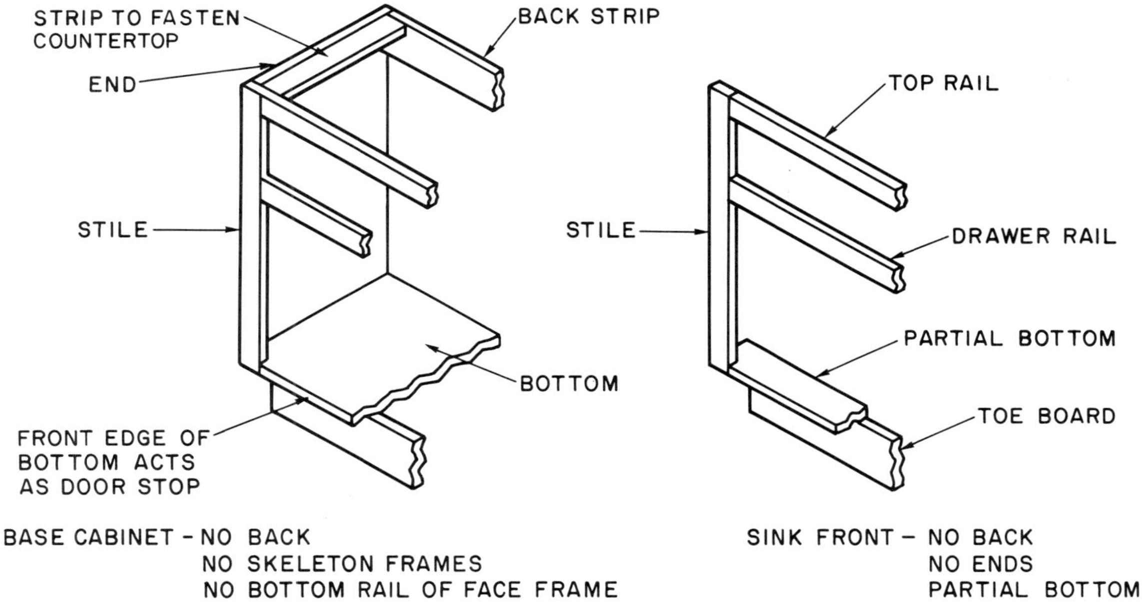

Fig. 22-14 Base kitchen cabinet construction methods

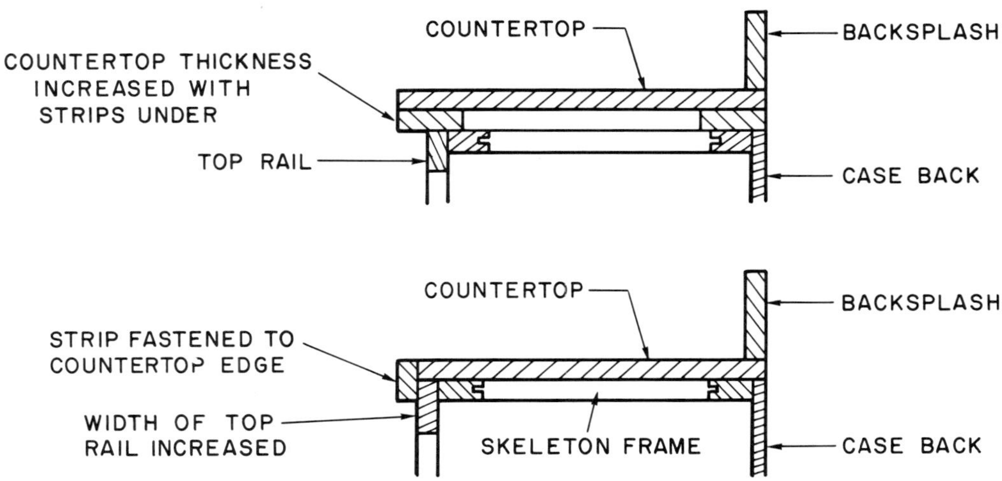

Fig. 22-15 Two methods of countertop construction

between two existing cabinets. The bottom in this case is usually not provided and is installed on the job.

Members of the face frame in lesser quality work are butted against each other. They are fastened together with power-driven corrugated fasteners on the inside of the frame. In some cases the bottom rail of the face frame is eliminated. The front edge of the bottom acts as the bottom rail of the face frame.

INSTALLING KITCHEN CABINETS

The cabinetmaker often is required to install kitchen cabinets. Cabinets must be installed in a straight, level, and plumb line. This requires skill because floors and walls are not level or plumb, especially in older buildings.

When installing a line of base and wall units, many installers prefer to mount the wall units first so that work does not have to be done over base units.

Installing Wall Units

The first step is to locate the bottom of the wall units 52 inches from the floor. This leaves a 16-inch space from the base unit countertop.

1. Using a level on the floor, find the low point between each end of the line of cabinets to be installed.
2. Measure up 52 inches from the low point and mark the wall.
3. Using a level and straightedge, draw a level line from the mark across the wall. The bottom of the wall units are installed to this line.

In a wood-frame wall, the studs must be located. Screws are driven into the studs to fasten the cabinets to the wall.

1. Just above the level line previously marked on the wall, lightly tap a hammer on and across the wall. When a solid sound is heard, you have located a stud.
2. To make sure, drive a finish nail in at the point the solid sound is heard. Drive the nail above the line so the nail holes are later covered by the cabinets.
3. If a stud is found, mark the location with a pencil.
4. Measure 16 inches in both directions from the first mark to locate the next studs. Drive a finish nail to test for solid wood. If studs are not found at 16-inch intervals, then tap the wall with a hammer to locate each stud.
5. At each stud, use a level and draw a plumb line down below the line for the bottom of the wall cabinets. Projecting below the wall units makes it easier to locate the studs when installing both wall and base units.

To install the wall unit:

1. Place the unit on a stand that holds it near the line of installation.
2. Level the unit up with wood shingles until the bottom of the unit is on the line of installation.
3. Test the front edge of the unit with a level for plumbness. If the unit is not plumb, shim it between the wall and its back edge with wood shingles until it is plumb.
4. Set dividers for the farthest distance the back edge of the unit is away from the wall.
5. Scribe the back edge by riding the dividers against the wall and marking the back edge of both end panels to the contour of the wall.
6. Take the cabinet down from the stand. Cut the back edges with a handsaw to the scribed line. A handsaw rather than a saber saw is used because a saber saw cuts on the upstroke and splinters out the face side while a handsaw cuts on the downstroke.
7. Place the cabinet back into position.
8. Drill holes through the mounting strips into the studs.

9. Fasten the cabinet in place with wood screws. Screws should be of sufficient length to hold the cabinet securely.

On masonry walls, first drill holes through the mounting strips, place the cabinet in position and mark the location of the drilled holes on the wall. Remove the cabinet and drill the holes into the masonry wall for lead inserts. Replace the cabinet and screw in place.

Adjacent cabinets are installed in the same manner. The back edges of these cabinets are scribed so their face frames are flush with the cabinet previously installed. Adjacent cabinets are fastened to each other by means of screws or bolts through the ends or through the stiles of the face frame.

Installing Base Cabinets

Before installing the base cabinets, draw a level line 16 inches plus the thickness of the countertop below the line previously drawn for the location of the wall units. This will be the location of the top of the base units without the countertop.

1. Place the first unit to be installed against the wall in the desired location.
2. Shim the bottom with wood shingles until the top is level across its length and width.
3. Adjust the dividers to correspond with the amount the top of the unit is above the line. Scribe this amount across both ends and toeboard by riding the dividers on the floor.
4. Cut both ends and toeboard to the scribed line.
5. Draw a line on the back between the cut ends and cut the back.
6. Place the unit back in position. The top edge should correspond with the layout line. If the back edges do not fit the wall, scribe them in a manner similar to wall units.
7. Scribe and fit adjacent units like the first unit. Fasten them to the wall and each other in the same manner as wall units.

Installing Countertops

After the base units are fastened in position, the countertop is laid on top of the units and against the wall.

1. Move the countertop, if necessary, so that it overhangs the same amount over the face frame of the base cabinets.
2. Adjust dividers for the difference between the amount of overhang and the desired amount of overhang. Scribe this amount on the backsplash if it has a scribing strip.
3. Cut the backsplash to the scribed line and fit it to the wall.
4. Fasten the countertop to the base cabinets with screws up through the top skeleton frame of the base units. Use a stop on the drill bit so you do not drill through the countertop.

In some cases, backsplashes are not built with scribing strips. To fit the backsplash to the wall, hold the countertop in the desired position. Press the backsplash against the wall at intervals and mark its outside face on the countertop. Remove the countertop and fasten the backsplash to the countertop on the marked lines. Fasten the countertop and backsplash in position.

Another method is to leave off the laminate on the face of the backsplash. Fasten the countertop in position. Hold the backsplash down tight on the countertop and nail it to the wall through its face. Then laminate the face of the backsplash on the job after it has been fastened in position. The disadvantage of this method is that it is difficult to remove the backsplash if the countertop has to be replaced.

ACTIVITIES

1. Look at manufacturers' catalogs and study the many different styles of kitchen cabinets. Note their dimensions and instructions for installaton.
2. Construct the base unit, wall unit, and countertop described in this unit. Be sure to develop a plan of procedure before starting any work.

3. Install the kitchen units constructed in #2. Doors and drawers for these cabinets will be made in later units.

4. Inspect the kitchen cabinet units in your home. How are they constructed? Note the dimensions and space between wall and base units.

UNIT REVIEW

Completion

1. The vertical distance between the base unit and wall unit is ________ inches.
2. The top shelf of the wall unit should not be over ________ feet from the floor.
3. The surface of the countertop is ________ inches from the floor.
4. The countertop is ________ inches wide and has a ________ inch high backsplash.
5. The countertop overhangs the base unit by ________ inch.
6. The drawer opening height in a base unit is ________ inches.
7. Typical wall units are usually ________ inches deep and ________ inches high.
8. A wall unit above a range is ________ inches high; above a sink it is ________ inches high; and above a refrigerator it is ________ inches high.
9. The vertical members of a face frame are called ________. The horizontal members of a face frame are called ________.
10. In both the wall and base units, the top edge of the bottom rail and the top surface of the bottom shelf are ________.

Multiple Choice

1. A projection is made on the backsplash and cabinet end panels
   a. to allow for shrinkage.
   b. to allow for damage in handling.
   c. to allow for scribing.
   d. to allow for mistakes.
2. The countertop is doubled up around the edges
   a. to give a heavier appearance.
   b. to make it stronger.
   c. to drive screws into.
   d. to save material.
3. Frames are installed inside the base unit
   a. to fasten the top of the cabinet back.
   b. to fasten the top of the cabinet ends.
   c. to fasten the countertop to the base unit.
   d. all of the above.
4. The face frame is usually made up of
   a. 1″ x 2″ stock.
   b. 1″ x 3″ stock.
   c. 1″ x 4″ stock.
   d. 1″ x 5″ stock.
5. Cabinets are mounted to masonry walls by
   a. fastening wood screws to the studs.
   b. driving screws into lead inserts.
   c. fastening with finishing nails.
   d. gluing and clamping in place.

6. To find the line of installation for the bottom edge of wall units, measure up from the low point on the floor
   a. 36 inches.
   b. 48 inches.
   c. 52 inches.
   d. 82 inches.

7. The depth of the dadoes in the end panels of the base unit is
   a. 1/4 inch.
   b. 3/16 inch.
   c. 1/8 inch.
   d. 5/16 inch.

8. A base unit is 36 inches wide and uses 3/4-inch plywood for end panels with a 3/16-inch dado. What is the length of the bottom platform?
   a. 34 1/2 inch
   b. 34 5/8 inch
   c. 34 3/4 inch
   d. 34 7/8 inch

9. To check a panel for squareness,
   a. use a combination square.
   b. measure the diagonals.
   c. use a framing square.
   d. all of the above.

10. When making a countertop, the top and backsplash are not joined until
   a. they have been sanded.
   b. they are installed.
   c. they are laminated.
   d. they are planed flush.

unit 23 Drawer Construction

OBJECTIVES

After studying this unit, the student will be able to:

- design a drawer to fit a face frame opening.
- construct overlay, lipped, and flush drawers.
- install drawers using common drawer guides.
- describe types of drawer pulls.

KINDS OF DRAWERS

Drawers are classified according to the type of drawer front. There are three basic kinds of drawers: the overlay drawer, the lipped drawer, and the flush drawer.

Overlay Drawer

The overlay drawer has a front that laps over the opening by 3/8 inch on all sides, figure 23-1. It is widely used on plastic laminated kitchen cabinets and desks.

The overlay drawer is easy to install because it does not require fitting in the opening. The four sides and the bottom of the drawer are assembled. The overlay front is then fastened to the false front, usually with screws from the inside.

Fig. 23-1 Overlay drawers

Lipped Drawer

The lipped drawer also covers the opening, figure 23-2, usually by 3/8 inch on all sides. It does not have a false front. Its front is rabbeted deeper on the ends than the top and bottom to

Fig. 23-2 Lipped drawers

receive the drawer sides. Usually the ends and edges of the front are rounded over to give a more pleasing appearance.

The lipped drawer is also easy to install because no fitting is required. However, it takes a little more time to make the front because of the extra milling. An advantage of overlay and lipped drawers is that the fronts act as stops for the drawer.

Flush Drawer

The flush drawer has a solid front that fits into and flush with the opening, figure 23-3. This drawer is a little more difficult to install because it has to be fitted in the opening. A 1/16-inch clearance (about the thickness of a dime) must be maintained between the drawer front and the opening.

In many cases, the edges and ends of the flush drawer front are shaped. The sides of the drawer are made up into the one-piece front. Stops are applied in the cabinet so the drawer will close flush.

DRAWER SIZES

Drawer dimensions are usually given as width, height, and depth in that order. The *width* of the drawer is the distance across the drawer opening. The *height* is the vertical distance of the opening. The *depth* is the distance from the front to the back.

Fig. 23-3 Flush drawers

Usually drawer fronts are made of 3/4-inch plywood or solid wood. The design must be in keeping with the cabinet. Sides and backs are generally 1/2-inch thick solid wood. Sides are made thicker if they are to be grooved for certain types of drawer guides. The drawer bottom is usually made of 1/4-inch plywood or hardboard. Small drawers may have 1/8-inch hardboard bottoms.

DRAWER JOINTS

Front and Side Joints

Typical joints between the front and sides of a drawer are the dovetail, lock, and rabbeted joints, figure 23-4.

The dovetail joint is used in high-quality furniture. It is difficult to make, but is the strongest. The lock joint is simpler to make and is still very strong. Though the rabbeted joint is the easiest to make, it is the weakest joint to use. It must be strengthened with screws, nails, or staples and glue.

Back and Side Joints

Typical joints between the drawer back and sides are the dovetail, dado and rabbet, dado, and butt joints, figure 23-5.

The dovetail joint is used in high-quality drawer construction. The dado and rabbet joint is easy to make and is usually set in from the ends of

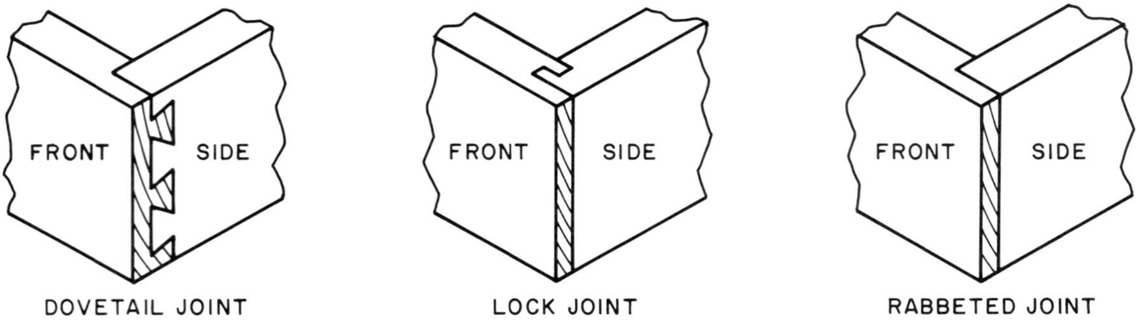

Fig. 23-4 Typical joints between drawer front and side

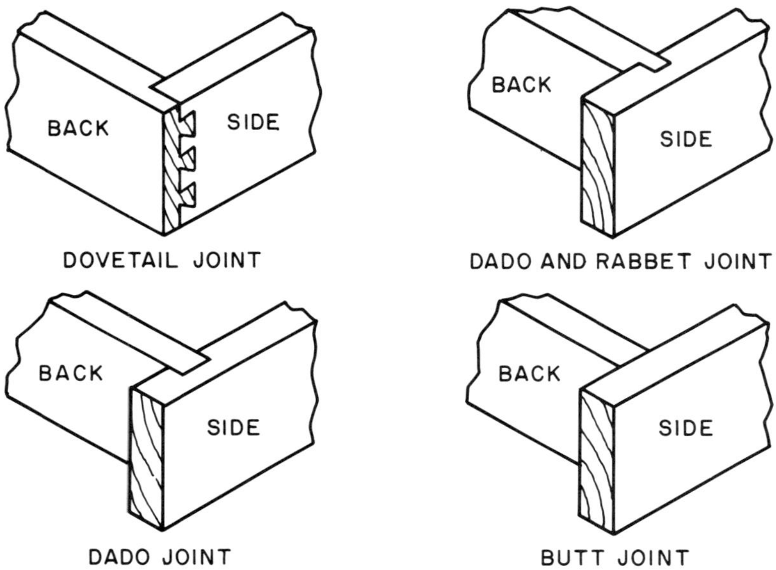

Fig. 23-5 Typical joints between drawer back and side

the sides. The dado joint is widely used. It requires less time to make because the ends of the backs need no rabbeting and are just squared off. This joint must be strengthened with fasteners in addition to glue. Care must be taken to hold the pieces in place while fastening.

Bottom Joints

In high-quality drawer construction, the bottom is fitted into a groove on all four sides, figure 23-6. An alternate method is to cut the back narrower, assemble the four sides, slip the bottom in,

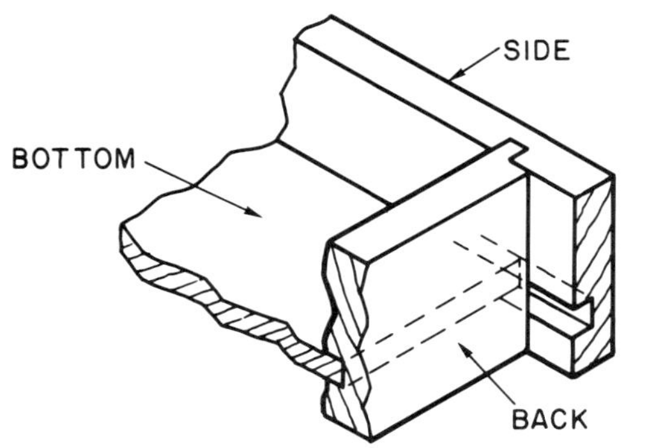

Fig. 23-6 Draw bottom fitted in groove at drawer back

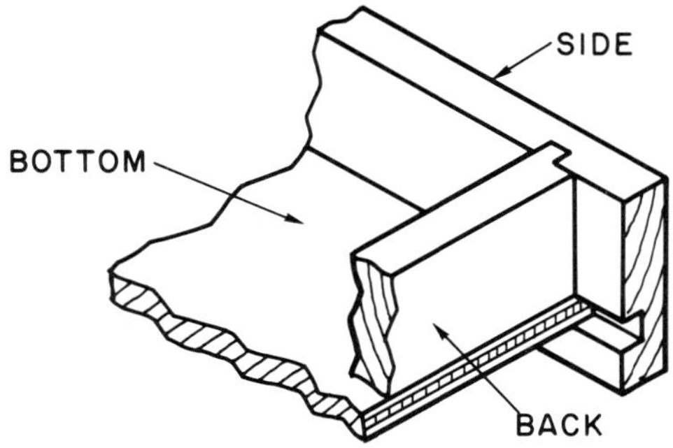

Fig. 23-7 Drawer bottom fastened to bottom edge of drawer back

and nail the bottom edge to the back of the drawer, figure 23-7. However, after a while the fasteners may work loose and cause the drawer bottom to sag.

On larger drawers, glue blocks are used between the sides of the drawer and the bottom. These blocks keep the sides from bowing out.

DRAWER GUIDES

There are many ways of guiding drawers in and out of the opening, figure 23-8. The type of drawer guide selected affects the size of the drawer and its parts. The drawer must be guided sideways and vertically. It also must be kept from tilting down when it is opened.

Wood Guides

Probably the simplest type wood guide is the center strip. It is installed in the bottom center of the opening from front to back, figure 23-9. The strip projects 1/4 inch above the bottom of the opening. The bottom of the back of the drawer is notched for the thickness of the strip. The drawer rides on the strip and is guided both vertically and sideways. An additional strip of wood, called a *kicker,* is installed above the drawer to keep it from tilting downward when the drawer is opened.

Another type of wood guide is the grooved center strip which is grooved on both edges, figure 23-10. This strip is placed in the center of the opening from front to back. A matching strip is fastened to the drawer bottom. This method keeps the drawer from tilting in addition to guiding it.

Grooved side guides are commonly used in casework construction. Grooves are made in the sides of the drawer. Strips of hardwood are fastened to the inside of the cabinet. The grooves in the drawer sides fit on the strips, figure 23-11. The sides of the cabinet act as side guides in this case, and the strips act as vertical guides and kickers. Care must be taken to mount the strips in the correct position. Wax applied to these strips, as well as all other wood guides, make the drawer run smoother.

Wood guides can also be made by installing *rabbeted side strips* on each side of the drawer opening. These strips are installed at the bottom of the opening from the front to the back of the cabinet, figure 23-12. The drawer sides fit into and slide along the rabbeted pieces. Sometimes these guides are made up of two pieces instead of being rabbeted. A kicker above the drawer is necessary with this type of guide.

Metal Drawer Guides

There are many different types of metal drawer guides available. Some have a single track mounted on the bottom center of the opening. Others may be mounted on the top center. Side metal guides mount either on the top or bottom of the drawer sides and opening, figure 23-13.

Instructions for installation differ with each type. When using commercially made drawer

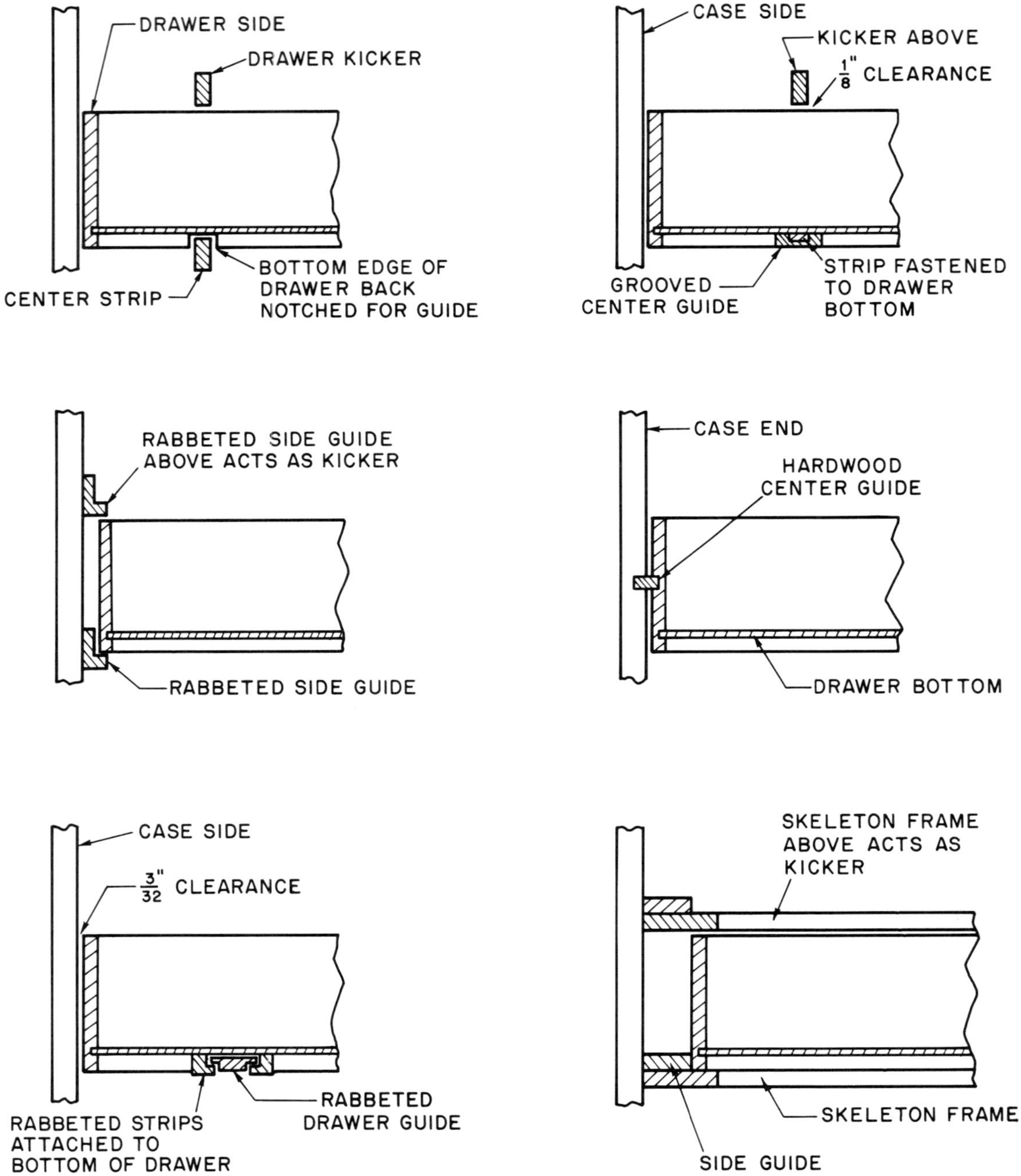

Fig. 23-8 Installing wood drawer guides

guides, read the instructions first before making the drawers. Allowances must be made for the installation of metal guides. Drawers usually have to be made smaller to accommodate the hardware.

Fig. 23-9 Simple center wood drawer guide. The back of the drawer is notched to run on the guide.

Fig. 23-10 Grooved center wood drawer guide

Fig. 23-11 Grooved drawer sides fit into guides.

DRAWER PULLS

Generally, drawers are opened and closed by means of pulls or knobs. They come in many different styles and designs, figure 23-14. They are made of metal, plastic, wood, procelain, or other materials.

Sometimes the pull is recessed in the bottom edge of the drawer front. This *stopped cove cut* is made with either a router or shaper using a cove-cutting router bit or a cove-shaped cutter. Recessed drawer pulls can only be made on overlay drawer fronts, figure 23-15.

Drawer pulls are usually fastened by drilling holes through the drawer fronts. The drawer pull is inserted and screwed from the inside of the

Fig. 23-12 Rabbeted wood side guides

Fig. 23-13 Commercial type side drawer guides

Fig. 23-14 Pulls and knobs *(Amerock Corp.)*

drawer, figure 23-16. The holes are drilled slightly larger than the screws in case the holes are slightly off center.

MAKING AN OVERLAY DRAWER

Cutting the Parts

From 1/2-inch solid stock, cut the drawer sides, back, and front. The width should be at least 1/8 inch less than the height of the drawer opening. This width may be less if commercial metal drawer guides are used. The length of the sides should be at least 1/2 inch less than the depth of the cabinet. The length of the front and back must allow clearance between the drawer sides and opening.

From 1/4-inch plywood or hardboard, cut the drawer bottom to size. Allow 1/16-inch clearance in the grooves. Do not cut the bottom oversize.

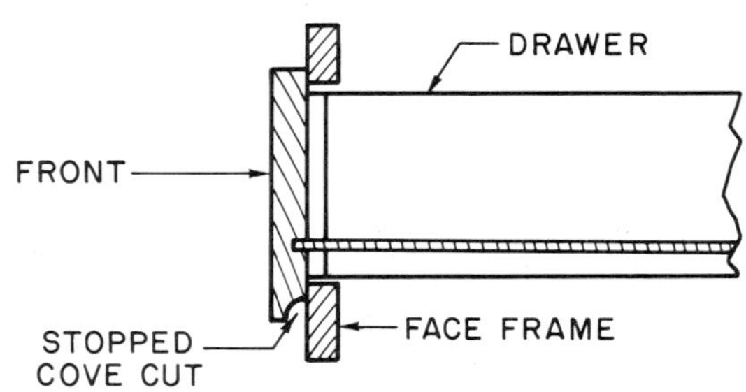

Fig. 23-15 Stopped cove cut

From 3/4-inch solid stock, cut the overlay front. It should be 3/8-inch wider and longer than the false front.

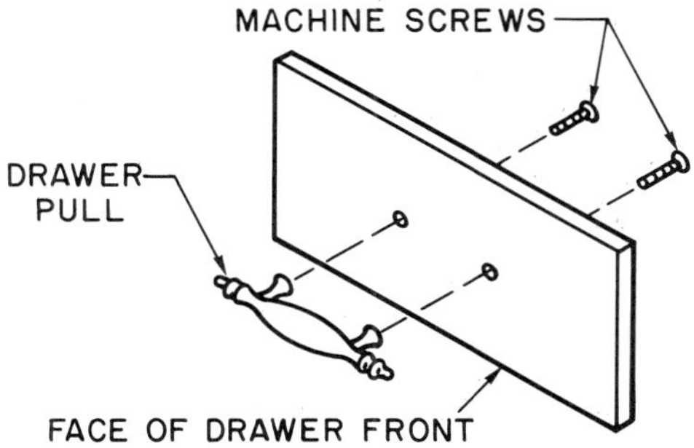

Fig. 23-16 Fastening drawer pulls

Machining the Parts

Dado the sides for the back of the drawer 1/4 inch deep. Rabbet for the front of the drawer the same depth, figure 23-17. Making these cuts the same depth allow the front and back to be cut to the same length.

Smooth and round off the top edges of the drawer sides, front, and back.

Make 1/4-inch deep grooves on all the parts for the drawer bottom. If the groove is too deep, it will weaken the sides, front, and back. If the groove is too shallow, the bottom may fall out after assembly. The groove should be wide enough to allow the bottom to slide in freely. If the groove is too narrow, assembly will be difficult. The distance from the bottom edge should be 3/8 inch to the groove, figure 23-18.

When cutting the grooves, alternate the side pieces so that both right and left sides will be obtained. If not carefully done, it is possible to end up with all right-hand sides or all left-hand sides.

Assembling the Parts

Glue and fasten the sides to the back with 4d coated box nails. Slip in the bottom. (Do not apply glue to the bottom.) Fasten the false front to the sides in the same manner as the back. Place the overlay front on the drawer and tack in position with brads. It should overlay all sides of the opening by 3/8 inch. Make sure the front is aligned. Adjust if necessary. Fasten the front to the drawer by driving screws in from the inside of the drawer.

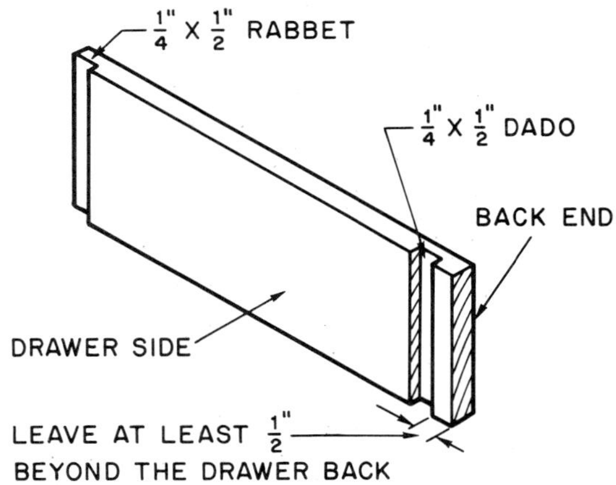

Fig. 23-17 Cuts in the side of an overlay drawer for the front and back

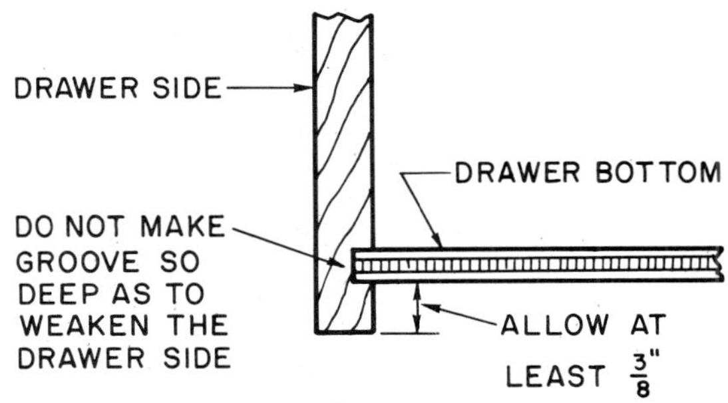

Fig. 23-18 Location of groove for drawer bottom

MAKING A LIPPED DRAWER

The sides, back, and bottom of a lipped drawer are measured the same as for the overlay drawer. However, the lipped drawer does not have a false front. The sides of the drawer are made up to the lipped front. The front is cut to its overall size so that it overlaps the opening by 3/8 inch on all sides.

Make a rabbet 3/8″ x 3/8″ on the bottom edge. On the top edge make a rabbet 3/8″ x 1/2″ to allow 1/8-inch clearance at the top. On each end make rabbets 3/8″ x 15/16″ to allow for the overlap, the thickness of the drawer sides, and side clearance, figure 23-19. Round over the ends and edges of the drawer front.

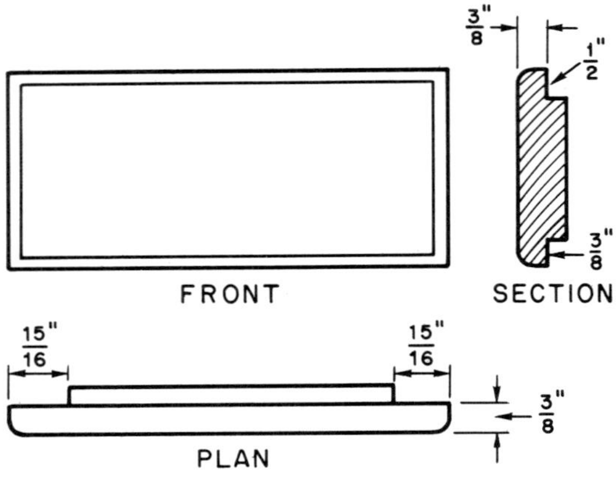

Fig. 23-19 The lipped drawer front

The drawer is assembled in a manner similar to the overlay drawer.

MAKING A FLUSH DRAWER

The sides, back, and bottom of a flush drawer are measured the same as for the overlay and lipped drawers. The front must fit exactly in the drawer opening.

Cut out the front to the overall width and height of the drawer opening. Rabbet the two ends of the drawer front to receive the sides of the drawer and to allow the proper side clearance. Make the necessary cuts to fasten the sides to the front according to the type of joint to be used. Allow the necessary side clearance according to the type of drawer guide to be used.

After the drawer is assembled, try it in the opening. Fit the drawer front to the opening by hand planing until the proper joint is obtained.

ACTIVITIES

1. Inspect the drawer construction of several pieces of furniture. Note the type of drawer, the joints used, and the method used to guide the drawer.
2. Make a display to show the different ways of guiding drawers with wood guides.
3. Construct overlay, lipped, or flush drawers for the base unit constructed in *Unit 22 Kitchen Cabinets.* Install the drawers. Explain your choice of type of drawer, joints, door pulls, and drawer guides.

UNIT REVIEW

Multiple Choice

1. The overlay drawer front overlaps the opening by
   a. 1/4 inch. c. 1/2 inch.
   b. 3/8 inch. d. 1 inch.
2. When selecting joints to be used in drawer construction, an important factor is the
   a. type of drawer guide. c. width of the drawer.
   b. kind of wood used. d. strength of the joints.
3. A drawer front that is rabbeted on both edges and both ends is the
   a. overlay front. c. flush front.
   b. lipped front. d. lapped-over front.
4. When cutting the bottom for a drawer, allow a clearance of _________ in the grooves.
   a. 1/32 inch c. 3/32 inch
   b. 1/16 inch d. 1/8 inch

5. Drawer bottoms are usually
   a. 1/8 inch thick.
   b. 1/4 inch thick.
   c. 3/8 inch thick.
   d. 1/2 inch thick.

6. A strip of wood placed above the drawer to keep it from dropping down when it is opened is called a
   a. kicker.
   b. top guide strip.
   c. hold-down strip.
   d. vertical guide strip.

7. A joint that can be used in place of the dovetail joint because it is easier to make is the
   a. rabbet and dado joint.
   b. dado joint.
   c. lock joint.
   d. rabbeted joint.

8. The groove for the bottom of the drawer should be made with a
   a. slide fit.
   b. tight fit.
   c. loose fit.
   d. none of the above.

9. Drawer dimensions are given in this order:
   a. height, width, depth.
   b. width, depth, height.
   c. width, height, depth.
   d. depth, height, width.

10. Overlay and lipped drawers
   a. use less material than flush drawers.
   b. do not require stops.
   c. take less time to make than flush drawers.
   d. do not require fitting.

Questions

1. Which type of drawer has a false front?
2. Which type of drawer has a front that fits inside the opening?
3. Which type of drawer needs stops installed inside the cabinet?
4. Which type of joint is used most in high-quality drawer construction?
5. Which joints are used to join the sides and the back of a drawer?
6. What is the purpose of installing glue blocks between the sides and bottom of larger drawers?
7. Why may wax be applied to wood drawer guides?
8. In one method of drawer construction, the bottom is fitted into grooves on all four sides. What happens if this groove is too deep? If it is too shallow?
9. How is the front of a flush drawer fitted to the opening?
10. How are recessed drawer pulls made?

unit 24
Cabinet Doors

OBJECTIVES

After studying this unit, the student will be able to:

- classify cabinet doors according to the method of construction and installation.
- make solid and paneled cabinet doors.
- cover door edges.
- install overlay, lipped, flush, and sliding doors.

CABINET DOORS

The cabinetmaker has the option of buying ready-made cabinet doors from a mill. These are available in a variety of styles. However, the cabinetmaker must often design and construct a door for a project.

Cabinet doors are classified according to their construction and installation.

Construction

Doors are constructed as solid, flexible, or paneled doors.

Solid doors are made of plywood, hardboard, particleboard, or glued-up solid lumber, figure 24-1. Designs are often grooved into the door with a router, or molding may be applied to give the door a more attractive appearance.

Flexible doors are made of thin strips glued together on a canvas back or held together with special edge joints, figure 24-2. They are used on roll-top desks and other cabinets when the door must slide around a corner.

Paneled doors have an exterior framework of solid wood and a center containing one or more panels, figure 24-3. The panels may be solid wood, plywood, hardboard, metal, plastic, glass, or some other material and come in many different designs. The exterior framework can be shaped in a number of ways also.

Fig. 24-1 Solid flush doors and drawers

Installation

Cabinet doors can be installed as overlay, lipped, flush, sliding, or folding doors.

Overlay doors cover the opening, usually by 3/8 inch on all sides, and swing on overlay hinges, see figure 24-3. *Lipped doors* are rabbeted over the opening and swing on offset hinges, figure 24-4. *Flush doors* fit inside the opening and swing on either surface hinges or butt hinges, see figure 24-1.

Fig. 24-2 Flexible door

Fig. 24-3 Paneled overlay doors

Sliding doors roll on tracks of metal or plastic, figure 24-5.

MAKING SOLID DOORS

To make a solid door from plywood, particleboard, or hardboard, lay out the dimensions of the door on the stock and cut to size. Use a sharp, 60 or 80-tooth, triple-chip carbide blade to produce a smooth, splinter-free edge. If no edge treatment is needed on the door, simply smooth the sharp corners with sandpaper. The door is ready for hanging.

To make a solid door from glued-up lumber, first cut out the pieces to rough size. Glue up the pieces. If no edge treatment is needed, use joints that will not show at the edges and ends, such as doweled, butt, or shaped joints. After the glue has set, cut the stock to size. Scrape and sand all surfaces. Smooth sharp corners with sandpaper.

Applying Designs to Solid Doors

Designs may be grooved into the surface of the door with a door routing jig and templates,

Fig. 24-4 Paneled lipped doors and drawers

Fig. 24-5 Solid sliding doors

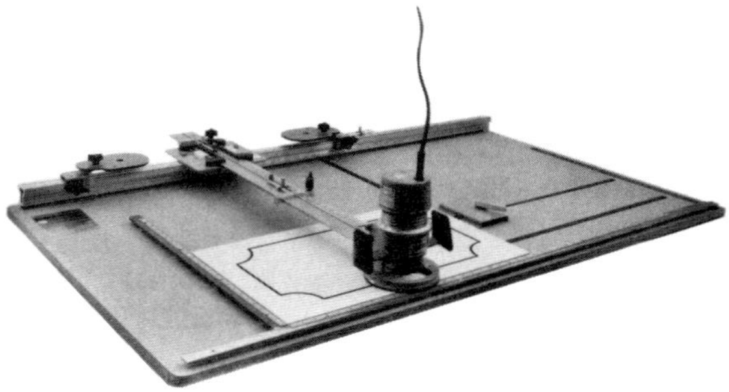

Fig. 24-6 Door routing jig and template

figure 24-6. The door is held in an adjustable clamp. Then the router is carefully guided along the template to make the various cuts.

If a door routing jig is not available, a frame can be made to hold the door, figure 24-7. A simple way to make this French provincial design is to remove the round base of the router and replace with an hourglass-shaped base. (Instructions for making this base are given in *Unit 29 Pattern Details.*)

1. With a small veining bit in the router, start routing about 1/8 inch deep along one side.
2. Bring the router into the corner until the base hits the outer side.
3. Rotate the router as in figure 24-8 until the cut is complete.

Another way to enhance the design of a solid door is to apply molding to it, figure 24-9. Molding comes in a variety of shapes and sizes (see *Unit 36 Molding*). It is applied to the door face with glue and brads.

MAKING FLEXIBLE DOORS

Flexible doors, such as those found in roll-top desks, are often called *tambour doors.* Flexible doors may be installed to roll or slide up and down or sideways. Grooves are routed in the interior of the case for the ends of the door to slide in.

Fig. 24-7 Frame for routing door

The flexible door is made by cutting the slats to the size desired. The thickness and width of the slats are determined by the width of the groove made in the case and the radius of the curve around which the door must slide.

1. Clamp the number of slats required together tightly on a flat surface.
2. Apply three coats of contact cement to the slats and to a piece of canvas large enough to cover the assembled slats.
3. Press the canvas to the slats.
4. Cut off the excess canvas and rub the ends of the slats with paraffin.

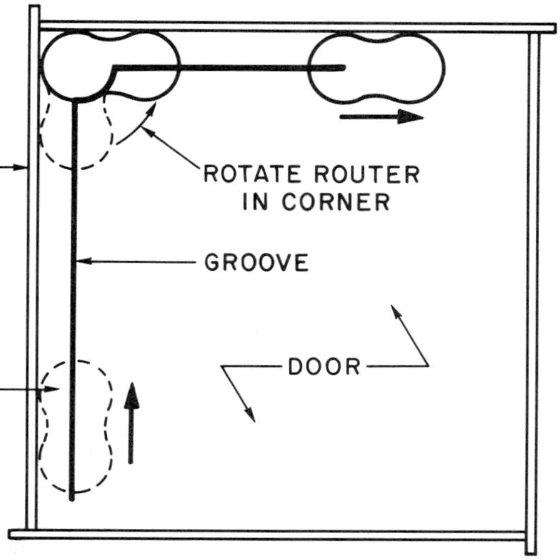

Fig. 24-8 Rotating the router in the corner makes a curved cut.

Fig. 24-9 Molding applied to a solid door face

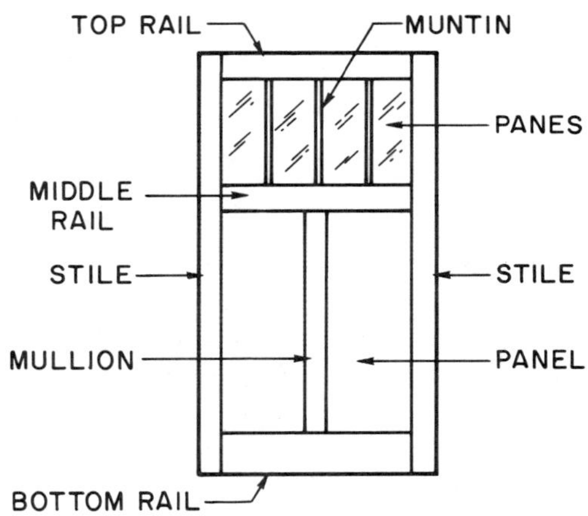

Fig. 24-10 Paneled door and its parts

MAKING PANELED DOORS

Although paneled doors take more time to make, they are preferred because of the wide range of designs possible. This type of door also has less tendency to warp than solid doors.

The paneled door consists of a frame and panels, figure 24-10. The frame is made of solid wood. The outside vertical pieces are called *stiles*. The horizontal pieces are called *rails*. Vertical dividing strips are called *mullions*. Small pieces that divide lights *(panes)* of glass are called *muntins*. The shape of the inside edges of the frame is called the *sticking*. *Panels* cover the area inside the framework.

Frames are either grooved or rabbeted to hold the panel, figure 24-11. If there is danger of the

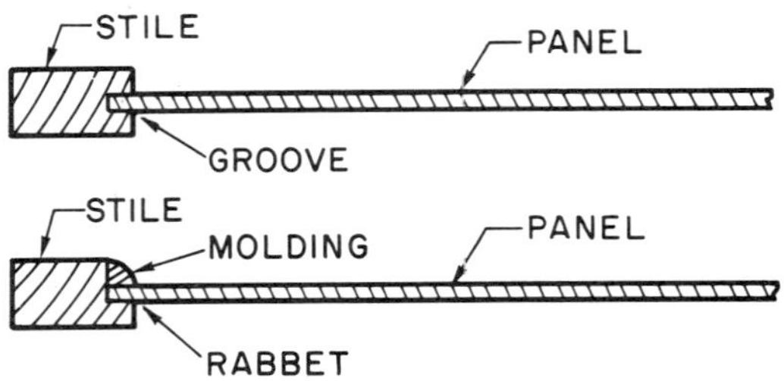

Fig. 24-11 Grooved and rabbeted frames for panels

panel being broken, such as a glass panel, the frame is rabbeted. The panel is laid in the rabbet and held in place with molding. Otherwise, the frame is grooved to hold the panel. Mortise-and-tenon or doweled joints hold the frame together.

The sticking may be square-edge or molded, figure 24-12. Panels may be flat or raised on one or both sides. They may be made of materials other than wood.

Making the Frame

To make the frame, cut all pieces to the desired thickness, width, and length. Remember to allow extra length for making the joints. For square-edge sticking, the inside edges of all parts of the frame are grooved or rabbeted. The joints are then made.

For small shops that have limited equipment, it is better to make frames with square-edge sticking. Molding is applied to the inside edges of the frame to give the same effect as molded sticking.

When all the parts are cut, the frame is assembled dry to make sure all joints fit properly, then taken apart. The panel is cut to size and shaped. If the frame is grooved, all parts are assembled and glued and clamped together. If the frame is rabbeted, the frame only is assembled. Then the panel is laid in and molding applied to hold it in place.

Making Panels

Straight or plain panels of plywood, hardboard, or solid stock are made by cutting the panel to the correct size. The thickness of the panel edges must be the same as the groove in the pieces of the

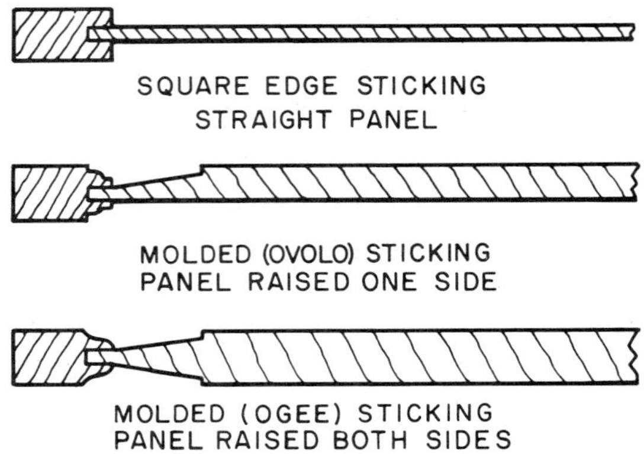

Fig. 24-12 Kinds of sticking and panels

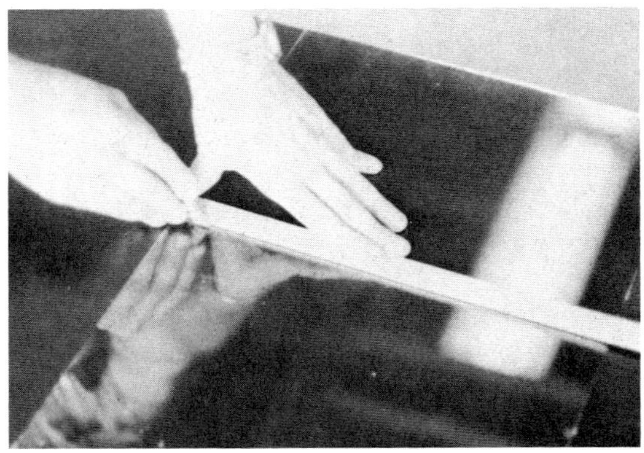

Fig. 24-13 Scoring the glass (gloves are removed for clarity)

Fig. 24-14 Breaking the glass along the scored line (gloves are removed for clarity)

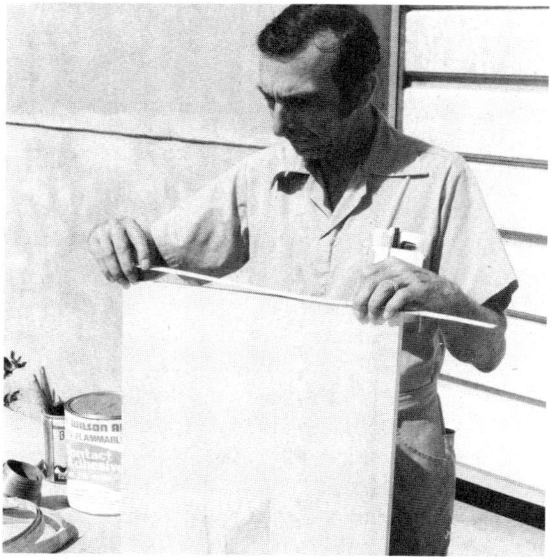

Fig. 24-15 Applying veneer strips to door edges

frame. For a panel that is thicker than the width of the groove, the edges of the panel are rabbeted to fit into the groove.

Raised panels are cut on a shaper, jointer, table saw, or radial arm saw. Many different effects can be achieved. Raised panels with curved edges can only be made by using the shaper.

Cutting Glass for the Panels

Panes of glass must often be cut for the panels.

Caution: Always wear gloves when cutting glass. Remove broken or unused pieces from the work area.

1. Lay the glass on a clean, smooth surface. Brush some mineral spirits along the line of cut.
2. Hold a straightedge on the line to be cut. Draw a glass cutter along the line and make a sharp, uniform *score* (cut), figure 24-13. Do not go over the scored line because it will dull the glass cutter. The line must be scored along the whole length the first time with no skips. Otherwise the glass may not break where desired.
3. Move the glass so the scored line is even with the edge of the workbench.
4. Apply downward pressure on the overhanging glass. If the glass is properly scored, it will break along the scored line, figure 24-14.
5. Slip the glass into the rabbet. Apply molding to hold it in place.

COVERING EDGES

On high-quality cabinetwork, end grain of lumber is either completely hidden or exposed as little as possible. End grain shows all around plywood doors because of the alternate direction of the plies. On solid wood doors, both ends also need to be covered to hide the end grain.

Using Plastic Laminates

Doors edges can be laminated in the same manner as kitchen countertops (see *Unit 25 Plastic Laminating*). The laminate is first applied to the door edges and trimmed off. The faces of the door are then covered and the excess laminate trimmed flush.

Using Veneers

Another method is to apply a thin veneer of the same material as the door to all the edges, figure 24-15. These veneers may be purchased in rolls of different kinds of wood. Veneers may also be made on the job by slicing thin strips slightly wider than the thickness of the door from the scrap material left over after making the doors.

The veneers are applied to the door edges with contact cement. Coat both the underside of the strips and the door edges with the cement and let dry. Apply the strips to the edges of the door, making tight joints at the corners. Trim the excess with a sharp knife and sandpaper.

Plastic Edge Moldings and Tapes

Rolls of plastic tape are available to cover edges of plywood and other core material. These tapes come in a variety of colors and wood grain and are backed with a heat-sensitive adhesive. They are applied by simply pressing the tape with a heating iron on to the edge and trimming flush, figure 24-16.

Plastic edge moldings are widely used to cover edges. A channel type, figure 24-17, fits completely over the edge. Another type has a grooved tongue that presses into a saw cut made in the edge, figure 24-18. These moldings are better used to cover the exposed edges of shelves or round tabletops where there are no corner joints.

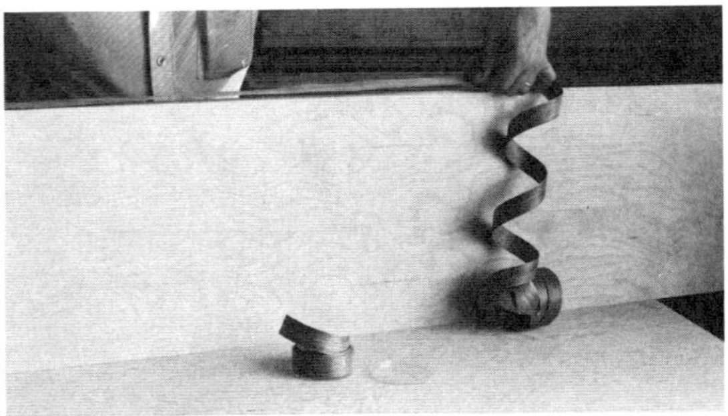

Fig. 24-16 Covering plywood edges with plastic tape

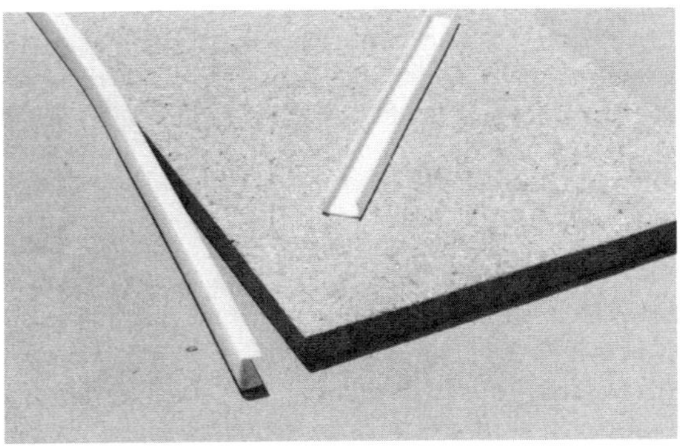

Fig. 24-17 Channel-type plastic edge molding

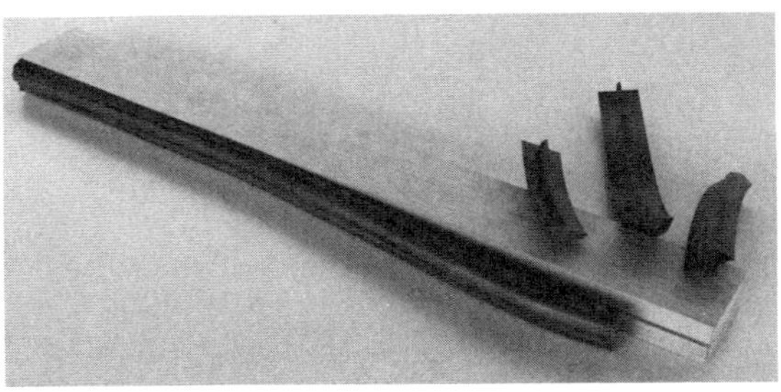

Fig. 24-18 Tongued plastic edge molding

Banding the Edges with Wood

Strips of wood of the same material as the door are used often to band the edges. The strips may be thin, from 1/4 to 1/2 inch and applied with glue and brads. The corners are mitered to avoid exposure of any end grain.

Strips of screen molding may also be used and are applied in the same manner as flat strips. The molding should be the same width as the thickness of the door edge and have tight-fitting mitered joints at the corners.

Wide strips of square edge stock are also used to band edges. These strips are joined to the edge with a tongue and grooved joint, figure 24-19. They are usually glued in place and held by three-way edging clamps.

HANGING CABINET DOORS

Overlay Doors

An overlay door is simply laid over the opening and hung in place. The door is cut to size so that it overlaps the opening by usually 3/4 inch on all sides. Hinges are applied to the door, the door is set into position, and overlay hinges are fastened to the frame. Use a straightedge to line up the door properly.

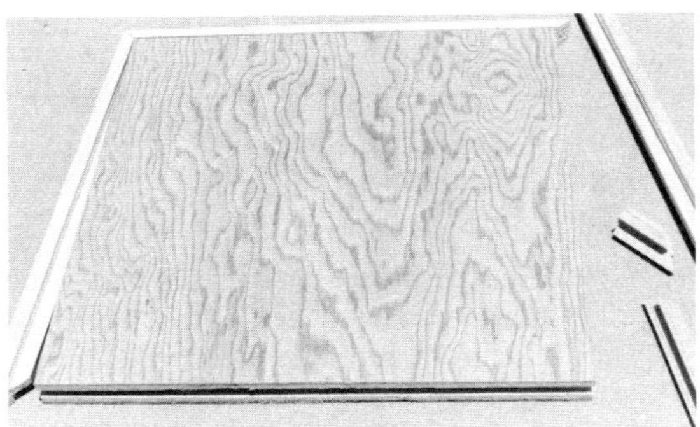

Fig. 24-19 Tongue-and-grooved edge band

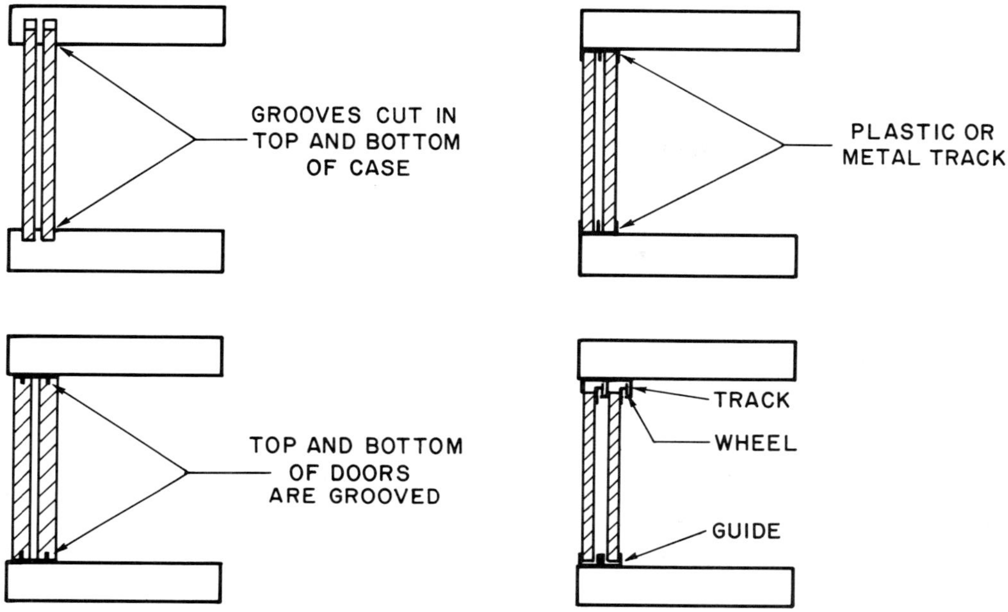

Fig. 24-20 Ways of installing sliding doors

Lipped Doors

Lipped doors are made by rabbeting out all edges of the door. Hinges for lipped doors are offset (bent) for a 3/8" x 3/8" rabbet on the hinge side of the door. On the other three sides make the rabbet 3/8" x 1/2" to allow clearance between the door and frame. Apply the hinges to the door. Line up the doors on the openings and fasten the hinges to the frame. Keep the hinge side against the opening and center the doors top to bottom.

Flush Doors

To hang a flush door, it first must be jointed in the opening. Cut the door oversize. Hand plane it until a clearance of 1/16 inch is obtained all around the door. Shim the door securely in position. If surface hinges are used, simply fasten them to the door and frame while the door is held in position. The pin of the hinge is centered on the joint. If butt hinges are used, proceed in the same manner as hinging the chest cover in *Unit 21 Casework Construction.*

Sliding Doors

There are a number of ways to install sliding doors, figure 24-20. One method is to cut two grooves in the top and bottom of the case near the front edge. The grooves in the top are made deeper so that the doors can be installed after the case is assembled and also removed when necessary. To install the doors, the top edge of the door is pushed all the way into the top grooves. The door height is made so that when dropped into the bottom groove, the door still remains in the top groove. The doors are removed in a similar manner. Paraffin is rubbed on the bottom edges of the doors to make them slide easier.

Sliding doors are also installed by using metal or plastic track. The upper track, installed at the top of the case, has grooves deeper than the lower track. The height of the doors is such that when the door is all the way in the groove of the top track, it just clears the bottom track. Metal or plastic track is available to accommodate doors from 1/4 inch to 3/4 inch in thickness.

Hardware is also available to hang sliding doors from the top. A double track is installed in the top of the case. Wheels that ride in the track are installed on the doors. A guide is installed on the bottom of the case to keep the door bottoms in line.

Sliding doors often have grooves cut into the bottom and top edges that ride in tracks with ridges. Sometimes ball-bearing wheels are installed in the bottom groove of the door for easy operation.

Special sliding door hardware is available for installing sliding glass doors.

Folding Doors

Folding doors are installed with hardware specially designed for this purpose. The hardware consists of a track which is installed at the top of the case or opening. Adjustable pins are inserted on the top and bottom of one door near the edge where it pivots. Hinges hold the required number of folding doors together. A spring-loaded guide is installed in the top edge of the last door and rides in the track at the top.

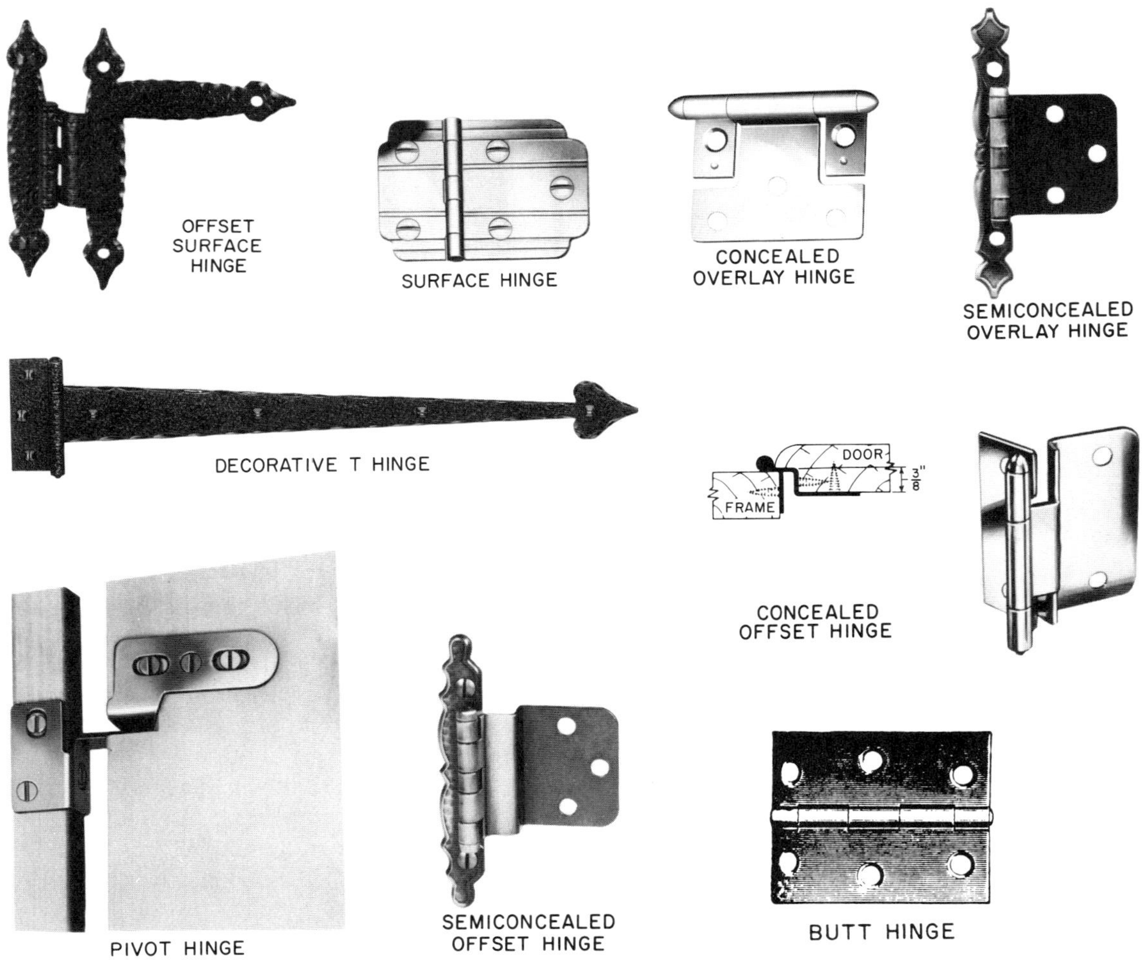

Fig. 24-21 Cabinet door hinges *(Amerock Corp.)*

HINGES

Hinges are made in many styles and shapes. If the kind of hinge is not specified, select a design that blends well with the cabinet being constructed. Some types of hinges are the surface, butt, offset, overlay, pivot, and continuous or piano hinge, figure 24-21.

The *surface hinge* mounts on the exterior surface of the door and frame. It is made straight for flush doors or offset for lipped doors. This type of hinge is used when it is desirable to show the hardware, such as early American furniture.

The *butt hinge* is used on flush doors when little hardware must show. When installed, only the pin of the butt hinge shows when the door is closed. These hinges require a little extra time to install.

The *overlay hinge* is used on overlay doors. The *offset hinge* is used on lipped doors. Both of these are available in *semiconcealed* or *concealed* types. Only one leaf is exposed with the semiconcealed type. One leaf is fastened to the back

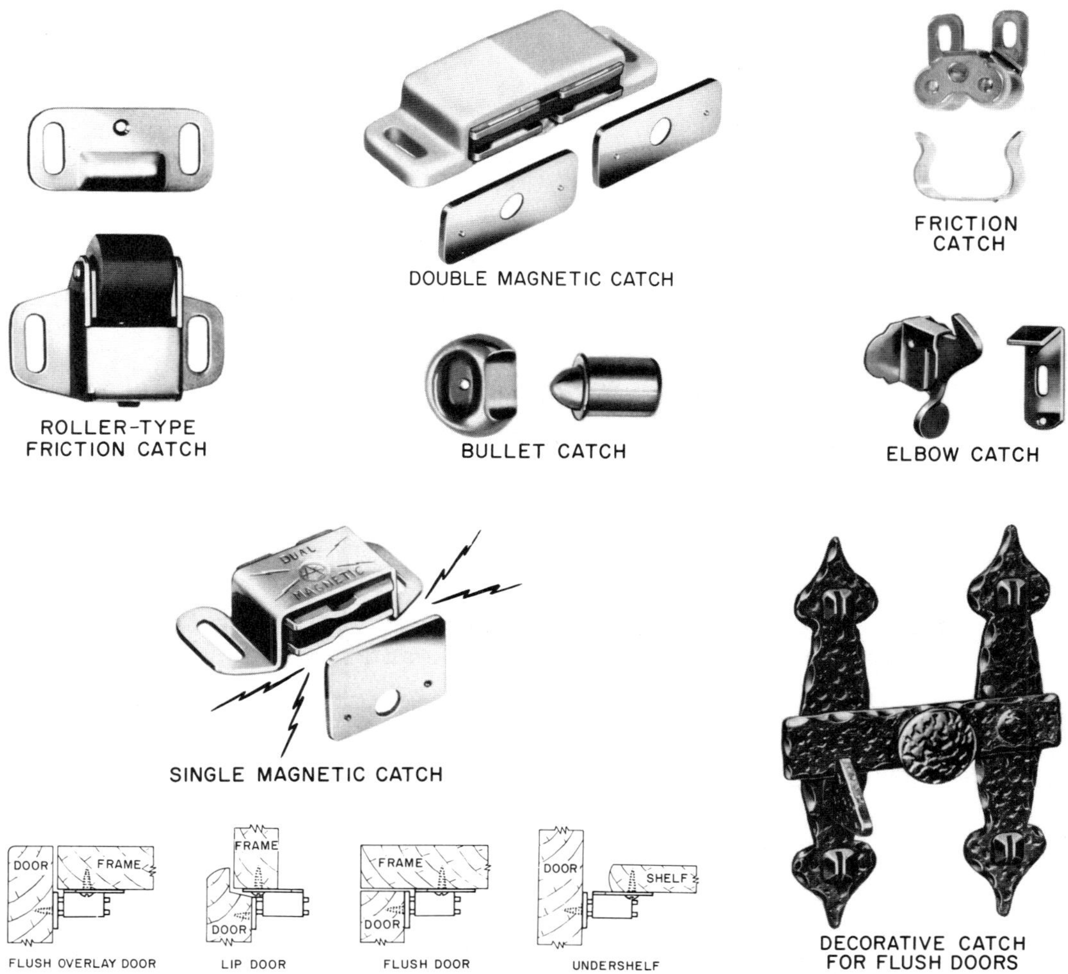

Fig. 24-22 Catches *(Amerock Corp.)*

of the door and the other is screwed to the face of the frame. The concealed type is completely hidden except for the hinge pin.

The *pivot hinge* is used on overlay doors. It is fastened to the top and bottom of the door and to the inside of the case. It is used frequently when there is no face frame on the case. The doors completely cover the face of the case.

The *continuous* or *piano hinge* is a one-piece hinge that usually extends the whole length of the door. It is installed like a butt hinge, and only the hinge pin is exposed. This type of hinge is used when the door is subjected to heavy use.

CATCHES

Some hinges are self-closing and therefore eliminate the need for installing catches to hold the door closed. Others require catches, figure 24-22. There are many kinds of catches available for holding doors.

Catches should be placed in the most out-of-the-way position possible. For instance, they are placed on the underside of shelves instead of on top.

Magnetic catches are used widely. They are available in single or double magnets of varying holding power. An adjustable magnet is attached to the inside of the case and a metal plate to the door. Other types of catches are the *roller type* and the *friction* type.

Elbow-type catches are used to hold one door of a double set. It must be released by reaching in back of the door. These are used when one of the doors is locked against the other.

Bullet catches are spring-loaded and fit into the edge of the door. When the door is closed, the catch fits into a recessed plate mounted on the frame.

ACTIVITIES

1. Study several pieces of furniture containing cabinet doors. Note the method of construction and installation. What type of hinges and catches are used? List at least five different types.
2. Make a solid door of plywood and solid lumber.
3. Edge band a solid door with veneer and with wide, thin strips of solid lumber.
4. Make a paneled door with doweled joints, rabbeted edges, and square-edge sticking. Set in a plain panel held in place with molding.
5. Make a paneled door with blind mortise-and-tenon joints, grooved edges, and square-edge sticking. Set in a panel raised on one side.
6. Design, construct, and install cabinet doors for the wall and base units made in *Unit 22 Kitchen Cabinets.* Explain your choice of construction and installation.

UNIT REVIEW

Completion

1. Three types of doors classified according to their construction are the ________, ________, and ________ doors.
2. Five types of doors classified by the method of installation are the ________, ________, ________, ________, and ________ doors.
3. Designs may be grooved into the surface of a solid door with a router and a ________.
4. The frame of a paneled door is either ________ or ________ to hold the panel.

5. The shape of the inside edge of the frame in a paneled door is called the ________.

6. Raised panels with curved edges should be made using the ________.

7. Three ways of covering the edges of doors are by using ________, ________, or ________.

8. A flush door must be ________ in the opening.

9. A ________ hinge is frequently used when there is no face frame on a case.

10. ________ catches are used to hold one door of a double set.

Multiple Choice

1. Small strips dividing lights of glass are called
   a. mullions. c. rails.
   b. muntins. d. stiles.

2. The first step in hanging a flush door is to
   a. level the door securely. c. level the hinges.
   b. mark the location of the hinges. d. joint it in the opening.

3. Overlay and lipped doors are easier to hang than flush doors because
   a. no hinge gains are necessary. c. no jointing is necessary.
   b. no leveling is necessary. d. no catches are necessary.

4. Overlay and lipped doors usually lap over the opening by
   a. 1/4 inch. c. 1/2 inch.
   b. 3/8 inch. d. 3/4 inch.

5. The type of hinges commonly used for lipped doors are
   a. butt hinges. c. offset hinges.
   b. surface hinges. d. overlay hinges.

6. Paneled doors are preferred over solid doors because they
   a. have less tendency to warp. c. take less time to make.
   b. take less material to make. d. are easier to hang.

7. Joints used on the frame of a paneled door are usually
   a. mitered joints. c. butt joints.
   b. mortise-and-tenon joints. d. half-lap joints.

8. The overlay door
   a. covers the opening. c. fits flush in the opening.
   b. is rabbeted in the opening. d. slides in the opening.

9. Hinges used on flush doors may be
   a. offset or overlay hinges. c. overlay or surface hinges.
   b. butt or surface hinges. d. butt or offset hinges.

10. A one-piece hinge that usually extends the whole length of a door that is subjected to heavy use is a
    a. semiconcealed butt hinge. c. concealed overlay hinge.
    b. pivot hinge. d. piano hinge.

unit 25 Plastic Laminating

OBJECTIVES

After studying this unit, the student will be able to:

- describe plastic laminates, sheet sizes, and thickness.
- laminate a countertop with a backsplash.
- laminate kitchen cabinet faces.
- laminate and cut scribing strips.

PLASTIC LAMINATES

Plastic laminate is a very tough material. It is widely used for surfacing countertops, kitchen cabinets, and many other kinds of cabinetwork, figure 25-1. It can be scorched by an open flame, but resists heat, alcohol, acids, and stains. Another advantage of plastic laminate is that no finishing is required. It also cleans easily with mild detergent.

Laminates are known by such trade names as Formica®, Micarta®, Texolite®, Wilson Art®, Melamite®, and many others. They are manufactured in many colors and designs including many wood grain patterns. Surfaces are available in gloss, satin, textured, and other finishes. The distributor supplies samples or chips of the different colors and finishes to help the customer decide which to use.

Thickness

Generally two thicknesses of laminates are widely used: thick and thin.

Thick laminate is about 1/16 inch thick. It is used on horizontal surfaces such as countertops, tables, dressers, and desk tops.

Thin laminate is about 1/32 inch thick. It is used for vertical surfaces such as the sides and fronts of kitchen cabinets. This is because vertical surfaces take less wear than horizontal surfaces. Thin laminate makes a more pleasing appearance because of the thin edge line it presents when trimmed. It is also less expensive than the thick laminate.

Fig. 25-1 A laminated kitchen desk unit

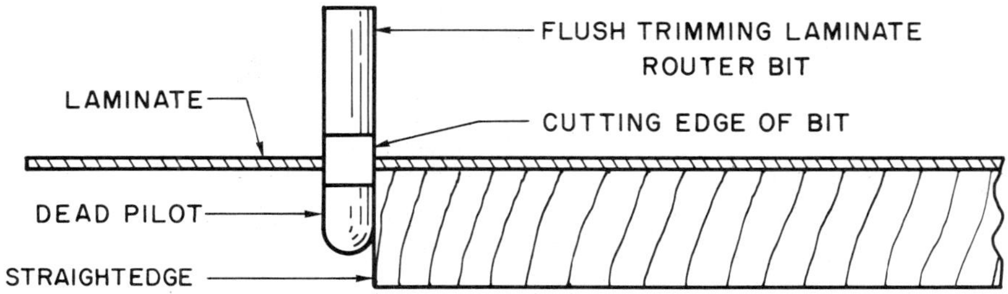

Fig. 25-2 Using a router and straightedge to cut laminate to rough size

A thinner laminate, called *backer laminate,* is also available. It is used to cover the inside of doors and the underside of tabletops to give a balanced construction to the core.

Widths and Lengths

Plastic laminate sheets come in widths of 24, 30, 36, 48, and 60 inches and lengths of 5, 6, 8, 10, and 12 feet. Sheets are usually 1 inch wider and longer than the size indicated.

Most distributors cut sheets in half through their width or length. This increases the range of sizes. Since the material is relatively expensive, it is wise to carefully plan and order the most economical sizes.

LAMINATING A COUNTERTOP

Inspecting the Surface

Before laminating a countertop, make sure all surfaces are flush. There should be no indentations where the pilot of the router bit will ride. Check for protruding nailheads and points. Plane or sand surfaces that are not flush. Fill in any holes and sand smooth. Drive nailheads flush, fill, and sand (see *Unit 27 Finishing*).

Cutting Laminate to Rough Size

There are a number of ways to cut laminate. Whatever method is used, cut the pieces 1/4 to 1/2 inch wider and longer than the surface to be covered. Laminate must be handled carefully because it is very brittle. It may crack if dropped or handled roughly.

Fig. 25-3 The Mar-Bel® Slitter is designed solely to cut plastic laminates. *(Mar-Bel)*

One method of cutting laminate is to use a straightedge and router with a flush trimming bit, figure 25-2. This method is used frequently by installers on the job and in the shop. It is easier to run the cutting tool across a large sheet than to move a large sheet across the cutting tool. Also, the router bit leaves a smooth edge.

Laminates can also be cut with other tools. The Mar-Bel® mica slitter cuts laminates up to 36 inches and produces an extremely smooth, chip-free edge, figure 25-3. The table saw can produce a smooth edge cut with a 60-tooth, triple-chip carbide blade. Laminate may also be cut with a portable circular saw, saber saw, or band saw. However, these tools will not give a clean, chip-free edge.

Laminating Countertop Edges

Contact cement is used for bonding plastic laminates. There are a number of reasons why a contact-bonded piece of laminate may fail to adhere:

- *Not enough cement is applied.* If the material is porous, like the edge of plywood or particleboard, a second coat is usually needed after the first coat dries. When enough cement has been applied, a glossy film will appear over the entire surface when dry.
- *Not enough time is allowed for the cement to dry.* Both surfaces must be dry before contact is made. To test for dryness, lightly press your finger to the surface. Although it may feel sticky, if no cement remains on your finger, the cement is dry.
- *Allowing the cement to dry too long.* If contact cement dries for over two hours, depending on the humidity, it will be too dry. To correct this condition, merely apply another coat of cement.

The front edge and two ends of a countertop can be laminated at the same time if one end of the two end pieces of laminates is cut perfectly square.

1. Apply a liberal coat of contact cement to the edges of the countertop and to the back side of the laminate with a brush or roller.

> **Caution:** Apply contact cement in a well-ventilated area. Avoid inhaling the fumes.

2. After the cement is dry, apply the laminate to the front edge of the countertop, figure 25-4. Position it so that the bottom and top edge can be trimmed and the ends overhang. A permanent bond is made when the two surfaces touch. A mistake in positioning means removing the bonded piece – a time-consuming, frustrating, and difficult job.
3. After the front edge is laid in place, roll out the entire surface with a rubber roller not over 3 inches wide. Or, tap the surface with a hammer and a block of soft wood. Do not tilt the roller or block as this will break the overhanging edges or ends.
4. Apply laminate to the ends in the same manner as the front edge piece. Make sure that the square ends butt up firmly against the overhanging ends of the front edge piece. This makes a tight joint between the two pieces.

Fig. 25-4 Apply the laminate to the front edge of the countertop.

Trimming Laminated Edges

Two routers are used for trimming laminates. One is set up with a flush trimming bit and another with a bevel trimming bit, figure 25-5.

The *flush trimming bit* is used to cut the edges of the first piece of laminate bonded down. The *bevel trimming bit* is used to trim the edge of laminate that is bonded against another piece of laminate. The router base is adjusted so that the bevel trimming bit is exposed gradually until there is just enough to trim the laminate flush with the first piece without cutting into it. The overhanging edge is trimmed flush with, but does not cut into the side of the laminate against which the pilot of the bit rides.

Some trimming bits have ball-bearing pilots, while others have dead or solid pilots that turn with the bit. When bevel trimming with a solid pilot, the laminate must be lubricated to keep it from being burned as the pilot rides against it. Usually paraffin or wax is rubbed on to the surface, or vegetable fat, such as Crisco®, is brushed on.

To trim laminated edges:

1. Using the bevel trimming bit, trim the end overhanging the front edge piece.

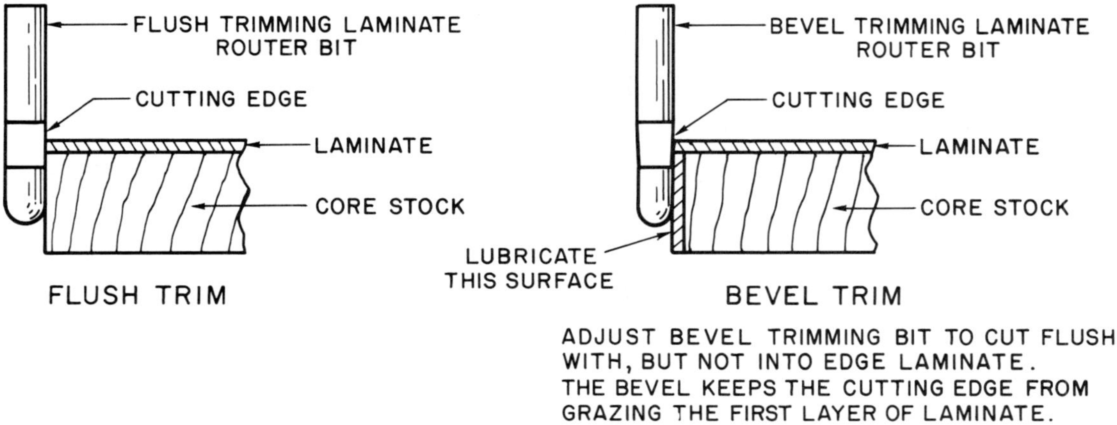

Fig. 25-5 Flush and bevel trimming laminate bits

2. Then, using the flush trimming bit, trim off the top and bottom edges of both front and end edge pieces, figure 25-6.
3. Belt sand or file the top and bottom edges flush with the surface.
4. Knock off the sharp edge on the bottom with a file or sandpaper.
5. Keep a sharp edge on the top to assure a tight joint with the countertop laminate.
6. Any exposed edges not being joined should be slightly rounded off.

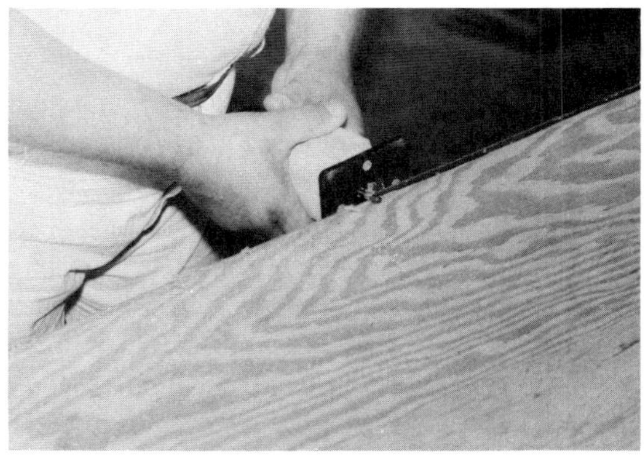

Fig. 25-6 Flush trimming the edge band

Laminating The Countertop Surface

1. Apply contact bond cement to both the countertop and the back side of the laminate. Make sure enough cement is applied and allowed to dry.
2. To position large pieces such as countertops, thin strips of wood about 3/8" x 3/4" are placed a foot apart on the surface. Or preferably, use metal venetian blind slats with the crowned side up.
3. Lay the laminate to be bonded on the strips or slats and position correctly, figure 25-7.
4. Bond on one end. Gradually remove the slats one by one until all slats are removed. The laminate is now positioned correctly, and no costly errors in positioning occurs.

Fig. 25-7 Position the laminate on the countertop using metal venetian blind slats.

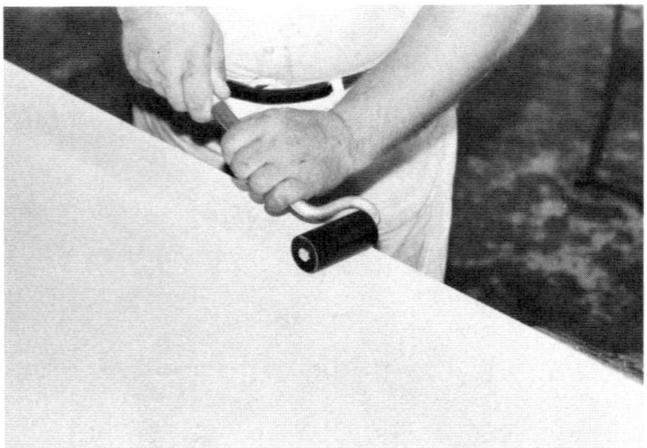

Fig. 25-8 Roll out the laminate to assure a good bond.

5. Roll out the laminate, figure 25-8.
6. Trim the back edge with a flush trimming bit. Trim the ends and front edge with a bevel trimming bit, figure 25-9.
7. Touch up with a mill smooth file where necessary. Clean off any excess cement with lacquer thinner and a rag.

> **Caution**: Lacquer thinner is extremely flammable. Use it in a well-ventilated area. There should be no open flame nearby and no smoking. Avoid inhaling the fumes.

In cases where the countertop must be laminated in two or more pieces, the joint must be made tight to present a good appearance. To make this joint:

1. Clamp the two pieces of laminate with their edges in a straight line on some strips of 3/4-inch stock.
2. Butt the ends together or leave a space less than 1/4 inch.
3. Using one of the strips as a guide, run the flush trimming bit through the joint. Keep the pilot of the bit against the straightedge. This is an easy method of producing a tight joint, figure 25-10.

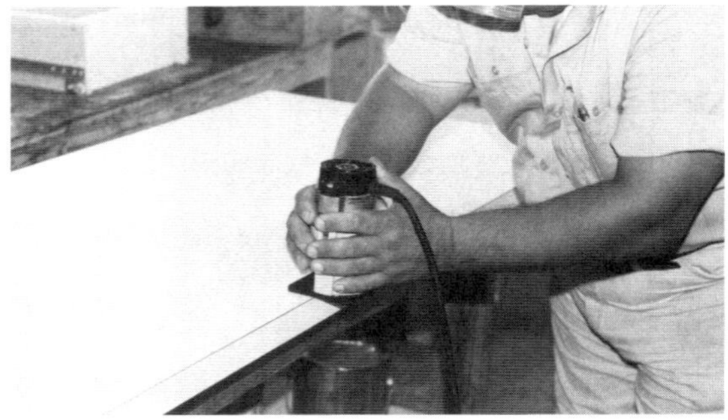

Fig. 25-9 Bevel trim the countertop laminate.

Fig. 25-10 The router with a flush trimming bit is used to make joints between pieces of laminate.

4. Press seam-filling compound into the joint with a narrow putty knife. Wipe off excess compound with a solvent.

Laminating Backsplashes

If backsplashes are to be part of the countertop, laminate these pieces separately in the same manner as the countertop. Then fasten it to the countertop by screwing or nailing up through the bottom of the countertop into the backsplash. Use a little caulking compound in the joint. This makes a tight and attractive joint, figure 25-11.

LAMINATING KITCHEN CABINET FRONTS

To laminate kitchen cabinet fronts, first laminate the two ends of the cabinet in the same manner as described in laminating countertops. Trim the bottom, top, front, and back edges flush. Belt sand

Fig. 25-11 Applying the backsplash

or file the flush edges true. Clean off excess cement with solvent or lacquer thinner.

Laminating the Stiles

To laminate the front face of the cabinet, first laminate the stiles. Trim the inside edges and top and bottom ends flush. Soften all sharp corners, except where joints are to be made, with a file or sandpaper. Bevel trim the edges that overlap the ends of the cabinet and touch up with a mill smooth file.

Laminating the Rails

One of the most difficult jobs in laminating the front frame of a kitchen cabinet is fitting the rail pieces to the stile pieces. However, by the following method, the job becomes a little easier and almost foolproof.

1. Apply enough cement to the rails and to the back of the laminate. Let dry.
2. Position the laminate on the rails. Contact in the center of the rail, but do not allow 3 or 4 inches of each end to make contact.
3. Slip a block of 1/2-inch material, about 6 inches long and laminated on one edge, under the end of the rail laminate. Hold the laminated edge of the block tightly against the edge of the stile laminate.
4. Using a flush trimming bit and the laminated edge of the block as a straightedge, trim the end of the rail pieces, figure 25-12.

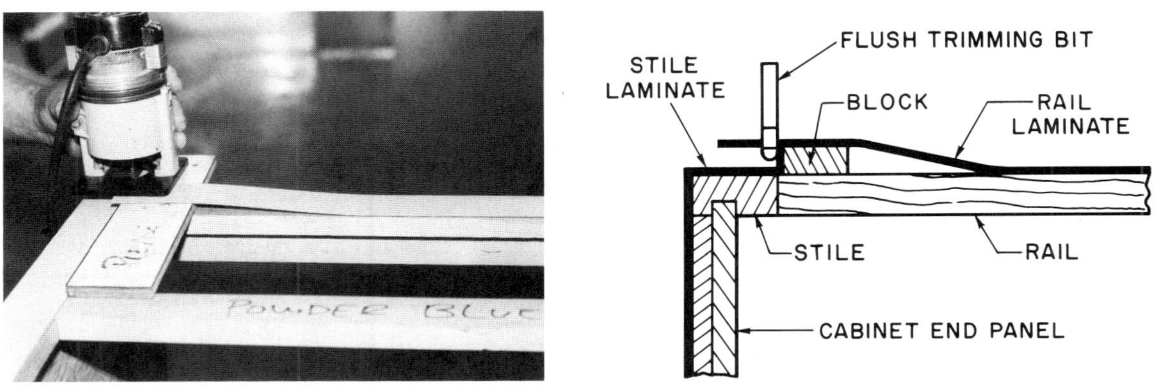

Fig. 25-12 Method of fitting the rail laminate to the stile laminate

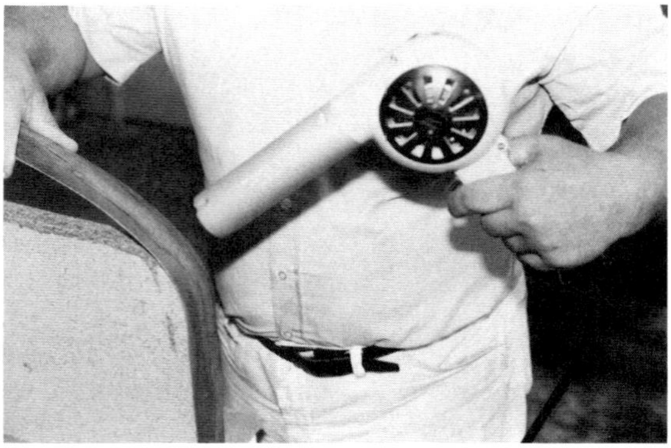

Fig. 25-13 Heating and bending laminate with a heat gun

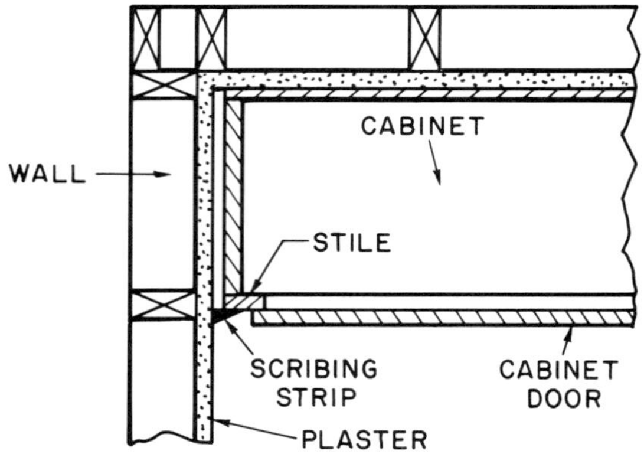

Fig. 25-14 Location of scribing strip

5. Remove the block. Buckle up the rail pieces and make the joint. Then contact the rest of the rail. Do the same on the other end and with the rest of the rails. This results in a perfect fit with all stiles and rails.
6. Trim the top and bottom edges with a flush trimming bit. Soften all sharp, exposed corners with a file or sandpaper.

If the Betterly® underscribe router is used, no guide block is necessary. Its specially constructed base is made to be guided by the edge of the stile laminate.

LAMINATING CURVED SURFACES

If the edge of a countertop has a curve, the laminate can be bent to the desired radius by heating with a heat gun, figure 25-13. If a heat gun is not available, then a small propane torch can be used. Do not let the laminate get too hot or it will blister.

Heat the laminate carefully and bend until the desired radius is obtained. Allow the laminate to cool while holding it in the shape desired. After it has cooled, it will retain the shape in which it was held.

Caution: Keep fingers away from the heated area of the laminate. Remember that the laminate retains heat for some time.

LAMINATING DOORS AND DRAWER FRONTS

Doors and drawer fronts are laminated in the same way as countertops. The edge laminate is applied first and trimmed off flush. The face laminate is then applied and bevel trimmed all around. The trimmed edges are filed where necessary and all sharp corners slightly rounded off.

SCRIBING STRIPS

In many cases when laminated kitchen cabinets are installed, irregularities on the wall or ceiling make it difficult to obtain a tight joint with the cabinets. Strips used to cover these joints are called *scribing strips*. These strips are made in the shop and installed on the job, figure 25-14.

1. Laminate one side of 3/4-inch clear pine stock. Use a laminate that matches the cabinets.
2. Using a 60-tooth, triple-chip carbide blade, rip the laminated stock into 1 3/16-inch strips. Rip with the laminated side up.
3. Set the saw at a 30-degree angle and rip the strips to approximately 1 1/8 inches wide.
4. Set the saw at a 20-degree angle. With the laminated side against the fence, rip the stock as thick as possible, yet producing a feather edge.

ACTIVITIES

1. Collect brochures and chips describing plastic laminates.
2. Study the latest government bulletins on the use of contact cement.
3. Experiment bending plastic laminates with and without heating to determine their bending radii under these conditions.
4. Laminate a kitchen wall or base unit. Laminate the end panels, face frame, and door and drawer fronts.
5. Laminate the countertop for the kitchen cabinet units made in *Unit 22 Kitchen Cabinets.*

UNIT REVIEW

Multiple Choice

1. Before beginning to laminate a surface, make sure that
   a. the surface is securely fastened.
   b. the wood grain is not raised.
   c. all surfaces are flush and smooth.
   d. the work area is neat and clean.
2. To get a chip-free edge when cutting laminate, use a
   a. router.
   b. portable electric circular saw.
   c. saber saw.
   d. band saw.
3. When trimming laminates, the router is pulled
   a. clockwise.
   b. counterclockwise.
   c. against the rotation of the bit.
   d. with the rotation of the bit.
4. Rough sizes of laminate should be cut oversize by
   a. 3/4 to 1 inches.
   b. 2 to 3 inches.
   c. 1/4 to 1/2 inch.
   d. any of the above.
5. Laminate bonding will fail if
   a. not enough cement is applied.
   b. the cement is not allowed to dry.
   c. the cement has dried too long.
   d. all of the above.
6. When the laminate has been positioned and contact has been made,
   a. trim the edges immediately.
   b. roll out the surface.
   c. clean off excess cement.
   d. none of the above.
7. When using a bevel trimming bit with a solid pilot,
   a. keep the router base on the surface.
   b. do not lubricate the laminated edge.
   c. lubricate the laminated edge.
   d. bear heavily against the edge.
8. Large pieces of laminate are positioned before bonding by
   a. placing strips on the surface.
   b. using two people to hold the laminate.
   c. sighting carefully by eye.
   d. a positioning machine.

9. When laminating kitchen cabinet faces, the
   a. stiles are laminated first.
   b. cabinet ends are laminated first.
   c. rails are laminated first.
   d. back is laminated first.

10. When fitting laminated pieces between the stiles of a kitchen cabinet front, use
   a. tin snips.
   b. a hacksaw and square.
   c. a block plane.
   d. a router and straightedge.

Questions

1. How does a customer pick a laminate? What choices are available?
2. How thick is thick laminate? Thin laminate?
3. Why would thin laminate not be used on a horizontal surface such as a countertop?
4. How is laminate cut to size on the job?
5. How does the cabinetmaker know when enough contact cement has been applied to a surface?
6. If contact cement is allowed to dry too long, what may be done to correct the condition?
7. Why must plastic laminate be positioned carefully?
8. If the countertop must be laminated in two pieces, how is the joint made?
9. How are curved areas laminated?
10. What should be done to sharp, exposed edges that are not being joined?

unit 26
Tables

OBJECTIVES

After studying this unit, the student will be able to:

- make common shaped legs and apply designs to their surfaces.
- describe the commonly used methods for joining legs to tables.
- construct and fasten tabletops, rails, legs, and stretchers.
- make, hang, and support drop leaves in tables.

TABLES

A simple table consists of four legs, four rails, and a tabletop, figure 26-1. Sometimes lower rails, called *stretchers,* are added to give extra support. Drawers and shelves may also be added to the table, depending on its use.

Dining, writing, and game tables range from 29 to 32 inches high. Coffee tables are usually from 14 to 18 inches high. Typewriter tables are usually 26 inches high.

LEG SHAPES

The *straight square leg* is the easiest to make, figure 26-2. It is made from either solid lumber or by gluing strips together to form the desired thickness. To make this type of leg, two edges of the stock are straightened and squared. Then the remaining two edges are cut to desired size. Finally, the leg is cut to length.

Tapered square legs are sometimes tapered on all four sides or on the two inside edges, figure 26-3. The top end of the tapered leg is left straight for a short distance in order to join it to the table rail (horizontal piece under the tabletop).

To make a tapered leg, cut the stock to the overall size as described for straight legs. Measure down the desired distance from the top of the leg and square a line around the leg. This line marks the start of the taper. Mark the bottom end of the leg for the amount of taper. Cut the taper.

Tapered round legs may be round from top to bottom or a portion of the top may be left square, figure 26-4. Tapered round legs are made on a wood lathe. (Review *Unit 12 Wood Lathe.*)

Other round legs are *turned* to cut many different designs, figure 26-5. Much experience using the wood lathe is required when more than one leg of a complicated design is needed. In mills that turn out quantities of turned legs, automatic wood lathes are used. Once properly set up, this machine duplicates every leg perfectly in a short time.

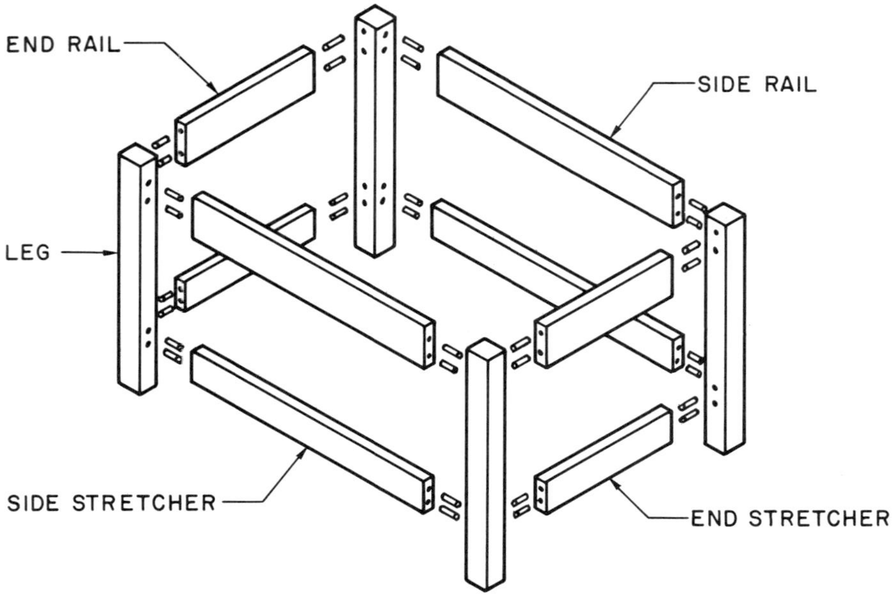

Fig. 26-1 Typical construction of a table without top

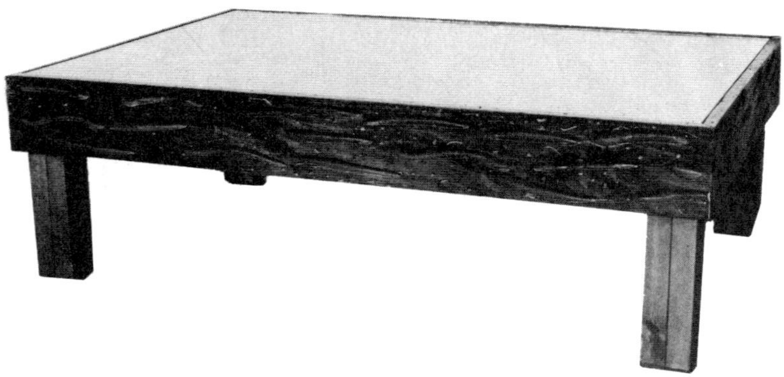

◁ Fig. 26-2 Straight square legs

▷ Fig. 26-3 Tapered, square leg. Note the bottom part is tapered while the upper part is left square. The middle part of these legs are decoratively shaped.

Fig. 26-4 Tapered round legs

Applying Designs on Legs

The appearance of straight and tapered square legs is improved by applying designs on the surfaces of the legs. Cutting shallow, spaced saw cuts along the length of the leg is an easy design to make, figure 26-6.

Reeds and flutes may be machined on square legs by using a shaper, figure 26-7. (See *Unit 10 Shaper and Overarm Router.*) *Reeds* are rows of convex cuts along the length of the legs. *Flutes* are concave cuts, figure 26-8. Reeds and flutes are usually stopped on one or both ends of the leg. This means they do not go all the way through.

METHODS OF JOINING LEGS

It is important that legs be joined securely to produce a rigid and strong table. There are many ways to join legs. One consideration to make is whether or not it will be a removable or permanent joining.

Joining Removable Legs

One method of attaching removable legs to a tabletop is by means of a hanger bolt and a metal plate, figure 26-9. The metal plate has a threaded hole in the center. It is fastened to the underside of the tabletop. A hanger screw is inserted in the

Fig. 26-5 Turned legs

Fig. 26-6 Shallow-spaced saw cuts provide a design on this table leg.

Fig. 26-7 Making flutes on a square leg

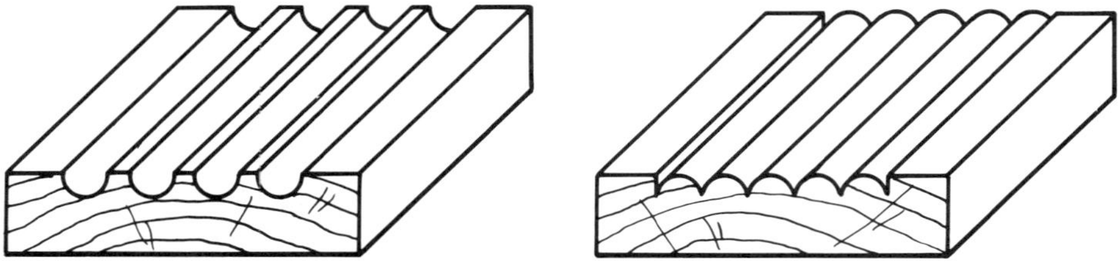

Fig. 26-8 Reeds and flutes

Fig. 26-9 Attaching a leg with a hanger bolt and metal plate

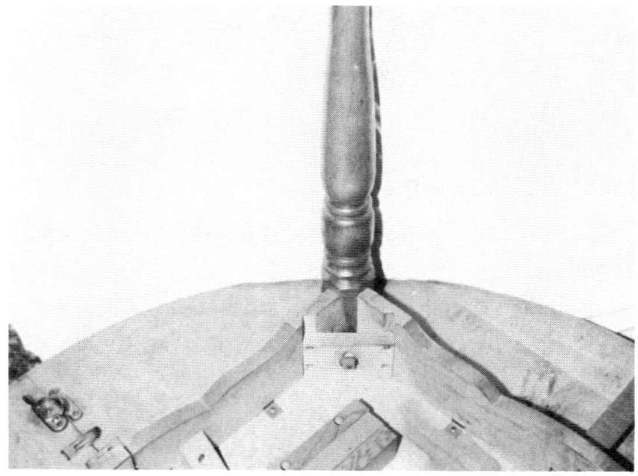

Fig. 26-10 Removable table leg held by hanger bolt and corner block

center of the top end of the leg. The leg is then screwed into the metal plate. The plates are sometimes made so that the leg will be straight or at an angle. This type of joining is useful on low tables, such as coffee tables or end tables. On higher tables, it does not provide enough strength or rigidity.

Another method of attaching removable legs is used for higher tables that must be very strong. In this method, the table rails butts up against the two flat surfaces at the top of the leg. Corner blocks with a hole in the center are fastened to the table rails. A hanger bolt is inserted in the inside corner of the leg and fastened with a nut and washer. By tightening the nut, the leg is drawn tightly against the ends of the table rails, figure 26-10. If the legs become loose in periods of dry weather or in a heated room, the nuts are simply tightened.

Fig. 26-11 Dowel jointed leg and rail

Joining Permanent Legs

On small tables, permanent joints are made between the leg and the rail. These are often doweled joints strengthened with corner blocks, figure 26-11. The dowel joint is preferred because it is strong and is quicker and easier to make than others.

The strength of the joint depends on the amount of contact between the rail and the leg. The rail should be wide enough to provide rigidity to the joint. The width of the rail depends on the size and use of the table. A rail that is too wide may interfere with a person's legs. The rail may be narrowed down if the piece has light use and is in keeping with the design.

Another permanent joint that is used frequently on tables is the blind mortise-and-tenon joint. A mortise is cut in the leg and a tenon cut on the rail, figure 26-12. The joint is glued and clamped and strengthened with a corner block. The rail may be centered on the leg or set off-center. Bottom edges of the rails are left straight or shaped in many different ways.

Fig. 26-12 Leg and rail joined with mortise and tenon

TABLETOPS

Construction

Tabletops are made of solid stock, edge-banded plywood, laminated particleboard, or combinations of each. Tops may be cut in many different shapes: square, rectangular, round, oval, or octagonal (8-sided). The cabinetmaker must know how to lay out any shape.

Solid stock tabletops are made by gluing pieces edge to edge. Pieces should be no wider than 6 inches. Usually doweled joints or shaped-edge joints are used. The edges and ends of the top may be left square or shaped in a number of ways, figure 26-13. Sometimes the edges are banded with strips of various width to hide the end grain.

Many tabletops are made from hardwood plywood, such as birch plywood. Hardwood plywood comes with many different kinds of face veneers. Some widely used veneers are oak, mahogany and walnut. The edges of plywood tabletops must be treated to hide the end grain. They are treated similar to covering the edges of doors (see *Unit 24 Cabinet Doors*).

There are probably more plastic laminated tops being made than any other kind. There are several reasons for its popularity. The most important is that it is practically indestructible. It is available in many beautiful wood grain patterns, colors, and designs; no finishing is required; and it is relatively inexpensive and easy to apply.

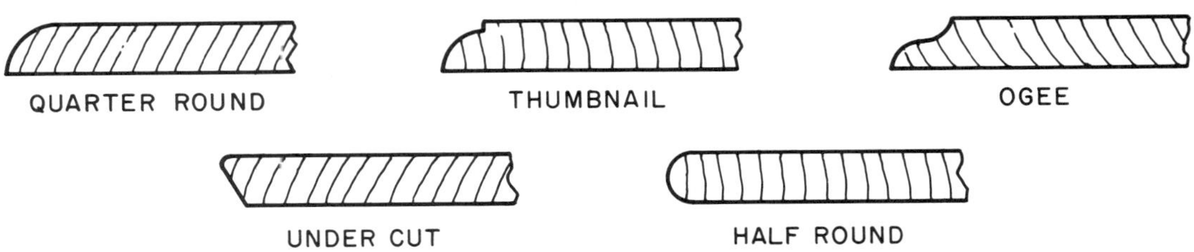

Fig. 26-13 Typical shaped tabletop edges

Usually particleboard or plywood is used for the core of laminated tops. Their edges are covered with the same laminate as the top or in other ways as described for doors. It may also be banded with solid stock before the laminate is applied. The laminate is then applied and the edges shaped, leaving the wood and the thin edge of the laminate exposed, figure 26-14.

Fastening Tops

There are a number of ways to fasten tabletops, figure 26-15. It is important that the fasteners

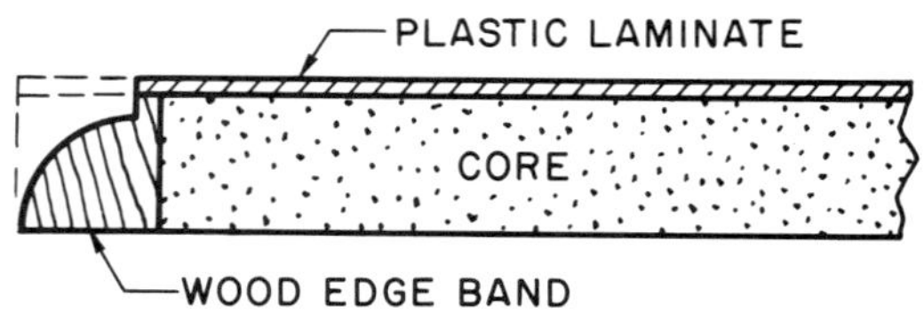

Fig. 26-14 Shaped edge of a laminated top

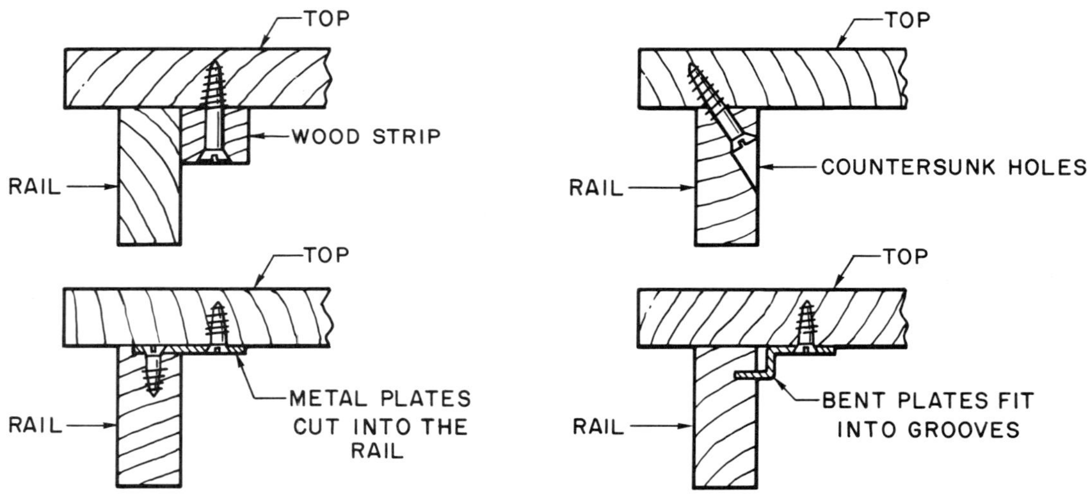

Fig. 26-15 Methods of fastening tabletops

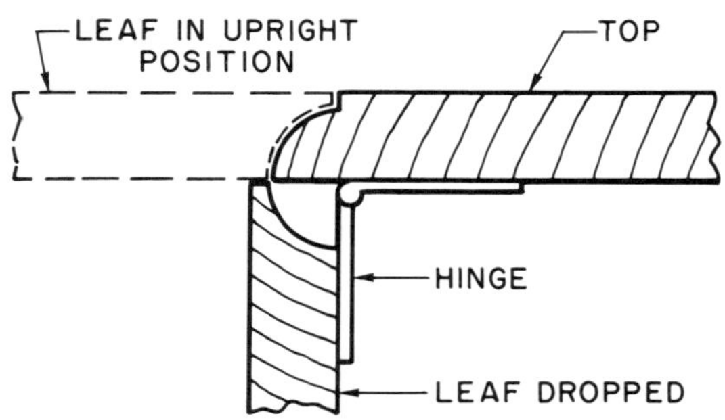

Fig. 26-16 Drop leaf hung on a shaped edge

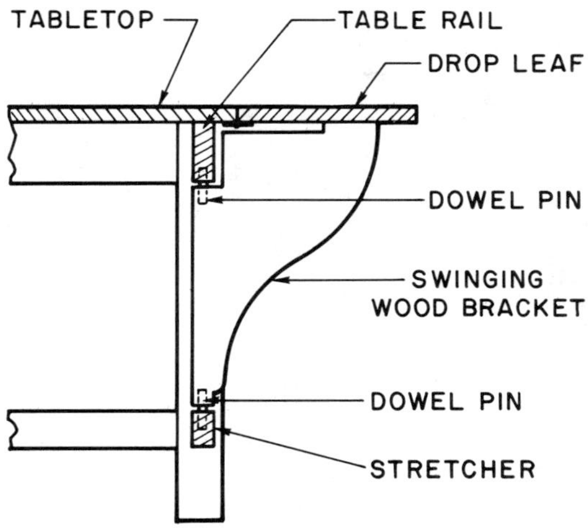

Fig. 26-17 Swinging wood bracket is used to support drop leaves. The bracket swings on dowel pins inserted in the rail and stretcher.

are hidden. The fastening method must also allow for swelling and shrinking. Glue should never be used between the top and the rails.

When drilling for or driving screws, put a stop on the drill to insure that the drill will not pierce the top. Select screws of sufficient length, but not so long that they will go through the stock.

DROP LEAVES

Construction

Drop leaves for tables are made of the same material and design as the main tabletop. Sometimes the hinged edge is squared. Other times the hinged edge fits the contour of the shaped edge of the main top, figure 26-16. To make this joint, matching shaper cutters are needed.

Hinges with wide flaps are usually used to hang drop leaves. These keep the screws used to fasten the hinges farther in from the edge.

Supports

There are many different ways of supporting the drop leaf in an upright position. Supports may be wood that swing or slide or metal folding brackets, figures 26-17, 26-18, and 26-19.

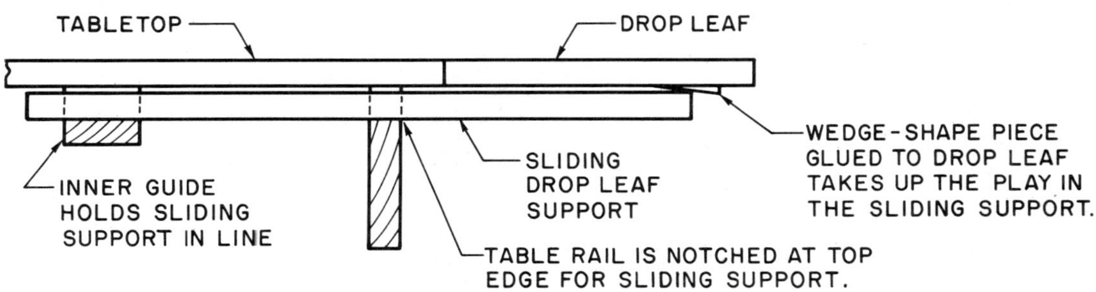

Fig. 26-18 Wood-sliding drop support

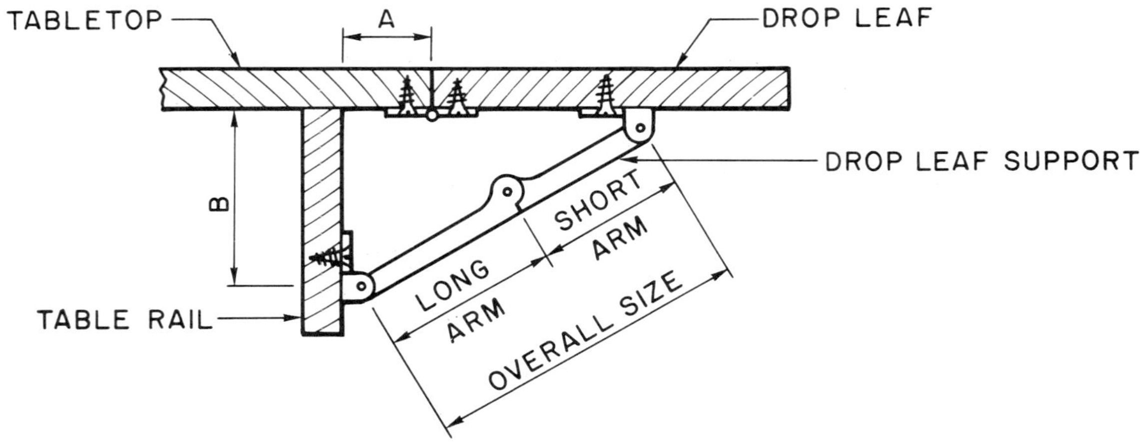

A	6" BRACKET B	8" BRACKET B	10" BRACKET B
1/2"	3 1/2"	3 1/2"	5"
1"	3 1/4"	3 3/16"	4 11/16"
1 1/2"	2 13/16"	2 7/8"	4 7/16"
2"	2 1/2"	2 1/2"	4"
2 1/2"	2 1/8"	2 1/8"	3 9/16"

Fig. 26-19 One type of metal drop leaf support. Directions for installation must be followed carefully.

MAKING A SMALL TABLE

The small table in figure 26-20 requires legs, rails, and a tabletop.

Legs and Rails

1. Cut the four legs to their overall dimensions.
2. Taper the legs. Leave 3 inches of the top end of each leg straight.
3. Cut the rails to exact size.
4. Bore holes for 3/8-inch dowels centered on the legs and on the ends of the rails.
5. Cut and glue the dowels into the legs. Cut the dowels about 1/8-inch short to keep them from bottoming out before the joint is tight.
6. Assemble the legs and rails without glue. Make any necessary adjustments to insure a tight fit.

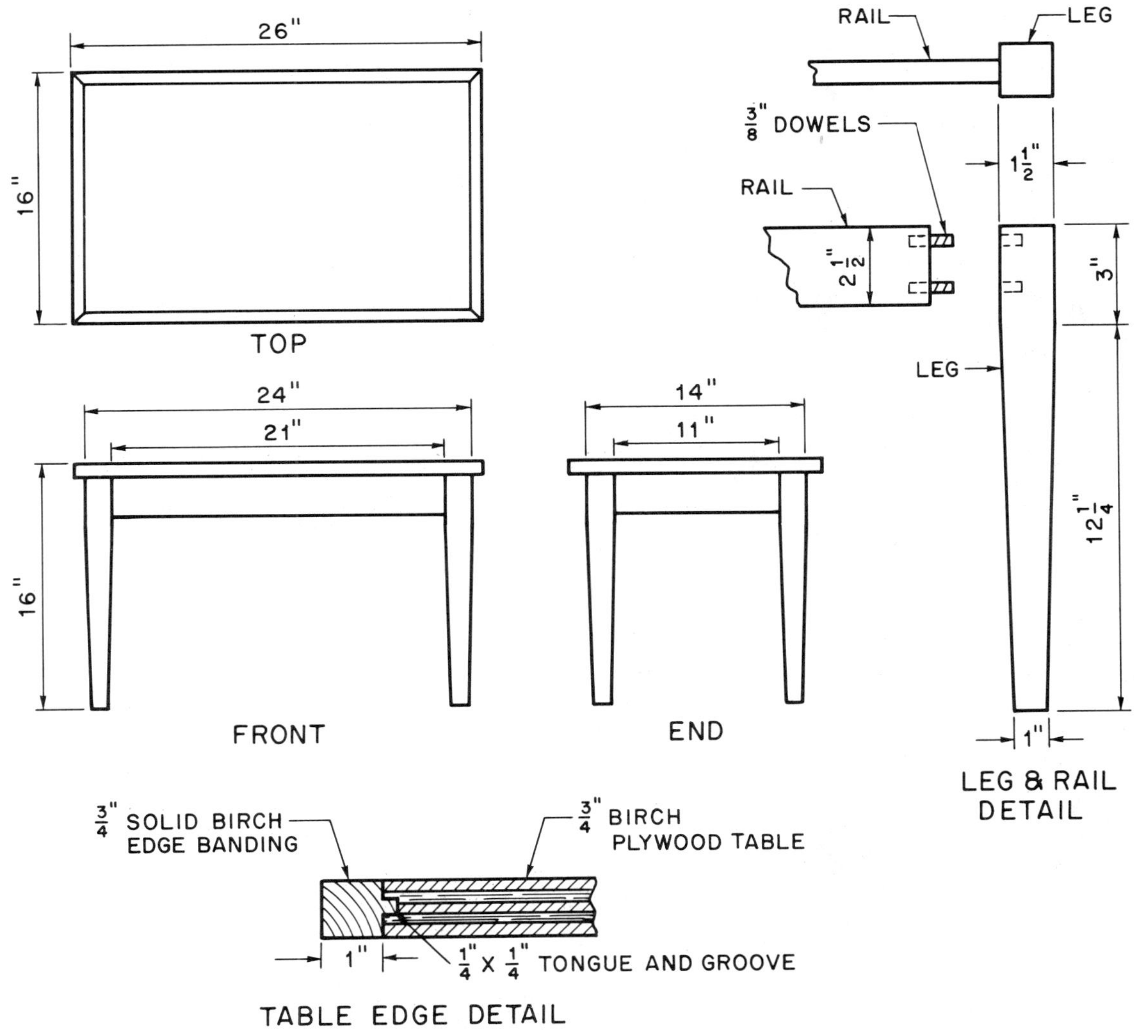

Fig. 26-20 Small table

7. Take apart, apply glue, and clamp together. Use blocks behind the clamp to avoid damaging the work.

Tabletop

1. Cut the plywood top to its overall size. Keep the face side up when cutting. Use a blade that will produce a splinter-free edge.
2. Cut the edge banding strips to rough lengths.
3. Groove the edges of the tabletop.
4. Tongue the edge banding strips to fit the grooves made in the tabletop.
5. Miter the edge banding strips around the tabletop. Make snug fits and glue and clamp in place with edge banding clamps.
6. Fasten the top to the leg and rail.
7. Sand all parts in preparation for finishing. Slightly round off all exposed sharp corners.

ACTIVITIES

1. Visit furniture stores and study magazines, catalogs, and books to note the different styles of tables and their construction.
2. Apply reeds and flutes to square straight legs, square tapered legs, and round tapered legs.
3. Design a small, round coffee table with four turned legs mounted on an angle and a laminated tabletop.
4. Design an end table with a shelf and four tapered square legs. Explain your choice of materials and construction methods.

UNIT REVIEW

Multiple Choice

1. Tabletops must be fastened in such a way as to
   a. keep them from swelling or shrinking.
   b. allow for swelling and shrinking.
   c. provide rigidity to the table.
   d. be fastened permanently to the rails.
2. Hinges with wide flaps are used to hang drop leaves on a table because
   a. they are easier to apply than other kinds.
   b. the flap will go under the rail.
   c. the screws can be kept farther in from the edge.
   d. they are stronger than other kinds.
3. When making a doweled joint between table legs and rails, the dowels are cut about 1/8 inch short to
   a. allow for excess glue.
   b. keep from bottoming.
   c. allow for wood chips in the bottom.
   d. allow for compression of the air.
4. When cutting a plywood top to size, keep the face side up to
   a. avoid marring the face as it is pushed through the saw.
   b. see the line you are cutting.
   c. avoid splintering the face edges.
   d. avoid kickback.
5. Average heights of dining room, writing, and game tables range from
   a. 24 to 26 inches.
   b. 26 to 28 inches.
   c. 29 to 32 inches.
   d. 34 to 36 inches.
6. Coffee table heights usually range from
   a. 8 to 10 inches.
   b. 10 to 14 inches.
   c. 14 to 18 inches.
   d. 18 to 24 inches.
7. Sometimes stretchers are added to a table
   a. to extend its height.
   b. to extend its length.
   c. to add a decorative effect.
   d. to give it extra support.
8. Convex cuts along the length of a table leg are called
   a. flutes.
   b. tapers.
   c. reeds.
   d. stops.

9. The strength of a leg joint depends on
   a. whether or not stretchers are used.
   b. the amount of contact between the rail and leg.
   c. the length of the rail.
   d. the shape of the leg.
10. When gluing pieces edge to edge to make a solid stock tabletop, the pieces should be no wider than
    a. 1 inch.
    b. 2 inches.
    c. 4 inches.
    d. 6 inches.

Questions

1. Name four shapes of table legs.
2. Name three ways of applying designs to table legs.
3. What types of joint should be used if the table leg must be attached at an angle and still be removable?
4. When a removable leg is fastened with a corner block and hanger bolt, why may the joint loosen? How is it tightened?
5. Why is a doweled joint preferred over a blind mortise-and-tenon joint for fastening permanent legs?
6. What determines the width of a rail?
7. Why are plastic laminated tabletops so popular?
8. How are drop leaves supported in an upright position?
9. What will happen if the dowels are cut too long when making a doweled joint between a table leg and rail?
10. Why should glue never be used to fasten the tabletop to the rails?

unit 27
Finishing

OBJECTIVES

After completing this unit, the student will be able to:

- select the proper kind and grit abrasive to sand wood.
- describe types of wood bleaches and bleach a surface.
- describe kinds of stains and stain a surface.
- describe paste wood fillers and fill a surface.
- describe kinds of sealers and seal a surface.
- describe topcoating materials and topcoat a surface by brushing and spraying.

SANDPAPER AND SANDING

Sandpaper is the common name for *coated abrasives.* Sandpaper does not use sand as an abrasive; nor is its backing necessarily paper. (It may also be cloth.) The terms *sandpaper* and *sanding* are still used in industry, however, for the material and the operation.

Kinds of Sandpaper

There are three popular kinds of sandpaper used for wood finishing. *Aluminum oxide* is a light brown, all-purpose, tough sandpaper. It is used to hand or machine sand wood, metal, plastics, and other material. *Silicon carbide* is a very tough, black abrasive used to sand wood or metal. It is a popular choice for sanding between coats of finish. *Garnet* is a natural reddish-brown material used throughout the furniture industry for finish sanding.

There are other types of sandpaper, but these three are most commonly used. Aluminum oxide is the toughest and most durable. Other sandpapers may be less expensive, but they do not last as long.

Grits

Sandpaper grits are designated either by a grit number or by an "0" series number. The grit numbers range from No. 12 (coarsest) to No. 600 (finest). The "0" series number 10/0 is equivalent to a 400-grit number, while a 1/0 is the same as a 80-grit sandpaper. See the chart in figure 27-1 for other grits.

Sanding Procedure

Most sanding of production work is done by machines. Custom work is finished by hand sanding.

To sand by hand:

1. Examine the piece and note all major defects.

GRIT NO.	0 SERIES
400	10/0
360	——
320	9/0
280	8/0
240	7/0
220	6/0
180	5/0
150	4/0
120	3/0
100	2/0
80	1/0
60	1/2
50	1
40	1 1/2
36	2
30	2 1/2
24	3
20	3 1/2
16	4

Fig. 27-1 Grits of Coated Abrasives

2. Smooth the wood with a cabinet scraper to remove planer marks and ripples.
3. Select the finest grit abrasive that will do the job. Use a sanding block with a piece of hard felt, cork, or rubber glued to the bottom.
4. Always sand with the grain, figure 27-2. Use straight strokes. Avoid a twisting or circular motion.
5. As soon as the surface has no cuts or scratches deeper than those produced by the grit, go to the next finer grit.
6. Between sandings, brush the surface clean. A lump of coarse grit under a fine-grit sandpaper will groove a smooth surface. Clean loaded spots off the sandpaper by knocking out or brushing. Loaded spots will pick up grit and scratch the surface.
7. To obtain a smoother surface, sponge with water to raise the grain. Do not soak the surface because too much water can damage thin veneers. Let the surface dry and sand with the same grit as last used.
8. Finish sanding bare wood with a 280-grit or 220-grit abrasive.

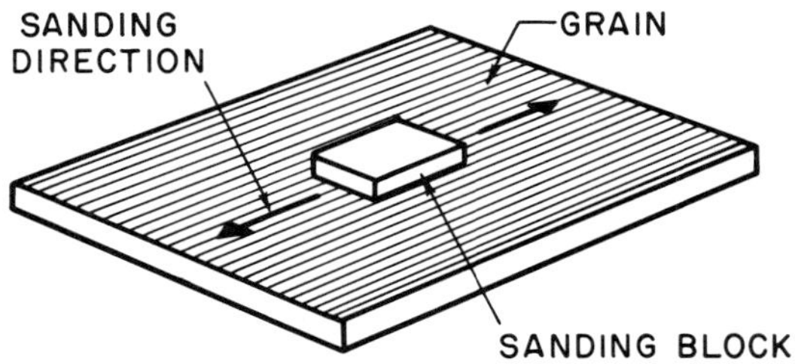

Fig. 27-2 Always sand with the grain.

BLEACHING

Bleaching is done to lighten the wood, remove undesirable stains, or achieve a more uniform color. Light-colored woods are the easiest to bleach. Darker, close-grained woods are harder to bleach. Douglas fir and yellow pine cannot be bleached satisfactorily.

Household bleach is sometimes used to bleach wood. This is the type used to wash clothes. Household bleach is used when a stronger bleach is not needed. It is effective on woods such as gum, maple, and walnut.

Oxalic acid can be purchased from drug stores and is not expensive. It comes in a crystal form that dissolves in hot water. It is a mild bleach but does a good job on woods such as ash, chestnut, and oak.

Hydrogen peroxide is the same type bleach used to lighten hair, but on wood a stronger solution is used. It is more expensive but produces satisfactory results on most woods.

Commercial wood bleaches are usually two-solution bleaches that are mixed before application. They give the best results because they are made especially to bleach wood. They work on most woods and will bleach wood as light as needed.

Bleaching Wood

Caution: Follow the manufacturer's directions carefully. All bleaches contain strong chemicals. Wear rubber gloves and protect clothing with an apron. Keep liquids away from mouth and eyes.

1. Apply bleach to wood with a brush or sponge.
2. Allow to dry until the desired effect is reached.
3. Apply a neutralizer according to directions. White vinegar is used to neutralize or stop the action of oxalic acid for instance.
4. After rinsing, let the surface dry thoroughly.
5. Sand lightly to remove the raised grain.

STAINING

Staining changes the natural color of the wood and brings out the beauty of the grain pattern. The soft part of the wood absorbs more color than the harder part and shows up darker. A more uniform color is obtained, if desired, by applying a thin coat of sealer prior to staining. It must be remembered that stains are not a finish. Finishing coats must be applied over them.

Kinds of Stains

There are many different kinds of stains. Water, alcohol, and oil stains are the most popular.

Water stains are made by mixing dyes with water. The dye is dissolved in hot water according to the shade desired. Water stains are easily applied with a brush. Keeping a wet edge obtains a uniform color with no blotches. The color can be darkened by applying a second or third coat. Water stains have the disadvantage of raising the grain so a light sanding is necessary after using them.

Alcohol stains, also called non-grain-raising stains, are made by mixing dyes with wood alcohol. They are similar to water stains and are applied in the same manner. Alcohol stains dry very fast, making them more difficult to use. Extra care must be taken to keep a wet edge. Spraying this type of stain overcomes the difficulty of keeping a wet edge. Alcohol stains do not raise the grain and no sanding is needed after staining.

Oil stains are made by adding color to linseed oil and turpentine. The stain is brushed on the surface and then wiped with a clean, soft rag before it dries. Oil stains are easy to apply and do not raise the grain. However, they are more expensive than water or alcohol stains, do not penetrate as deeply, and take longer to dry. Oil stains are best for the beginner to use because there is no danger of lapping a dry edge.

PRIMARY COLORS — RED, YELLOW, BLUE	
COMBINATION	RESULT
Red and Yellow	Orange
Red and Blue	Violet
Yellow and Blue	Green
Blue and White	Light Blue
Yellow and White	Ivory
Black and White	Gray
Red and Black	Brown
Green and Black	Olive
Violet and Black	Plum
Blue, Green, and White	Turquoise

Fig. 27-3 Color matching

Mixing Colors

Stains are available in a wide range of colors. From the three primary colors – red, yellow, and blue – other colors can be produced. See the color chart in figure 27-3 for the proper color combinations.

FILLING

Open-grained woods like oak, mahogany, and walnut have large pores that need to be filled before finishing coats can be applied. *Filler* gives the surface of the wood a smooth, glass-like surface after it has dried hard and been sanded.

Wood paste filler is made with silex (a finely ground rock) mixed with linseed oil, drier, and thinned with turpentine. It is available in different colors to match the wood being filled. It also comes in a natural, off-white color that can be tinted to match the color of the stain being used. Sometimes, for special effects, the filler is applied with a color different than the stain.

Applying Paste Filler

1. Thin the paste filler with turpentine until it is the consistency of heavy cream.

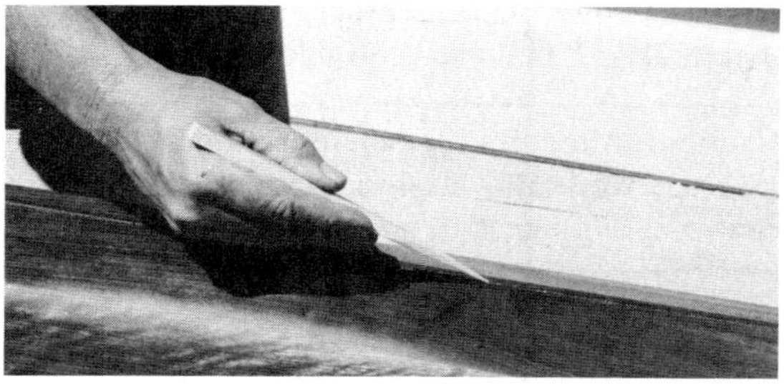

Fig. 27-4 Remove excess filler from corners with a sharp piece of wood.

2. Brush it on with the grain. Do a small area at a time so the filler does not set too hard. If set too hard, it is a difficult and time-consuming job to remove.
3. When the filler loses its shine, wipe it off across the grain. Use a coarse-weave cloth, like burlap, because it absorbs the excess easily.
4. Make sure every bit of excess filler is removed from the surface. Wipe across the grain because filler will be pulled out of the pores by wiping with the grain. Remove excess filler from corners with a sharp pointed piece of soft wood, figure 27-4.
5. Allow the filler to dry completely, preferably overnight. The final finish may not dry completely if it is applied before the filler has completely dried.
6. Sand lightly with very fine sandpaper to remove any excess filler on the surface.

Crack Fillers

Crack fillers are used to fill checks, nail holes, and dents that are too deep to be removed by sanding. Crack fillers include plastic wood and stick shellac. They may be purchased in a variety of colors to match the wood being finished.

After nailheads are set below the surface, the holes are filled with *plastic wood,* figure 27-5. The paste-like filler is pushed into the hole with a putty knife.

Fig. 27-5 Nail holes are concealed with plastic wood.

Stick shellac comes in stick form. It is applied by melting the stick with a flame or hot knife blade and pushing it into the dent.

Crack fillers are then sanded down level with the surface and allowed to dry thoroughly. When possible, it is best to apply fillers before staining the wood.

SEALING THE SURFACE

Once the wood has been sanded, stained, and filled, it is ready for the first coat of finish. The purpose of the first coat is to seal the surface and form a bond between the wood and the top coats of finish.

The sealer coat is a thinned out coat of the same material as the finish coats. For a shellac finish, thin 1 part of pure white shellac with 8 parts of denatured alcohol. For a polyurethane finish, reduce it half and half with the recommended solvent. For a lacquer finish, use a lacquer sanding sealer. For painted surfaces, use the recommended undercoater.

Applying the Sealer

One of the worst mistakes that is made in the finishing process is to apply a heavy coat of sealer. Sealers may be applied by brush or spray. It is a better practice to apply two light coats than a single heavy one. A light, two-coat finish will always dry faster than a single heavy coat. A heavy coat may trap air or dust in pores causing pinholes in the finish. Also, a heavy coat of sealer may react with the stain or filler and cause a milky appearance in the sealer.

The sealer should dry thoroughly. It then sands easily to a dry powder without clogging the sandpaper. Sanding with 220-grit abrasive gives a level and hard foundation for the finishing coats.

TOPCOATING

Clear Finishes

The most popular types of clear finishes are the *synthetic resins.*

The most widely used synthetic finish is the *polyurethane type.* It is used whenever a clear, hard, chemical-resistant finish is needed. It is a superior replacement for varnish. Polyurethane finish can be applied easily by brush or spray, dries in about four hours, and can be recoated in one hour. It is available in high-gloss and satin finishes and does not darken with age. Polyurethane cannot be used for exterior work. *Spar varnish* is still used as a clear finish on work exposed to the weather.

Lacquers are widely used in the furniture finishing industry. It is an interior finish that has replaced shellac and varnish on factory-made furniture. This is because varnish takes too long to dry and shellac is sensitive to heat, alcohol, and water.

Most lacquers are extremely fast drying and must be sprayed. *Brushing lacquer* is available when it is not practical to spray. Lacquer finishes dry dust-free, clear, and hard. It also darkens wood the least of all the finishes. Lacquers are available in gloss, satin, and flat finishes.

Shellac is not as widely used for finish coats. In recent times, lacquer has replaced shellac because it dries as fast. However, it still maintains its popularity as a sealer. Shellac is available in orange and white and is thinned with alcohol. It is easy to apply when thinned properly, but a poor job will result if it is applied too thick.

> **Caution:** Most finishing materials are highly flammable. Work in a well-ventilated area. Do not smoke or work near an open flame.

Penetrating Oil Finishes

Teak and walnut are often given a penetrating oil finish. This enhances the beauty of the grain as well as protects the wood.

Penetrating oils are easy to apply. All that is needed is a smooth, clean surface. They are applied with the palm of the hand or a clean cotton-cloth and rubbed into the wood. Each additional coat increases the gloss.

Tung oil, also known as *china wood oil,* is one of the most durable of all natural oils. It is superior to linseed oil because it will not darken with age.

Danish oil seals and finishes the wood in one application and is not greasy. Water spots, burns, and scratches are easily removed by buffing with steel wool and Danish oil.

Brushing on Clear Finishes

1. Use a good natural bristle brush. Nylon brushes may dissolve in lacquer. The bristles may fall out of cheap brushes and spoil the job.

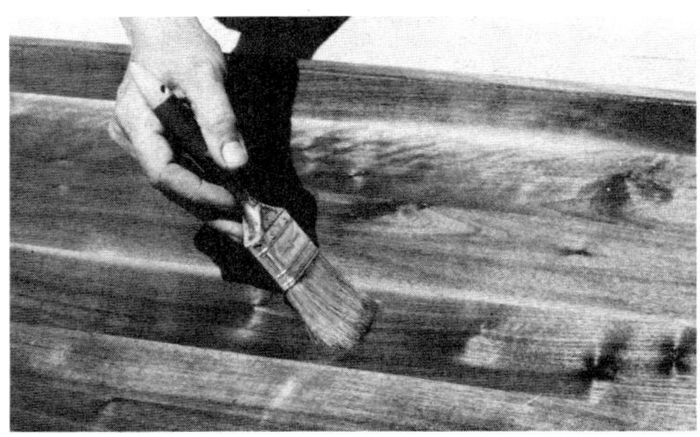

Fig. 27-6 Use light strokes when brushing on a clear finish.

2. Dip the brush not more than half the length of the bristles into the finish. Tap it lightly on the side of the can to prevent dripping.

3. Apply the finish with a full brush and flow onto the surface. Use light strokes. Do not force the bristles against the work, figure 27-6.

4. Work rapidly. Keep a wet edge, working from the dry to the wet. Lift the brush up at the end of the stroke. Smooth out with very light strokes with the grain as quickly as possible. Do not continue to dab once the finish is smooth.

5. Allow the first coat to dry thoroughly. Sand lightly with 320-grit sandpaper.

6. Apply the second coat. When dry, sand again with the same grit.

7. Clean the surface and apply a coat of paste wax.

8. To obtain a highly polished surface, rub the final coat with a pumice stone and rubbing oil using a felt pad.

9. After completing the pumice rubbing, dip a clean pad in water and then into rottenstone. Rub the surface in one direction. Continue until the surface is as smooth as possible.

Spraying Clear Finishes

Before attempting to finish any cabinetwork by spraying, practice using the spray gun. Practice on some cardboard cartons to get the feel of the gun and how to adjust it. Most of the time, poor spraying jobs result from a dirty spray gun. Spray guns should be cleaned immediately after each use.

Air pressure that is too high causes a textured finish called *orange peel.* Pressure for a typical spray gun should not exceed 40 pounds. Follow the manufacturer's direction for the correct air pressure.

Try to avoid tipping the gun at too sharp an angle. This may cause the gun to suck air instead of fluid and give a sputtering spray. Keep the air vent clear in suction-type guns.

Holding the gun too far away from the surface also causes orange peel. With fast-drying finishes, the fluid starts to dry before it reaches the surface and prevents the finish from flowing

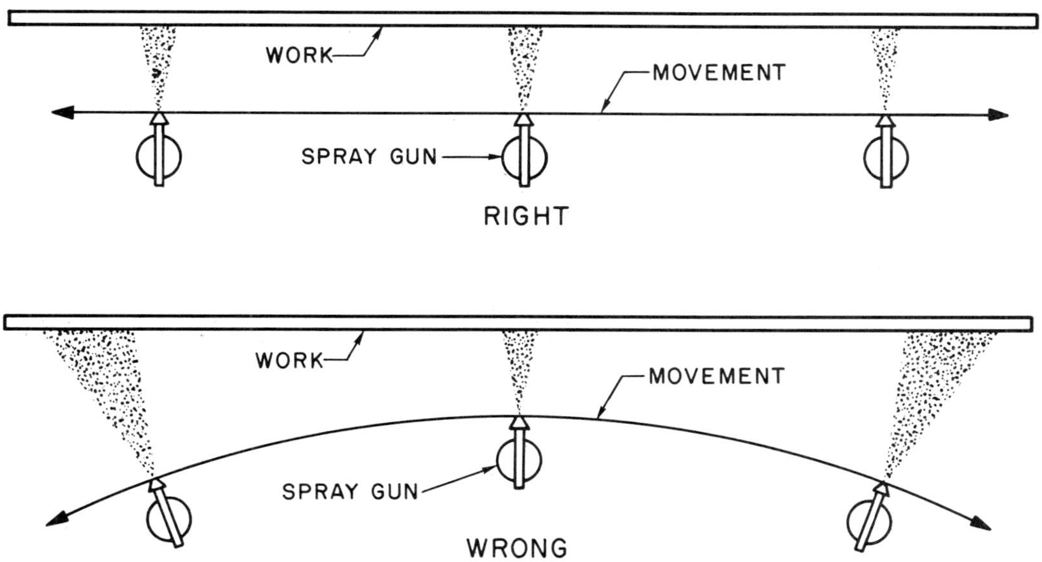

Fig. 27-7 Spray with the gun at right angles to the work.

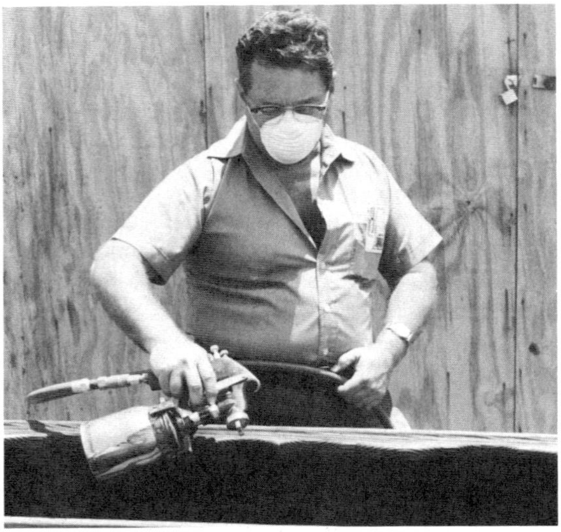

Fig. 27-8 Wear a mask if spraying must be done outside a spray booth.

out smoothly. Under most conditions, the nozzle is held from 6 to 8 inches from the work.

When spraying, keep the gun at right angles to the work, figure 27-7. Move it back and forth with a straight motion, not a curving motion. Start the spray off the work and end it off the work. Overlap the spray to assure complete coverage.

Caution: Spraying should be done in a spray booth. Wear a mask if spraying is done outside the booth, figure 25-8, to prevent inhaling the spray. Inhaling the vapor of fast-drying finishes is extremely dangerous and can be fatal.

A good spray job requires little rubbing between coats. With knowledge of the correct spraying techniques and the problems that cause a poor spray job, excellent results are possible.

ACTIVITIES

1. Make a display of different kinds of coated abrasives and outline the characteristics of each.
2. Make a chart showing the recommended type of wood bleach to use for different kinds of wood.
3. Experiment mixing colors to match a stain on a piece of furniture.
4. Finish the casework and furniture constructed in this section:
   - Bookcase made in *Unit 21 Casework Construction*
   - Chest made in *Unit 21 Casework Construction*
   - Wall and base units made in *Unit 22 Kitchen Cabinets*
   - Table made in *Unit 26 Tables*
   - Additional casework and furniture constructed as part of the activities in these units.

     Decide which techniques to use to finish each piece. Explain your choices. Try to use the following technique at least once. (Note that all of these techniques would not necessarily be used to finish any one piece.) Include:

     - Sanding
     - Bleaching
     - Filling the surface
     - Sealing the surface
     - Staining
     - Applying a clear finish
     - Applying a penetrating oil finish

UNIT REVIEW

Multiple Choice

1. Water stains
   a. do not penetrate deeply.
   b. do not bring out the grain.
   c. raise the grain.
   d. do not give a uniform color.
2. The recommended material to remove excess filler is
   a. burlap.
   b. cotton.
   c. paper towels.
   d. felt.
3. Before sealing the surface, wood paste filler must be allowed to dry at least
   a. 1 hour.
   b. 6 hours.
   c. overnight.
   d. a week.
4. Shellac must be thinned with
   a. mineral spirits.
   b. turpentine.
   c. alcohol.
   d. linseed oil.
5. The most common cause of a poor spray job is
   a. tipping the gun at a sharp angle.
   b. a dirty spray gun.
   c. holding the gun too far away.
   d. air pressure too high.
6. Under most conditions, the spray nozzle is held
   a. 2 to 4 inches from the work.
   b. 4 to 6 inches from the work.
   c. 6 to 8 inches from the work.
   d. 8 to 10 inches from the work.
7. A 1/0 grit sandpaper is equivalent to a
   a. 60-grit number.
   b. 80-grit number.
   c. 100-grit number.
   d. 120-grit number.
8. To obtain a smoother surface in the sanding operation, sponge with water to
   a. remove all traces of grit.
   b. see defects more clearly.
   c. bring out the natural color.
   d. raise the grain.
9. Two woods that do not bleach satisfactorily are
   a. ash and chestnut.
   b. gum and maple.
   c. Douglas fir and yellow pine.
   d. mahogany and walnut.
10. To obtain a more uniform color when applying stains, use
    a. a spray gun.
    b. a thin coat of sealer first.
    c. an oil stain.
    d. bleach before staining.

Questions

1. What are three popular kinds of sandpaper?
2. How are sandpaper grits designated?
3. Describe the proper sanding motion to use.
4. When bleaching wood, how is the action of the bleach stopped?
5. What safety precautions must be taken when working with bleach?

6. What is the purpose of wood paste filler?
7. What will happen if a final finish is applied before the filler is completely dry?
8. What is the purpose of the first coat of finish?
9. Why are two light coats of sealer applied to a surface instead of one heavy coat?
10. How is a highly polished, clear finish obtained?

section 5

Patternmaking

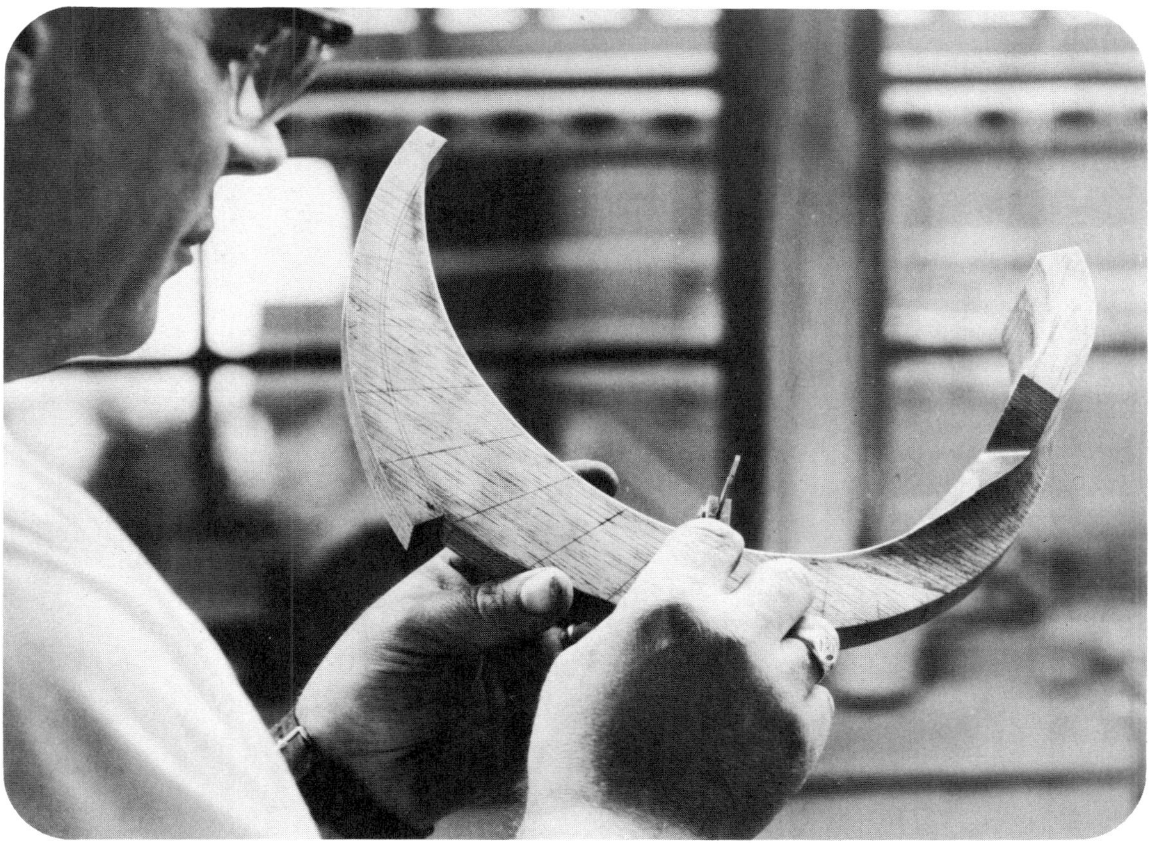

unit 28 Patterns and the Sand Mold

OBJECTIVES

After studying this unit, the student will be able to:

- define patterns and describe kinds of patterns.
- describe the parts of a flask and tell how sand molds are formed.
- describe foundry practices of casting by sand molding.

PATTERNMAKING

Patternmaking is the art of making *patterns* used to form molds for casting. The pattern is the basic model for a casting. It is used to shape an impression in the sand called a *mold.* The *casting* is made by pouring molten metal or plastic into this mold. The mold shapes the material as it cools and hardens. The pattern is designed to make the molding process fast and efficient.

For a simple casting, the pattern will closely resemble the casting, figure 28-1. For more complex castings that have holes and interior openings, the pattern will look much different than the finished casting. In all cases, the pattern is slightly larger than the casting because molten metal shrinks as it cools.

Patterns are made of different materials, but the master pattern is usually made of wood. Woods like cherry, mahogany, and sugar pine are good for patternmaking because they are close-grained and easy to work.

Patternmaking is one of the few trades where much of the work is still done with hand tools. The patternmaker must know how to shape wood with extreme accuracy to keep the costs of machining the casting made from the pattern to a minimum, figure 28-2.

METRICS IN PATTERNMAKING

The metric system of measurement is used extensively in patternmaking shops and the foundry. The units in *Section 5 Patternmaking* will therefore include metric equivalents for all inch measurements. A list of inch-to-millimetre conversions is in the appendix.

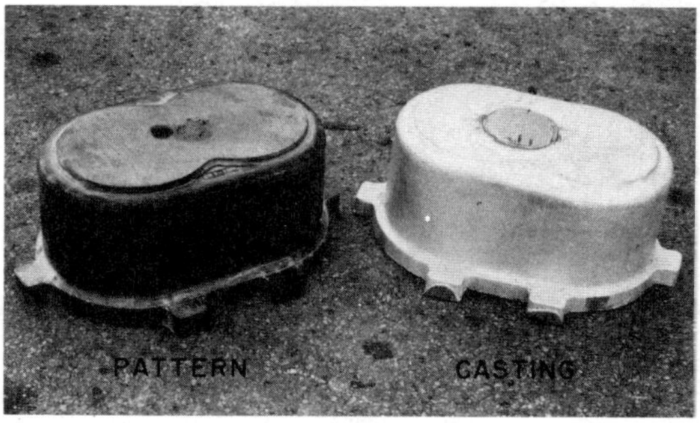

Fig. 28-1 Pattern and casting

Fig. 28-2 Patternmakers must know how to shape wood with extreme accuracy. *(Berkshire Pattern & Woodworking Shop)*

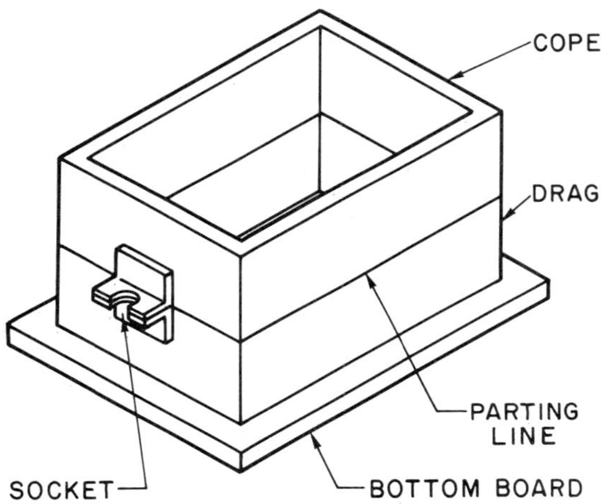

Fig. 28-3 A flask

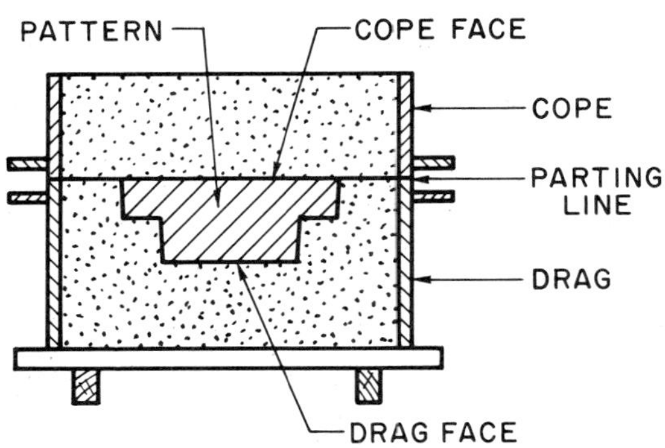

Fig. 28-4 The cope face and drag face of a pattern

THE SAND MOLD

In order to make good patterns, the patternmaker must understand how molds are made and used. The most common type of sand mold uses sand that will not crumble after the impression of the pattern is made. This kind of sand is called *green sand* because it has more clay in it than might be found in sand at the beach, for example. When the clay is moistened, the sand becomes sticky. This is called *tempering.* Foundries generally use green sand, but school shops often use oil-tempered sand.

A mold is made by ramming green sand around a pattern that is placed in a box. This box has no top or bottom and is called a *flask.* The flask consists of a top half called the *cope* and a bottom half called the *drag,* figure 28-3. The joint between the cope and the drag is called the *parting line.* The two sections are fitted with *pins* and *sockets* so that they are detachable.

Patterns have a *parting surface* that coincides with the parting line of the cope and drag, figure 28-4. The surface of the pattern that is exposed when the cope is removed is called the *cope face.* This is also the parting line of the pattern. The surface of the pattern that leaves its impression in the drag is called the *drag face.*

Making a Mold

1. Put the cope face of the pattern down on the molding board. Place the drag upside down on the molding board, figure 28-5. Dust a fine, dry *parting compound* on the pattern so that the sand does not stick to it.

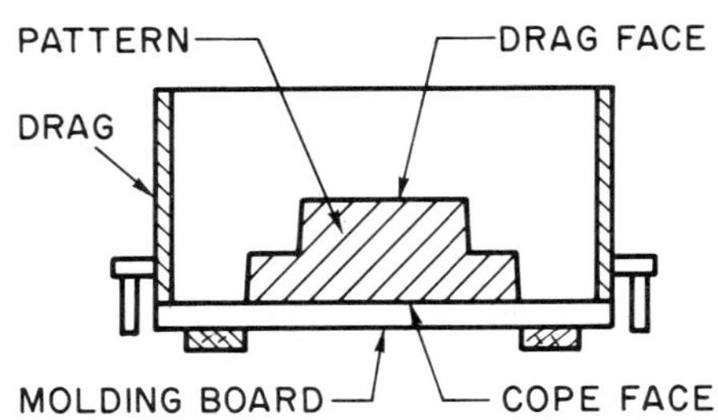

Fig. 28-5 The pattern is positioned in the drag.

Fig. 28-6 Ramming the drag *(Schmidt Aluminum Castings)*

Fig. 28-7 Scraping off the drag with a strike-off bar *(Schmidt Aluminum Castings)*

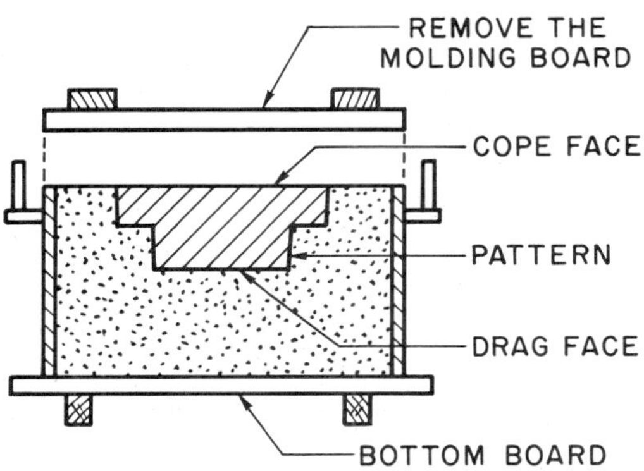

Fig. 28-8 Turn the drag over

2. Sift sand through a sifter, called a *riddle,* over the pattern until it covers the pattern by about 2 inches (50 mm).
3. Press the sand by hand around the pattern and the sides of the drag. Fill the remaining space in the drag with unriddled sand.
4. Pack the sand tightly with a *rammer,* figure 28-6. Scrape off excess sand with a strike-off bar, figure 28-7.
5. Place a bottom board over the drag. While holding the two boards and the drag firmly together, turn them over. Remove the molding board to expose the pattern, figure 28-8.
6. Smooth out any roughness on the parting surface with a trowel. Dust parting compound over the surface to prevent the cope from sticking to the drag when it is set in place.
7. Set a tapered pin, called a *sprue plug,* in line with the center of the pattern and about 2 inches (50 mm) away from it, figure 28-9. The sprue will provide the opening through which the molten material is poured.
8. Place one or more straight pins, called *riser pins,* on the opposite side of the pattern from the sprue plug. Molten material cools more slowly in the riser hole than in the sprue hole. The riser hole feeds the castings as the casting cools and shrinks. This process is called *venting the mold.*

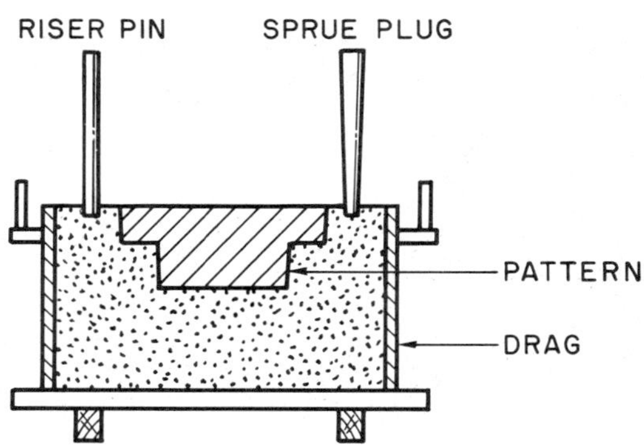

Fig. 28-9 Locating the sprue plug and riser pin

Fig. 28-10 Riddling sand over the pattern *(Schmidt Aluminum Castings)*

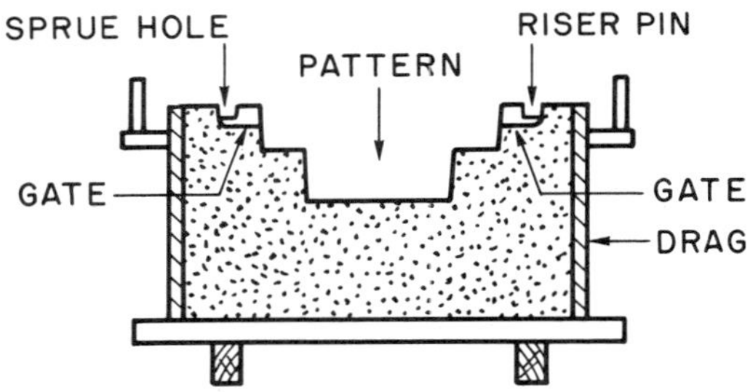

Fig. 28-11 Cutting gates

Fig. 28-12 Pouring a casting in the foundry *(Schmidt Aluminum Castings)*

9. Place the cope over the drag and riddle sand over the pattern, figure 28-10. Pack the sand firmly around the pattern, the sprue plug, and the riser pins. Finish filling the cope with unriddled sand, ram, and strike off the excess.
10. Draw out the sprue plug and riser pins. Lift off the cope section carefully and set aside where it will not be damaged.
11. Draw the pattern from the sand by lightly driving a large pin, called a *draw spike,* into the pattern to loosen it. Lift the pattern from the sand. Repair any breaks in the impression with a trowel.
12. Cut a channel, called a *gate,* through which the material can flow from the sprue hole to the mold. Use a *gate cutter.* Cut gates from the mold to the riser pins, figure 28-11.
13. Blow all loose sand from the mold and dust the surface with parting compound. Replace the cope section carefully.
14. The casting is now ready to be poured, figure 28-12. The molten material is poured in the sprue hole. The sand mold is destroyed after making the casting, but any number of molds can be made with a single pattern.

KINDS OF PATTERNS

The *solid* or *one-piece pattern* is the simplest of all patterns. It may be made of many pieces, but they are fastened together to make one solid piece, figure 28-13.

Some patterns, because of their shape, cannot be drawn from the mold unless they are made

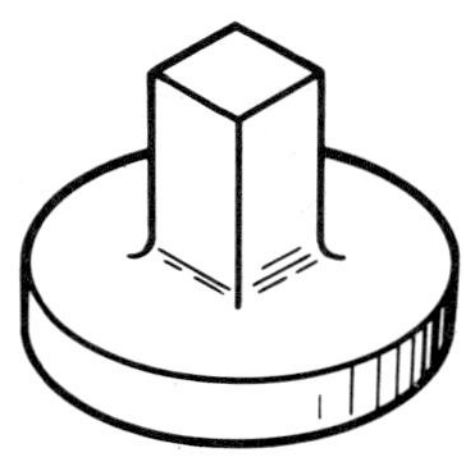

Fig. 28-13 Solid or one-piece pattern

in two or more pieces. These are called *split* or *parted patterns.* Parted patterns have tapered dowel pins in the joint. The dowels keep the pieces in line and allow them to be separated, figure 28-14.

Patterns with *loose pieces* have projecting parts. These projecting parts prevent the loose pieces from being drawn from the mold with the rest of the pattern. Dowels with a slide fit hold the loose pieces in place, figure 28-15.

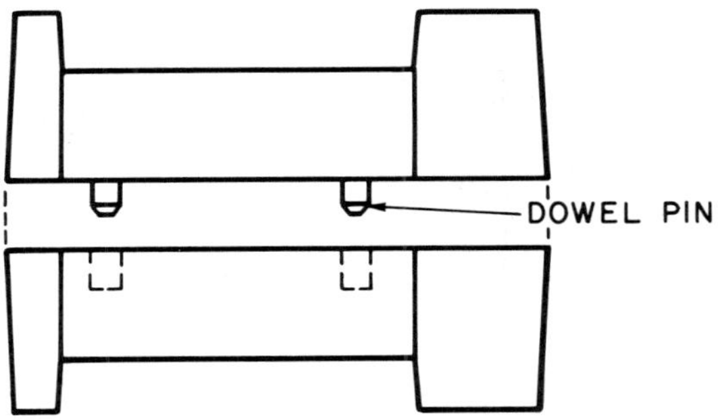

Fig. 28-14 Split or parted pattern

THE FOUNDRY

Foundries do both mass-production and custom work. In mass-production work where many castings of the same kind are made, assembly line techniques and automatic equipment are used. Conveyors move sand to the molder. A mold-making machine releases sand into the flask and hydraulic presses ram the sand tightly around the pattern, figures 28-16 and 28-17. The pattern is removed and the casting is poured. When the casting cools, it is placed in a shaker where the sand is shaken loose.

Large foundries usually handle production work as well as custom work with automatic equipment. In small shops where only one or a few castings are made, castings are generally made by hand. The processes described in this section apply to single, handmade castings.

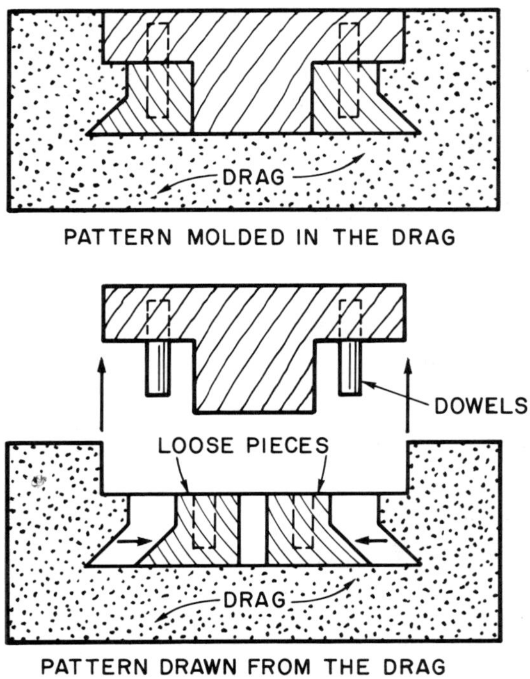

Fig. 28-15 Pattern with loose pieces

Fig. 28-16 Filling the cope with sand on a mold-making machine *(Schmidt Aluminum Castings)*

Fig. 28-17 Ramming the sand with a hydraulic press *(Schmidt Aluminum Castings)*

ACTIVITIES

The following activities can be done while you visit a foundry:

1. Observe both production and custom methods of making castings.
2. Follow a particular job through from the molding to the finishing of the casting.
3. Grasp green sand firmly in your hand and see how it keeps its shape.
4. Visit the foundry's patternmaking shop and observe how various machine tools and hand tools are used to make patterns.

UNIT REVIEW

Completion

1. To make a casting, molten material is poured into an impression in the sand called a ________.
2. The object that is used to make the impression in the sand for the casting is called a ________.
3. The art of making castings is called ________.
4. The most common type of sand used in the molding process is called ________ sand.
5. The box that contains the sand is called a ________.
6. The bottom section of the box is called the ________.
7. The top section of the box is called the ________.
8. The joint between the two sections of the box is called the ________.

9. When making a mold the first step is to place the pattern on a ________.

10. Sand is sifted over the pattern through a ________.

Multiple Choice

1. The surface of the pattern that is exposed when the cope is removed is called the
   a. cope face.
   b. drag face.
   c. parting face.
   d. flask face.

2. The surface of the pattern that leaves its impression in the drag is called the
   a. cope face.
   b. drag face.
   c. parting face.
   d. flask face.

3. Sand is scraped off the drag with a
   a. shrink.
   b. trowel.
   c. strike-off bar.
   d. scraper.

4. The opening through which the molten material is poured is called a
   a. gate.
   b. vent.
   c. sprue.
   d. riser.

5. The channels that connect the mold to the sprue and riser holes are called
   a. gates.
   b. vents.
   c. plugs.
   d. spikes.

6. As the casting cools and shrinks, it is fed molten material by the
   a. gate.
   b. vent.
   c. sprue.
   d. riser.

7. After the pattern is removed from the sand mold, the surfaces are dusted with
   a. parting compound.
   b. dust.
   c. powder.
   d. dry sand.

8. Loose pieces on patterns are held in place with dowels that have a
   a. press fit.
   b. slide fit.
   c. loose fit.
   d. tight fit.

9. Parted patterns use dowels that
   a. have a taper.
   b. keep the pieces in line.
   c. allow the pieces to be separated.
   d. all of the above.

10. Patterns have a parting surface that coincides with
    a. the drag face.
    b. the flask.
    c. the parting line.
    d. all of the above.

unit 29 Pattern Details

OBJECTIVES

After studying this unit, the student will be able to:

- explain shrinkage and the use of shrink rules.
- state the allowances made for machine finishing.
- explain the purpose of draft and lay out centerlines to provide draft on a small pattern.
- describe the materials, tools, and methods used for applying fillets.
- describe how patterns are sealed and explain the color code.

SHRINKAGE

Most metals expand when heated to a liquid and shrink when they change back to a solid. If the pattern is made the same size as shown in the drawing, the casting made from the pattern will be undersize. In order for the casting to shrink to the right size, the pattern is made larger than specified.

Patternmakers use *shrink rules* so that they do not have to calculate the allowance from shrinkage, figure 29-1. These rules look the same as ordinary rules, but they have larger graduations. A two-foot shrink rule with a scale of 1/4-inch shrinkage per foot is 1/2 inch longer than the standard rule. There are different scales because metals do not shrink alike. Some of the metals used for casting shrink as follows:

- Aluminum – 1/4 inch per foot
- Brass – 3/16 inch per foot
- Cast Iron – 1/8 inch per foot
- Steel – 3/16 inch per foot

Metric shrink rules are not available as yet. When working in metric, the patternmaker must use the percentage of shrinkage:

- 1/8 inch shrinkage per foot = 1 percent
- 3/16 inch shrinkage per foot = 1.56 percent
- 1/4 inch shrinkage per foot = 2.08 percent
- 1/2 inch shrinkage per foot = 4 percent

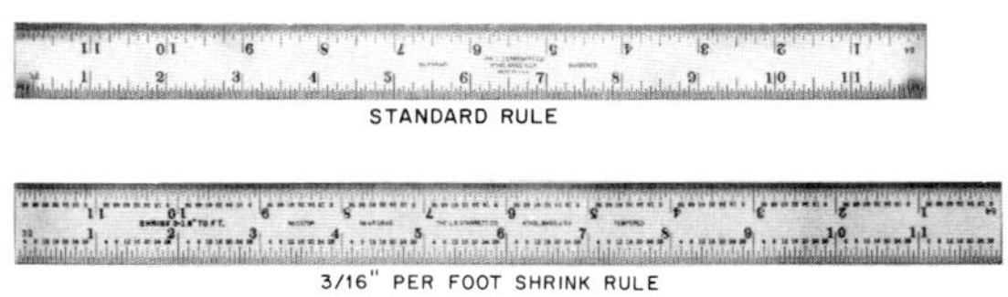

Fig. 29-1 Shrink rules are longer than standard rules. *(L.S. Starrett Co.)*

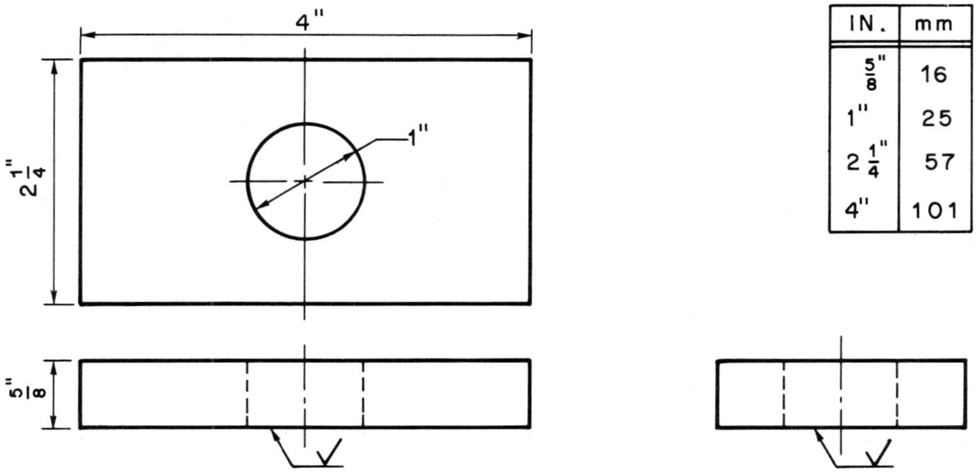

IN.	mm
5/8"	16
1"	25
2 1/4"	57
4"	101

Fig. 29-2 The symbol √ indicates the bottom surface is to be finished.

MACHINE FINISH

Sometimes parts of a casting must be machined. An allowance must be made on the pattern to provide the extra material that will be removed by machining. The symbol "√" on a line in the drawing indicates the surface to be finished. In figure 29-2, the bottom surface is to be finished. The pattern is made 3/4 inch (19 mm) thick instead of 5/8 inch (16 mm) as indicated in the drawing. This allows 1/8 inch (3 mm) for finishing.

The amount allowed for machining the casting varies from 1/16 inch to 1/4 inch (1.5 to 6 mm). This depends on the way the casting is made, the kinds of machines used to do the finishing, and the amount of finish required. Usually a 1/8-inch (3-mm) allowance is made for iron and steel, and a 1/16-inch (1.5-mm) allowance is made for brass and aluminum.

DRAFT

Draft is a slight taper added to the vertical surfaces of a pattern. It allows the pattern to be removed or *drawn* without damaging the mold. The part of the pattern that lies in the drag is laid out to the given dimensions on the drag face. Then, the pattern is enlarged from the drag face toward the cope face, figure 29-3. This insures that the casting will not be smaller than specified. If there are holes in the pattern, they are laid out to size on the cope face so that the holes are not smaller than required.

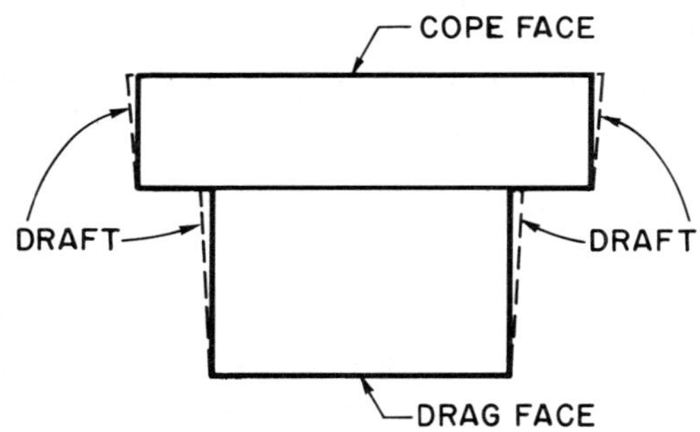

Fig. 29-2 Draft is a slight taper added to the vertical surfaces of a pattern.

The amount of draft varies according to the pattern. Patterns are usually made so that their position in the mold requires the least amount of draft. Draft varies from 1/64 to 1/16 inch per inch (0.4 to 1.5 mm per 25 mm). A tight draft is given the pattern when close tolerances are required. More draft is provided whenever possible without interfering with critical dimensions.

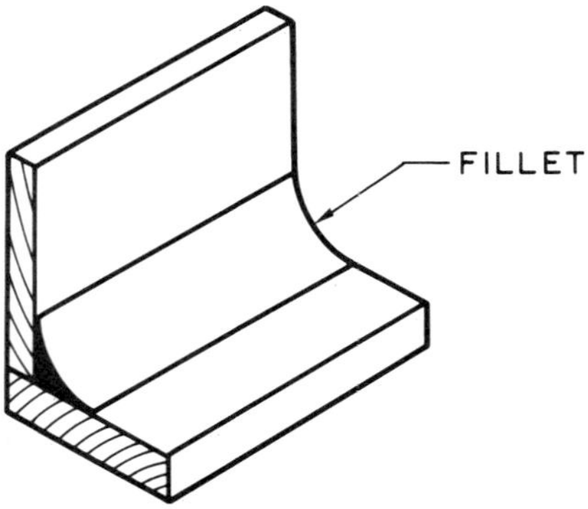

Fig. 29-4 Fillets round off sharp inside corners.

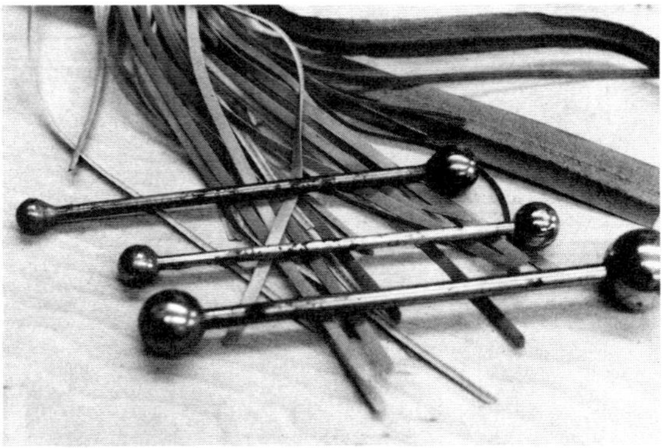

Fig. 29-5 Leather fillets and fillet irons *(Schmidt Aluminum Castings)*

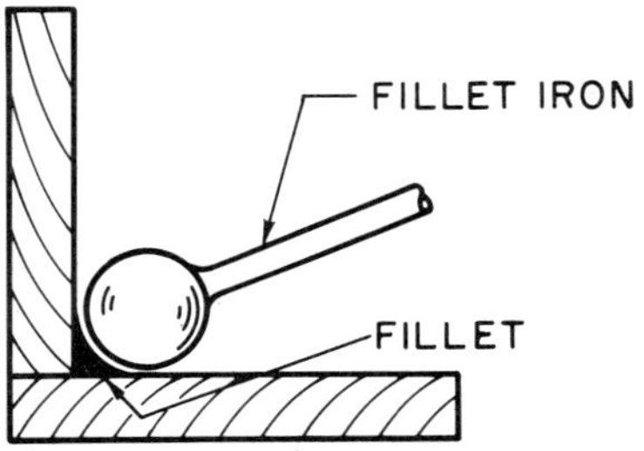

Fig. 29-6 A fillet iron of the desired diameter shapes the fillet.

APPLYING FILLETS

Fillets are concave pieces placed on sharp inside corners of the pattern, figure 29-4. They round off sharp inside corners, improve the appearance, and add strength to the casting. They also allow the pattern to be drawn from the mold with less danger of damage to the sand. Fillets are made of leather, wax, and wood.

Leather Fillets

Leather fillets are most widely used because they are very pliable, figure 29-5. They come in 4-foot (1219-mm) lengths of different sizes.

Leather fillets are usually fastened in place with glue. Then they are rubbed with the ball end of a fillet iron. The shape of the fillet iron makes the rounded corner desired, figure 29-6. Excess glue is removed immediately with a damp cloth. The featheredges of the fillet should be well bonded.

Leather fillets are sometimes applied to a finished surface. In this case glue will not make a good bond, and shellac is used instead. The back sides of the fillet and the corner are coated with heavy shellac and allowed to become sticky. The fillet is applied and rubbed in place with the fillet iron. Excess shellac is wiped off with a rag dampened in alcohol.

Wax Fillets

Wax fillets are used only when there is no danger of the fillet being melted by hot sand. They come in threads of various diameters, or they can be made on the job with a beeswax gun.

Wax of the proper diameters should be used. To make a fillet with a 1/4-inch (6-mm) radius, for instance, a 1/8-inch (3-mm) diameter wax thread is used. The thread is laid in the corner and worked in place to a featheredge with a heated fillet iron. Excess wax is removed with a rag dampened in mineral spirits.

Caution: The fillet iron is heated only enough to form the wax, not to melt it. Care should be taken in handling the heated equipment.

Wood Fillets

Wood fillets are usually used to make corners with large curves. The stock is first cut to a triangular shape and hollowed out on one surface.

The fillet is glued securely to the corner. Brads are then pushed below the surface of the fillet to join the fillet to the pattern. The holes are filled and sanded. Care must be taken not to damage the featheredge of the wood fillet.

FINISHING THE PATTERN

Sealing the Pattern

Patterns are sealed to help prevent swelling and warping caused by absorbing moisture. Usually shellac is used to seal the surfaces of patterns. The pattern is sanded smooth and the surface wiped free of dust. A wash coat of shellac (1 part shellac and 8 parts alcohol) is first applied. After it dries for about one hour, it is sanded smooth with fine sandpaper. Two more coats of slightly thicker shellac are then applied, sanding between coats when dry. The pattern is now ready for color coding.

Color Coding

The American Foundrymen's Association recommends a standard color code to help foundry workers identify different parts of patterns. A new code has been adopted, but some foundries still use the old code. These color codes are listed in figure 29-7.

LAYING OUT A PATTERN

The pattern for the router base shown in figures 29-8 and 29-9 is used in *Unit 24 Cabinet Doors* to make a French Provincial design on a solid cabinet door. The router base will be cast of aluminum and no surfaces need additional finishing.

PART	NEW CODE	OLD CODE
Main Body	Clear	Black
Core Prints	Black	Yellow
Finish	Red	Red
Seats for loose pieces	Aluminum	Red stripes on yellow

More information on coloring patterns may be obtained from the American Foundrymen's Association.

Fig. 29-7 Color Codes

Cut the Stock

1. Select a piece of clear, dry lumber suitable for patternmaking.
2. Plane one side to the required thickness of the object. Use the 1/4-inch shrink rule to test the thickness.
3. Joint one edge, ripping about 1/2 inch (13 mm) wider than required.
4. Square both ends leaving the stock longer than needed.
5. Mark one side *cope face* and the other *drag face.*

Lay Out the Centerlines

The next step is to lay out the centerlines. Centerlines must be laid out carefully on both faces of the pattern and must be exactly opposite each other. Use the marking gauge to score centerlines with the grain and a knife to score centerlines across the grain.

1. Using a marking gauge, score centerlines AA on both faces, figure 29-10. Ride the marking gauge along the same edge to mark both faces. In this way, even if the line is slightly off center, the centerlines will still be in line.
2. At the middle of the centerline (point 0), square centerline BB all around the stock. Use a square and a knife.

Fig. 29-8 Router base to make a French provincial design.

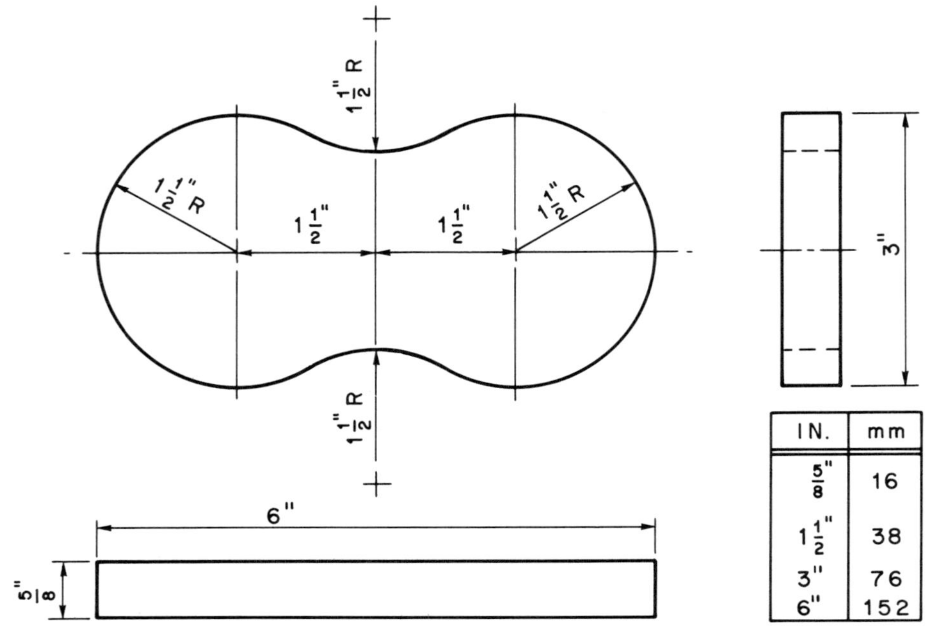

IN.	mm
5/8"	16
1 1/2"	38
3"	76
6"	152

Fig. 29-9 Router base

3. Set dividers for 1 1/2 inches (38 mm) using the 1/4-inch shrink rule. Strike arcs on both sides from points 0 so that they intersect centerlines AA at points E.

4. From points E, square lines with a knife around the stock making centerlines CC and DD.

5. With the dividers still set at 1 1/2 inches (38 mm) and using points E as centers, draw the circular ends on the drag face only, figure 29-11.

6. Open the dividers 1/64 inch (0.4 mm) more and strike similar ends on the cope face of the pattern. This accounts for the draft on the pattern.

7. It is necessary to scribe arcs whose centers lie outside the stock to complete this pattern layout. To locate these centers, square a line across a piece of scrap stock of the same thickness. Hold the scrap stock against the edge of the pattern material so that the squared mark lines up with centerline BB, figure 29-12.

8. Set the dividers at 1 1/2 inches (38 mm). Locate the center on the scrap block of the arc that is tangent to the circular ends of the pattern. Strike tangent arcs on the drag face.

9. Open the dividers 1/64 inch (0.4 mm) more and strike arcs on the cope face in a similar manner.

Cutting the Pattern

1. Band-saw the pattern with the cope face up to within 1/16 inch (1.5 mm) of the outline.

2. Using chisels and gouges, pare the edge by cutting from both sides toward the center of the edge until a true surface is obtained. Split the scribed lines when cutting. Do not leave stock outside of the line and do not cut the line off.

3. Test the edge surface frequently with a straightedge.

4. Sand the faces and edges.

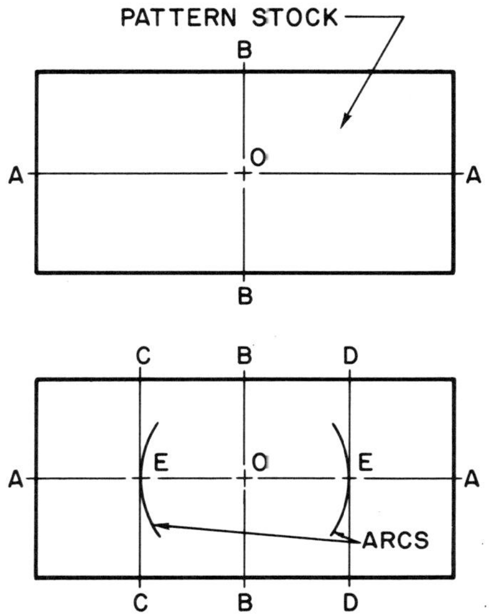

Fig. 29-10 Laying out centerlines

The pattern is now complete, except for a shellac finish and color coding. A draft of 1/32 inch per inch (0.8 mm per 25 mm) is provided, and shrinkage of 1/4 inch per foot is accounted for.

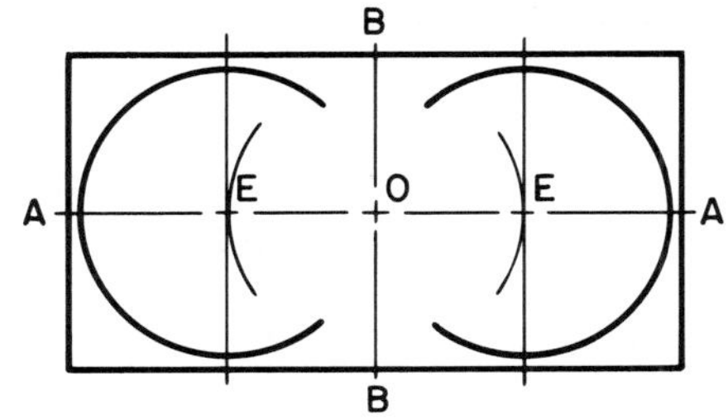

Fig. 29-11 Laying out the circular ends of the router base pattern

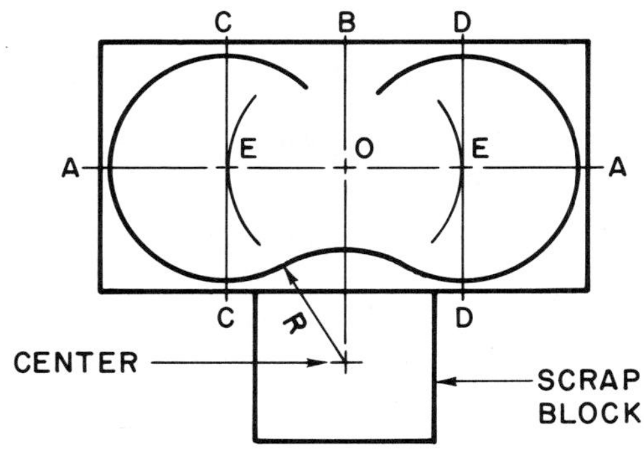

Fig. 29-12 Use a scrap block when the center of an arc is off the pattern stock.

ACTIVITIES

1. Visit a patternmaker's shop and observe the following:
    - the use of shrink rules.
    - pattern drawings specifying machine finish.
    - application of fillets.
    - provisions on the pattern for draft.
    - starting the layout of a pattern.
    - how patterns are finished and colored.
2. Make a chart showing the amount of shrinkage per foot for different metals.
3. Apply leather, wood, and wax fillets.
4. Practice drawing arcs tangent to each other.
5. Practice paring stock to a scribed line using chisels and gouges.
6. Finish the pattern for the router base with three coats of shellac as described in this unit.

UNIT REVIEW

Completion

1. Most materials ________ when heated to a liquid state.
2. The pattern is made ________ in order for the casting to be the required size.
3. Aluminum shrinks about ________; brass shrinks about ________; and cast iron shrinks about ________ when cooled from a liquid to a solid state.
4. A two-foot shrink rule with 1/4-inch shrinkage per foot is ________ inch shorter than the standard rule.
5. When parts of a casting are to be machine finished, ________ must be provided on the pattern.
6. The symbol ________ on a drawing indicates the surface to be finished.
7. An allowance of ________ inch to ________ inch must be made for machine finishing.
8. Fillets are made from ________, ________, or ________.
9. Draft is a slight ________ on the vertical surfaces of a pattern.
10. Patterns are sealed with shellac to prevent ________ and ________.

Multiple Choice

1. Draft enlarges the pattern toward the
   a. bottom face.
   b. cope face.
   c. drag face.
   d. flask face.
2. The part of the pattern that lies in the drag is usually laid out on the
   a. bottom face.
   b. cope face.
   c. drag face.
   d. flask face.
3. Draft varies for each inch from
   a. 1/64 inch (0.4 mm) to 1/16 inch (1.5 mm).
   b. 1/16 inch (1.5 mm) to 1/8 inch (3 mm).
   c. 1/8 inch (3 mm) to 3/16 inch (5 mm).
   d. 3/16 inch (5 mm) to 1/4 inch (6 mm).
4. Patterns are usually sealed with
   a. lacquer.
   b. shellac.
   c. polyurethane.
   d. linseed oil.
5. The pattern is color coded so that
   a. moisture cannot warp the pattern.
   b. foundry workers can identify the parts.
   c. sand cannot stick to the surface.
   d. foundry workers can tell it apart from other patterns.

6. The color specified in the recently adopted color code for core prints is
   a. black.
   b. red.
   c. yellow.
   d. aluminum.
7. Centerlines are scored across the grain with
   a. an awl.
   b. a sharp nail.
   c. a knife.
   d. a marking gauge.
8. Centerlines are scored with the grain with
   a. an awl.
   b. a sharp nail.
   c. a knife.
   d. a marking gauge.
9. To obtain a true surface when cutting the pattern,
   a. leave 1/32 inch (0.8 mm) outside the scribed line.
   b. test the edge surface when you are finished.
   c. split the scribed line.
   d. cut the scribed line off.
10. Test the edge surfaces of a pattern frequently with the
    a. marking gauge.
    b. outside calipers.
    c. straightedge.
    d. inside calipers.

unit 30 One-Piece Patterns

OBJECTIVES

After studying this unit, the student will be able to:

- describe the construction of a one-piece pattern.
- make a rectangular and round one-piece pattern.
- make a round, segmented one-piece pattern.
- find the miter cut for three or more equal-sided patterns.

ONE-PIECE PATTERNS

A *one-piece pattern* is a pattern that is molded with no partings or loose pieces. It may be made with many pieces, but when complete, all of the pieces are held securely together.

The pieces that make up the pattern are positioned so the grain direction gives the pattern the most strength. They are held together with glue, brads, nails, screws, dowels, or other fasteners. There is no need, as in cabinetmaking, to hide the fasteners as long as the surface is smooth. For instance, nails, screws, and brads are set below the surface and filled smooth. Dowels need not be bored blind as long as the exposed ends are sanded smooth.

Sanding is done in the direction that is most convenient, either with or across the grain. However, pattern edges are sanded along their length so the edges are not rounded over.

Finally, the pattern is constructed to be molded with the least amount of draw, figure 30-1. The patternmaker must decide on the parting line of the one-piece pattern and the direction it will be drawn from the mold.

Many castings from one-piece patterns can be seen right in the woodworking shop. The tabletop of the band saw, table saw, jointer, disc sander, scroll saw and other stationary machines were probably cast from a one-piece pattern. Many parts of woodworking and metal vises, hand planes, cabinet scrapers, and spokeshaves are also cast from a one-piece pattern. Examine some of these parts and try to determine their parting line and direction of draw from the mold.

MAKING A RECTANGULAR ONE-PIECE PATTERN

An isometric and exploded view of an aluminum bracket is shown in figure 30-2. Note that the pattern is made from two pieces of stock. These pieces are joined and two fillets round off the edges between the pieces.

The working drawing for the aluminum bracket is shown in figure 30-3. Notice that the

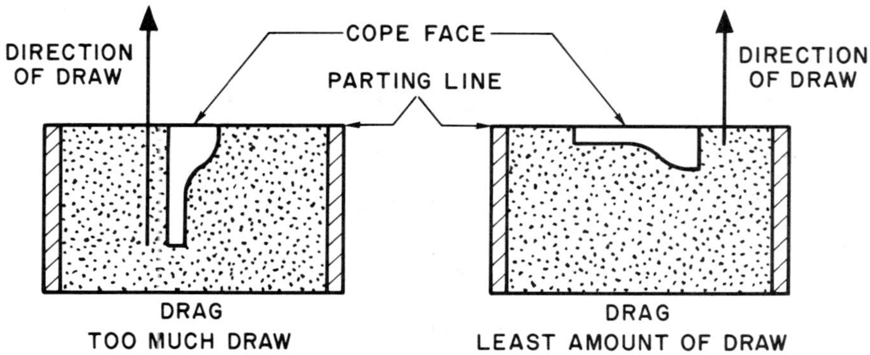

Fig. 30-1 Choose the parting line that provides the least amount of draw.

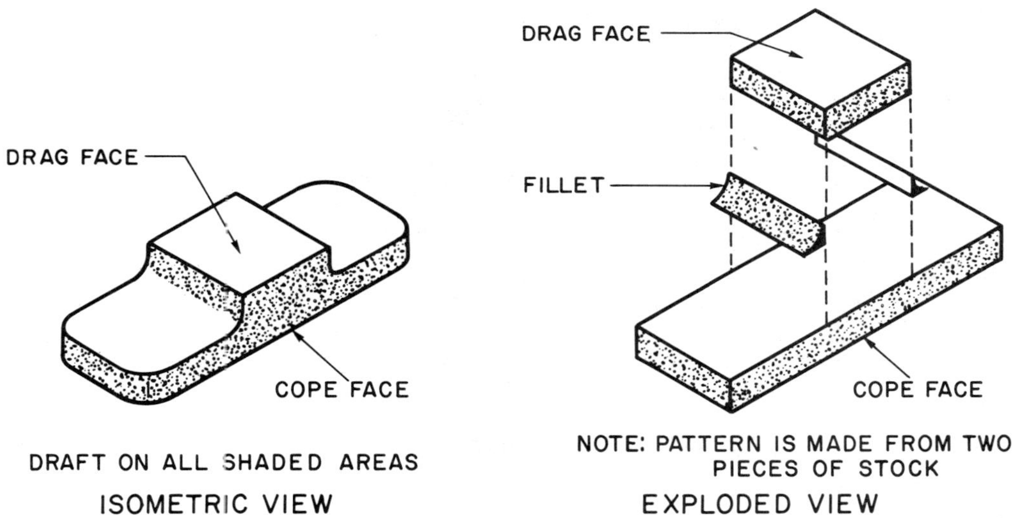

Fig. 30-2 Aluminum bracket

bottom surface is to be finished. The holes will be drilled after the casting is made, so the patternmaker does not have to worry about them. The bottom surface is the cope face and the parting line, figure 30-4. The pattern will be molded entirely in the drag.

Constructing the Top Section

1. Plane a piece of stock to a 1-inch (25-mm) thickness using the 1/4 inch per foot shrink rule for aluminum.
2. Lay out centerlines for the pattern in both directions on one surface marked *drag face*.
3. From the centerlines, measure out 1 1/2 inches (38 mm) in each direction to make a 3-inch (76-mm) square.
4. Band saw to within 1/16 inch (1.5 mm) of the scribed lines.
5. Use a disc sander to provide draft to the four edges.

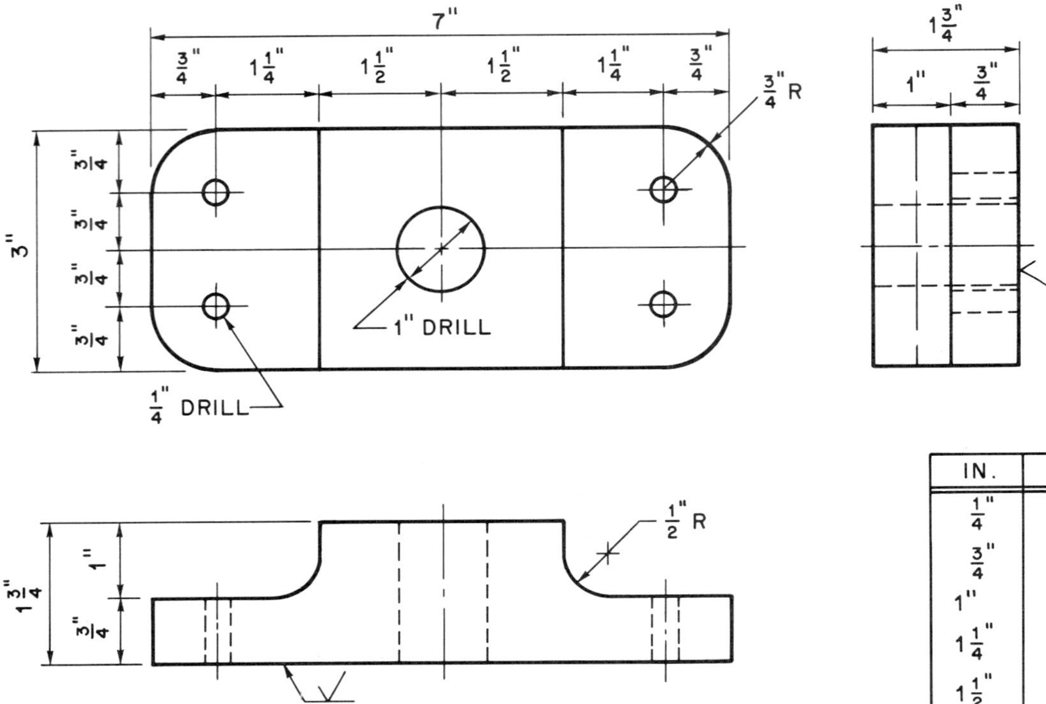

IN.	mm
1/4"	6
3/4"	19
1"	25
1 1/4"	32
1 1/2"	38
1 3/4"	44
3"	76
7"	178

Fig. 30-3 Working drawing of aluminum bracket

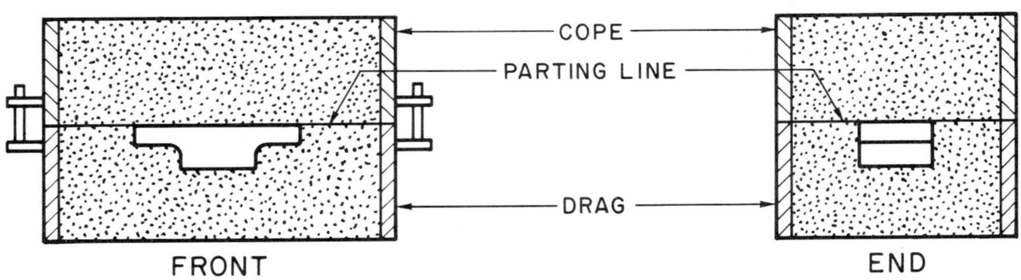

Fig. 30-4 Position of the pattern in the flask

6. Tilt the table up to sand a taper of 1/64 inch (0.4 mm) to the inch. Sand the edges of the block until the scribed lines are split. Using the disc sander eliminates the need for scribing centerlines on the cope face to enlarge the pattern.

Constructing the Bottom Section

1. Plane a piece of stock of sufficient width and length down to 7/8 inch (22 mm) thick. This allows 1/8 inch (3 mm) for finishing as indicated in the drawing.
2. Lay out centerlines in both directions on the drag face.
3. Using dividers, strike arcs 3 1/2 inches (89 (mm) in both directions from the center on the long centerline.
4. Scribe lines to mark the length and width of the pattern. The width should be 1/32 inch (0.8 mm) wider than indicated in the drawing to match the cope face of the top section, figure 30-5.
5. Set the marking gauge for 3/4 inch (19 mm). Mark in both directions at each corner to find the center of the arc.
6. Set the dividers from this center to the edge of the stock and scribe arcs at the four corners.
7. Cut the bottom section almost up to the scribed lines.
8. Disc sand with the table tilted down and the scribed lines up. Tilt the table to the same setting as for the top section. Sand until the scribed lines are split.

Joining the Sections

1. Scribe lines square across the width of the bottom section, 1 1/2 inches (38 mm) from the centerline. This marks the location of the top piece.
2. Place the cope face of the top piece on the drag face of the bottom piece. The edges of the top section must line up with the edges of the bottom section and the scribed lines.

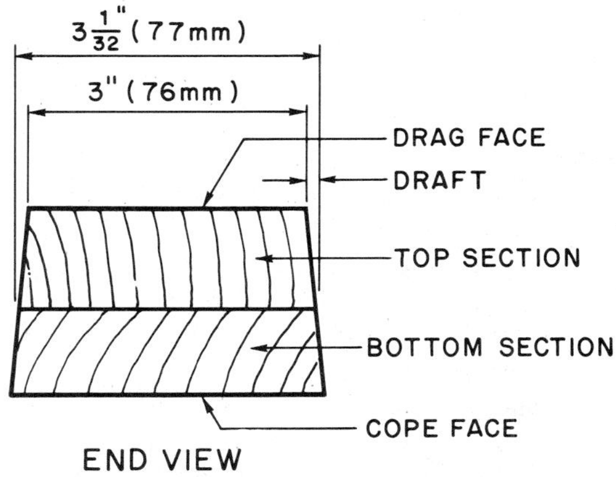

Fig. 30-5 The edges of both sections must be flush with each other.

3. Apply glue and clamp the two pieces together in the correct position. Protect the pattern by using scrap blocks under the clamp.
4. When dry, remove the clamps. Choose leather fillets to obtain a 1/2-inch (12.5-mm) radius between the two pieces.
5. Apply glue to the fillets and rub into place with a fillet iron.
6. Wipe excess glue off with a rag dampened with water.
7. Trim the ends of the fillet flush with a sharp knife.
8. Sand all surfaces smooth and slightly round off sharp edges except those on surfaces to be finished. Rounding over corners of surfaces to be machine finished may interfere with the finishing process.
9. Shellac and color the pattern. Color the bottom red and the main body with a clear finish.

MAKING A ROUND ONE-PIECE PATTERN

Patterns for casting round objects are usually made by building up the circle in layers with small curved pieces, like bricks in a circular wall, figure 30-6. This is called *segmental construction* and each piece is called a *segment.* (See *Unit 18 Making Curved Pieces.*)

The number of segments in a ring depends on the diameter of the ring and the width of the segments. Segments are made to provide maximum strength to the ring. However, long segments have too much cross-grain and are weak. Too many segments waste the patternmaker's time, figure 30-7. Rings up to 14 inches (350 mm) may contain four segments per layer. Rings over 14 inches may have six or more segments per layer.

The joints between the segments are placed midway on the segments of the layer below for strength, figure 30-8. When it is complete, the face and edges are trued by faceplate turning on a wood lathe.

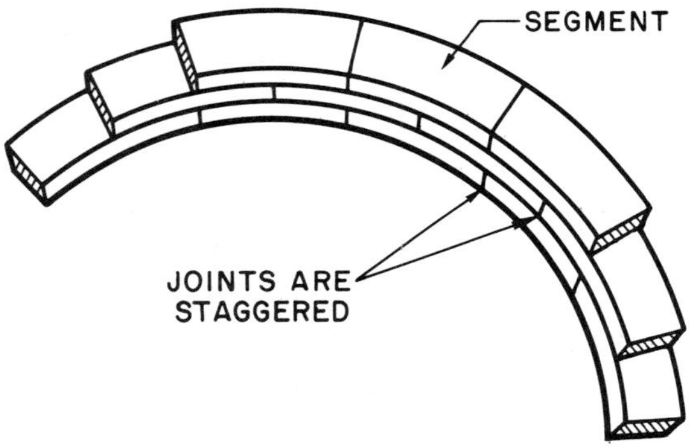

Fig. 30-6 Segmental construction

Laying Out The Segments

Make a pattern for the brass ring in figure 30-9 using four segments per layer.

1. On a smooth piece of scrap, lay out a circle 6 1/2 inches (165 mm) in diameter, figure 30-10(A).
2. Using the same center, scribe another circle 3 1/2 inches (89 mm) in diameter.
3. Draw two diameters AA and BB at right angles to each other.
4. Inscribe (draw inside) a square inside the inner circle so its corners touch the circle, figure 30-10(B).
5. Circumscribe (draw outside) another square outside the outer circle so the square's midpoints touch the circle.
6. Measure the right angle distance between the squares. Rip enough stock to this width to make three layers of segments.
7. Plane the ripped stock down to 3/4 inch (19 mm), which allows 1/4 inch (6mm) for truing.

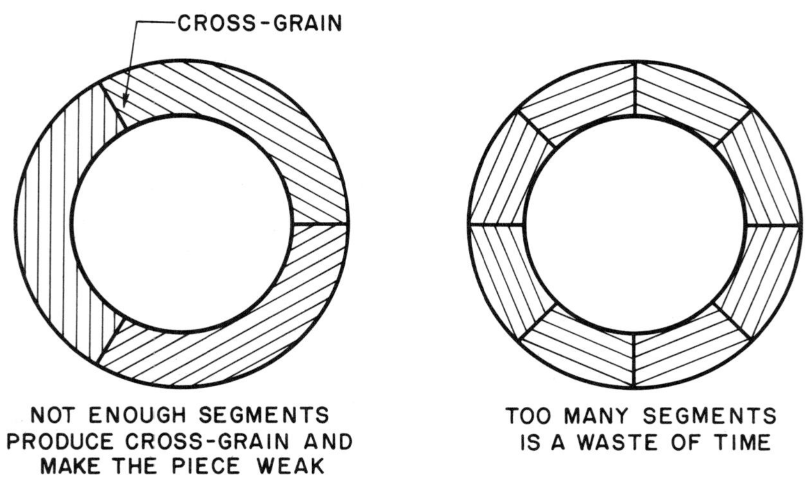

Fig. 30-7 Too little or too many segments

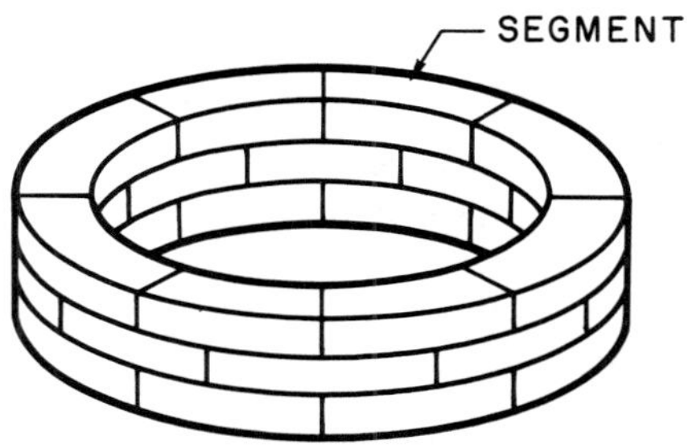

Fig. 30-8 Segmental construction must have at least three staggered layers.

8. Lay out the length of the segment and the miter cut of 45 degrees on each end from the full-size layout.

Laying Out the Curves

1. Set the dividers at 3 1/2 inches (89 mm). With the point of the dividers on a scrap block of the same thickness, scribe an arc that intersects both corners of the short edge of the segment, figure 30-11.
2. Set the dividers for 6 1/2 inches (165 mm). Without moving the segment or scrap block,

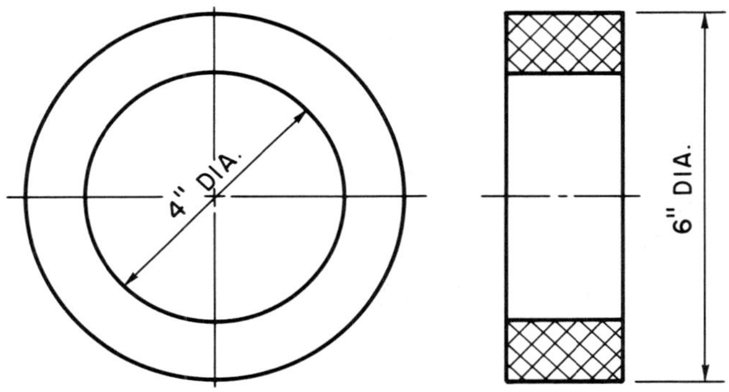

IN.	mm
4"	102
6"	152

Fig. 30-9 Brass ring

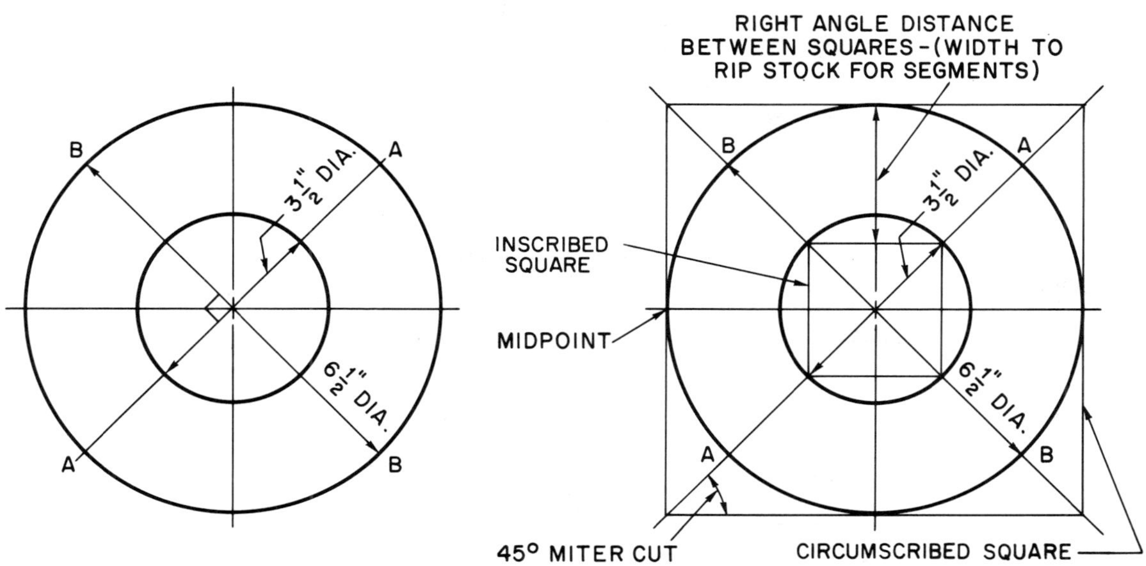

A B

Fig. 30-10 Layout for a brass ring

use the same center to scribe an arc tangent to the midpoint of the long edge of the segment.

3. Cut out both curves keeping outside of the scribed lines.
4. Sand the edges smooth down to the line.
5. Use this piece as a pattern to lay out, cut, and smooth the rest of the segments needed. Twelve are needed.

Assembling the Segments

1. Place a circle of segments on the full-size layout.
2. Join them tightly with brads to hold them in place.
3. Build up the layers using glue and brads and staggering the joints. Place fasteners so they will be clear of the turning chisels.
4. Remove from the layout. Clip off and sink any protruding brads.

Turning the Construction

1. Screw a scrap round piece of wood slightly smaller in diameter and about 3/4 inch (19 mm) thick onto the pattern. Center it carefully.
2. Center and screw the faceplate to the scrap piece.

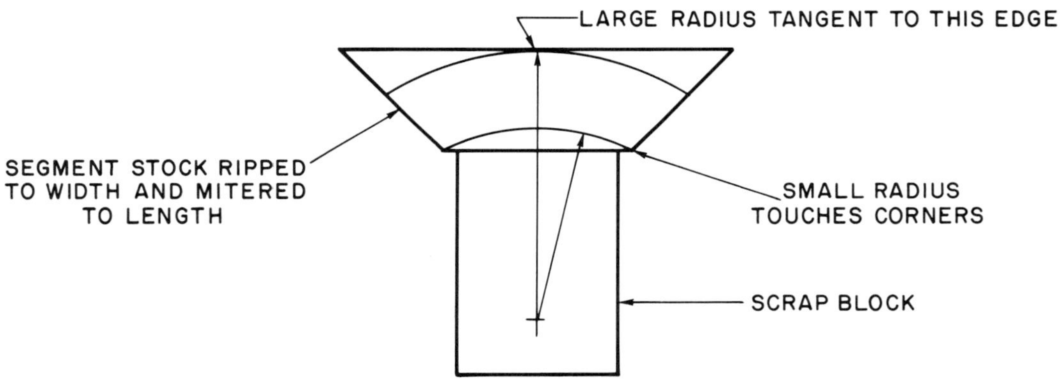

Fig. 30-11 Method of striking arcs on segment

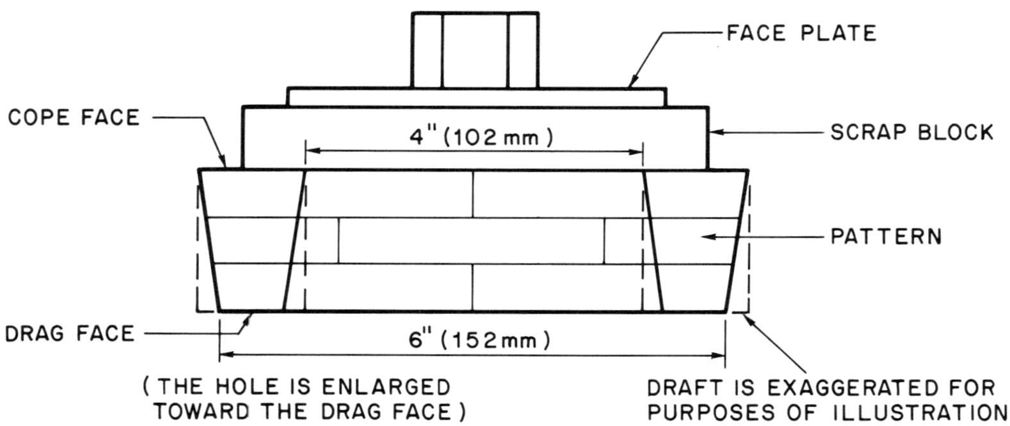

Fig. 30-12 Making draft by faceplate turning

3. Mount the assembly on the head center of the wood lathe.
4. Joint the face of the pattern to the required 1 3/4 inch (44 mm) thickness. Test around the edge with outside calipers.
5. Using inside calipers, turn the hole on the drag face to a 4-inch (102-mm) diameter, figure 30-12.

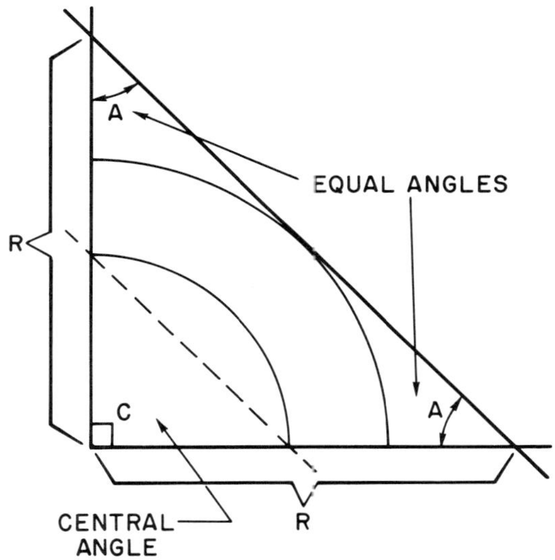

Fig. 30-13 The inside angles of a triangle total 180 degrees.

6. Close the calipers about 1/32 inch (0.8 mm) and taper the hole smaller toward the cope face. Take small cuts and test frequently in order to avoid taking out too much stock.
7. Using outside calipers, bring the circumference of the drag face to a 6-inch (152-mm) diameter.
8. Open the calipers about 1/32 inch (0.8 mm) and taper the outside larger toward the cope face.
9. Sand the piece while still mounted on the faceplate.
10. Remove the piece. Fill all screwholes and finish.

FINDING MITER ANGLES

The miter cut for the segments of the brass ring is 45 degrees. This is found by dividing a circle into four equal parts. Since there are 360 degrees in a circle, four equal parts make central angles of 90 degrees.

A line connecting the ends of two radii form an *isosceles triangle,* which is a triangle with two equal sides. Since the two radii of the triangle are equal, then the other two angles are equal, figure 30-13. The interior angles of a triangle total 180

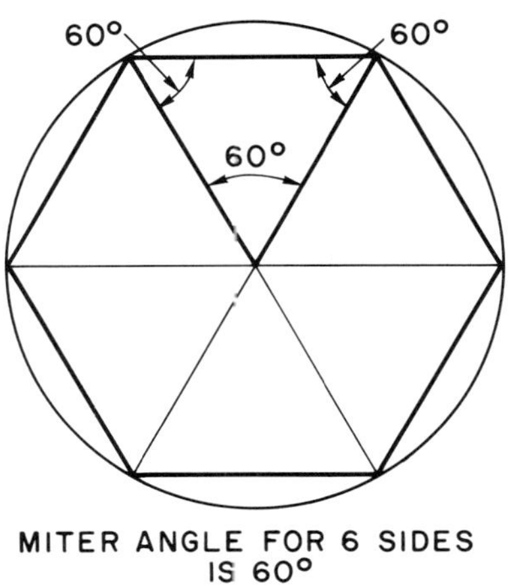

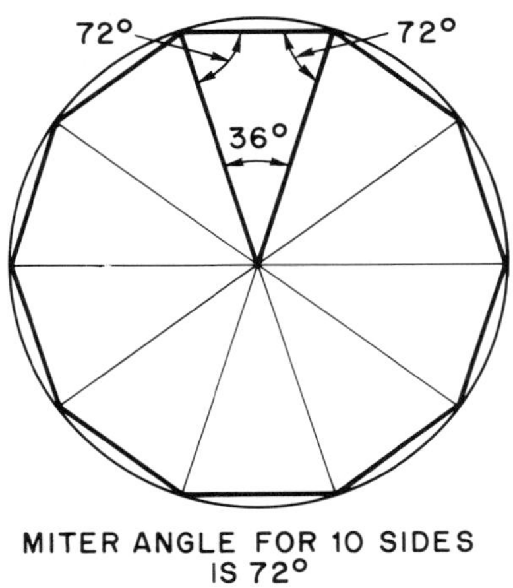

Fig. 30-14 Finding the miter angle for equal-sided shapes called ***polygons***

degrees. Subtract 90 degrees (central angle) from 180 degrees (total degrees in a triangle) and divide by 2 to find the remaining equal angles are 45 degrees. This is the miter cut for an object with four equal sides.

A *polygon* has three or more equal sides. To find the miter angle for a polygon, divide 360 degrees by the number of sides. Then subtract that number from 180 degrees and divide by 2, figure 30-14.

ACTIVITIES

1. Make a list of the parts of a table saw cast from a one-piece pattern. Give reasons for each part on the list.
2. Select a casting molded by a segment-constructed pattern. List the reasons for the selection.
3. Outline a method for laying out and cutting the segments for the casting in #2.
4. Draw three equal-sided shapes and find the miter angle for each one.
5. Make a pattern for a brass bracket. Use the same dimensions as those for the aluminum bracket in figure 30-2 except make it 4 inches (102 mm) wide with a corner radius of 1 inch (25 mm). Use the shrink rule for brass and allow for both top and bottom surfaces to be finished.

UNIT REVIEW

Questions

1. What is a one-piece pattern?
2. How is a one-piece pattern held together?
3. When making patterns, why must the grain direction be considered?
4. Why must the pattern be constructed with the least amount of draw?
5. What woodworking machine is used frequently to taper patterns for draft?
6. The edges of what surface cannot be sanded? Why?
7. What does the number of segments in a ring pattern depend on?
8. What do segments provide to the construction of a ring pattern?
9. When building up each layer of a ring pattern, how are the joints placed?
10. How many layers are required in segmental construction?

Multiple Choice

1. Sanding pattern surfaces is done
   a. across the grain only.
   b. with the grain only.
   c. on every surface.
   d. in the most convenient direction.
2. Sanding pattern edges is done
   a. across the grain only.
   b. along their width.
   c. along their length.
   d. in the most convenient direction.

3. Sharp edges on surfaces to be machine-finished should be
   a. rounded over.
   b. painted black.
   c. left as they are.
   d. sanded.

4. A hole in a pattern that lies in the drag is enlarged toward the
   a. top.
   b. bottom.
   c. cope face.
   d. drag face.

5. Segmental construction builds a circular object
   a. by bending long curved pieces.
   b. by cutting circles out of solid stock and gluing together.
   c. in layers with long curved pieces.
   d. in layers with small curved pieces.

6. Long segments of a segmental constructed pattern
   a. are weak.
   b. waste time.
   c. have less joints.
   d. save time.

7. With segmental construction, rings 14 inches (350 mm) or larger in diameter require
   a. at least four segments per layer.
   b. five segments per layer.
   c. six or more segments per layer.
   d. at least ten segments per layer.

8. Fasteners that join the segments are placed so that they
   a. protrude from the pattern.
   b. provide draft.
   c. are clear of the turning chisels.
   d. are staggered.

9. Diameters of holes are tested with
   a. outside calipers.
   b. inside calipers.
   c. dividers.
   d. a rule.

10. The miter angle for a pattern with six equal sides is
    a. 45 degrees.
    b. 60 degrees.
    c. 67 1/2 degrees.
    d. 75 degrees.

unit 31 Parted Patterns

OBJECTIVES

After studying this unit, the student will be able to:

- describe the parts of a parted pattern and explain their use.
- explain how the parts are matched.
- explain the advantages of parted patterns over solid patterns.
- make typical parted patterns.

PARTED PATTERNS

Patterns must be constructed to keep foundry labor to a minimum. Extra work due to poor pattern construction adds to the cost of each casting.

To simplify the making of a mold, patterns are often made in two or more parts that can be easily joined together and taken apart. This type of pattern is called a *parted pattern.* The parts take their name from the section of the flask in which they are molded, such as the cope part and the drag part of the pattern, figure 31-1.

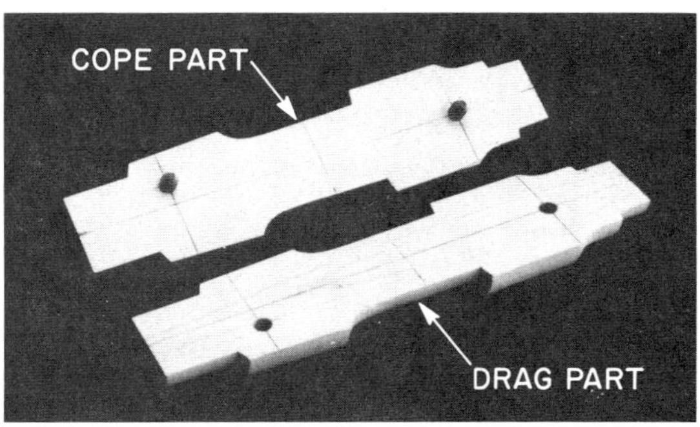

Fig. 31-1 A parted pattern

The advantage of a parted pattern over a solid pattern is that it allows a straight parting of the cope and drag. If part of a solid pattern is molded in the cope, the parting is irregular. An irregular parting makes lifting of the cope difficult, figure 31-2, and often mars or ruins the cope mold. With parted patterns, however, the cope half of the pattern is lifted off with the cope and then drawn from the mold in the usual way.

Matching the Parts

To hold the parts in their proper position, two or more *dowel pins* are used. The size and shape of the work determines the number and diameter of the dowel pins. They are carefully tapered on one end to allow a snug fit and an easy separation of the parts, figure 31-3. The length of the taper is equal to the diameter of the dowel pin. The end of the dowel pin is tapered down to 1/2 the diameter.

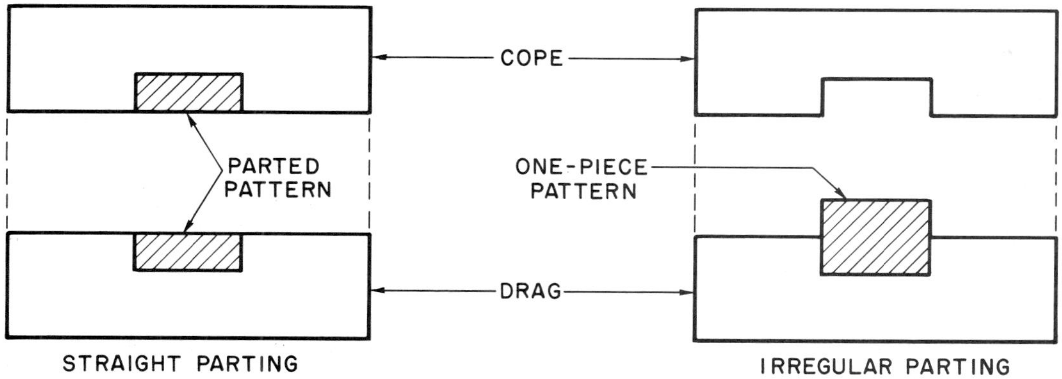

Fig. 31-2 Straight and irregular partings

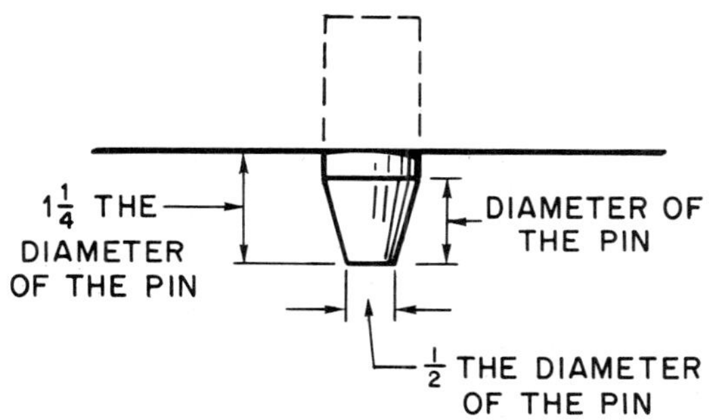

Fig. 31-3 Tapered dowel pin

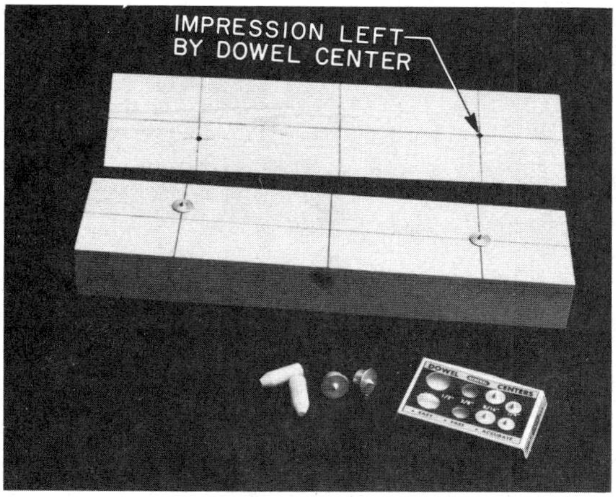

Fig. 31-4 Dowel centers

The dowels are set solid in the cope part and project about 1 1/4 times the diameter. If they are set in the drag part, they interfere with the molding of the drag.

To bore for dowel pins, mark their location on the parting surface of the cope part of the pattern. Bore holes all the way through the part. Clamp the two parts together. Place the bit through the cope part and bore into the drag part. Do not bore deeper than necessary, but bore deep enough so that the parting will come tight before the dowel pin bottoms in the hole.

Dowel centers, figure 31-4, are also used to mark centers for dowel pins. Locate and bore the holes in one part of the pattern. Insert the dowel centers in the bored holes. Accurately position the two parts and press them together. The dowel centers mark the center of the holes to be bored in the other part.

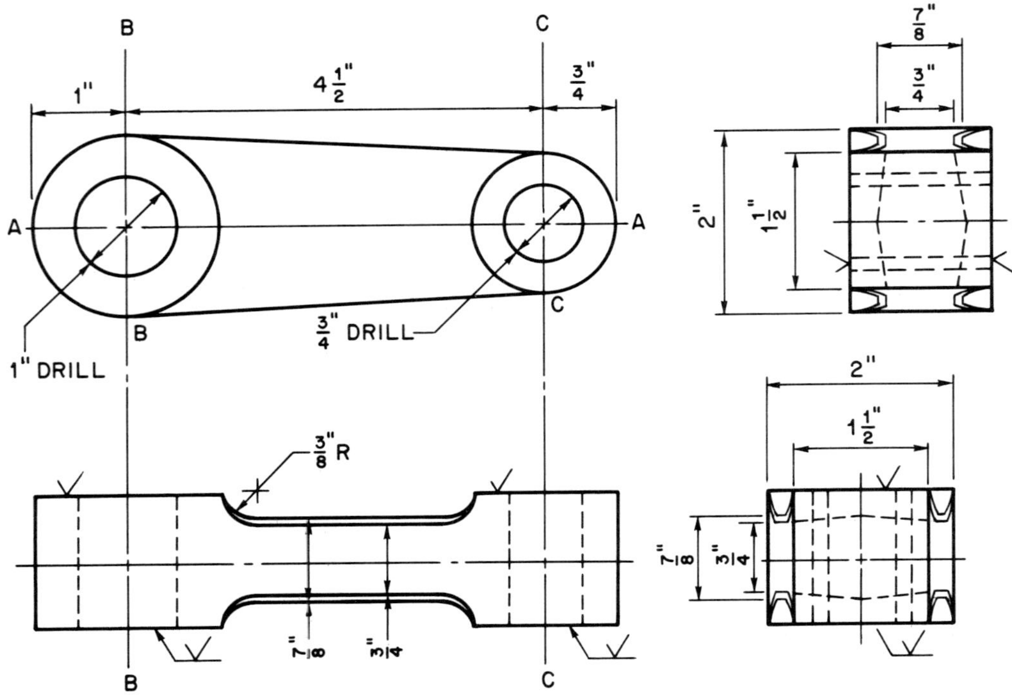

IN.	mm
3/8"	10
3/4"	19
7/8"	22
1"	25
1 1/2"	38
2"	51
4 1/2"	114

Fig. 31-5 Cast-iron arm

CAST-IRON ARM PATTERN

The cast-iron arm in figure 31-5 has an overall arm length of 6 1/4 inches (159 mm). The overall thickness is 2 inches (51 mm). The diameter of the larger hub is 2 inches (51 mm) and the smaller hub 1 1/2 inches (38 mm). The ends of both hubs are to be finished, and the web is thicker in the center than at the outside edges. No consideration is given to the holes by the patternmaker because they are to be drilled. The shrink rule for cast iron is used in all the measurements.

This pattern cannot be molded completely in the drag in any position without destroying the mold when drawn. It is not desirable for it to be molded partly in the cope. It must be a parted pattern. The best parting line is along centerline AA.

After determining the parting line, a plan of procedure is developed to construct the pattern. The pattern is built of ten pieces. Two pieces make up the web and eight semicircular pieces form the hubs, figure 31-6.

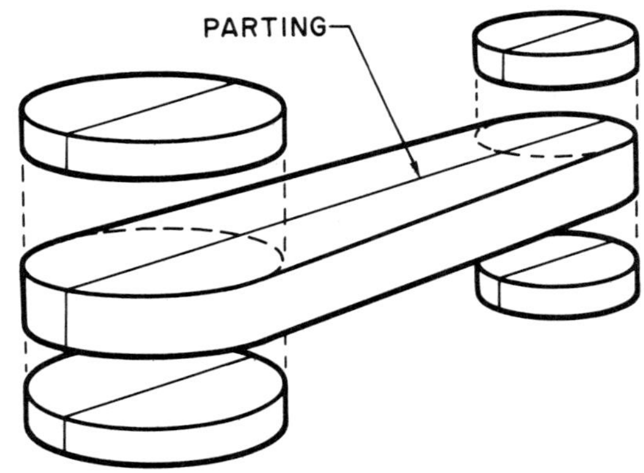

Fig. 31-6 Construction of the cast-iron arm pattern

Constructing the Web

To make the web,

1. Plane a piece of stock down to the overall thickness of 2 inches (51 mm). Rip the stock in two pieces and joint the ripped edges.
2. Clamp the jointed edges together. Lay out and bore for two 3/8-inch (10-mm) dowels on these edges.
3. Shape the dowel pins and glue them in the cope part of the pattern. Wipe off any excess glue and join the two parts together again. This joint is the parting line of the pattern.
4. To lay out the two hubs, scribe centerlines BB and CC 4 1/2 inches (114 mm) apart across the stock.
5. Set the dividers at 1 inch (25 mm). Scribe a circle for the large hub, using the intersection of centerline BB and the parting line as the center.
6. Scribe a circle with a 3/4-inch (19-mm) radius for the small hub on the other centerline CC.
7. Scribe straight lines on each side tangent to both circles.
8. Cut out the web leaving about 1/16 inch (1.5 mm) outside of the scribed lines.
9. Sand the square edges down to the lines.

Constructing the Hubs

Extra material is left on the faces of the hubs because they are to be finished.

1. Surface the hub stock to 3/4 inch (19 mm) allowing 3/16 inch (5 mm) for each face for finishing.
2. Scribe and cut out four semicircles with a 1-inch (25-mm) radius and four semicircles with a 3/4-inch (19-mm) radius.
3. Glue and clamp these semicircles to the parted sections of the web. Make sure the semicircular pieces are in perfect alignment.

Completing the Pattern

1. With a chisel and file, taper the web to the outside edges as indicated in the drawing.
2. Apply fillets in the corners between the web and hubs.
3. Bring the fillets down to a featheredge.
4. Sand the pattern smooth, rounding off all corners except the faces of the hub.
5. Finish the pattern in the usual way.

ALUMINUM PIN PATTERN

The pattern for the aluminum pin in figure 31-7 must be parted. If it is molded vertically with its round end in the drag, it cannot be drawn from the mold unless the square end is molded in the cope. Too much draft is required if it is molded vertically on either end.

The pattern is therefore parted, figure 31-8, and consists of two half-round sections and two triangular sections. The round section is turned on a wood lathe, and the triangular sections are cut out using the table saw. The shrink rule for aluminum is used in all measurements.

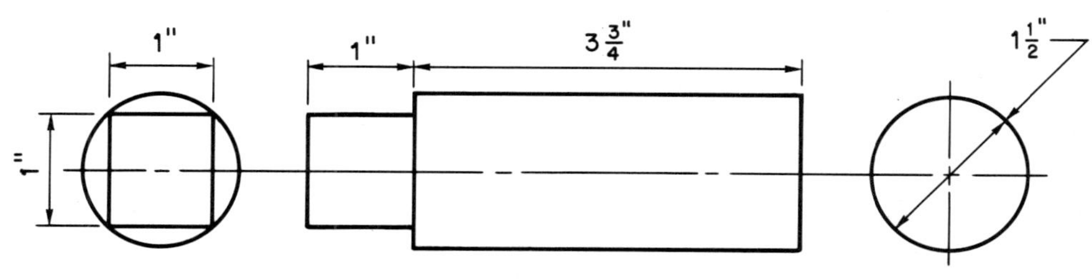

IN.	mm
1"	25
1 1/2"	33
3 3/4"	95

Fig. 31-7 Aluminum pin

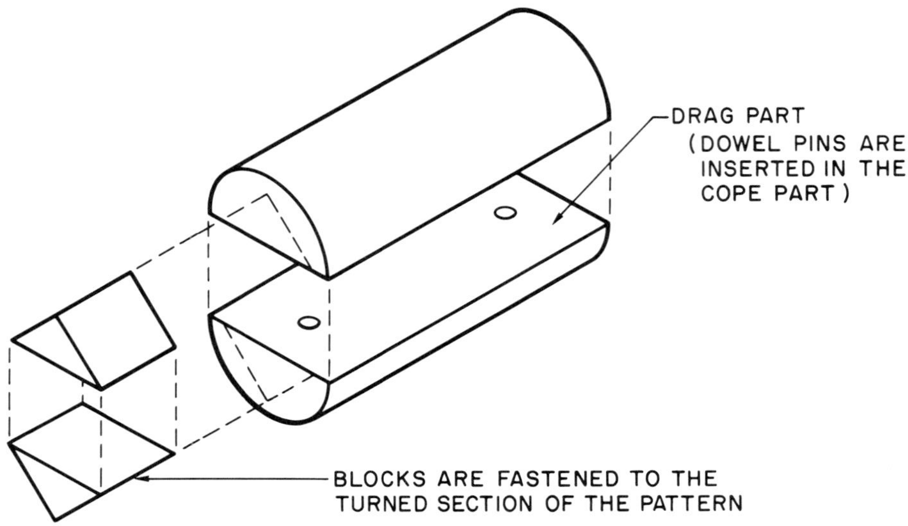

Fig. 31-8 The pieces of the aluminum pin pattern and its parting

Constructing the Round Section

1. Surface the stock to a thickness of 1 inch (25 mm) and rip it to a width of 2 inches (51 mm).
2. Cut two pieces 6 inches (152 mm) long. When the two pieces are joined, this allows 1/2 inch (13 mm) extra thickness and about 2 inches (51 mm) extra length for turning on the lathe.
3. Scribe a centerline down the length of the parting surfaces of both parts. Locate, bore, and glue two dowel pins along the centerline. If both parts are about the same, stagger the dowels so that the parts cannot be reversed, or use dowels of different diameters.
4. Join the pieces by gluing a piece of paper between the parting surfaces. Keep the ends and edges flush. Separate the parts later with a knife blade.
5. Less turning is required if the square piece is first made eight-sided or octagonal. Lay out an octagon on one end of the piece, figure 31-9. Cut off the corners to the lines.
6. Drill a small hole on each end and mount on the wood lathe.
7. With the parting tool, space cuts along the length of the piece. Using the outside calipers set at a little more than 1 1/2 inches (38 mm) with the shrink rule, test each cut.
8. With the gouge and skew, bring the piece down to the test cuts.
9. Sand smooth to the required diameter.
10. Open the dividers to the required length of the piece. Set the tool rest close to the work. While holding the dividers on the tool rest, lightly push them against the revolving stock.

> **Caution**: If you do not bring the tool rest close to the work, it may slip the dividers out of your grasp and injure you.

11. Turn the ends perfectly square using the parting tool and skew. Test the ends with a straightedge as turning progresses. Turn each end down to about 1/4 inch (6 mm). Turn the end at the tail center first. Turning the

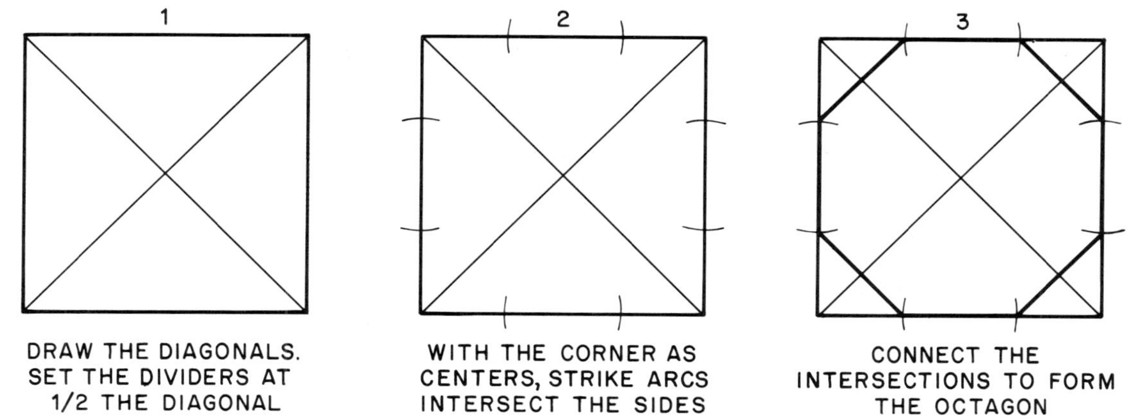

Fig. 31-9 Laying out an octogon

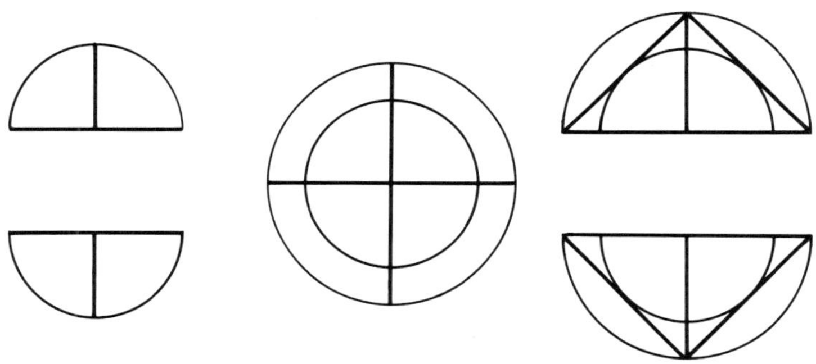

Fig. 31-10 Laying out the square end of the aluminum pin pattern

end at the head center first may cause the piece to break off when the tail end is cut.

12. Remove the piece from the lathe and cut off the waste with a backsaw. Trim smooth and flush with a disc sander.

Constructing the Square End

1. Separate the two halves by inserting a knife blade in the joint. Sand the parting surfaces smooth. Square the centerline across the ends.
2. Join the two pieces together and scribe a circle 1 inch (25 mm) in diameter on the end. Scribe lines at a 45-degree angle to the parting surface and tangent to the circle, figure 31-10. This marks the location of the square end.
3. From a piece of stock at least 1 inch (25 mm) thick, rip a strip of sufficient length at a 45-degree angle. Rip the stock to a 1-inch (25-mm) width and height.
4. Cut triangular pieces 1 inch (25 mm) long and square them on both ends. Fasten the pieces to the ends of the half-round section over the layout lines. Check the alignment before fastening.
5. With dowels, clamp and glue the pieces into place.
6. When dry, sand smooth and finish.

ACTIVITIES

1. List three parts of woodworking machinery in your shop that were cast from parted patterns. Locate the parting line, the cope part, and the drag part for each pattern. Give reasons for your choice.
2. Make a parted pattern to duplicate the tool rest on the wood lathe in your shop. Draw a layout for your work and write a plan of procedure. Use the shrink rule for cast iron and provide an allowance for finishing on the appropriate surfaces.
3. Visit a patternmaking shop and a foundry to watch the method of making and molding with parted patterns.

UNIT REVIEW

Questions

1. What is a parted pattern?
2. From what do the parts of a parted pattern take their name?
3. What holds the parts of a parted pattern in their proper position?
4. What are dowel centers used for?
5. What is the advantage of a parted pattern over a solid pattern?
6. When both parts of a pattern are about the same, how are the dowels placed so that the parts cannot be reversed?
7. How are the pieces of a parted pattern joined together in order to be turned?
8. What is done to a square piece so that less turning on the lathe is needed?
9. How is the length of a small cylinder marked that is to be turned on a wood lathe?
10. Which end of a cylinder on the wood lathe should be cut first?

Multiple Choice

1. Dowel pins used on parted patterns are tapered to a length of
   a. one-half the diameter of the pin.
   b. three-quarters the diameter of the pin.
   c. the diameter of the pin.
   d. one and one-quarter the diameter of the pin.
2. The ends of dowel pins are tapered down to
   a. one-eighth the diameter of the pin.
   b. one-quarter the diameter of the pin.
   c. one-half the diameter of the pin.
   d. the diameter of the pin.
3. Dowel pins are set solid in
   a. the cope part of the pattern.
   b. the drag part of the pattern.
   c. either part of the pattern.
   d. both parts of the pattern.

4. If a solid pattern is molded in the cope, the parting is
   a. straight.
   b. less difficult.
   c. ruined.
   d. irregular.

5. With parted patterns, the cope half of the pattern is
   a. lifted off with the cope.
   b. left in the drag.
   c. lifted off with the drag.
   d. not drawn from the mold.

6. To mark the length of a piece of stock turning on the lathe, use the
   a. outside calipers.
   b. dividers.
   c. skew.
   d. parting tool.

7. Extra material is left on the faces of the hubs of the cast-iron arm pattern because these surfaces are to be
   a. straightened.
   b. smoothed.
   c. finished.
   d. polished.

8. To lay out an octagon on a square, set the dividers at
   a. the length of the side.
   b. the length of the diagonal.
   c. one-half the length of the diagonal.
   d. one-half the length of the side.

9. When turning a cylinder on a wood lathe, first make test cuts along its length using a
   a. gouge.
   b. skew.
   c. parting tool.
   d. any of the above.

10. The parting of a pattern should
    a. require the least amount of draft.
    b. not make the parting surface irregular.
    c. make the lifting of the cope less difficult.
    d. all of the above.

unit 32 Cores, Core Prints, and Core Boxes

OBJECTIVES

After studying this unit, the student will be able to:

- describe when green sand cores and dry sand cores are used and how to secure them in the mold.
- determine the size and shape of core prints for horizontal and vertical dry sand cores.
- construct patterns for castings requiring a vertical or horizontal dry sand core and build the core box for each.

CORES

Sand cores are used to provide holes and other interior openings in castings. The core prevents molten material from flowing into the opening during the casting process. It is also easily knocked out after the casting cools.

If the opening can be made in the pattern, the core is formed by green sand during the molding process. This is called a *green sand core.*

Green sand cores are only molded if the diameter of a round core is larger than the casting it passes through. For a rectangular core, the shortest distance across the face of the core must equal or exceed the thickness of the casting, figure 32-1. For instance, a green sand core passing through a casting 3 inches (76 mm) thick must equal or exceed that dimension.

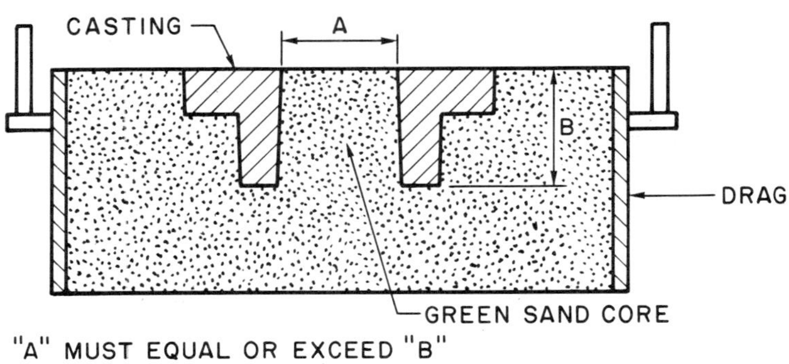

Fig. 32-1 Drag section

Fig. 32-2 Dry sand cores *(Schmidt Aluminum Castings)*

When green sand cores cannot be molded, *dry sand cores* are used. Dry sand cores are made separate from the pattern. Sand is mixed with a binder and pressed into a *core box* to form the desired shape. The core is then removed from the box and baked in an oven until it is dry and firm, figure 32-2.

After the pattern is drawn from the sand, the dry sand core is inserted to retain the shape of the opening during casting. Core boxes must therefore be made accurately so that the core will match the pattern, figure 32-3. Dry sand cores may be placed in the mold either horizontally or vertically.

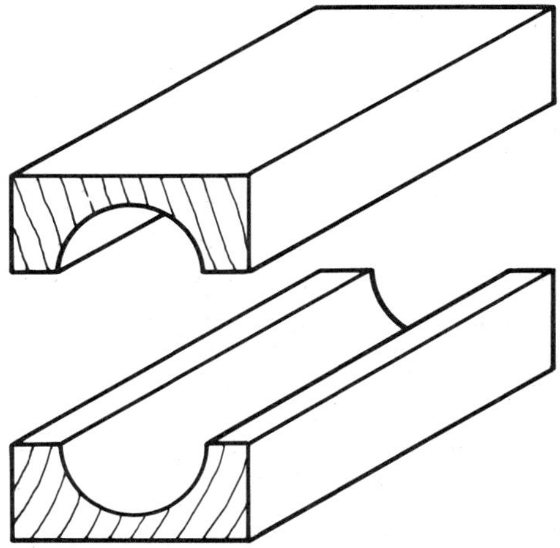

Fig. 32-3 Core box

CORE PRINTS

When using dry sand cores, an extension must be made on the pattern. This extension is called a *core print,* figure 32-4. The core print forms an impression in the sand to support the core. After the pattern is removed from the mold, the core is placed in the impression made by the core print. Like the mold, the core is destroyed in the casting process, but many cores can be made from the same core box.

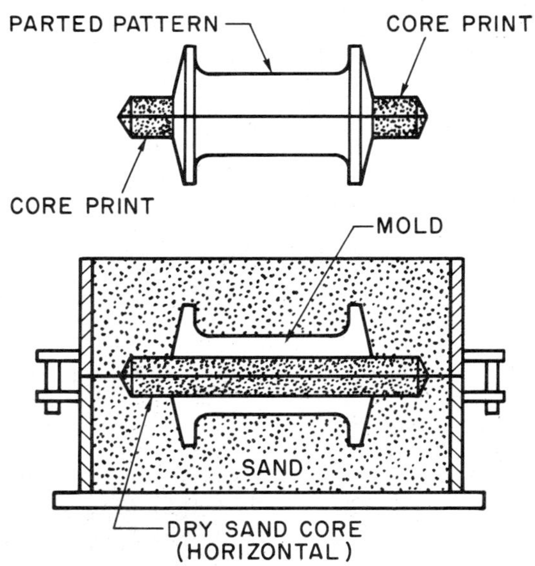

Fig. 32-4 The core print forms an impression in the sand to hold the core.

A vertical dry sand core may be supported at the drag end only if the diameter of a round core is one-half or more than the casting thickness. If the core is rectangular, the shortest distance across the end face must be one-half or more than the casting thickness, figure 32-5.

In other cases, the core must also be supported on the cope end. Core prints for the cope end are tapered and parted. The parting of the core

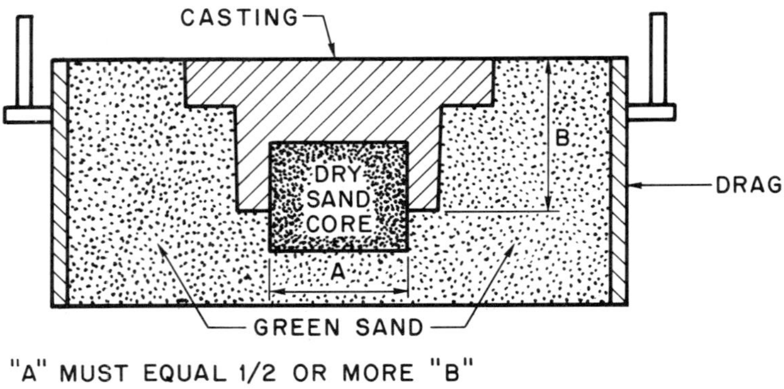

Fig. 32-5 Drag section. Vertical dry sand core secured at the drag end only

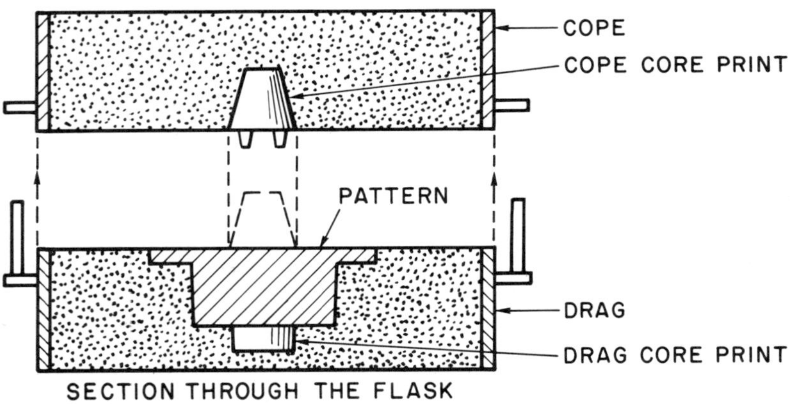

Fig. 32-6 Core prints for the cope end of a vertical dry sand core are tapered and parted.

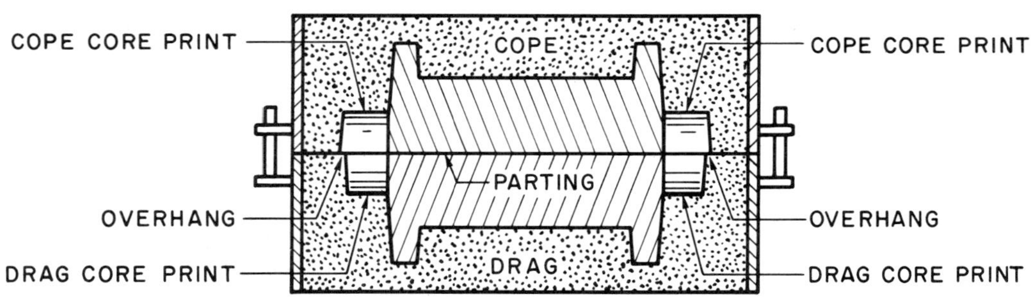

Fig. 32-7 The cope print overhangs the drag print to prevent the cope from touching the ends of the horizontal core.

print allows it to be lifted off with the cope, giving a straight parting of the mold. Dowels used to position the core print on the pattern are set solid in the drag face of the core print, figure 32-6. The tapered impression made by the core print centers the core without damaging the mold.

Horizontal dry sand cores may be round, rectangular, or other shapes. Core prints for round cores are parted through their diameter. Core prints for rectangular cores are parted through their diagonal if possible. Each half of the core is fastened solidly to the corresponding part of the pattern under most circumstances. The ends of the cope-half of core prints are made to overhang the drag-half by about 3/32 inch (2 mm). This is done to prevent the cope from touching the ends of the core when the mold is closed. The length of the core is made 1/16 inch (1.5 mm) less than the drag-half of the pattern, figure 32-7.

Core prints are made large enough to support the core. Their dimensions are set according to their location in the mold and how wide the core is. Core print sizes are determined according to the table in figure 32-8.

PATTERN FOR AN ALUMINUM STAND

The aluminum stand is an example of a casting that may be cored vertically, figures 32-9 and 32-10. Note that the shortest distance across the end of the opening is less than one-half the height of the casting. The core is therefore secured at both ends. The pattern must have core prints on both the drag face and the cope face, figure 32-11.

Constructing the Pattern

1. Cut the material for the base to the required thickness, width, and length. Scribe centerlines on both sides.
2. Cut the stock for the top section to the required size. Scribe centerlines on both ends.

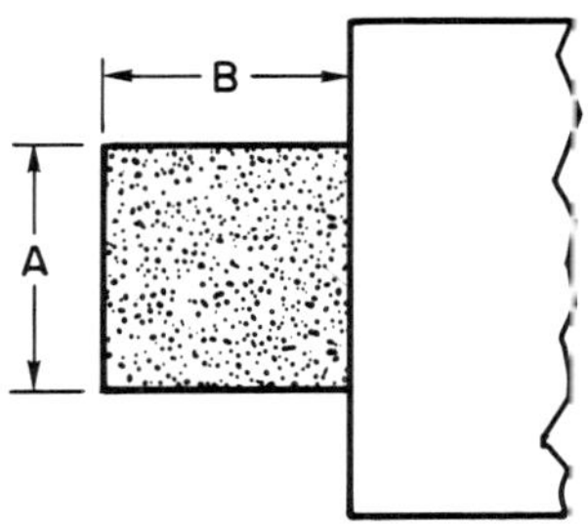

HORIZONTAL CORE PRINT

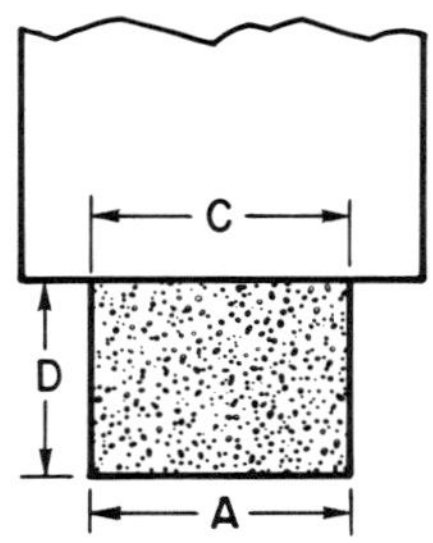

DRAG CORE PRINT

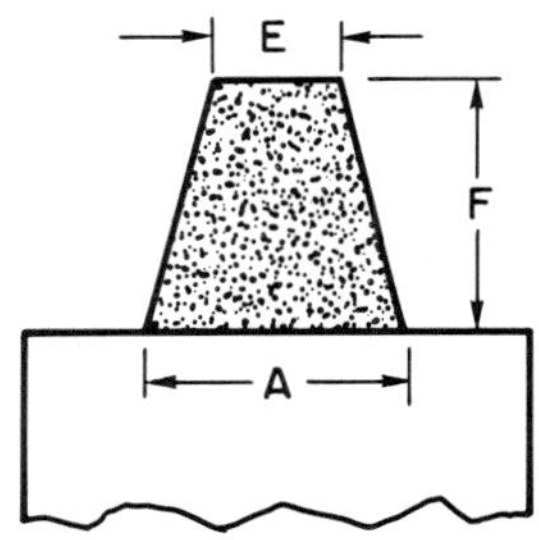

COPE CORE PRINT

A	B	C	D	E	F
1/2	1/2	17/32	1/2	1/4	1/2
5/8	5/8	21/32	1/2	5/16	5/8
3/4	3/4	25/32	1/2	3/8	3/4
7/8	7/8	29/32	5/8	7/16	7/8
1	7/8	1 1/32	3/4	1/2	1
1 1/4	7/8	1 9/32	7/8	5/8	1 1/4
1 1/2	1	1 17/32	/	3/4	1 1/2
1 3/4	1	1 25/32	/	1	1 1/2
2	1 1/8	2 1/32	/	1 1/4	1 1/2
2 1/2	1 1/8	2 17/32	/	1 3/4	1 1/2
3	1 1/8	3 1/32	/	2	1 1/2

SIZES IN INCHES

A	B	C	D	E	F
12.7	12.7	13.5	12.7	6.4	12.7
15.9	15.9	16.7	12.7	8.9	15.9
19.0	19.0	20.0	12.7	9.5	19.0
22.2	22.2	23.0	15.9	11.0	22.2
25.4	22.2	26.2	19.0	12.7	25.4
31.8	22.2	32.5	22.2	15.9	31.8
38.1	25.4	39.0	25.4	19.0	38.1
44.5	25.4	45.2	25.4	25.4	38.1
50.8	28.6	51.6	25.4	31.8	38.1
63.5	28.6	64.3	25.4	44.5	38.1
76.2	28.6	77.0	25.4	50.8	38.1

SIZES IN MILLIMETRES

Fig. 32-8 Table of core print sizes

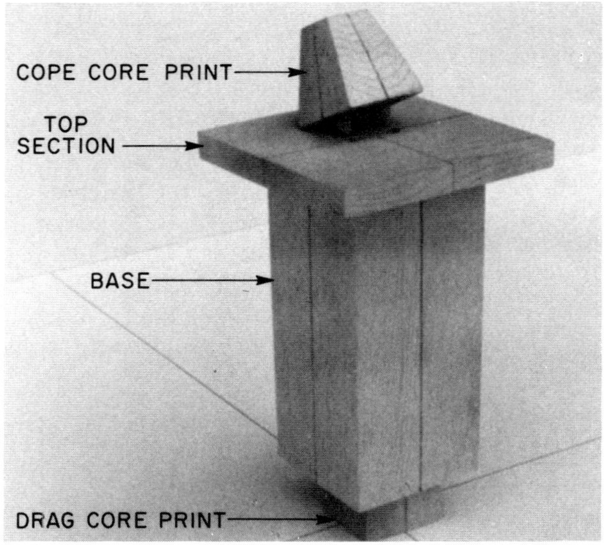

Fig. 32-9 Pattern for the aluminum stand

3. Taper the edges of the base and the sides of the top section for the proper amount of draft. Remember that draft enlarges the pattern from the drag face toward the cope face.
4. Scribe centerlines on all four sides of the top section.
5. Locate the top section on the base and fasten it with nails and glue.
6. Apply fillets to round the corners between the top section and the base.

Constructing the Core Prints

1. Machine the stock for the drag core print to the required size, see figure 32-8.
2. Scribe centerlines on the cope face. Provide draft on all four sides. Extend the centerlines down the sides of the core print.
3. Locate and fasten the core print with glue and nails to the drag face of the pattern.
4. Cut the stock for the cope core print to its overall size. Lay out and scribe centerlines on both ends and four sides.
5. Locate and bore holes for two dowels in the drag face of the core print and on the cope face of the pattern. Lay out the dowels carefully to insure the proper alignment of the core print on the pattern. Use dowels of different diameters to avoid reversing the core print on the pattern.
6. Taper the four sides of the cope core print to the necessary dimensions, see figure 32-8. Make sure that dimensions are measured from the centerlines.
7. Insert the dowels in the core print.
8. Sand and finish the pattern. Color both core prints and the surface covered by the loose cope core print black.

Constructing the Core Box

The core for the aluminum stand is made in halves and later glued together. Making a core in halves saves time in building the core box. Figure 32-12 shows the construction of the core box needed to make the dry sand core for the aluminum stand casting. Only those dimensions of a core box

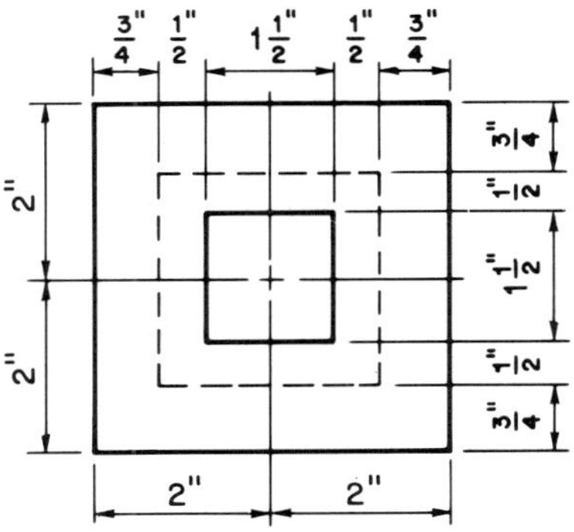

IN.	mm
$\frac{3}{8}''$	10
$\frac{1}{2}''$	13
$\frac{3}{4}''$	19
$1\frac{1}{2}''$	38
$2''$	51
$5\frac{1}{2}''$	140

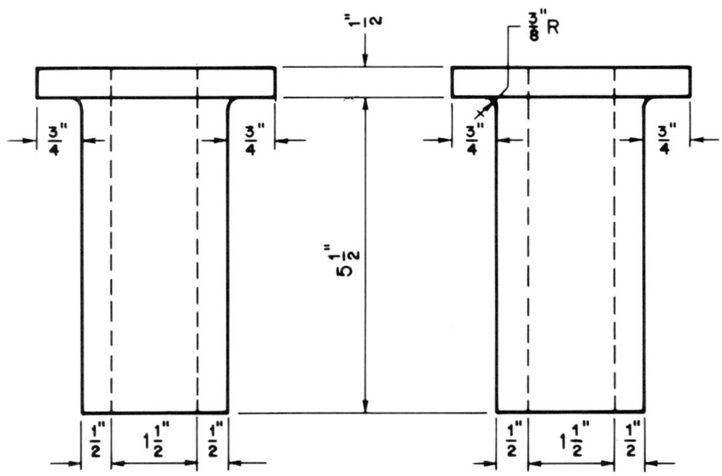

Fig. 32-10 Aluminum stand

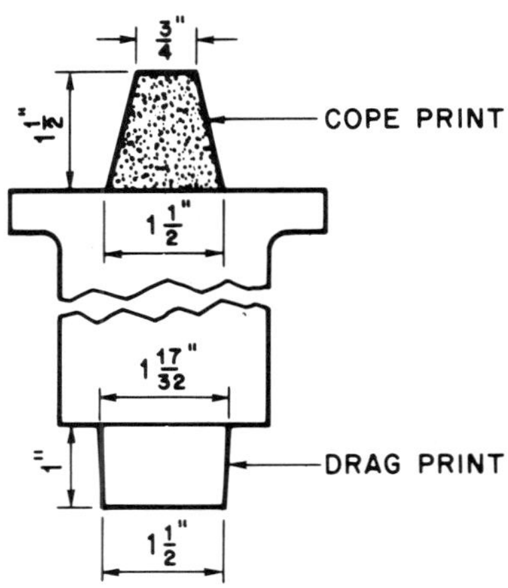

IN.	mm
$\frac{3}{4}''$	19.0
$1''$	25.4
$1\frac{1}{2}''$	38.1
$1\frac{17}{32}''$	39.0

Fig. 32-11 Cope and drag print sizes for the aluminum stand

that give shape to the core are important. Outside dimensions are not critical unless those surfaces help shape the core.

1. Cut a piece of stock that is longer than the total length of the core.
2. Cut one piece from this stock to the length of the straight part of the core plus the height of the drag print.
3. Cut the other piece to the height of the cope print, 1 1/2 inches (38 mm).
4. Scribe centerlines across the width and ends of each piece.
5. On each end draw a semicircle with a diameter equal to the width of the core ends, figure 32-13. In this case the drag end is 1 1/2 inches (38 mm) and the cope end is 3/4 inch (19 mm).
6. Scribe lines at a 45-degree angle to the face of the core box and tangent to the semicircles, intersecting the end centerlines. Join the end marks with straight lines across the face of the core box, figure 32-14.
7. Cut out the core cavity to the scribed lines. Cut out the end pieces for the core box.

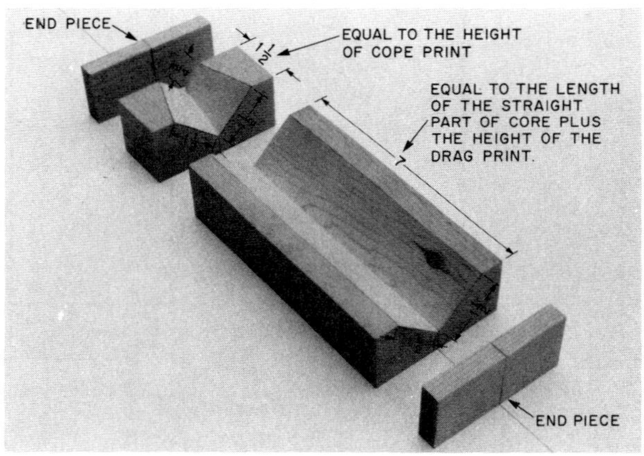

Fig. 32-12 Construction of the core box

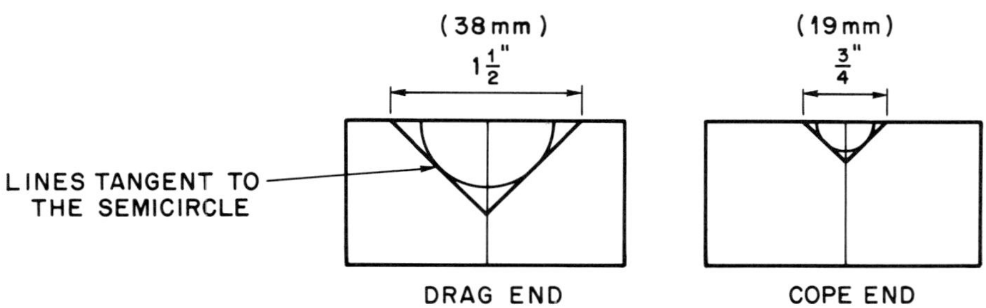

Fig. 32-13 Laying out the ends of the core box

8. Sand the cutout parts and join together by gluing, clamping, and nailing. Provide a slight draft on the inside end surfaces of the core box by slightly beveling the ends.

PATTERN FOR A CAST-IRON SLEEVE

The cast-iron sleeve in figure 32-15 is molded horizontally and requires a dry sand core. The pattern is parted diagonally and is constructed with ten parts, figure 32-16. Fillets are required between the body of the sleeve and the flanges. The core prints are attached permanently to the body of the pattern. The cope core prints are made slightly longer to overhang the drag core prints.

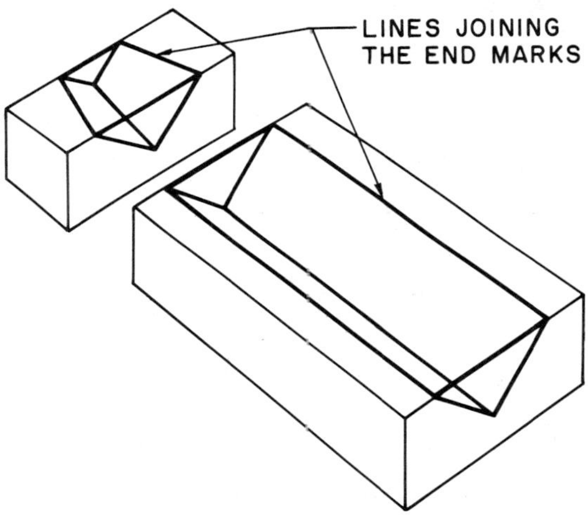

Fig. 32-14 Locating the cavity of the core box

1. Mill out the stock for the body of the sleeve and cut it in half along its length. Scribe centerlines along the length and across the width of the parting surfaces.
2. Bore holes and insert dowels in the cope part. Stagger their placement.
3. Join the pieces and cut to the exact length with both ends square.
4. Extend the centerlines across the ends. Scribe circles with a 2 1/2-inch (64-mm) diameter on both ends. Separate the pieces and draw 45-degree angles from the parting surface tangent to the semicircles.
5. Cut the pieces to the scribed lines.
6. Surface stock for the flanges to the required thickness.
7. Scribe a line square across the width. Scribe a semicircle with a 3 1/2-inch (89-mm) diameter on the stock. From the edges, scribe lines at a 45-degree angle tangent to the circle, figure 32-17.
8. Cut the piece to the lines. Make three other pieces exactly the same.

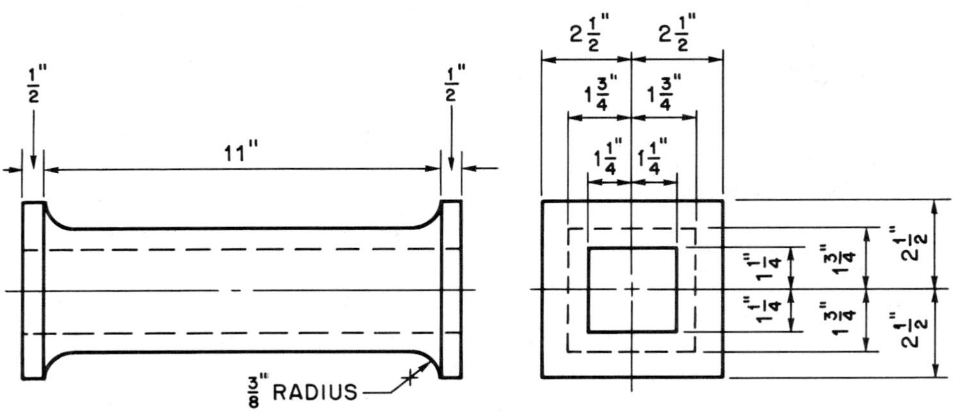

IN.	mm
3/8"	10
1/2"	13
1 1/4"	32
1 3/4"	44
2 1/2"	64
11"	280

Fig. 32-15 Cast-iron sleeve

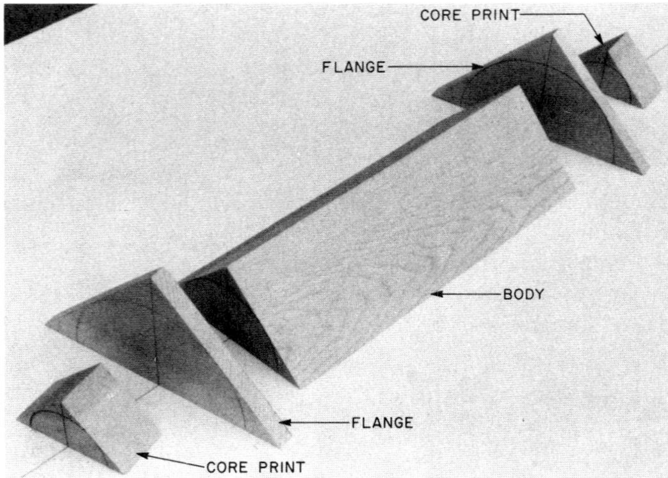

Fig. 32-16 Parted pattern for the cast-iron sleeve

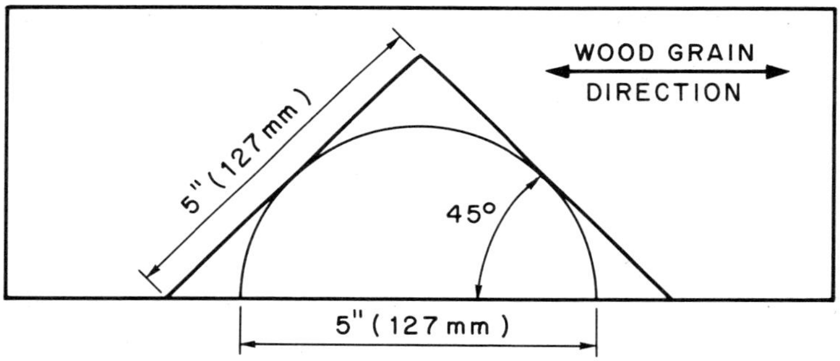

Fig. 32-17 Laying out the flanges

9. Cut the stock for the core prints in a manner similar to that for the flanges. Make the cope core prints 3/32 inch (2.4 mm) longer than the drag core prints.

10. Join the pieces of both halves with glue, clamps, and nails. Make sure all centerlines match.

11. Provide a slight draft on all surfaces parallel to the direction of draw.

12. Apply fillets where required.

13. Sand and finish the pattern. Color the core prints black.

CORE BOX FOR THE CAST-IRON SLEEVE

The core box to make the dry sand core for the cast-iron sleeve is parted. This allows the core to be made in one piece, figure 32-18.

1. Cut the stock to size and cut the parting surface.
2. Stagger dowels close to the edges.
3. Join the pieces together and cut the box to the same length as required for the core.
4. Lay out the ends as needed for the size of the core. Cut out the core cavity.
5. Sand the core cavity and finish. Number the core box with the same number as the pattern.

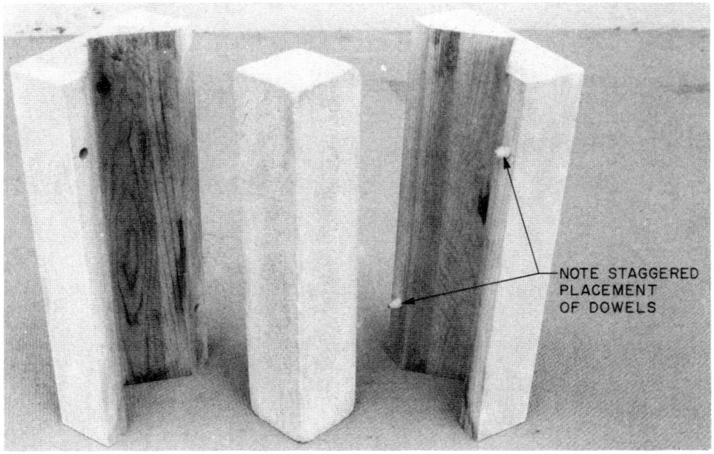

Fig. 32-18 Construction of the core box for the core

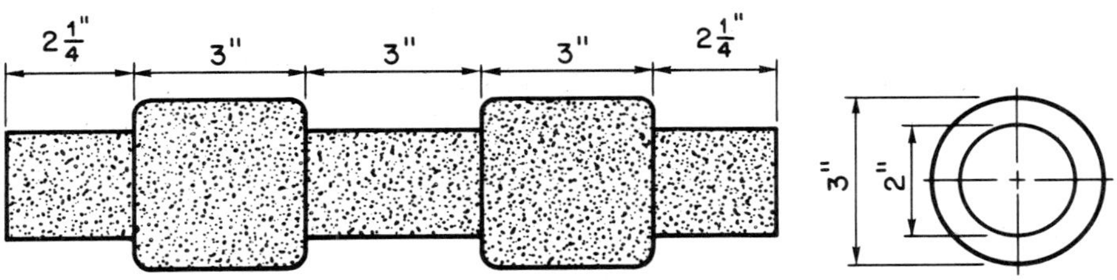

IN.	mm
2"	51
2 1/4"	57
3"	76

Fig. 32-19 Horizontal dry sand core

ACTIVITIES

1. List three castings that were probably made with a dry sand core and three that were made with a green sand core. Give reasons for your choices. Were the dry sand cores horizontal or vertical?
2. Construct the core box for the horizontal dry sand core in figure 32-19.

UNIT REVIEW

Questions

1. What kind of sand core is used when the diameter of the core is smaller than the casting it passes through?
2. What are dry sand cores shaped in?
3. How may a vertical dry sand core be supported, if the diameter is one-half or more than the casting thickness?
4. How are cope core prints for vertical cores made?
5. Why are dowels used on parted patterns and core boxes staggered?
6. Which dimensions of a core box are important?

7. Why are cope core prints for horizontal cores made slightly longer than drag core prints?
8. How much shorter is the length of a horizontal core than the length of the drag-half of the pattern?
9. What must be provided on the inside end surfaces of a core box?
10. In order to make the core in one piece, how is the core box constructed?

Multiple Choice

1. The height of a drag core print for a 2-inch (51-mm) diameter vertical core is
   a. 1 inch (25 mm).
   b. 1 1/4 inches (32 mm).
   c. 1 1/2 inches (38 mm).
   d. 2 inches (51 mm).
2. The height of a cope core print for a 2-inch (51-mm) diameter vertical core is
   a. 1 inch (25 mm).
   b. 1 1/4 inches (32 mm).
   c. 1 1/2 inches (38 mm).
   d. 2 inches (51 mm).
3. The length of the drag core print for a horizontal core 1 inch (25 mm) in diameter is
   a. 1/2 inch (13 mm).
   b. 5/8 inch (16 mm).
   c. 7/8 inch (22 mm).
   d. 1 inch (25 mm).
4. The cope core print for a horizontal core overhangs the drag print by about
   a. 1/32 inch (0.8 mm).
   b. 3/32 inch (2.4 mm).
   c. 3/16 inch (5 mm).
   d. 3/8 inch (10 mm).
5. Dowels in parted cope core prints for vertical cores are set solid in the
   a. end face.
   b. cope face.
   c. pattern face.
   d. drag face.
6. Core prints are colored
   a. red.
   b. yellow.
   c. white.
   d. black.
7. Core prints for horizontal cores are joined to the pattern
   a. loosely.
   b. tightly.
   c. with dowels.
   d. permanently.
8. If a cope core print for a vertical core has a drag face 1 inch (25 mm) square, the cope face is
   a. 1/4 inch (6 mm) square.
   b. 3/8 inch (10 mm) square.
   c. 1/2 inch (13 mm) square.
   d. 3/4 inch (19 mm) square.
9. Extensions made on patterns to form impressions in the mold to support cores are called
   a. core supports.
   b. core patterns.
   c. core prints.
   d. core boxes.
10. Dry sand cores placed in molds are used to
    a. form the exterior of the casting.
    b. provide extra support to the casting.
    c. provide a passage for the escape of gases.
    d. form interior passages in the casting.

unit 33 Stave Construction

OBJECTIVES

After studying this unit, the student will be able to:

- explain stave construction.
- lay out and construct a pattern with staves.
- lay out and construct a staved core box.

STAVE CONSTRUCTION

Staved work is a way to construct patterns and core boxes for long, round castings that exceed 8 to 10 inches (203 to 254 mm) in diameter. It gives a lighter, stronger pattern than solid wood. Staved work also retains its shape better than solid wood since the pattern has less tendency to swell, shrink, and warp.

A staved pattern is constructed by fastening strips, called ***staves,*** together around foundation pieces, called ***heads,*** figure 33-1. This construction is similar to that of a wooden barrel. Heads are spaced no more than 24 inches (610 mm) on centers. ***Braces,*** running the length of the pattern, are set into the heads. They strengthen the pattern by tying the heads together. Also they can be tapped lightly in different directions to loosen and help draw the pattern from the mold.

The number of staves required depends on the diameter of the work. Enough staves are laid out to leave at least a 3/4-inch (19-mm) gluing surface at the edge of each stave. The amount of gluing surface between the staves is important to obtain a good joint and to give the pattern strength, figure 33-2.

The thickness of the staves depends on the number used and the strength required. Usually 1-inch (25-mm) stock is suitable for ordinary work. The heads are 1 1/2 times thicker than the staves.

A STAVED PATTERN

A brass cylinder is shown in figure 33-3. Its length is 48 inches (1220 mm) and its outside diameter is 12 inches (305 mm). The inside diameter is 10 1/2 inches (267 mm), making the wall thickness 3/4 inch (19 mm). The size and shape of this casting requires a pattern of staved construction parted through its diameter. The required core box is made in a similar manner.

Laying Out the Staves and the Heads

1. On a piece of smooth plywood, lay out a semicircle with a radius of 6 inches (152 mm), figure 33-4.
2. Divide the semicircle into four equal parts by stepping it off with dividers.

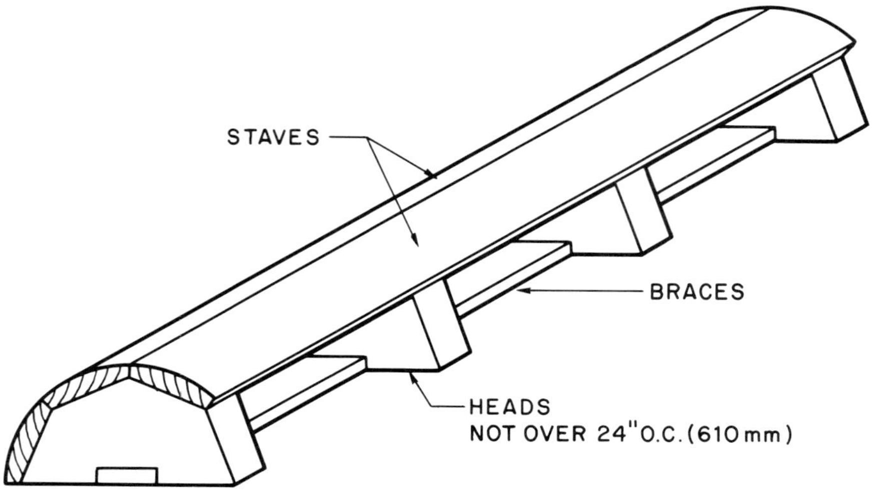

Fig. 33-1 Stave construction

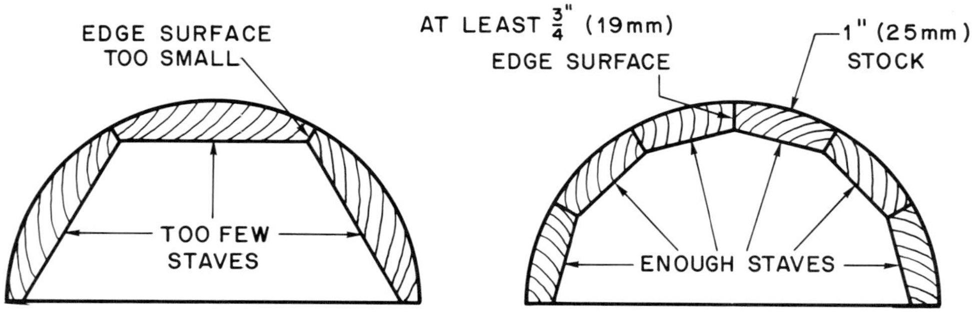

Fig. 33-2 Determining the number of staves

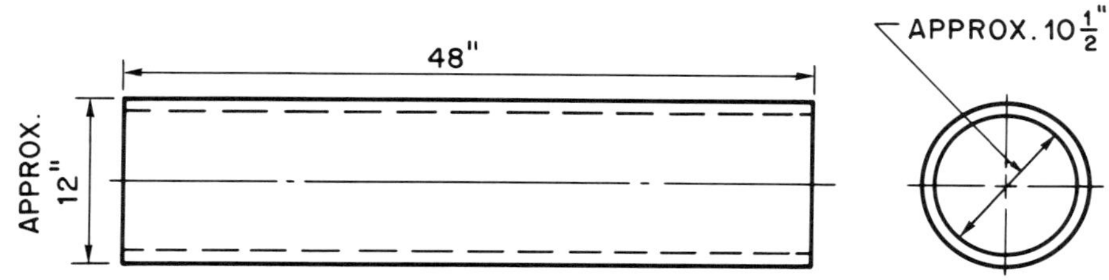

IN.	mm
$10\frac{1}{2}$"	267
12"	305
48"	1220

Fig. 33-3 Brass cylinder

3. Draw straight lines from the center to these points.
4. Draw the chord of one of the arcs.
5. Draw the line AB parallel to the chord about 1/8 inch (3 mm) outside the circle to allow for waste when turning.
6. Measure the thickness of the stave material by measuring at right angles to line AB. Draw line CD. ABCD outlines the stave material before turning.
7. If the stave edges are too thin, repeat the layout by dividing the semicircle into a greater number of parts.
8. With OC as a radius, draw a semicircle and connect points E, F, and G with straight lines. This lays out the shape of the heads. Make the heads about one and one-half times thicker than the staves.

Making the Heads

1. Mill out enough stock to make eight heads as required.
2. Lay out the shape on one head.
3. Band-saw to within 1/16 inch (1.5 mm) of the line and then disc sand to split the line.
4. Use this head as a pattern and trace its outline on the other material.
5. Saw and sand the other pieces in a similar manner as the first.
6. Working from centerlines, lay out and notch the heads for the braces with a handsaw and chisel.
7. Handsaw each end of the notch and chisel in from both sides of the head toward the center for the depth of the brace. To avoid breaking the stock below the scribed line when chiseling, lightly score the lines on both sides with the chisel. Then take out a little stock at a time, working down to the scribed lines.

Joining the Header and the Braces

1. Make two braces 3/4 inch (19 mm) thick by 48 inches (1220 mm) long plus enough to allow for drafting the ends of the pattern.
2. Lay both braces edge to edge. Square lines across the face of the braces to mark the location of the heads.
3. Glue and screw the braces to the heads, figure 33-5. Use two flathead screws 1 3/4 inches (44 mm) long. Set the screws slightly below the surface.
4. Scribe a centerline along the length of the braces.

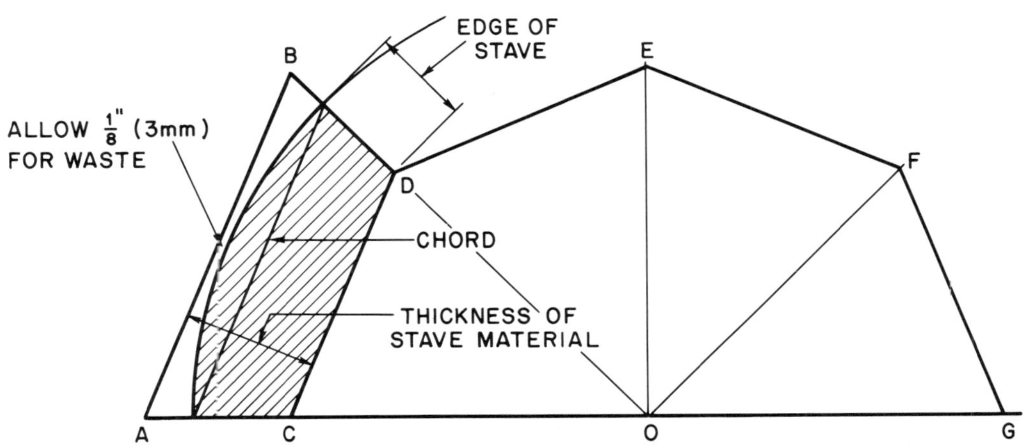

Fig. 33-4 Laying out the staves and heads

Making the Staves

1. From the layout, determine the number and size of the staves.
2. Plane suitable stock of extra length and width down to the required thickness.
3. On scrap stock, bevel the edges to the desired angle. Try them on the heads of the pattern until the necessary bevel and width is obtained to make an accurate joint.
4. Rip stock to width and bevel at the correct angle.
5. Cut the staves to the same length as the braces.

Fastening the Staves to the Heads

The assembly of the staves and heads is done on a smooth, straight piece of plywood that is longer and wider than the pattern.

1. Draw a centerline on the plywood and square lines across it which mark the length of the pattern, figure 33-6.
2. Fasten the heads to these lines with their centerlines matching those on the plywood. Screw up from the bottom of the plywood so that the pattern can be removed later, figure 33-7.

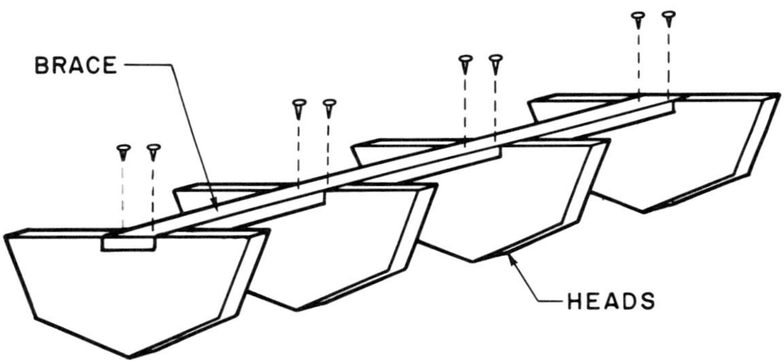

Fig. 33-5 Glue and screw the braces to the heads.

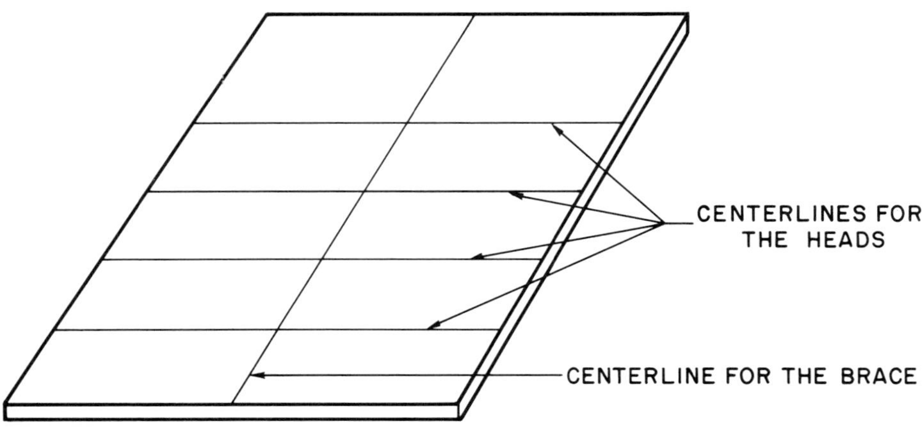

Fig. 33-6 Lay out the centerlines.

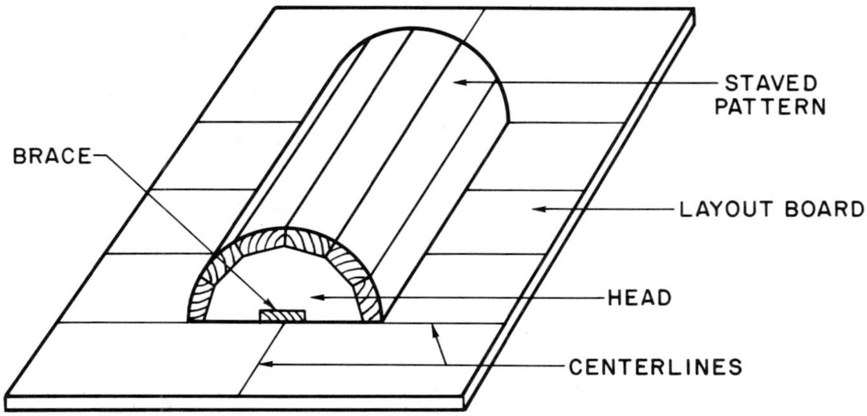

Fig. 33-7 Assembly of the staves and heads

3. Glue and nail each stave in place. Begin by fastening a stave on each side so that the edges are against the plywood board at the parting line.
4. Work up on each side toward the top. Make sure a tight joint is made between the staves.
5. Set the fasteners well below the surface to prevent the heads from being slipped off during the turning operation.
6. When both parts of the pattern are assembled, prepare the parting. Lay out and insert dowels in the cope half. Bore matching holes in the drag half.

Turning the Pattern

1. Glue the two parts together with paper between them in the parting.
2. Let the glue set and then mount the pattern on the wood lathe. Accurately line up the centers.
3. Make sure the tail center is locked securely in place and set the lathe to turn at a slow speed.
4. Turn the pattern down to 12 inches (305 mm). Test the diameter frequently along the length of the pattern with outside calipers.
5. While the pattern is still mounted on the lathe, fill in all nail holes and let dry.
6. Sand the pattern as it turns in the lathe.
7. Remove the pattern from the lathe and separate it at the parting line.
8. Scrape and sand the parting smooth.
9. Disc sand the ends of the pattern to provide the proper amount of draft and bring the pattern to the desired length at the cope and drag ends.

Making the Core Prints

1. Rip two pieces of stock large enough to make the core prints for both ends of the pattern. Glue them together with paper in the parting.
2. Mount the piece on the lathe and turn it to the desired diameter.
3. Remove it from the lathe and separate it along the glued parting.
4. Scrape and sand the parting smooth.
5. Cut and fit the drag core prints to the drag part of the pattern, lining up the centers. Provide draft on the ends of the prints.
6. Fasten the drag prints with dowels and glue and clamp them in place.

7. Cut, fit, and fasten the cope core prints in a similar manner to the cope part of the pattern.

8. Cut the cope core prints to overhang the drag core prints by 1/4 inch (6 mm).

9. Finish the pattern.

A STAVED CORE BOX

The staved core box is laid out and made in a similar manner as the pattern. The inside surfaces of the staves are concave, however, and the core box is the reverse of the pattern, figure 33-8. The heads are attached to the braces, and the staves are fitted and fastened into place one at a time.

Preparing the Heads, the Staves, and the Braces

1. Lay out a semicircle using the inside diameter of 10 1/2 inches (267 mm). Divide into equal parts, figure 33-9. Draw chord AB of one of the arcs.

2. Measure a distance equal to the thickness of the stave material at right angles to the chord. Draw a line CO from this point parallel to the chord. ABCD outlines a cross section of the stave.

3. To outline the core box head, set the dividers at OC and draw a semicircle. Connect points E, F, and G with straight lines.

4. Cut out and assemble the braces and heads in a manner similar to that of the pattern. The same number of heads are needed for the core box as for the pattern.

5. Cut out the staves, as described for the pattern, to their overall width and length. Bevel both edges of each stave as required.

Hollowing the Inside Surface of the Staves

The radius of the pattern was turned using the wood lathe. This cannot be done to make the curved inside of the core box. This curved surface is made by hollowing out or making cove cuts on the inside surface of each stave. The pieces are run

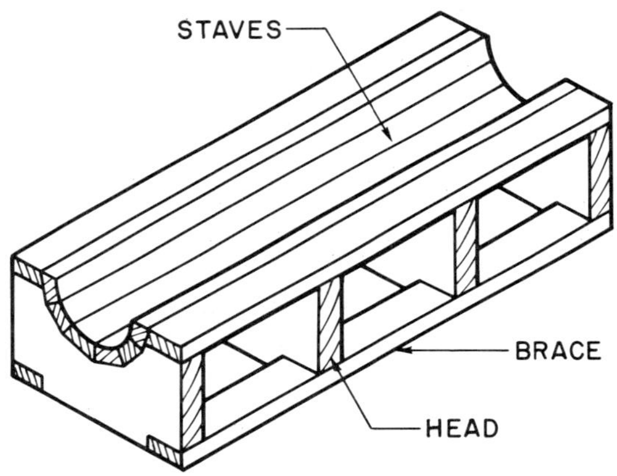

Fig. 33-8 Staved core box

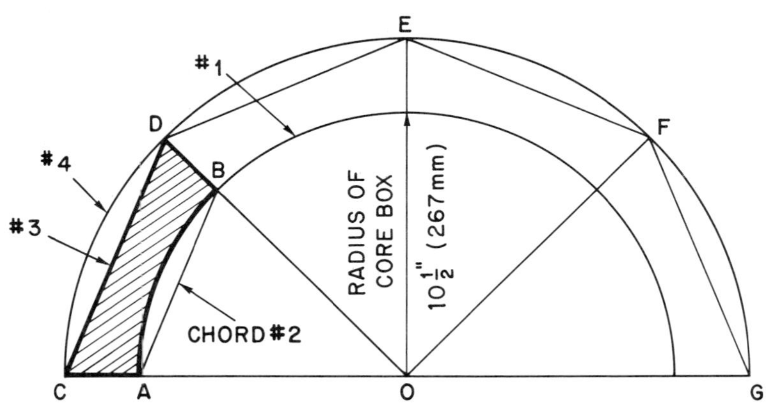

Fig. 33-9 Laying out the staves and heads for the core box

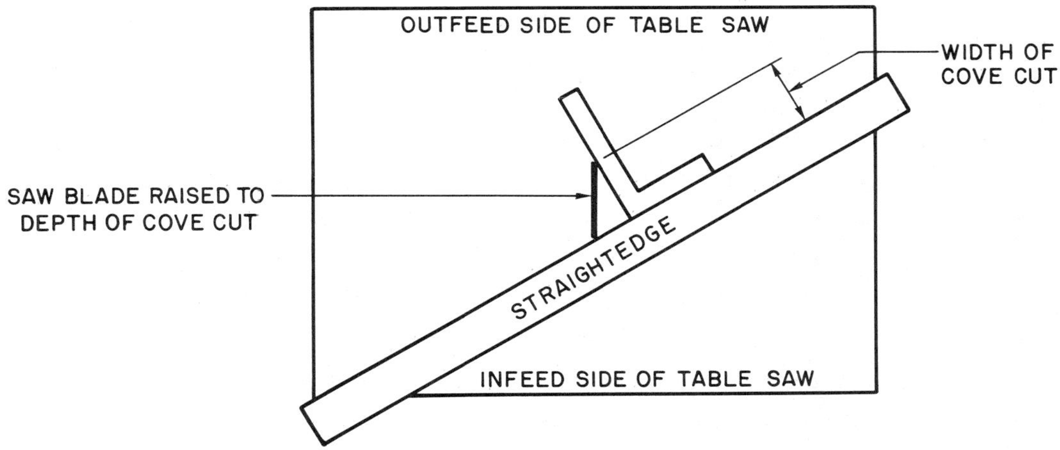

Fig. 33-10 Determining the straightedge angle for making cove cuts on the staves

diagonally across the table saw blade as described in *Unit 7 Table Saw.* The staves are guided by a straightedge clamped to the table saw top at an angle to the saw blade. The angle at which the straightedge is clamped determines the width of the cut.

To determine the straightedge angle:

1. Raise the saw blade to the overall depth of cut.
2. Lay the straightedge on the table and bring it up against the teeth of the blade on the infeed side.
3. Move the straightedge, keeping it against the teeth of the blade as a pivot, until a measurement taken at right angles from the straightedge to the outfeed side of the blade is equal in length to the width of the core cut, figure 33-10.
4. Clamp the straightedge in this position. Place the clamps on the side of the straightedge away from the blade.
5. Make a number of cuts on scrap pieces by raising the blade about 1/16 inch (1.5 mm) with each cut. Check the scrap piece on the layout and continue making cuts until the desired depth is obtained.
6. With the correct set-up, lower the blade again and run the staves through. Cut a small amount each time until the desired radius and width of cut is obtained.

Caution: Use a push stick because the blade is dangerously close to the top surface of the staves.

7. Fit and fasten the staves one at a time on to the heads of the core box.
8. Sand the core cavity smooth and finish.

ACTIVITIES

1. Observe the construction of staved patterns and core boxes in a patternmaking shop.
2. Make a scale model of a staved pattern and its core box and label the parts for display.
3. Make a conical (one end larger than the other) staved pattern and its core box.
4. Make a staved pattern and reinforce the joints with splines.

UNIT REVIEW

Questions

1. What diameter should a long, round casting be in order to use staved construction?
2. What are the advantages of staved patterns?
3. What are the parts used in staved construction?
4. What does the number of staves required for a pattern depend on?
5. What thickness of stock is suitable for staves on ordinary work?
6. How much thicker than the staves are the heads?
7. How far apart on center are heads spaced?
8. What pieces strengthen the pattern by tying the heads together?
9. How is the staved core box hollowed out?
10. How is the angle of the cove cut maintained?

Multiple Choice

1. Enough staves are laid out to provide an edge gluing surface of at least
   a. 1/2 inch (13 mm). c. 3/4 inch (19 mm).
   b. 5/8 inch (16 mm). d. 7/8 inch (22 mm).
2. If the edges of the staves are too thin, repeat the layout and
   a. divide the semicircle into more parts.
   b. divide the semicircle into less parts.
   c. increase the radius of the semicircle.
   d. decrease the radius of the semicircle.
3. To allow for waste in turning a pattern of staved construction, the staves are made about
   a. 1/16 inch (1.5 mm) thicker. c. 3/16 inch (5 mm) thicker.
   b. 1/8 inch (3 mm) thicker. d. 1/4 inch (6 mm) thicker.
4. Staves are cut to the same length as the
   a. braces. c. cope print.
   b. heads. d. drag print.
5. When the parts of the pattern are assembled, the parting is prepared by setting dowels in the
   a. braces. c. drag part.
   b. cope part. d. heads.
6. The diameter of the pattern turning on the lathe is tested with
   a. a flexible tape. c. outside calipers.
   b. inside calipers. d. a rule.
7. The inside surfaces of the staves for a core box are
   a. convex. c. concave.
   b. a chord. d. inside out.

8. Before the staves can be joined to the heads, the heads must be
   a. joined to the braces.
   b. notched for the braces.
   c. fastened to a plywood layout.
   d. all of the above.

9. To set up the angle of the straightedge when making staves, measure the length of the cove cut at right angles from the straightedge to
   a. the infeed side of the saw blade.
   b. the outfeed side of the saw blade.
   c. the center of the saw blade.
   d. the highest tooth of the saw blade.

10. To make a cove cut, the stave is run across the table saw blade
    a. in one pass for speed.
    b. a number of times, at the desired radius.
    c. back and forth carefully.
    d. a number of times, cutting a small amount each time.

unit 34 Match Plates

OBJECTIVES

After studying this unit, the student will be able to:

- describe a match plate and give reasons for using it.
- build a pattern on a match plate.

MATCH PLATES

A *match plate* is used for fast, multiple molding of small patterns and to mold fragile patterns. When a fragile pattern is built into a match plate, the pattern can withstand the foundry molding process better. Small patterns are built into a match plate so that more than one object can be cast from the same mold.

A match plate consists of a wood or metal plate fitted with sockets to fit over the pins of the drag part of the flask, figure 34-1. The drag and cope parts of the pattern are attached solidly to the match plate. The drag part is made upon one side of the match plate and the cope part upon the other. This eliminates the parting of the pattern.

The gates, riser, and sprue locations are also made on the match plate at the same time as the pattern. This eliminates the need of cutting them in by hand when molding.

MAKING A MATCH PLATE

Select a smooth, straight piece of exterior grade plywood for the plate. The thickness of the plywood depends on the size of the match plate. For ordinary work, 1/2 inch (13 mm) is sufficient.

Cut the plywood to the outside dimensions of the flask, leaving enough material on both ends to bore sockets to fit the pins of the drag. Scribe matching centerlines on both sides, figure 34-2. Dimensions for placing the pattern are taken from these centerlines.

MAKING A PATTERN FOR THE MATCH PLATE

Figure 34-3 shows a brass porthole ring with an outside diameter of 8 1/2 inches (216 mm). The total thickness is 5/8 inch (16 mm). It has a 6 1/2-inch (165-mm) diameter hole through 1/4 inch (6

Fig. 34-1 A match plate

mm) of its thickness and 5 1/2-inch (140-mm) hole through the rest of the casting.

The shape of this casting is ideally suited to be molded on a match plate. The pattern is parted where the two inside diameters meet. The cope side of the match plate holds the pattern with the small inside diameter. The drag side holds the pattern with the large inside diameter. The core for the ring is made of green sand. The match plate is made to mold two rings at the same time.

Fig. 34-2 Making the match plate

Laying Out the Pattern

1. Scribe a circle 8 1/2 inches (216 mm) in diameter, figure 34-4. With the same center, scribe circles 6 1/2 inches (165 mm) and 5 1/2 inches (140 mm) in diameter.
2. Divide the circle into four equal parts.
3. On one quarter draw the chord AB of the inside arc.
4. Draw a line CD parallel to the chord AB and tangent to the outside circle. This outlines the stock for one-quarter of the ring for the drag part of the pattern.
5. On another quarter of the circle draw the chord EF on the middle circle.
6. Draw line GH parallel to chord EF and tangent to the outside circle. This outlines the

IN.	mm
1/4"	6
3/8"	10
5/8"	16
5 1/2"	140
6 1/2"	165
8 1/2"	216

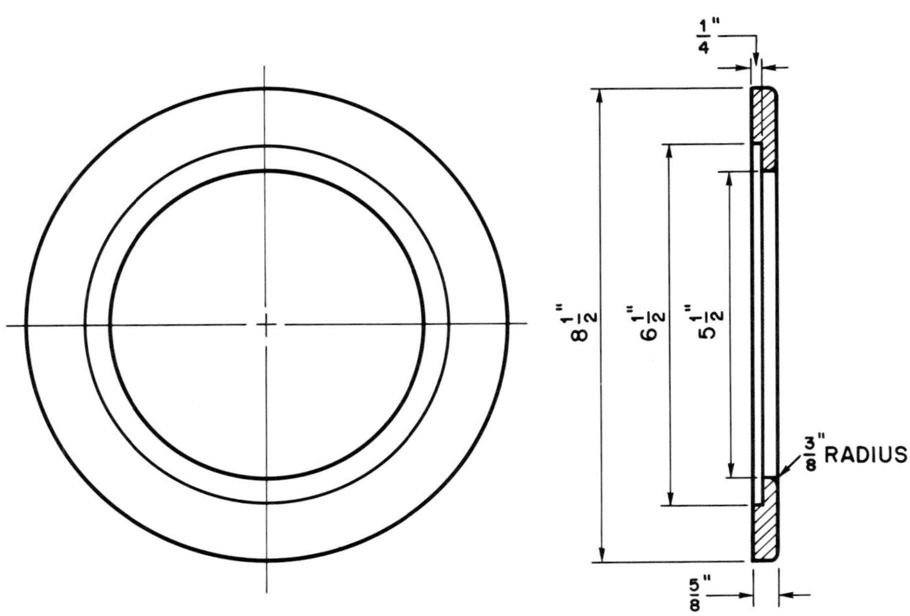

Fig. 34-3 Brass porthole ring

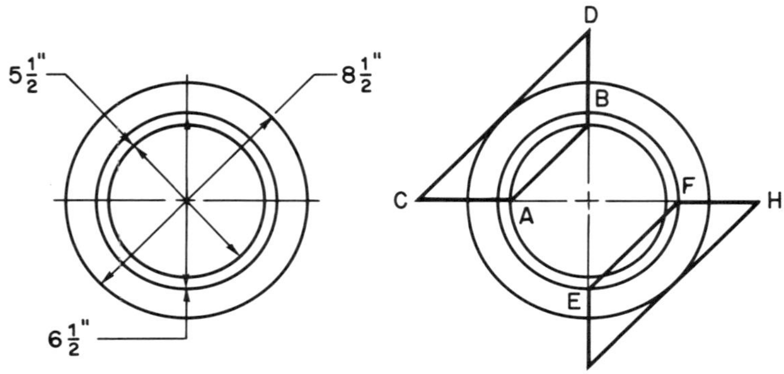

Fig. 34-4 Laying out the ring

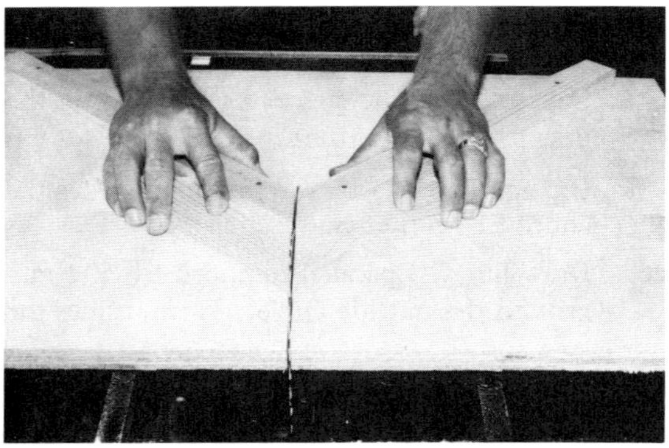

Fig. 34-5 The end of each piece is cut at a 45-degree angle.

Fig. 34-6 Clamping up the ring segments with band clamps

stock for one-quarter of the ring for the cope part of the pattern.

Making the Pieces of the Pattern

1. For the cope part of the pattern, plane sufficient stock for two rings to 3/8 inch (10 mm) thickness.
2. For the drag part, plane sufficient stock for the two rings to 1/4 inch (6 mm) thickness.
3. Rip the stock to the required width and length as determined from the layout.
4. The end of each piece is cut at a 45-degree angle, figure 34-5.
5. Join four pieces for each of the rings with a splined joint. A single saw cut about 1/8" x 1/8" (3 mm x 3 mm) on the end of each piece is sufficient for the spline. Use 1/8-inch tempered hardboard or solid wood. If solid wood is used, the grain of the spline must run at right angle to the joint. Do not make the splines so wide as to keep the joint from coming up tight.
6. Clamp and glue the stock to make two circles for the drag part of the pattern and two circles for the cope part of the pattern. Use band clamps to clamp the segments, figure 34-6.
7. Lay the assemblies on the layout and scribe the inside and outside circumferences, figure 34-7.

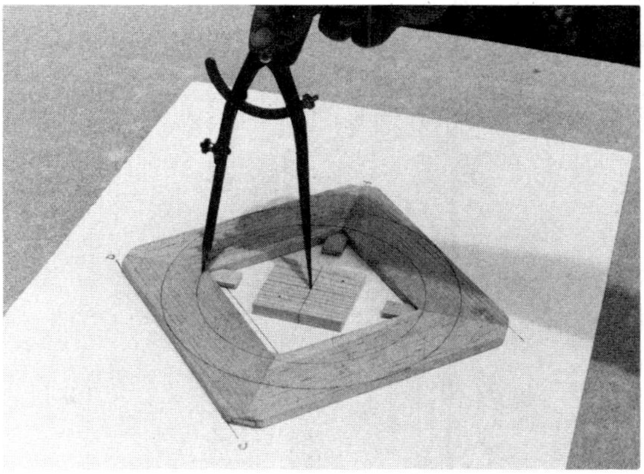

Fig. 34-7 Scribing the inside and outside diameters

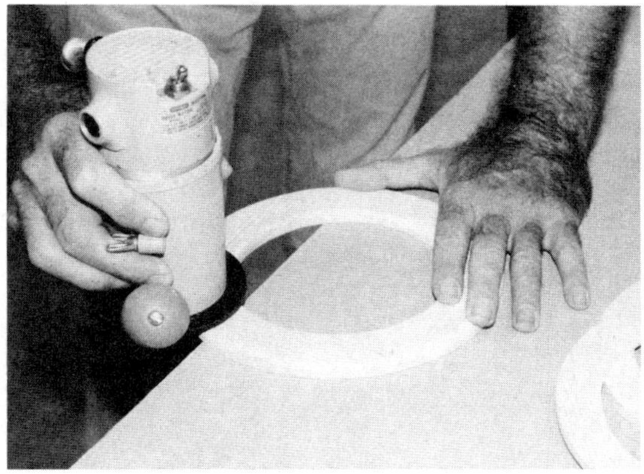

Fig. 34-8 Round over the edges of the cope part.

8. Cut outside the scribed lines enough to allow for sanding. Use a band saw for the outside circle. Use a scroll saw or saber saw for the inside circle.
9. Sand both the inside and outside circumferences to the scribed lines. Sand at a slight angle to provide draft.
10. Round over the inside and outside edges of the cope part of the pattern. Use a router with a rounding-over bit, figure 34-8.

MOUNTING THE PATTERN ON THE PLATE

1. Fasten the cope parts of the pattern to the drag parts with small brads, keeping the outside circumferences flush.
2. Mark out and drill three equally spaced, 1/4-inch (6-mm) holes through both rings on a drill press. Drill through the face until the drill point comes through. Turn the stock over and drill from the other side so the stock does not splinter.
3. Separate the rings and mount the cope part on one side of the match plate using glue and screws. Mount the two rings in line with the long centerline and an equal distance from the short centerline, figure 34-9.

Fig. 34-9 Mounting the cope part of the pattern on the match plate

4. For perfect alignment, bore holes through the match plate in line with the previously bored holes in the cope part of the pattern.
5. Insert dowels through the holes. Lay the drag part of the pattern over the dowels on the drag face of the match plate.
6. Fasten the drag part of the pattern in position with glue and screws.
7. Sand the dowels flush with the surface of the pattern. Fill in and sand all recessed screws and brads.
8. Center a gate between the rings on the cope face. Fit and fasten a block of the same

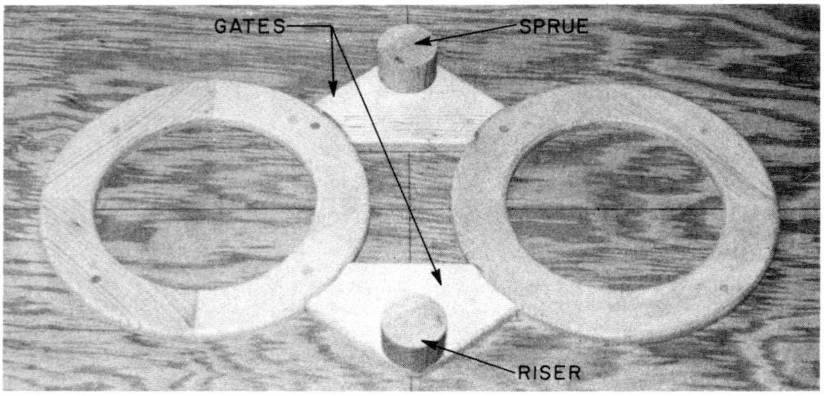

Fig. 34-10 The pattern needed to mold the gate, sprue, and riser

thickness against the rings to mold the gate and sprue, figure 34-10.

9. On the opposite side of the rings, fit a block to mold the riser.

10. Round over and provide draft on the edges of the blocks.

11. Finish the pattern and plate with three coats of shellac.

ACTIVITIES

1. On your field trip to a foundry, observe sand molding using match plates. Take notes to explain the molding process and the reasons for using match plates.
2. Make a pattern on a match plate to duplicate a small casting from one of the woodworking machines in the shop.

UNIT REVIEW

Questions

1. What is a match plate made of?
2. Where are the parts of a pattern placed on the match plate?
3. Why are fragile patterns built into a match plate?
4. Why are small patterns built into a match plate?
5. What does the use of a match plate eliminate from the pattern?
6. How are the drag and cope parts of a pattern attached to a match plate?
7. When are gates, riser, and sprue locations added to a match plate?
8. For ordinary work, how thick should a match plate be?
9. How is the match plate fitted to the drag?
10. How is the drag pattern aligned on the match plate?

Multiple Choice

1. A match plate is used
   a. to mold patterns with loose pieces.
   b. to make large molds.
   c. so parted patterns can be eliminated.
   d. none of the above.

2. When a match plate is used, the gates, riser, and sprue locations
   a. must be cut accurately in the mold.
   b. are not needed.
   c. are carved out of the match plate.
   d. are made on the match plate.

3. Dimensions for placing the pattern on the match plate are measured from the
   a. centerlines of the plate.
   b. edges of the plate.
   c. sockets of the plate.
   d. pins of the plate.

4. The cope and drag parts of the pattern are lined up on the match plate using
   a. centerlines.
   b. dowels.
   c. dividers.
   d. a rule.

5. If splines are made of wood,
   a. the grain must run at right angles to the joint.
   b. the cut must be small to keep the joint from coming tight.
   c. they must run with the grain of the segments.
   d. the cut must run with the grain.

6. If the splines are too wide,
   a. the cut will be off center.
   b. they will not be at right angles to the joint.
   c. jam them into the joint until they fit.
   d. the joints will not come tight.

7. To clamp up the porthole rings, use
   a. band clamps.
   b. bar clamps.
   c. C clamps.
   d. miter clamps.

8. To round over the edges of the cope part of the pattern, use a
   a. block plane.
   b. router.
   c. shaper.
   d. spokeshave.

section 6

Millwork

unit 35 Wood Mills and Machines

OBJECTIVES

After studying this unit, the student will be able to:

- define millwork and list common millwork products.
- describe millwork machinery.
- explain how to manufacture millwork.

MILLWORK

Millwork is shaped items of wood, figure 35-1. It is made in large quantities in woodworking mills on heavy production machinery. To mass-produce millwork, some of these machines are highly automated.

Most millwork is used in buildings and is installed by finish carpenters. Other items of millwork are parts that are shipped to other mills for more machining and assembly of a finished product.

A list of millwork products includes:

- Window frames and sashes
- Window screens and screen doors
- Blinds and shutters
- Doors
- Kitchen and bathroom cabinets
- Casework
- Chairs, tables, dressers, and furniture parts
- Stairs
- Columns and turned work
- Building products such as flooring, siding, shingles, moldings, interior trim, mantels, fencing, gates
- Special millwork made to order such as a special church window or entrance doors
- Specialized furniture or casework such as dental or doctor office and hospital furniture

Millwork is constructed and sold either set up or knocked down. *Setup millwork* is assembled and ready to be installed with little or no fitting. *Knocked-down millwork* is assembled by the finish carpenter on the job site. For example, many window frames are shipped in knocked-down form and later assembled. Knocked-down items of millwork save shipping space, and this form is used whenever practical.

KINDS OF MILLS

There are various kinds of woodworking mills. Mills vary in size and may employ a handful or hundreds of people. Some mills make only a few special items, while others make a full line of finished woodwork.

Most large mills specialize in a few items because very expensive machinery is required to produce parts economically in large quantities, figure 35-2. Smaller mills usually custom-make small quantities of millwork products to order. In this case, expensive production machinery is not required, figure 35-3. The making of various kinds of millwork described in this section is applied to a small mill without specialized production equipment.

MILLING OPERATIONS AND MACHINERY

Millwork Material

Millwork is made, in most cases, with well-seasoned, dried lumber. Items do not shrink or warp then after they are made. If wood is not thoroughly dry, it will not hold a finish, glue joints will fail, and veneers will not hold. Lumber used for millwork is reduced to about four to nine percent moisture content by kiln drying before it is processed.

Fig. 35-1 Typical millwork products

Fig. 35-2 A large mill uses expensive, specialized machines to mass-produce millwork. *(Woodworking and Furniture Digest)*

Fig. 35-3 A small mill uses conventional stationary machines to custom-make millwork.

Rough Mill

After the rough lumber is dried to the proper moisture content, it is brought to the *rough mill.* Here the rough stock is trued and straightened to uniform sizes in large quantities for later processing.

The stock is first cut to rough lengths using a *cutoff saw.* The cutoff saw shown in figure 35-4 will cut stock up to 6 inches thick and 20 inches wide. Then one face or side is straightened on a facer to remove any warp, figure 35-5. This machine is similar to a jointer except it has an arrangement that holds down and feeds the stock through the machine. A *single-surface planer* then planes the stock to the required thickness (see figure 6-20). Some mills have *double-surface planers* which plane both the top and bottom surface of the stock with one pass, figure 35-6.

Some machines, like the *Oliver Straitoplane®* in figure 35-7, combine the facing and planing operations and are widely used in millwork plants. This machine planes both sides of stock, reduces it to uniform thickness, and removes warp in one operation. It may also be used as a double-surface planer to cut both sides of lumber with one pass. The Straitoplane® planes stock up to 36 inches wide and 12 inches thick.

The stock is next cut to rough widths so it is slightly wider than the finished product. The best edge is first straightened on a jointer or an *edging*

Fig. 35-4 A production cutoff saw is used to cut stock to length. *(Oliver Machinery Co.)*

Fig. 35-5 The facer cuts one side of stock straight and true. *(Custom Cabinets)*

Fig. 35-6 Double-surface planer

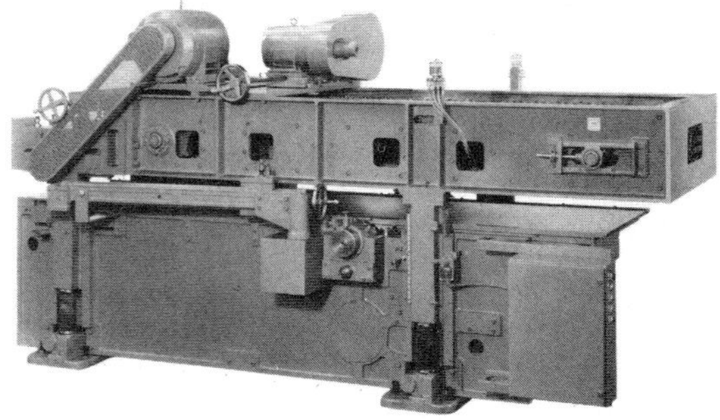

Fig. 35-7 The Straitoplane® faces and planes in one operation (guards are removed). *(Oliver Machinery Co.)*

saw, figure 35-8. Edging saws produce better edges for gluing or straightening than the jointer, figure 35-9. It is power-fed and removes a minimum of stock. Sawdust is the only waste.

After the edges have been straightened, the stock is ripped to a rough and uniform width. Since the best edge was straightened, ripping removes more of the defected edge. A *single* or *gang table ripsaw* is used for ripping, figures 35-10 and 35-11. In most production mills, chain-feed saws are used so the operator does not have to hand-feed the stock. This type of saw is safer for the operator to use and results in high production and reduced labor costs.

Once the stock has been cut to uniform lengths, it is stacked and stored until called for by the machining department for further processing.

Machining

The *machining department* of a mill is responsible for making the parts of the finished product. These parts are then sent to sanding, assembling, and finishing departments for the construction of the finished product.

The machining department must cut the pieces of rough stock to their finished size and shape. Holes are bored in the stock and dowels inserted. Carved work, turnings, and joints of all kinds are made in this department.

A wide variety of automatic equipment is used to do this work accurately and quickly. There is a great demand for machine operators who are

Fig. 35-8 The edging saw produces straight edges with a minimum of waste. *(Ekstrom Carlson)*

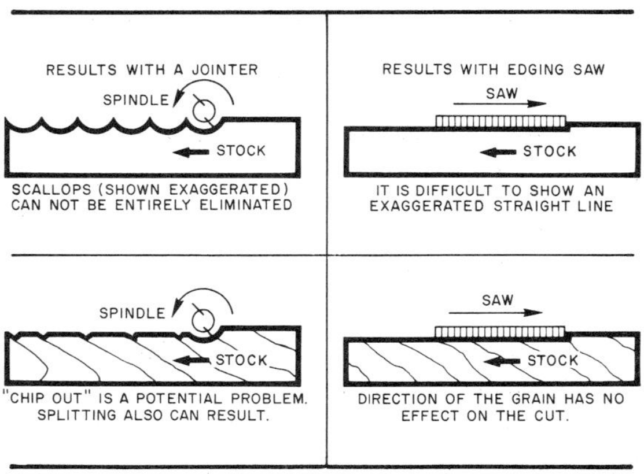

Fig. 35-9 The edging saw produces a better edge than a jointer. *(Ekstrom Carlson)*

Fig. 35-10 Single straight-line ripsaw *(Diehl Machines Division)*

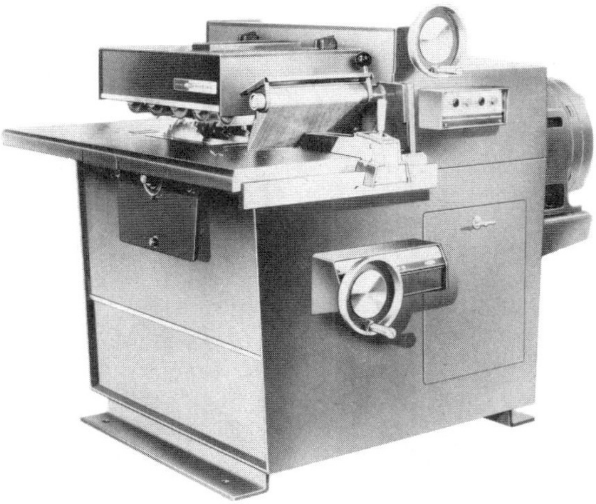

Fig. 35-11 Gang ripsaw *(Diehl Machines Division)*

Fig. 35-12 Feeding panels into a hot press *(Weyerhaeuser)*

able to set up and maintain the equipment needed in the machining department.

Large presses are used to bond veneers (thin layer of expensive hardwood) to plywood panels to make furniture parts. Some of these are *hot presses* that apply heat to bond the veneers, figure 35-12. *Cold presses* are also used, but involve a longer process, figure 35-13. However, fast-setting adhesives reduce the press time.

The *molder* or *sticker* is widely used to shape parts, figure 35-14. This machine is discussed in greater detail in *Unit 36 Molding.*

Horizontal and vertical *boring machines* save time and money also. Many have multiple spindles that can be moved to bore holes in specific locations, figures 35-15 and 35-16.

One of the most valuable production woodworking machines is the *double-end tenoner,* figure 35-17. It is actually two tenoners joined together. One machine is fixed while the other is adjustable for different lengths of stock. The double-end tenoner does much more than make tenons. It trims and shapes edges and ends as well as performs numerous edge and face machining operations. It also has vertical cope spindles and cutting heads to make special shapes.

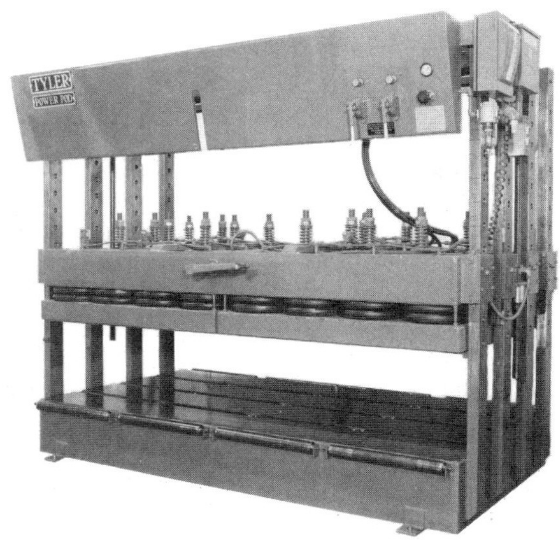

Fig. 35-13 Cold press *(Tyler Machinery Co. Inc.)*

Fig. 35-14 One of the many types of molders used in millwork plants *(Mattison Machine Works)*

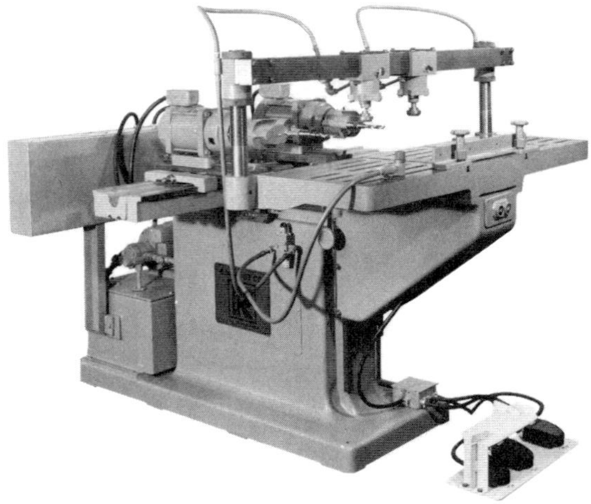

Fig. 35-15 Multispindle horizontal boring machine *(B.M. Root Co.)*

Many kinds of *shapers* are used in mills to machine wood parts (see *Unit 10 Shaper and Overarm Router*). The *double-spindle shaper* has two spindles that rotate in opposite directions, figure 35-8. The operator may use either spindle to cut with the grain as the direction of the grain in the stock changes. Cutting with the grain results in a much smoother cut not always possible with a single-spindle shaper.

An *automatic shaper* is similar to a hand-fed shaper except it has a power sprocket. The pattern that holds the piece to be shaped is fitted with a chain around its edge, figure 35-19. The sprocket drives the pattern along its chain, and a pressure roller holds the pattern against the sprocket.

Many other production machines are used in the machining department. *Drawer front dovetailers* make dovetail joints on drawer parts in large quantities, figure 35-20. The *dado machine,* figure 35-21, performs high-volume dadoing of stock for the mobile and modular home industry.

The *overhead pin router* is an extremely versatile machine in the woodworking industry, figure 35-22. Because the cutter is above the table, it can pierce the work as well as shape it. It can mortise, groove, dado, carve, and perform many other operations. Some have floating heads that travel up and down with the shape of the work.

Another machine used to shape parts is the *double-miter and cutoff saw,* figure 35-23. This

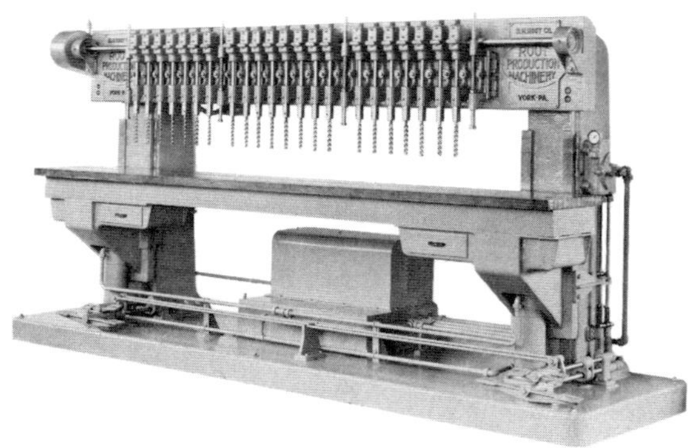

Fig. 35-16 Multispindle vertical boring machine *(B.M. Root Co.)*

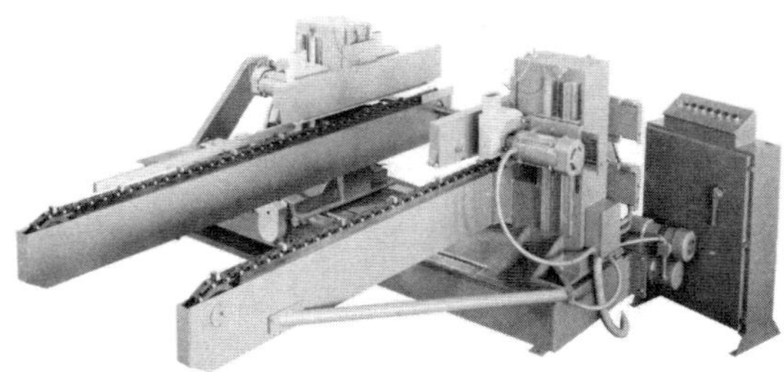

Fig. 35-17 Double-end tenoner *(Tyler Machinery Co., Inc.)*

Fig. 35-18 Double-spindle shaper *(Oliver Machinery Co.)*

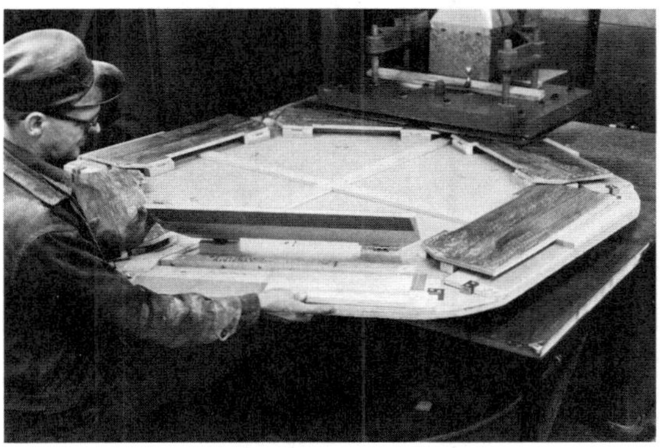

Fig. 35-19 An automatic shaper with a turntable to hold several parts *(Woodworking and Furniture Digest)*

Fig. 35-20 Dovetailer *(Tyler Machinery Co., Inc.)*

Fig. 35-21 Dado machine *(Tyler Machinery Co., Inc.)*

Fig. 35-22 Overhead pin router

machine cuts pieces to length with either or both ends cut at simple or compound angles.

The *automatic shaping lathe,* figure 35-24, turns duplicate pieces by a manual or hopper feed. A machine of this type produces such things as bowling pins in large quantities.

Some mills specialize in producing carved wood parts. In these plants, carved legs, panels, drawer fronts, and other pieces are made for furniture. One machine used to produce these parts is the *multiple-spindle woodcarving machine,* figure 35-25. This machine may hold as many as thirty-two carving spindles.

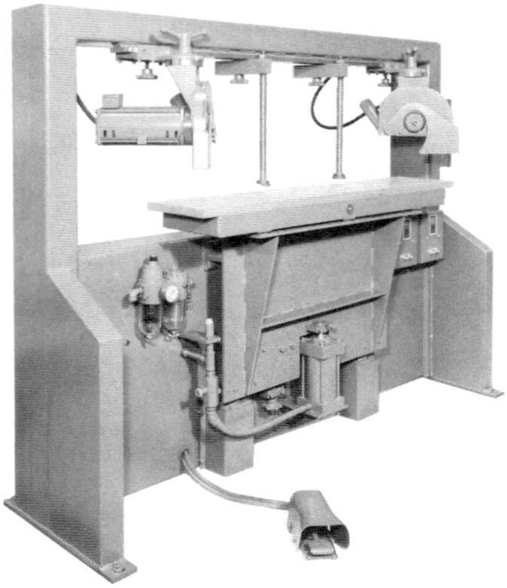

Fig. 35-23 Double-miter and cutoff saw *(Tyler Machinery Co., Inc.)*

Fig. 35-24 Automatic shaping lathe *(Mattison Machine Works)*

Fig. 35-25 Parten® multiple-spindle woodcarving machine *(Parten Machinery)*

Fig. 35-26 A large, high-volume drum sander *(Boise-Cascade Corp.)*

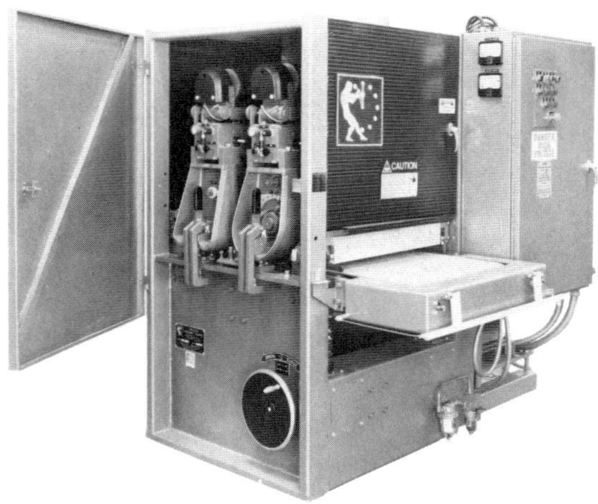

Fig. 35-27 The wide-belt sander is sometimes called an abrasive planer. *(Timesavers, Inc.)*

Fig. 35-28 A stroke sander is used for final flat sanding of wide surfaces.

Fig. 35-29 An edge belt sander being used to sand a curved edge *(Custom Cabinets)*

Fig. 35-30 The Taylor® clamp carrier is used for production clamping of wood parts *(James L. Taylor Mfg. Co.)*

Sanding

After the pieces are shaped, they usually need to be smoothed. Sanders range from very small to huge machines. They are generally designated as drum, belt, or disc sanders.

The *drum sander* is a large, heavy-duty production machine used to sand large flat items such as doors and plywood panels, figure 35-26. The size of the drum sander is determined by the width and number of drums. Each drum holds progressively finer abrasives to smooth the stock further as it passes through the machine.

Another type of sander widely used in woodworking is the *wide-belt sander,* figure 35-27. Sometimes called an *abrasive planer,* this machine reduces stock to its final thickness while smoothing the surface.

Most sanders are designated for specific work. The *stroke sander,* figure 35-28, is used for the final flat sanding of wide surfaces such as tabletops. The *edge belt sander,* figure 35-29, is used to sand the edges of stock with the grain.

Disc sanders are discussed in *Unit 14 Sanding Machines.*

Assembly and Finish

The sanded machined parts are then sent to the assembly and finishing departments. Machinery for gluing, clamping, and fastening are usually large and power-operated, figure 35-30. Finishing the assembled parts is accomplished by either spraying, dipping, or rolling.

ACTIVITIES

1. Visit a large wood mill and observe the sequence of operations and the machinery used to mass-produce millwork. List the types of millwork produced.

2. Visit a small mill and observe the method of producing custom-made parts. List the types of millwork produced.

3. Make a flow chart to produce an item of millwork such as a paneled door. Show the sequence of operations and the machinery used with each operation. Use the most efficient machinery available.

4. Make a list of the operations that can be performed using a double-end tenoner.

UNIT REVIEW

Multiple Choice

1. Select the item listed below that is not considered millwork.
   a. Kitchen cabinets
   b. Balusters
   c. Rough lumber
   d. Doors

2. Lumber used for millwork is reduced to a moisture content of about
   a. two to seven percent.
   b. three to eight percent.
   c. four to nine percent.
   d. five to ten percent.

3. After stock is cut to rough lengths, one side is first straightened on a
   a. facer.
   b. jointer.
   c. planer.
   d. shaper.

4. A machine that combines the facing and planing operations is called a
   a. facer.
   b. double-surface planer.
   c. Straitoplane.
   d. double-end tenoner.

5. To straighten the edges of stock, jointers are used. Another machine used to straighten edges is the
   a. planer.
   b. Straitoplane.
   c. molder.
   d. edging saw.

6. A machine similar to the shaper, except that the cutting tool is above the work, is called
   a. an automatic shaper.
   b. a contour profiler.
   c. an overhead pin router.
   d. a dado machine.

7. To make turnings in large quantities, many mills use
   a. an automatic shaper.
   b. a multiple carver.
   c. overhead pin router.
   d. an automatic shaping lathe.

8. A machine that cuts pieces to length at a simple or compound angle is called
   a. an automatic shaping lathe.
   b. an edging saw.
   c. a Straitoplane.
   d. a double-miter cutoff saw.

9. A multiple spindle woodcarving machine may hold as many as ________ carving spindles.
   a. twelve
   b. twenty-four
   c. thirty-two
   d. forty-eight

10. A machine that is sometimes called an abrasive planer is the
    a. single-surface planer.
    b. double-surface planer.
    c. drum sander.
    d. wide-belt sander.

Questions

1. What is knocked-down millwork? Why is it used whenever practical?
2. What may happen if the wood used in millwork is not thoroughly dry?

3. What is done to the stock at the rough mill?
4. Which machine is used to cut the stock to rough length?
5. What is the name of the machine that planes both the top and bottom surfaces of the stock with one pass?
6. What are the advantages of using an edging saw?
7. What machines are used to rip stock to a rough and uniform width?
8. What is done to the stock in the machining department?
9. What is the advantage of using a double-spindle shaper?
10. What is the edge belt sander used for?

unit 36
Molding

OBJECTIVES

After studying this unit, the student will be able to:

- describe commonly manufactured molding and give examples of their use.
- make moldings using a molding machine.
- grind shaper knives to the proper shape.
- make different kinds of molding using limited equipment.

MOLDING

Molding is a shaped strip of wood used as an ornamental edging in buildings. The surface of the wood strip is cut with alternating curved or straight shapes. The pleasing effects of light and shade on these surfaces help achieve a finished appearance, figure 36-1. Molding is also used on furniture to give a simple looking piece a definite style.

Molding includes all the finished woodwork used to cover spaces around openings in the interior of buildings. Sometimes called *interior trim,* these pieces are used around doors and windows, between ceilings and walls, and between walls and floors. Molding called *exterior trim* is used on the exterior of a building to give a finished and decorative appearance. Exterior moldings are used at the roof overhang (cornice), over and around doors and windows, and many other places.

Figure 36-2 shows the typical types of moldings. Note that the term *mold* is used instead of molding when a specific molding is described.

Fig. 36-1 Molding gives a dramatic effect to this entrance hall.

- *Crown, bed,* and *cove molds* cover large angles such as those between the ceiling and walls and on cornices. They are made up to five inches wide.
- *Casings* trim around doors and windows and close the space between the wall and the door or window frame.
- *Back bands* are added to casings to give them a heavier and different shape.
- *Stools* are used to finish the inside of window sills. They are rabbeted at an angle to fit the inclined top surface of the window sill.
- *Aprons* are placed under the stools to close off and decorate the space between the window sill and the wall.
- *Stops* guide or hold windows and doors in place.
- *Base* or *baseboards* have one, two, or three members. A three-piece member consists of the baseboard, the base mold, and the base shoe.
- *Half rounds* cover joints between panels on flat walls and also to edge such things as tabletops.
- *Quarter rounds* cover joints on inside corners and are also used for edging purposes.

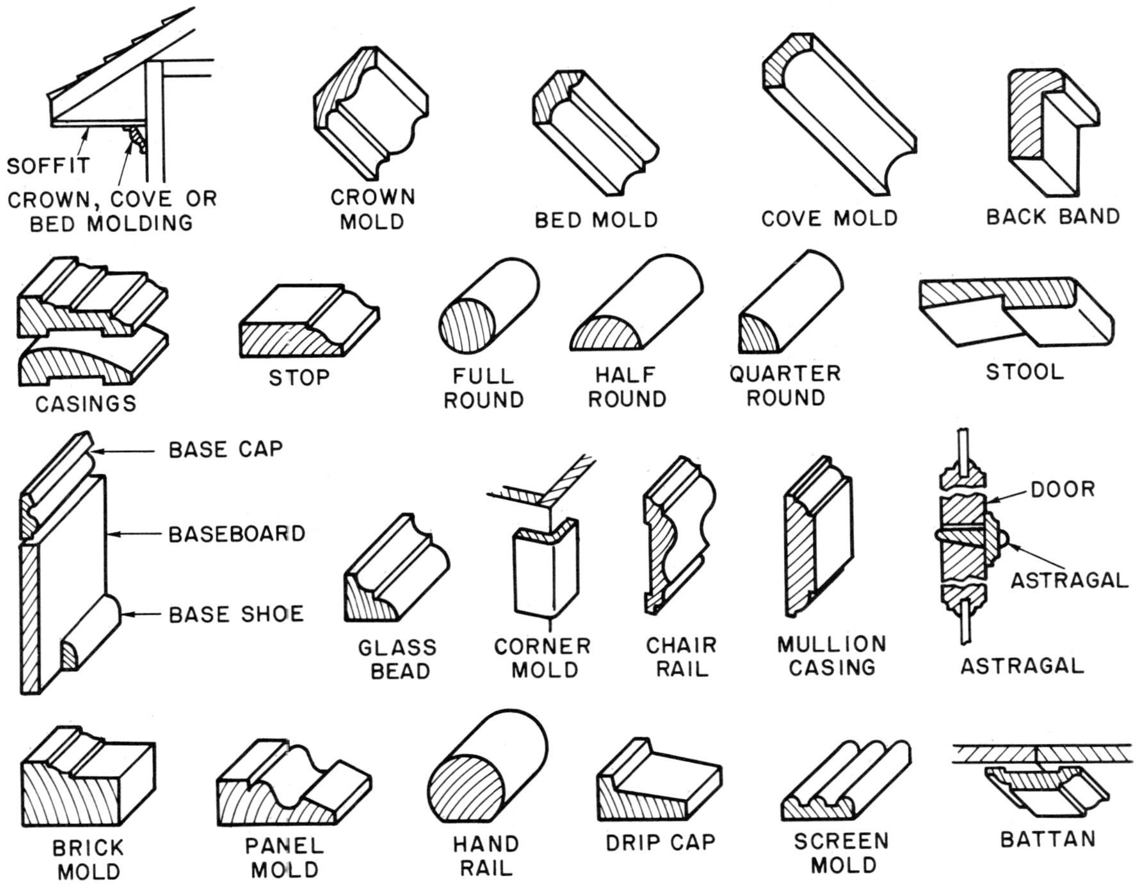

Fig. 36-2 Kinds of molding

- *Corner molds* decorate and cover exposed outside corners of walls and other places.
- *Thresholds* are placed under doors to cover joints where flooring of different kinds come together.
- *Panel molds* hide joints between panels on walls and ceilings.
- *Drip caps* are placed on the top of outside casings of doors and windows to keep water from entering between the wall and casing. The small groove in the bottom edge causes water to drip off and prevents it from flowing back to the window.
- *Brick molds* are used extensively on masonry window and door openings. They cover the space between exterior window or door frames and the masonry or siding opening.
- *Glass beads* are small moldings used to hold glass or other material in place in window or door frames.
- *Screen molds* cover and hold the edges of screening in place. It also covers the edges of shelves, tabletops, and similar objects.
- *Astragals* decorate and make a watertight joint between double doors.
- *Battans* conceal the line where two parallel boards or panels meet.

Molding comes in various sizes and patterns. Molding sizes are indicated by three measurements: thickness x width x length. Thickness and width are measured from the extreme points. Lengths vary from 3 to 20 feet. They usually run in strips of 8, 12, 14, and 16-foot lengths and are cut into shorter lengths as required. The millworker must know how to set up and operate the machinery to produce the required molding.

MOLDER

The *molder* or *sticker* shapes edges of wood pieces. It can be a simple machine with one cutterhead that *sticks* or shapes only one edge with each pass. Or it can have as many as nine cutting heads and make all the cuts in one operation, figure 36-3. Some models can handle stock up to 11 inches wide and 6 1/2 inches thick at speeds up to 140 feet per minute. The larger models, in addition to molding operations, also plane and make guide grooves in the stock.

The most popular molder for medium-sized work is the *four-sided molder,* figure 36-4. It operates like a double-surface planer. As the stock is fed into the machine, first the upper surface is shaped, then one side, then the other, and finally the lower surface. The shape of the cuts made depends on the shape of the cutting knives. This

Fig. 36-3 The number of cutterheads on a molder may vary from one to nine. *(Diehl Machines Division)*

Fig. 36-4 Four-way molder *(Anderson Lumber Co.)*

Fig. 36-5 Standard and special shapes of molder and shaper cutters. The storage cabinet is humidity controlled to prevent the cutting knives from rusting. *(Anderson Lumber Co.)*

machine will shape a straight piece of lumber into any form or design on one, two, three, or four sides in one operation.

Setup and Operation

Actual on-the-job experience is necessary to become familiar with the setup of a molder. Once this type of machine is set up, it takes little experience to operate. What distinguishes millworkers from operators is that they not only know how to operate woodworking machinery, but they also know how to setup and maintain the equipment. Setting up a molder is almost a trade in itself.

The setup and operation discussed in this unit applies to the four-sided molder. First the cutting knives are installed in each cutterhead and adjusted so they cut evenly. The cutterheads each hold two pair of knives. These pairs may have different shapes, figure 36-5. Only one pair is installed if this is enough to do the job.

Knives are adjustable in both directions. The cutterheads are rotated by hand against a piece of stock until each knife just touches the stock. Sometimes knives are installed so that one does the cutting and the other trails for balance.

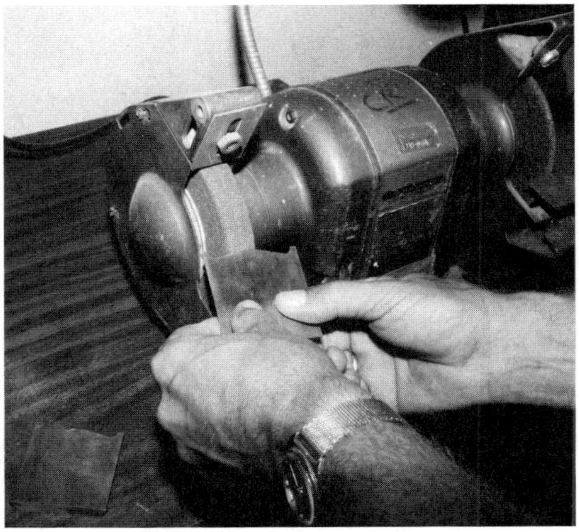

Fig. 36-6 Grinding shaper knives

Each cutterhead is adjusted to cut the desired amount of stock. Feed rollers and tables are adjusted to the proper setting. Guides are moved into position to hold the stock as it moves through the machine.

Once all adjustments are made, the machine is started. The stock, which is cut to overall size, is fed into the molder. The feed rollers pull the stock through while the knives cut one, two, three, or four sides as desired. More adjustments may be necessary after the first pass.

Coves, quarter rounds, bases, base shoes, stops, and casings are easily made using a shaper if a molding machine is not available. A description of the shaper and how to operate it is found in *Unit 10 Shaper and Overarm Router.*

Shaper cutter knives of the desired shape must be installed to do the work required. Many shapes are standard, and knives are available ground and sharpened to these shapes. For special shapes, the millworker grinds the knives to the desired shape.

GRINDING SHAPER KNIVES

To grind the knives to the proper shape, figure 36-6:

Fig. 36-7 Dressing a grinding wheel (guard removed for clarity)

1. From a plain shaper steel bar of the desired width, cut two pieces to length. The bars should be long enough so the knives are held by as much of the shaper collar as possible.
2. Since the bars are too hard to cut with a hacksaw, cut the bars to length by grinding on a narrow grinding wheel. Be careful not to overheat the metal when grinding. Overheating causes the metal to lose its temper.

Caution: Wear eye protection at all times when making shaper knives.

3. After cutting the bar to length, mark the outline of the desired shape on the side of a thin strip of plywood.
4. Grind the bars to the desired shape. Test it against the outline. Cool the metal frequently while grinding.
5. Different thicknesses and shapes of grinding wheels may be needed. Shape the edges of the grinding wheel as desired with a grinding wheel dresser, figure 36-7.
6. As the bar is ground to shape, bevel the back edge. This prevents it from contacting the stock when it is cut. The amount of bevel depends on the radius of the cut. Bevel the cutting edge no more than is required to provide clearance.

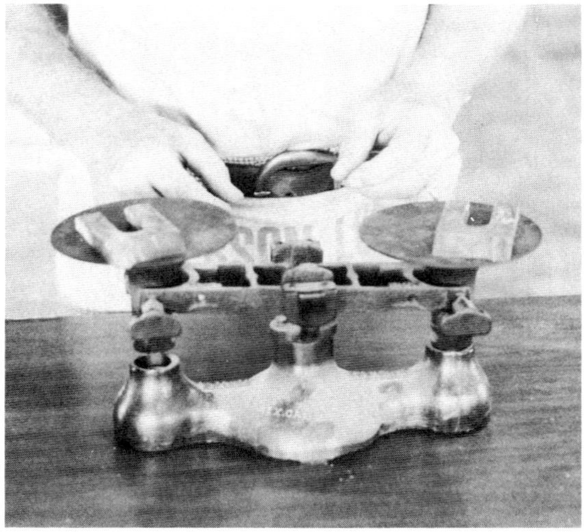

Fig. 36-8 Balancing cutting knives for equal weight *(Anderson Lumber Co.)*

7. After the knives are ground to shape, weigh them on balancing scales, figure 36-8. If one weighs more than the other, grind some metal from the heavier one until both knives weigh the same.
8. Sharpen the knives on an oilstone.
9. Install the knives in the shaper.

Balancing the knives prevents vibration when rotating at high speeds on the shaper spindle. Vibration may cause the knives to work free and fly out, possibly causing serious injury. Three or four-wing shaper cutters made in one piece are safer and should be used whenever possible. However, this type of cutterhead is very difficult to grind to special shapes with the usual shop equipment.

Knives for molders are also ground to special shapes using the method described here.

MAKING MOLDING ON THE SHAPER

1. Adjust the fence for the depth of cut.
2. Adjust the spindle up or down as necessary to produce the desired shape.

Fig. 36-9 Making molding using a shaper. Notice the feather boards used to keep the stock from vibrating.

3. Lay the stock against the fence and place the shaper hold-down on the top edge. Clamp as many featherboards or hold-downs as necessary on the side and top of the stock to keep it from vibrating as the stock is fed through the machine, figure 36-9.
4. With the spindle rotating, feed the stock in against the rotation of the knives while holding the stock against the shaper fence. Usually this operation requires two persons – one feeding the stock and one pulling the stock through the machine.
5. Feed the stock with an even rate of speed for maximum smoothness of cut.

MAKING MOLDING ON THE ROUTER

Small moldings may also be made with a router. Moldings such as quarter rounds are made with a rounding-over bit. Small cove molds are made with a cove-cutting bit. An ogee bit can be used to produce that shape on the edge of stops or casings.

The router is mounted underneath the table so the spindle projects through the tabletop, figure 36-10. This arrangement produces a small shaper. A straightedge clamped to the tabletop is used as a fence, or router bits with pilots may be used to control the depth of cut. When feeding the stock through the router, make sure the direction of feed is against the rotation of the blade.

Fig. 36-10 The router is mounted underneath the table to cut molding.

Fig. 36-11 Making corner mold using the table saw (guard is removed for clarity)

MAKING MOLDING ON THE TABLE SAW

Some moldings can also be made on the table saw. Corner molds, for instance, are made by rabbeting square stock to the desired depth, figure 36-11. After rabbeting the stock, the exposed edges are rounded over with a hand plane and sandpaper.

Wide cove molds can also be made. The stock is cut by making a number of passes at an angle over the table saw blade. This method is described in *Unit 7 Table Saw.* The radius of the cove cut is limited by the diameter of the saw blade, but quite a wide cove molding can be cut. After the cove cut is made, the edges of the stock are trimmed to width at a 45-degree angle. Only short runs are made on the table saw because of the labor involved. A disadvantage of this method is that the curved cut must be smoothed by scraping and sanding because the saw blade produces a rough surface.

ACTIVITIES

1. Obtain samples of different kinds of molding. Mount on a display board and identify each kind.
2. Make a list showing the uses of each kind of molding displayed.
3. Visit a mill that has a molder. Observe the setup of the machine and the manufacture of molding.
4. Grind, balance, and sharpen a pair of shaper knives to cut a specified shape.
5. Make a chart of the standard shaper knives available in your shop. Cut out sample moldings on the shaper.
6. Make a quarter round using the router.
7. Make a wide cove mold using the table saw.
8. Make a corner mold on the table saw.

UNIT REVIEW

Completion

1. ________ guide or hold windows and doors in place.
2. ________ are used to trim around doors and windows.
3. To finish the inside of window sills, ________ are used.
4. ________ are used extensively on masonry window and door openings.
5. ________ make a watertight joint between double doors.
6. ________, ________, and ________ molds cover large angles such as those on cornices.
7. A three-member baseboard consists of the baseboard, ________, and ________.
8. ________ keep water from entering between the outside wall and window casing.
9. Small moldings used to hold glass in place in windows and door frames are called ________.
10. ________ conceal the line where two parallel boards meet.

Multiple Choice

1. Shaping an edge on a molder is often referred to as
   a. bending the edge.
   b. rabbeting the edge.
   c. trimming the edge.
   d. sticking the edge.

2. The edges of a grinding wheel are shaped as desired with a(n)
   a. astragal.
   b. grinding apron.
   c. grinding wheel dresser.
   d. sticker.

3. To cut lengths from a plain shaper steel bar, use a
   a. hacksaw.
   b. grinding wheel.
   c. cutting torch.
   d. thin file.

4. The length of a single shaper cutting knife is determined by the
   a. width of the knife.
   b. kind of cut to be made.
   c. temper of the metal.
   d. diameter of the shaper collar.

5. When grinding shaper or molder cutting knives, overheating will cause
   a. the metal to become soft.
   b. the metal to lose its temper.
   c. the metal to lose its edge.
   d. all of the above.

6. After shaper or molder knives have been ground to the desired shape, they are
   a. dressed.
   b. tempered.
   c. balanced.
   d. sharpened.

7. To make sure the cutting edges of shaper or molder knives are the same distance from center, check by
   a. measuring with calipers.
   b. rotating the spindle against a scrap piece of stock.
   c. making a trial cut on scrap wood.
   d. measuring in from the edge of the table.

8. To keep the stock from vibrating when cutting molding on the shaper, router, or table saw,
   a. use featherboards as necessary.
   b. hold the stock firmly.
   c. get more help.
   d. apply silicon lubricant to the tabletop and fence.

9. One small molding sometimes made with a router is the
   a. crown mold.
   b. drip cap.
   c. astragal.
   d. quarter round.

10. A molding that can be made on the table saw is the
    a. quarter round.
    b. astragal.
    c. casing.
    d. corner mold.

11. Molding sizes are indicated by the
    a. thickness x width x length.
    b. width x length x thickness.
    c. length x width x thickness.
    d. width x length.

unit 37
Screens

OBJECTIVES

After studying this unit, the student will be able to:

- describe the construction and sizes of window screens and screen doors.
- make a half-size window screen.
- make a screen door.

WINDOW SCREENS AND SCREEN DOORS

Screens are made in large quantities in mills. They are sold with either the wire mesh in place or open without the wire mesh. Well-seasoned, clear white pine or clear fir is used in the construction of most window screens or screen doors, figure 37-1.

Full-size window screens are usually made of 1 1/8-inch thick stock. Half-size screens that cover half of a double-hung window are usually made of 3/4-inch stock. Screen doors are usually 1 1/8 inches thick. The width of a screen door is the same as the main door, but the height is generally 1 inch greater.

Usually window screens and screen doors are *square stuck.* This means that the inside edges are not shaped. Through mortise-and-tenon joints are widely used on window screens. On screen doors, doweled joints are a popular choice.

The wire mesh of the screen may be applied on the surface of the frame. It may also be rabbeted and the mesh and molding sunk flush with the frame to make a neat finish, figure 37-2. Wire mesh is made of either aluminum, bronze, or plastic wire.

Fig. 37-1 Wood frame window and door screens *(Iroquois Millwork Corp.)*

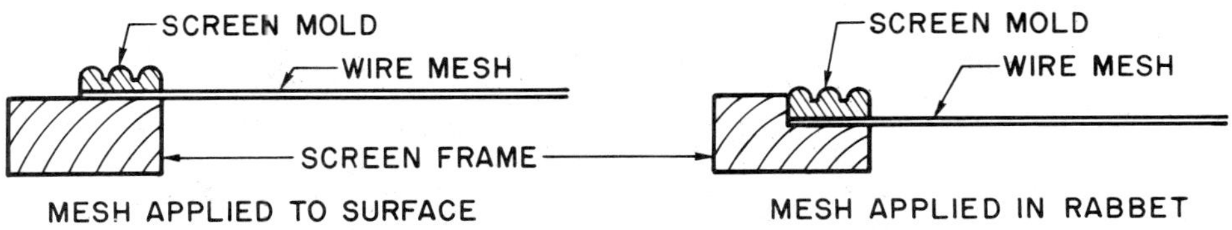

Fig. 37-2 Two methods of applying wire mesh to wood screens

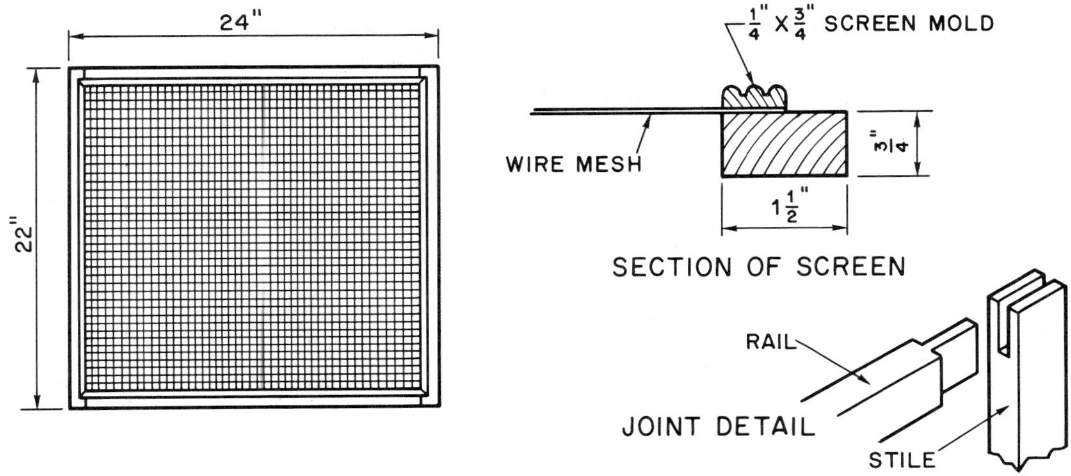

Fig. 37-3 Half-size window screen

MAKING A WINDOW SCREEN

A working drawing of a typical half-size window screen is seen in figure 37-3. The stiles and rails are 3/4" x 1 1/2". The frame is to be constructed with through mortise-and-tenon joints. The stiles are mortised and the rails are tenoned. The joints are clamped and glued and fastened with one wood screw in each joint. The strength of the frame depends on accurate, close-fitting joints. The wire mesh is stapled to the surface of the frame and covered with a screen mold.

Milling the Stock

When making large quantities of millwork such as window screens, it is essential that the stock be reduced to uniform size for accurate machining later. Rough stock must be square to insure straight pieces of uniform thickness and width. To do this, the stock is first reduced to thickness, then cut to width, then cut to length.

It is important that enough stock to do the job is milled at one time. Much time is wasted to set up the machinery again to cut more stock, and the second setup may not produce parts exactly the same as the first. It is also important not to waste material by milling out more stock than is required to do the job.

1. Determine the amount of stock required to produce the desired number of screens. Cut the stock to rough length.
2. Face one side and reduce it to 3/4 inch thickness.
3. Look the stock over carefully and select the best edge. Joint this edge, figure 37-4.
4. Rip to 1 1/2 inch width with the face side (better side) up, figure 37-5.

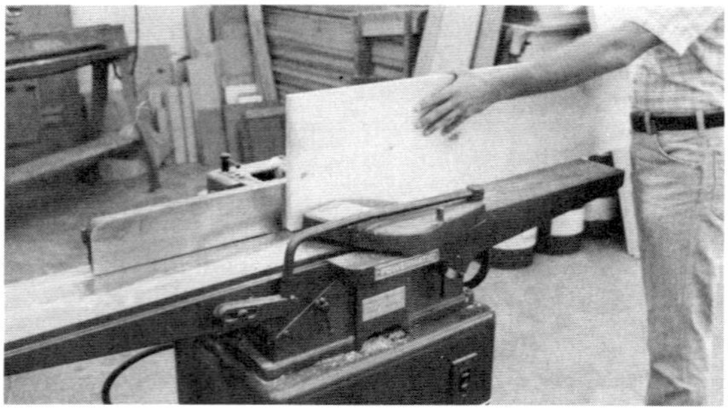

Fig. 37-4 Using a jointer to straighten the edges of the stock

Fig. 37-5 Ripping the stock to width with a power-fed ripsaw *(Custom Cabinets)*

5. Square the better end with the face side up. Remove only as much stock as necessary to square the end. This insures that there will be enough stock to cut the required length.
6. Set up stops as necessary and cut the required number of 22-inch and 24-inch pieces. Cut the other end of the stock with the same face side up, figure 37-6.

The reason for keeping the face side up on the ripping and squaring operations is that the back side edges and ends splinter out slightly. Clean cuts then show on the face side. The piece is assembled with the face of all exposed parts on the same side for a better-looking job.

After the parts have been cut to the required thickness, width, and length, additional machining is necessary. The stiles of the screens are mortised, and the rails are tenoned.

Mortising the Stiles

Through mortises are easily made on a production basis with a double-end tenoner. This machine also cuts to the finished length. Once the machine is set up properly, the stock is fed to the desired length. Mortises are made in both ends with one pass through the machine.

A single-end tenoner may also be used. However, two passes are required to machine both ends of the pieces.

If tenoners are not available, a table saw can be used. Set the ripping fence to leave on 1/4 inch.

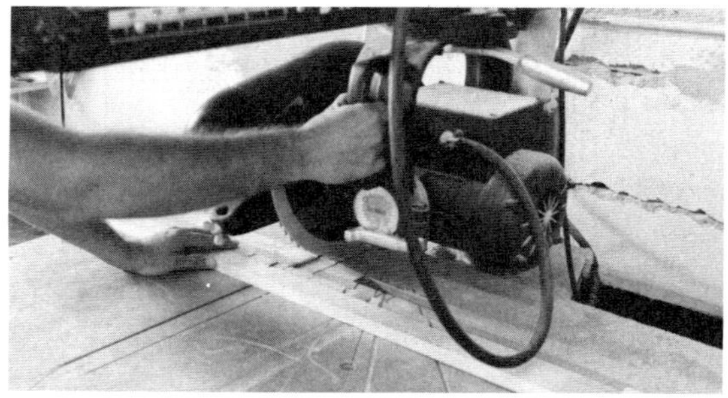

Fig. 37-6 Cutting the stock to length (guard is removed for clarity)

Raise the blade 1 1/2 inches above the table. Using a scrap piece of stock, stand it on end and run it through the saw. Turn the piece around and run it through again. Adjust the fence, if necessary, so that 1/4 inch is removed from the center of the stock. A tenoning jig makes this operation safer and more accurate and should be used if available.

Tenoning the Rails

The rails of the screen frames must be tenoned. Using a double-end or single-end tenoner is the fastest and most accurate method. This is described in *Unit 11 Mortiser and Tenoner.*

Tenons may also be made on the table saw in a manner similar to making mortises. Set the ripping fence so 1/4 inch is left in the center after the stock is passed through twice on end, figure 37-7. Cut the two shoulders of the tenons by using

Fig. 37-7 Making tenons using a table saw (guard is removed for clarity)

the miter gauge set at a 90-degree angle. Clamp a block to the ripping fence and set the fence to act as a stop to cut the tenon to the desired length.

Tenons may also be cut using a dado head on the table saw or radial arm saw. These are described in *Unit 7 Table Saw* and *Unit 8 Radial Arm Saw.*

Cutting the Screen Mold to Length

The screen mold is mitered on both ends at a 45-degree angle. These pieces can be cut to length and mitered on both ends in one operation by using a double miter and cutoff saw. Cut the required number of pieces to an overall length of 20 1/2 inches and 22 1/2 inches.

An alternate method of mitering screen mold to length is by using a mitering jig with the radial arm saw. Make a 45-degree cut on the molding using one side of the jig. With a stop set to the desired distance, make the mitered cut on the other end of the molding using the other side of the jig, figure 37-8. By using the mitering jig, the face side of the molding can be kept up for both cuts.

Another method of mitering the screen mold to length is by using the power miter box. Cut the required number of pieces to rough lengths and miter one end of each piece. Swing the saw to cut miters in the opposite direction. Set stops to cut the desired lengths and make mitered cuts on the other end of all the pieces, figure 37-9. Making cuts on one end of each piece first saves time by eliminating the need for swinging the saw back and forth if each piece were cut separately to length.

Fig. 37-8 Mitering screen mold using the radial arm saw and a mitering jig

Fig. 37-9 Using a power miter box to miter screen mold to length

Assembling the Frame

For fast, accurate assembly of the frame, build a jig on the work bench to clamp the frames and hold them square. Apply a water-resistant glue to the joints (plastic resin glue) and clamp the

Fig. 37-10 Stapling the wire mesh to the screen frame

parts in the jig. Drill and countersink holes and drive one 5/8 inch by #6 flathead screw in the center of each joint. Wipe off any excess glue with a damp rag.

Applying the Wire Mesh

After the frames are assembled, wire mesh is stapled on. Wire mesh comes in rolls and must be handled carefully to prevent putting any creases in the mesh. The mesh is applied so it is tight and flat with no wrinkles or sag.

1. Roll out the mesh on the frame. Let at least 3/4 inch extend beyond the inside of the frame.
2. With a hand stapler and 1/4-inch staples, staple one corner, figure 37-10.
3. Stretch the mesh to the second corner and staple that corner. Do not place the staples more than 3/4 inch from the inside edge of the frame. Otherwise, the screen mold will not cover the staples.
4. Staple along the first side every 3 inches.
5. Stretch the mesh to the adjacent side and staple the third corner. Staple along that side every 3 inches.
6. Stretch the screen in both directions and towards the fourth corner and staple it. Staple the two remaining sides, stretching the screen with each staple driven.

Fig. 37-11 Applying screen mold to the frame

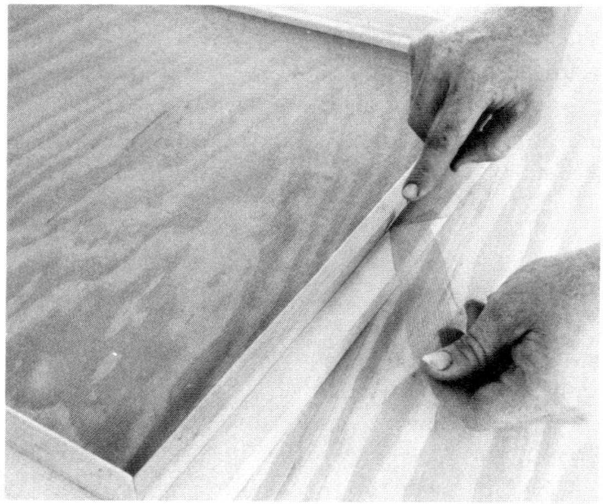

Fig. 37-12 Trimming the excess wire mesh with a sharp knife

Applying the Screen Mold

The screen mold is now applied using galvanized brads about 6 inches on center. The inside edge of the screen mold is kept flush with the inside edge of the screen frame, figure 37-11. The excess wire mesh is trimmed off flush with the outside edge of the screen mold with a sharp knife, figure 37-12.

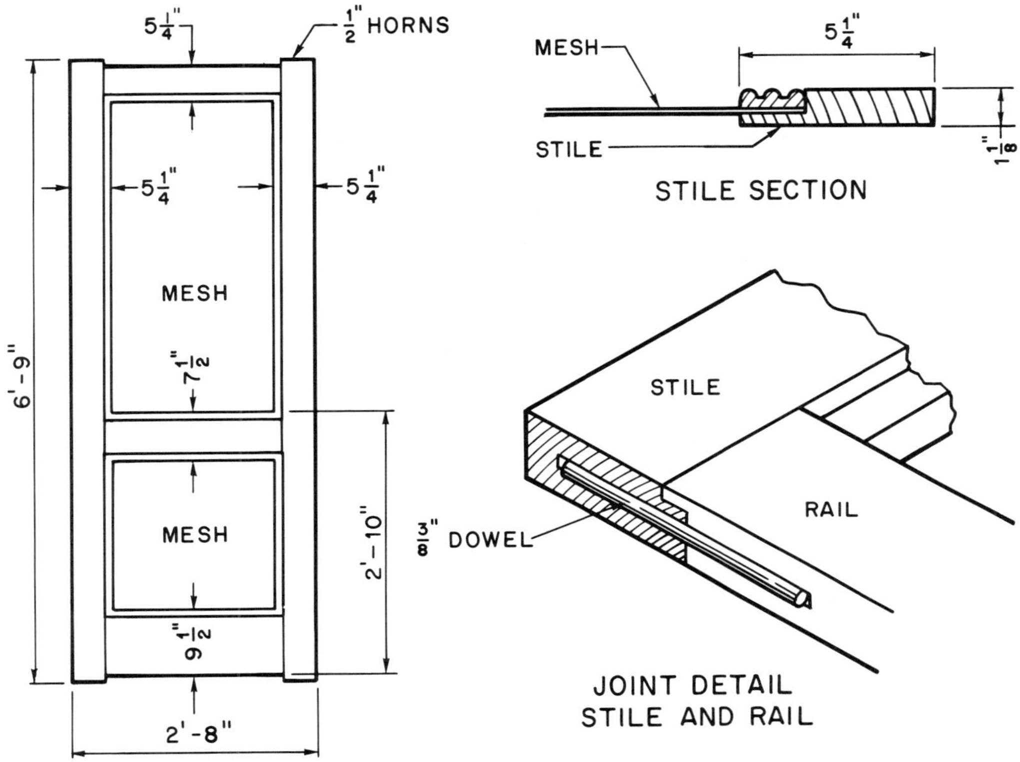

Fig. 37-13 Screen door

MAKING SCREEN DOORS

Figure 37-13 shows a typical screen door. In this case, the wire mesh and screen mold are set in a rabbet in the frame. The top rail is the same width as the stiles. The bottom rail is made wider to give the door more rigidity and to allow for planing. The lock rail (middle rail) is also wider than the top rail. Doweled joints are made between the rails and stiles.

Size and Location of Dowels

Most doors are made with doweled joints. The diameter of the dowel is determined by the thickness of the door, figure 37-14. A general rule of thumb is that the dowel diameter should be one-third the thickness of the door. For instance, if the door thickness is 1 1/8 inches, the dowel diameter should be 3/8 inch.

Two or more dowels are used to make a joint whenever possible. Dowels are spaced as far apart as possible for maximum strength. The outside dowel is centered not less than two dowel diameters from the outside edge. This permits fitting of the door without cutting into the dowel. The inside dowel is centered not less than one dowel diameter from the bottom of the groove or rabbet. A center dowel is spaced halfway between the other two dowels. On doors where space permits, dowel spacing does not exceed 3 1/2 inches.

The length of the dowel is determined by the width of the door stile. Generally, dowels extend into the stile not less than two-thirds of the stile width, but not so far that they are exposed when the door is fitted. Dowels extend an equal length into the rails of the door.

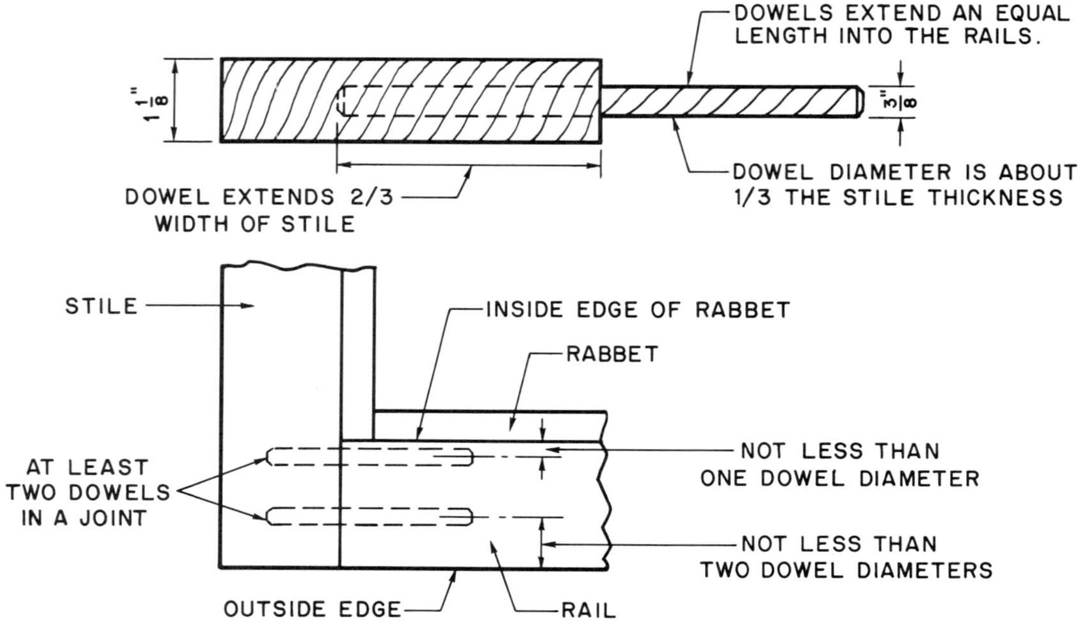

Fig. 37-14 Location and size of dowels used in jointing stiles and rails

Milling the Stock

Sufficient material of uniform thickness and width must be milled out to make the required number of screen doors.

The stiles are cut to a length of 6′-10″. This length allows the stiles to project 1/2 inch beyond the top and bottom rail. These projections, called *horns*, protect the edges of the rails during storage and transportation. The horns are later sawed off when the doors are fitted into their frames.

The rails are cut to a length of 23 inches. This allows enough length for the rails to be fitted into the rabbets of the stiles.

Rabbeting the Stiles and Rails

Make a rabbet 3/4 inch wide and 5/16 inch deep on one edge of each stile and the top and bottom rails. Rabbet both edges of the lock rail. Rabbet the rail ends to fit into the rabbet of the stile edges. Using a scrap piece of stock of the same size as the rail stock, cut the rabbet to the desired width and depth on both ends of the rails.

Laying out the Dowels

1. Lay the stiles side by side with the rabbeted edges facing each other.
2. Lay off the overall height of the door leaving horns on each end.
3. Mark the width of the top rail, the bottom rail, and the lock rail.
4. Mark the centerline of the dowels on the edge of the stiles. Set a combination square and mark the center of the thickness of the stile at the dowel locations, figure 37-15.
5. Locate the dowel centers on the rails. Hold the rails in their respective position against the stiles and transfer the dowel centers from the stiles to the rails. Center punch all dowel centers.
6. Bore the dowel holes.

In mills, dowel holes on stiles are bored with vertical boring machines. In the small custom shop, the drill press is used.

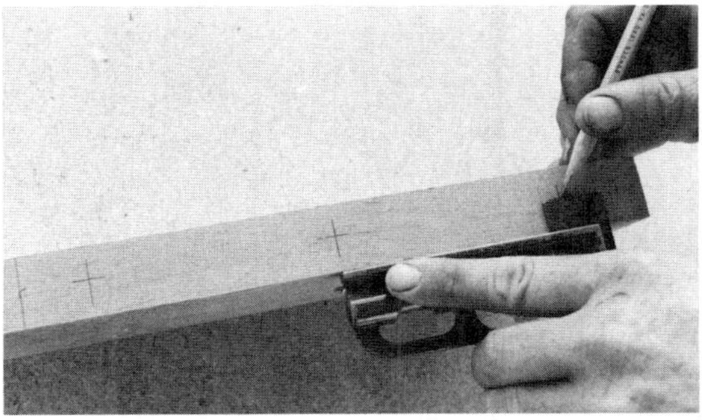

Fig. 37-15 Using a combination square to mark the centerline of dowels

Assembling the Door Frame

Dowels are usually inserted with glue into the stiles. The shoulders of the joints are then glued and the parts assembled by using bar clamps across each rail. Clamps are placed alternately on each side of the door to prevent the door from cupping. In production mills, special door clamping machines are used to speed the process.

After the glue has set, the frame is removed from the clamps and sanded. The wire mesh and screen mold is then applied in a similar manner as described for window screens.

ACTIVITIES

1. Obtain a catalog from a millwork manufacturer and study the styles of window screens and screen doors available.
2. Make a flow chart explaining how to efficiently produce window screens in large quantities with the equipment available in your shop.
3. Give a report on the advantages and disadvantages of the different types of screen wire mesh available.
4. Write a plan of procedure for making a full-size window screen with the equipment available in your shop. How would this procedure differ if you were making the window screen in a large woodworking mill?
5. From the plan of procedure written in #4, make a full-size window screen.

UNIT REVIEW

Completion

1. Full-size window screens are usually made of ________ inch thick stock. Half-size window screens are usually made of ________ inch thick stock.
2. Screen doors are usually made of ________ inch thick stock.
3. The height of a screen door is usually made ________ inch greater than the main door.
4. ________ joints are widely used on window screens.
5. On screen doors, ________ joints are a popular choice.
6. Wire mesh used on screens is made of either ________, ________, or ________.
7. The lumber used to make screens is usually well-seasoned, clear ________ or ________.
8. When using through mortise-and-tenon joints on window screens, the ________ are mortised and the ________ are tenoned.

9. Before rough stock is reduced to thickness, one side is first ________.
10. Before rough stock is cut to width, the stock must first be ________.

Multiple Choice

1. The strength of a screen frame depends on
   a. the type of wire mesh used.
   b. the width of the screen mold.
   c. the uniform size of the frame.
   d. accurate, close-fitting joints.
2. To make through mortises on a production basis for window screens, a first choice of machine is the
   a. double-end tenoner.
   b. radial arm saw.
   c. table saw.
   d. shaper.
3. To make tenons on a production basis, a first choice of machine is the
   a. double-end tenoner.
   b. table saw.
   c. shaper.
   d. radial arm saw.
4. For mitering both ends of pieces to a specified length on a production basis, use a
   a. mitering jig on the radial arm saw.
   b. double miter and cutoff saw.
   c. power miter box with stops.
   d. table saw with mitering jig.
5. When stapling wire mesh to the screen frame, staples are placed from the inside edges of the frame no more than
   a. 3/4 inch.
   b. the width of the screen mold.
   c. the width of the staple.
   d. half the width of the stile.
6. Excess wire mesh is trimmed off with
   a. tin snips.
   b. scissors.
   c. sharp knife.
   d. wire cutters.
7. The diameter of dowels used for making joints in screen doors should be
   a. 1/4 the thickness of the door.
   b. 1/3 the thickness of the door.
   c. 1/2 the thickness of the door.
   d. 3/4 the thickness of the door.
8. The outside dowel in a door frame is centered from the outside edge not less than
   a. two dowel diameters.
   b. the thickness of the door.
   c. half the thickness of the door.
   d. one dowel diameter.
9. The bottom rail of a screen door is made
   a. the same width as the stiles.
   b. narrower than the stiles.
   c. the same width as the top rail.
   d. wider than the stiles.
10. The projection that protects the edges of screen door rails during storage and transportation is called a
    a. stile.
    b. screen mold.
    c. horn.
    d. lock rail.

unit 38 Doors and Door Frames

OBJECTIVES

After studying this unit, the student will be able to:

- describe the basic parts and sizes of doors and door frames.
- describe the two commonly used methods of door construction.
- make an interior door using blind mortise-and-tenon joints.
- make an exterior door with raised panels using doweled joints.
- make an exterior door frame.

DOORS AND DOOR FRAMES

A typical exterior door is shown in figure 38-1. Its parts are similar to those of a cabinet door (see *Unit 24 Cabinet Doors*).

The parts of an exterior door frame are shown in figure 38-2.

Jambs form the sides and top of the door frame. The side members are called *side jambs* and the top member is called a *head jamb.* Door jambs may be plain or rabbeted to provide stops for the door. Stops are applied separately to plain jambs. Jambs may be rabbeted on one edge or on both edges. In double-rabbeted jambs, one rabbet receives the main door while the second receives a storm or screen door. The standard depth of rabbets in door jambs is 1/2 inch. The width of the rabbet is made for standard thicknesses of doors.

The *sill* of an exterior door frame is the horizontal bottom member. Sills are shaped in a number of ways according to the custom of the geographical location and the swing of the door, figure 38-3. Exterior doors in residences usually swing into the room. Exterior doors in schools and public buildings are required by law to swing outward to permit quick opening in case of emergencies.

The *casing* is generally furnished with and applied to a setup door frame. The top horizontal member is called a *head casing.* The vertical members are cut between the head casing and sill and are called *side casings.* The bottom edge of side casings is cut on a bevel to fit against the sill and provides a watertight joint. The upper end of the side casing is usually rabbeted into the head casing to keep rain and snow from entering the space between the wall frame and the door frame, figure 38-4.

When molding casings are used, a mitered joint that cannot be rabbeted must be used at the head. Quality construction calls for the miter to be splined to prevent rain and snow from entering through the joint.

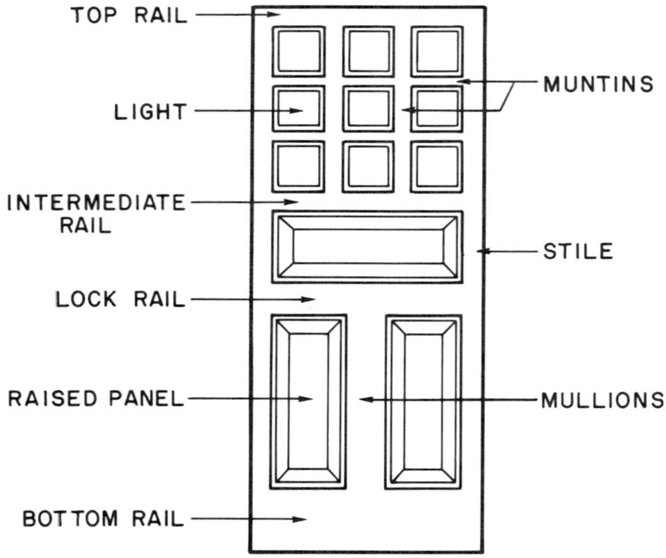

Fig 38-1 Parts of a typical exterior door

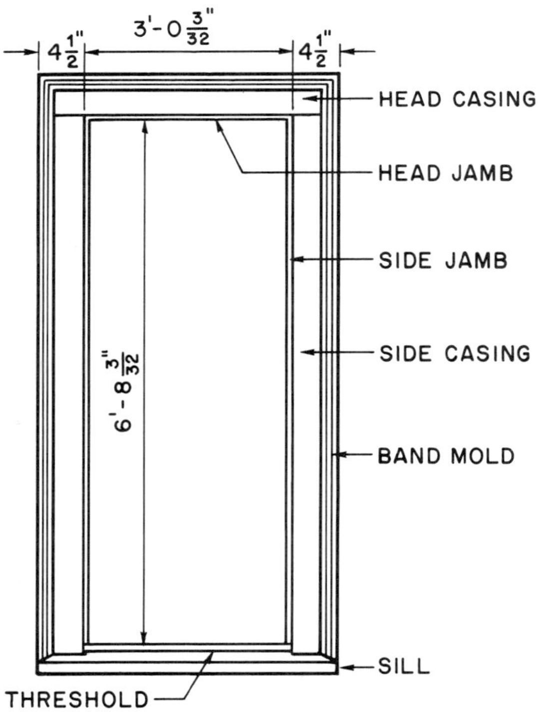

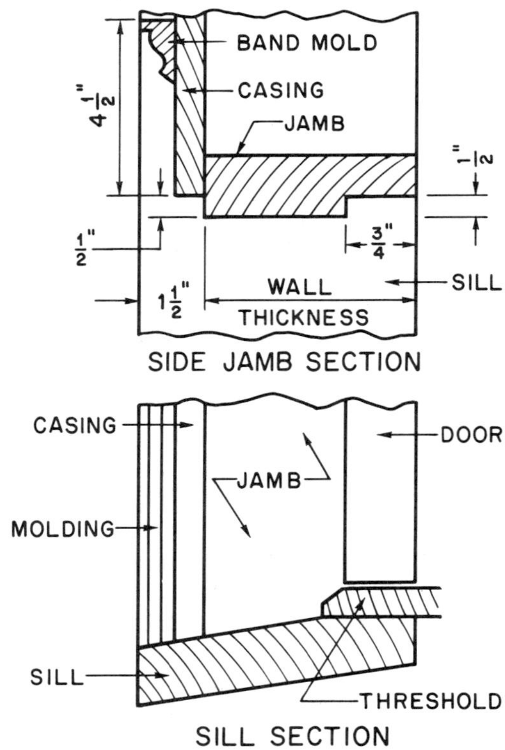

Fig. 38-2 Exterior door frame for a 3'-0" x 6'-8" x 1 3/4" door

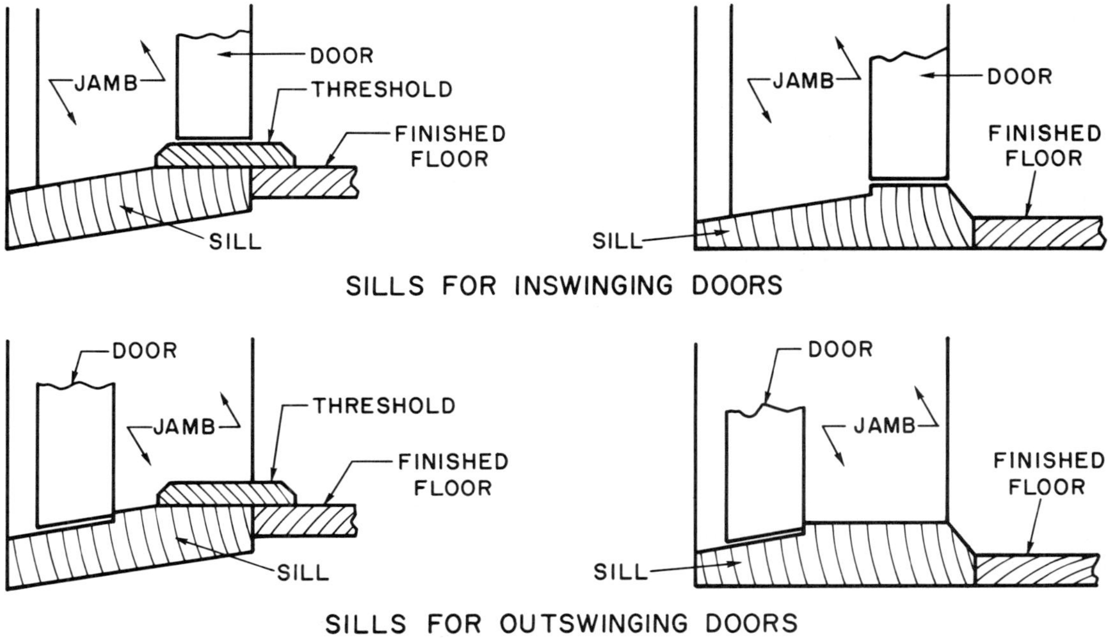

Fig. 38-3 Sill shape varies according to the swing of the door and custom of geographical locations.

Door Sizes

Since door sizes are standard, mills are able to mass-produce these sizes at a great savings to the buyer. Odd-sized doors must be custom-made at a considerably higher cost.

Door sizes are stated by width, height, and thickness in that order. The width and height is always listed in feet and inches and the thickness is always given in inches.

Standard thickness for exterior main doors is 1 3/4 inches. Interior doors are 1 3/8 inches thick. Storm and screen doors are usually made 1 1/8 inches thick.

Exterior doors are 3′-0″ wide and 6′-8″ high. Service or rear entrance doors may be narrower, usually 2′-6″ wide. Interior doors are 2′-6″ wide and 6′-8″ high. However, doors to bathrooms and closets may be 2′-4″ wide. Linen and broom closets may even be as narrow as 1′-4″ wide.

Stiles, top rails, and intermediate rails are usually 5 inches wide. Lock rails are 8 inches wide,

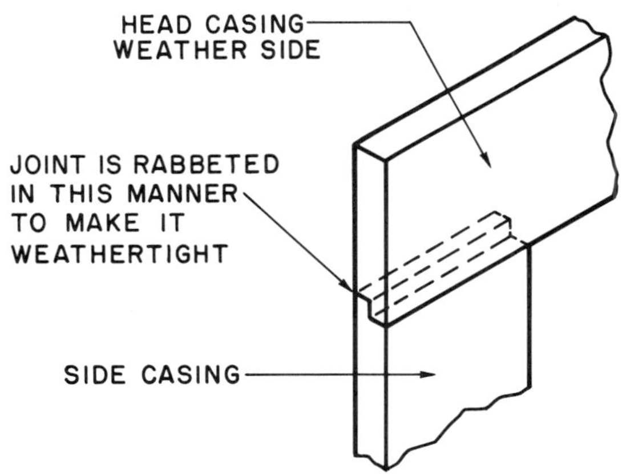

Fig. 38-4 A rabbeted joint is made between side and head casings of a door frame.

while bottom rails are usually 9 1/2 inches wide. Raised panels are made 3/4 inch thick. Solid flat panels range from 1/4 inch to 1/2 inch thick.

Rabbeted door jambs are usually 1 1/2 inches thick and 5 1/4 inches wide. Plain door jambs are

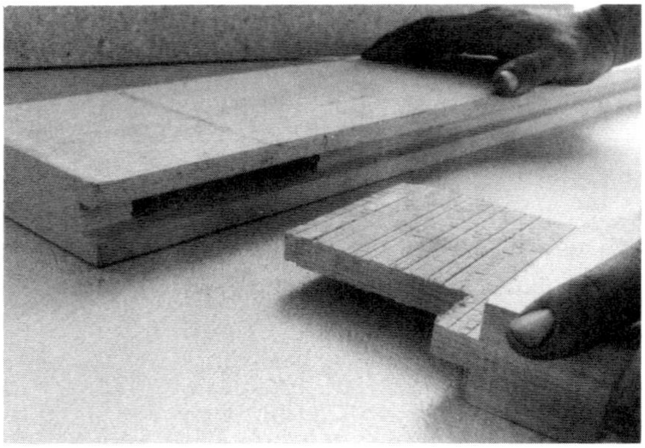

Fig. 38-5 The blind mortise-and-tenon joint is used in door construction.

usually made of 3/4 inch thick stock. The width of door jambs varies with standard wall thicknesses. Sills are usually made of 1 1/2 inch thick hardwood stock. Their width also varies according to wall thickness, but are made about 1 1/2 inches wider than the jambs.

Door Construction

The basic types of door construction use either the blind mortise-and-tenon joint or the doweled joint. Doweled joint construction used on doors is explained in *Unit 37 Screens.*

The blind mortise-and-tenon joint does not show when the joint is assembled, figure 38-5. Several factors determine the size and location of a mortise.

1. For a joint of maximum strength, a mortise is made about one-third the thickness of the stock if possible.
2. The mortise is made, when possible, to the same size as the groove in the stiles and rails that receives the panel. This makes it easier to make the joint. It also allows the haunch of the tenon to fit in the panel groove.
3. The inside edge of a mortise must line up with the bottom of the panel groove. The outside edge of the mortise is made not less than two mortise widths from the outside edge of the door.
4. The depth of the mortise is determined by the width of the stile. For the strongest job, mortises are made to a depth of about two-thirds the width of the stiles.
5. Mortises for wide lock or bottom rails are interrupted at intervals to prevent weakening of the stile by a long mortise. Tenons are made to fit these mortises by relishing at intervals.
6. Tenons are cut about 1/4 inch shorter than the depth of the mortises. This prevents the tenon from bottoming out before the shoulders of the tenon come up tight. If the tenon strikes the bottom of the mortise before the shoulders touch the stile, an open and weak joint is the result.
7. The haunch of a tenon is made slightly smaller than the depth of the panel groove to allow the joint to come up tight.

MAKING AN INTERIOR DOOR

A working drawing of an interior door is shown in figure 38-6. The door is to be constructed with blind mortise-and-tenon joints. The panels are 1/2-inch fir plywood set into 3/8-inch deep grooves in the stiles and rails.

Milling the Stock

The first step in making doors or any other items of millwork is to mill sufficient stock to uniform thickness, width, and length. A layout rod, as described in *Unit 20 Layout and Planning,* is a valuable aid to determine the actual sizes of the door parts.

1. Lay out a rod for the door to be constructed. Write a cutting list. Mark the face side of all parts as you mill.
2. Cut the stiles 6′-9″ long. This allows for a 1/2-inch horn on each end.

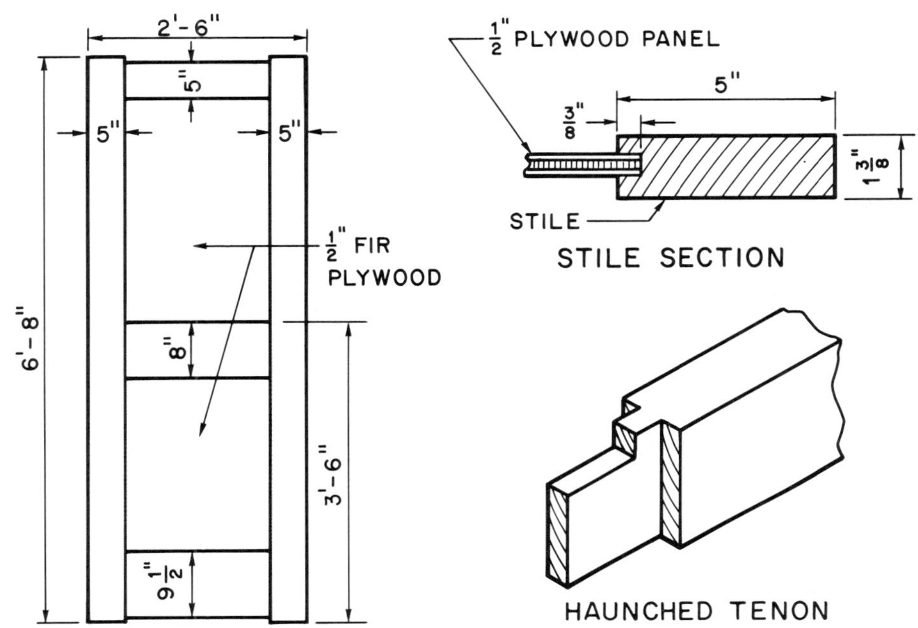

Fig. 38-6 Working drawing of an interior door

3. Cut each rail 26 inches long. This allows for a 3-inch tenon on each end.
4. Cut each panel 1/8 inch narrower than the required width of 21 inches. This will prevent the panel edges from bottoming out before the joints between the stiles and rails come tight.
5. Lock rails are centered about 38 inches from the bottom of the door. This is the same as the recommended height of the centerline of a lockset. Therefore, cut the top panel 1/8 inch shorter than 34 inches and the bottom panel 1/8 inch shorter than 25 1/2 inches.

Fig. 38-7 Mortising the stiles for an interior door with a hollow chisel mortiser

Machining the Parts

1. On the inside edge of the stiles, top rail, and bottom rail, cut a 3/8-inch groove to receive the panel. Cut the same size groove on both edges of the lock rail.
2. Lay out the mortises on the stiles for the top, lock, and bottom rails. Interrupt the mortises for the lock and bottom rails to prevent weakening the stiles. Mortise to a depth of 3 1/4 inches, figure 38-7.
3. Tenon the ends of the rails 1/2 inch thick by 3 inches long.

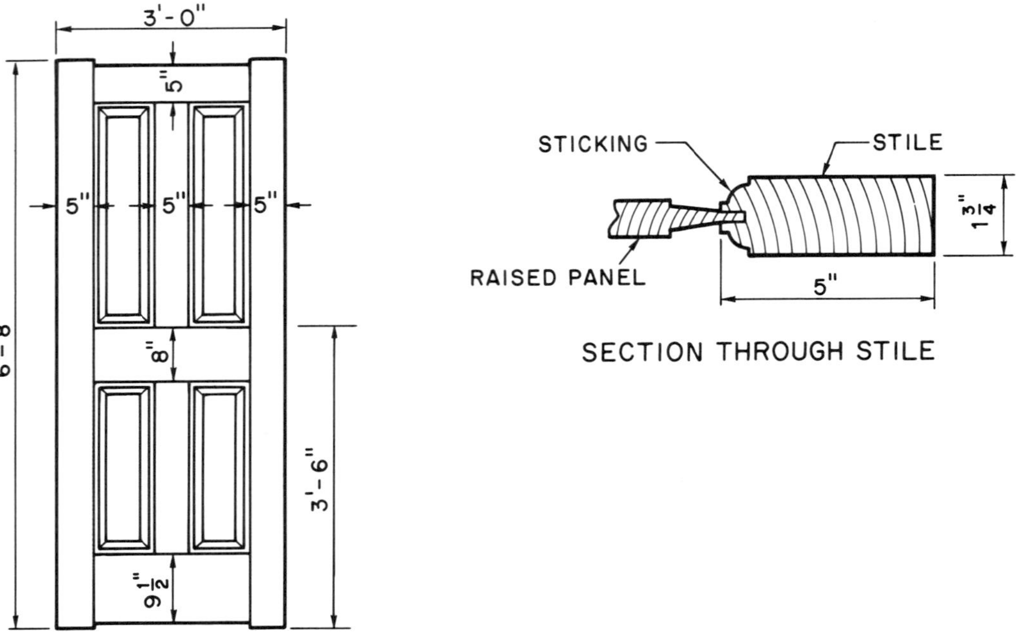

Fig. 38-8 3'-0" x 6'-8" x 1 3/4" Exterior door

4. Haunch the tenons on the top and bottom rails.
5. Relish the tenons on the lock rail and bottom rail.

Assembling the Door

1. Assemble the parts with the face side up.
2. Apply glue to the joints and clamp with bar clamps alternately across each rail.
3. After the glue has set, remove the clamps and sand the joints flush.

MAKING AN EXTERIOR DOOR

Figures 38-8 and 38-9 show working drawings for an exterior door and the door's construction. It is made with doweled joints, molded sticking, and raised panels. This door also has two mullions to reduce the width of the raised panels.

1. Lay out a rod for the door.
2. Glue up enough stock to make the raised panels (see *Unit 24 Cabinet Doors*).

Fig. 38-9 Construction of the exterior door in figure 38-8 *(C.F. Morgan Millwork)*

3. Cut all pieces to overall finished size. Allow enough for the horns of the stiles. Make sure the rails and mullions are long enough to be coped (fitted) to the molded sticking.

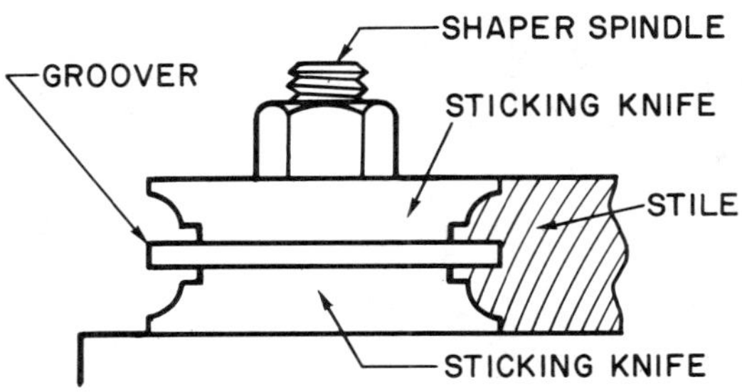

Fig. 38-10 Sticking a stile and cutting the panel groove

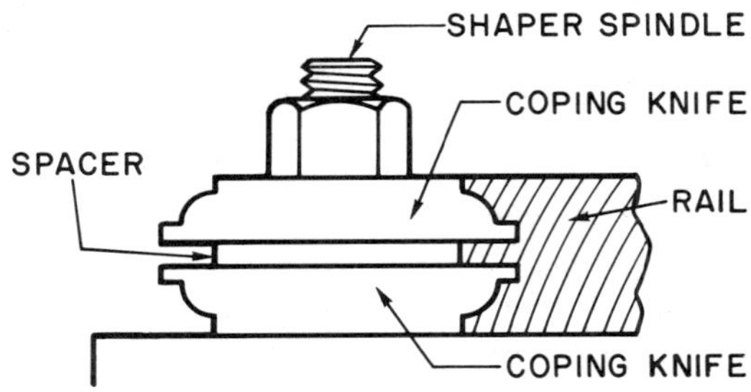

Fig. 38-11 Coping a rail. Cutting blades to make the cope must match the sticking on the stiles.

4. Cut the molded sticking and panel groove at the same time by running the stiles, rails, and mullions through a molder or shaper, figure 38-10. Set up the knives to cut a matching sticking on both sides of the pieces. Set a grooving cutter between the knives to cut the desired width and depth of the groove.
5. Cope the ends of the rails and mullions. If a shaper is used, hold narrow pieces securely square. The cutting blades used to make the copes must match the sticking on the stiles and rails, figure 38-12.
6. Make the raised panels (see *Unit 24 Cabinet Doors*). If a double-end tenoner is used, make frequent checks and adjustments, figure 38-13.
7. Determine the size and location for the dowels (see *Unit 37 Screens*). If a multiple boring machine is not available, bore the holes one at a time on a drill press. Save time by making jigs and setting stops when many pieces with the same dowel hole location must be bored.
8. Insert the dowels in the holes with glue.
9. Glue the joints and parts. Clamp together along the rails and mullions. Do not glue the panels.

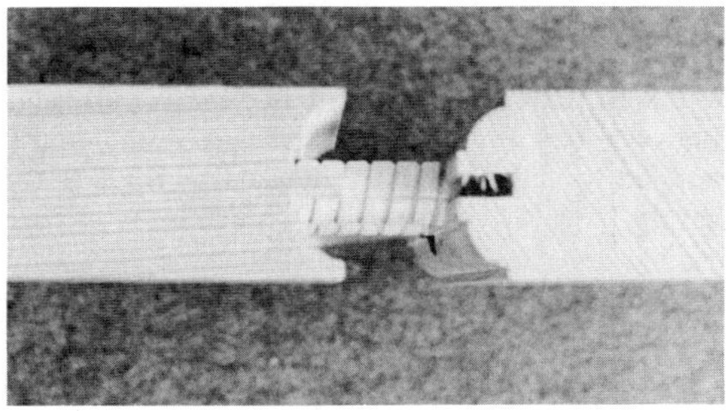

Fig. 38-12 Joint between rail and stile of an exterior door *(C.F. Morgan Millwork)*

Fig. 38-13 Raised panels are made efficiently with a double-end tenoner. *(Woodworking and Furniture Digest)*

MAKING AN EXTERIOR DOOR FRAME

The size of the door frame depends on the size of the door to be hung.

- Lay out the door frame 3/32 inch wider and higher than the door to be hung. This allows for a joint between the door and the frame.
- Allow for the thickness of the threshold of the door if necessary.

- Lay out the side jambs 6 inches longer than the door. This allows for horns that are later cut off when the frame is set in place.
- Lay out the head jamb 3/32 inch longer than the width of the door.
- Lay out the sill 3/32 inch longer than the width of the door plus twice the width of the outside casing.

An exterior door frame is shown in figure 38-2. A cutaway view of an exterior door and frame is shown in figure 38-14. The jambs are single rabbeted for a 1 3/4-inch thick door, figure 38-15. The sill has a pitch of 2 inches in 12 inches. The side jambs are dadoed to receive the head jamb and sill, figure 38-16. A rabbeted joint is provided between the side and head casings. A molding is applied to and flush with the outside edges of the casing.

Machining the Stock

1. Rabbet the jamb stock. Once set up, a molder can run large quantities in a short time. If a

Fig. 38-14 Cutaway of an exterior door and frame *(C.F. Morgan Millwork)*

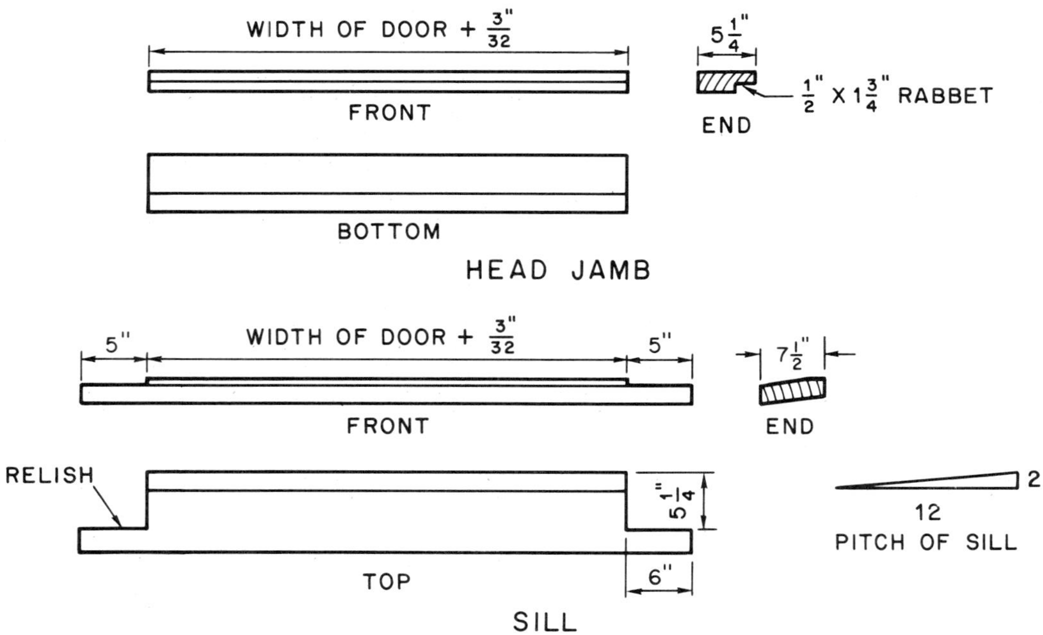

Fig. 38-15 Head jamb and sill construction

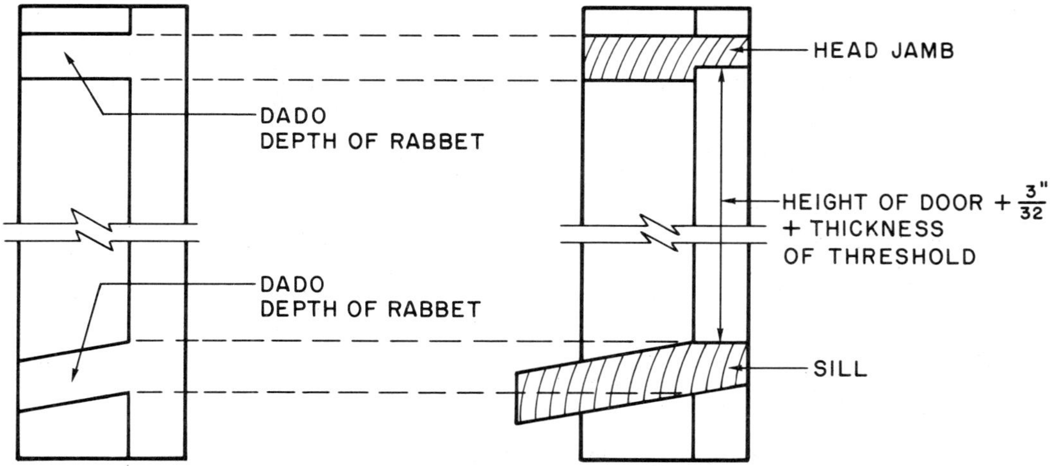

Fig. 38-16 Side jambs are dadoed to receive the head jamb and sill.

shaper is used, the stock must be hand-fed. Do not use a table saw to rabbet because it leaves a rough surface. Exposed rabbets must be cut smooth.

2. Bevel both edges and the top side of the sill stock. A molder may be used. If the beveled cuts are made on the table saw, smooth the cut edges on a jointer.
3. Cut the head jambs square on both ends.
4. Cut the sills to length. Relish each end so the extending portion fits under the bottom end of the side casings.
5. Space the side jamb dadoes to provide an opening which is 3/32 inch more than the height of the door. This allows for a joint between the sill and the door. Allow for the threshold.
6. Make the dado the same depth as the jamb rabbets. Cut them slightly shallow by machine to avoid marring the rabbet on the side jambs. Then cut the dadoes to the correct depth by hand. Do not score the exposed side of the side jamb rabbet.
7. Cut the head casing to length with both ends square. Make stop rabbets to receive the side casings, figure 38-17.

Fig. 38-17 Making stop rabbets on the head casing (guard removed for clarity)

8. Miter the top ends of the side molding. Bevel the bottom ends at the same angle as the bottom ends of the side casings. Miter the head molding to length.

Assembling the Frame

The frame is assembled by fastening the side jambs to the head jamb and sill. In case the dado is slightly wide, either or both the head jamb and sill is wedged toward the inside of the frame and fastened in place. The joint on the inside of the frame is exposed and must be tight. The joint on the outside of the frame is later covered.

1. Apply the side casings to the side jambs leaving a 1/2 inch reveal. A ***reveal*** is the amount of side jamb edge that is exposed. Apply the casings to the edge of the side jambs opposite the rabbet for in-swinging doors.
2. Fasten the head casing to the head jamb leaving the same reveal as on the side jambs. Make sure joints between side and head casings are tight.
3. Apply the side molding flush with the outside edges of the side casings. Fasten the head molding flush with the top edge of the head casing. Make sure the mitered joints are tight.

ACTIVITIES

1. Make a display of different styles and sizes of doors.
2. Compare the advantages and disadvantages of mortise-and-tenon joints and doweled joints in door construction. Consider the time involved to construct the door and the amount of material used.
3. Draw a cross section of sills showing their shape for inswinging and outswinging doors.
4. Design an interior door using molded sticking and mortise-and-tenon joints.
5. Design an exterior door using a molded sticking and doweled joint.
6. Make a door frame for the exterior door designed in #5. Hang the door in the frame.

UNIT REVIEW

Completion

1. The stiles of a door are the outside ________ members extending the full height of the door.
2. Horizontal members of a door extending the full length between stiles are called ________.
3. ________ are long vertical members of a door extending between rails and separating panels.
4. A part of a door filling the space between rails and stiles is called a ________.
5. A thick panel whose edges are tapered to fit in grooves made in stiles or rails is called a ________.
6. A pane of glass or an opening for a pane of glass is called a ________.
7. The ________ is the shape of the inside edges of rails and stiles.
8. ________ form the sides and top of a door frame.
9. The bottom member of a door frame is called a ________.
10. The top member of a door frame is called the ________.

Multiple Choice

1. The standard depth of a rabbet in a door frame is
   a. 1/4 inch.
   b. 3/8 inch.
   c. 1/2 inch.
   d. 5/8 inch.

2. The width of the rabbet in door frames is made
   a. 1 1/8 inches.
   b. 1 3/8 inches.
   c. 1 3/4 inches.
   d. for standard door thicknesses.

3. Exterior doors for residences usually
   a. swing out.
   b. swing in.
   c. slide.
   d. fold.

4. The top, horizontal casing of a door frame is called the
   a. side casing.
   b. head casing.
   c. head jamb.
   d. drip cap.

5. The bottom end of side casings in a door frame is usually cut on a bevel to fit against the
   a. jamb.
   b. sill.
   c. header.
   d. stop.

6. The joint between flat side and header casings is
   a. splined.
   b. rabbeted.
   c. butted.
   d. mitered.

7. To provide a weathertight joint between molded side and header casings, the joint is
   a. splined.
   b. rabbeted.
   c. butted.
   d. mitered.

8. Door sizes are stated in this order:
   a. height, thickness, and width.
   b. thickness, width, and height.
   c. width, thickness, and height.
   d. width, height, and thickness.

9. For a mortise-and-tenon joint of maximum strength, the mortise is made about
   a. 1/8 the thickness of the stock.
   b. 1/4 the thickness of the stock.
   c. 1/3 the thickness of the stock.
   d. 1/2 the thickness of the stock.

10. The outside edge of a mortise in a door stile is made from the outside edge of the door not less than
    a. 3/4 inch.
    b. one mortise width.
    c. two mortise widths.
    d. necessary to fit the door.

unit 39 Sash and Window Frames

OBJECTIVES

After studying this unit, the student will be able to:

- describe the parts of a sash and window frame.
- determine the overall size of a sash from its glass size and number of lights.
- cut glass to a specified size.
- make a 3-light basement sash and a window frame for it.

SASH

A window consists of a window frame and sash. The *sash* is that part of the window that holds the glass. It is set in a *window frame.* A window may contain more than one sash. The sash may be stationary, slide vertically or horizontally, or swing in or out on any edge.

The parts of a sash are similar to a door, figure 39-1. *Stiles* are the vertical outside pieces of a sash. *Rails* are the horizontal pieces of the framework. Rails are further identified according to their location and design. *Muntins* are the small vertical or horizontal strips that divide the glass in a sash. *Lights* are the openings in the sash that are filled with glass.

Sash Sizes

Sash are classified by the size and number of lights contained in the unit. Thus, a sash may be called a 10″ x 12″, 3-light sash. This means that the sash contains three lights with each light measuring 10″ x 12″ (see figure 39-1). A sash may be divided into a great number of lights if desired and if the sash is large enough.

When stating the size of the sash, always give the width of the light first, then the height, and then the number of lights. The same applies to windows even though they contain more than one sash.

Standard finished thicknesses of sash are 1 1/8 inches and 1 3/8 inches. In special cases, sash are made thicker. The width of stiles and top rails usually extends 2 inches outside of the glass. The bottom rail is made 2 3/4 inches outside of the glass. Muntins are usually 3/16 inch thick between lights of glass and are spaced about 1/32 inch more than the glass size. Therefore, the overall width of a 10″ x 12″, 3-light sash is 34 1/2 inches. Its height is 16 3/4 inches. Sizes of parts may vary and depends on geographical location and custom.

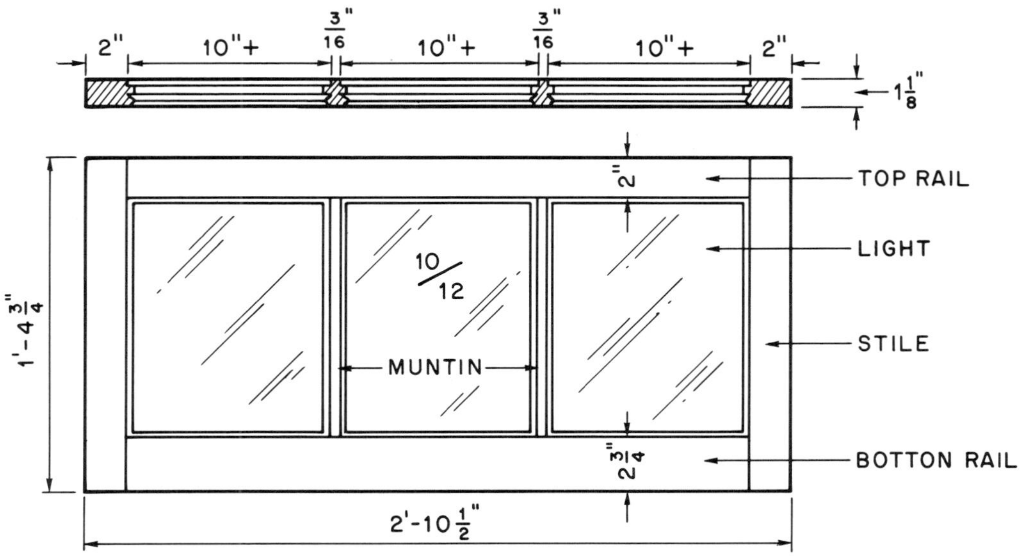

Fig. 39-1 A 10" x 12", 3-light basement sash and its parts

Making a Sash

The stiles and rails of the sash in figure 39-1 are to be joined with through mortise-and-tenon joints. The muntins are joined to the rails with blind mortise-and-tenon joints. The ends of the muntins are coped and tenoned on both ends. Figure 39-2 shows the shape of the sash parts.

1. Make a layout rod for the sash. Write a cutting list, figure 39-3.
2. Square up enough stock to overall length to make the stiles, rails, and muntins. If many sash are to be made, multiply the number of pieces required for one sash by the number of sash to be made.
3. Make a through mortise on the stiles of the sash, figure 39-4. To size the mortise, follow the rules for making mortises outlined in *Unit 38 Doors and Door Frames.*
4. Tenon the rails. Cope the shoulders of the tenon to fit against the stuck (shaped) edge of the stiles. The shape of the coping knives

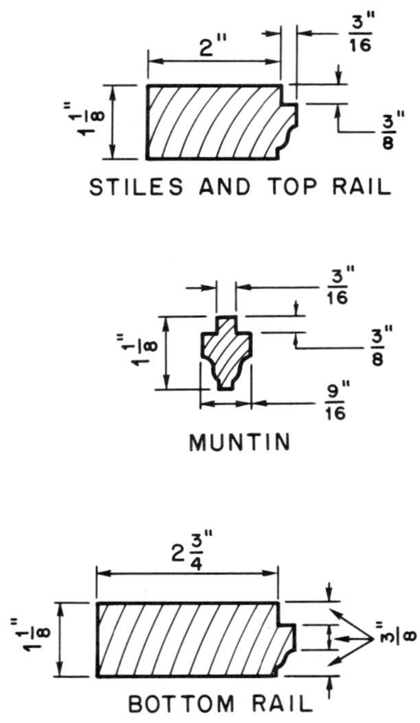

Fig. 39-2 Detail of the sash parts

NAME	NO. OF PIECES	THICKNESS	WIDTH	LENGTH
STILE	2	1 1/8"	2 1/4"	16 3/4"
TOP RAIL	1	1 1/8"	2 1/4"	34 1/2"
BOTTOM RAIL	1	1 1/8"	3"	34 1/2"
MUNTIN	2	1 1/8"	3/4"	13"

Fig. 39-3 Cutting list for the sash in figure 39-1

used must match the knives used to stick the edges of the stiles.

5. Using the same setup except for length, tenon the muntins. The length of the tenons on the muntins are cut shorter than on the rails because they fit into a blind mortise, figure 39-5.
6. Make 3/8-inch mortises on the top and bottom rails to receive the tenons of the muntins, figure 39-6.
7. Stick the inside edges of the rails and stiles to the desired shape, figure 39-7. Stick both edges of the muntins.
8. Loosely fit the parts together, clamp, and square. A sash clamp or clamping jig speeds the process.
9. Bring the joints tightly together. Drive steel dowels or headless nails into each corner. Countersink the steel dowels below the surface of the wood.
10. Remove the clamps. Sand the sash by running through the drum sander.

GLAZING GLASS

Outside sash are made so the glass is held in place with glazier points and glazing compound, figure 39-8. *Glazier points* are small, triangular or diamond-shaped pieces of metal. They are driven into the sash parts to hold the glass in place. The glass is installed with the convex or crowned side out. Glass set in this manner is not as apt to break when installed.

1. Brush off loose dust and dirt from the sash.
2. Apply a thin bed of glazing compound to the opening.

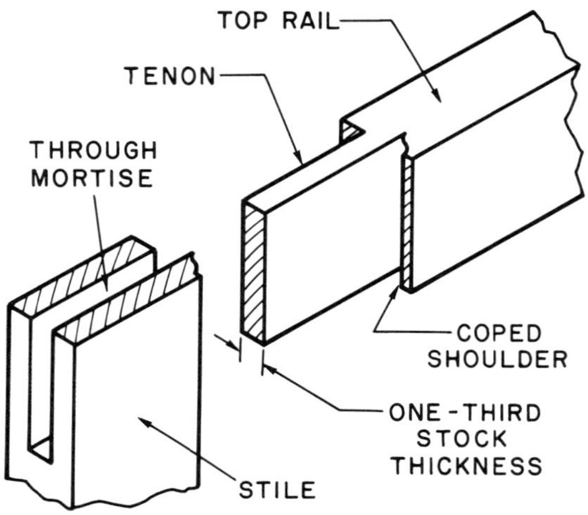

Fig. 39-4 Detail of the joint between the stile and rail

3. Lay the glass in the compound convex side out.
4. Fasten the glass in place with glazier points. Do not let the glazier point driver strike the glass or the glass may break. Special glazier point driving tools will prevent glass breakage.
5. If a special driving tool is not available, drive the points with a chisel or putty knife. Slide the tool across the glass while driving the glazier points.

WINDOW FRAMES

The parts of a window frame are similar to that of a door frame. The sides are called the *side jambs.* The top member is called a *head jamb.* The bottom member is called the *sill,* or sometimes called a *stool.* Usually *casings* are applied to the outside edges of jambs just as on door frames.

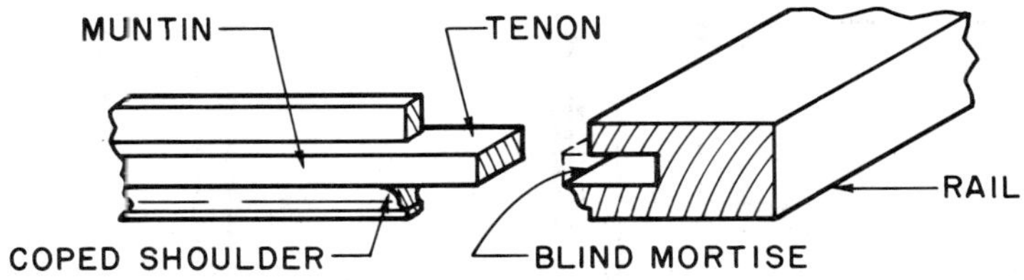

Fig. 39-5 Detail of the tenon made on the ends of the muntin

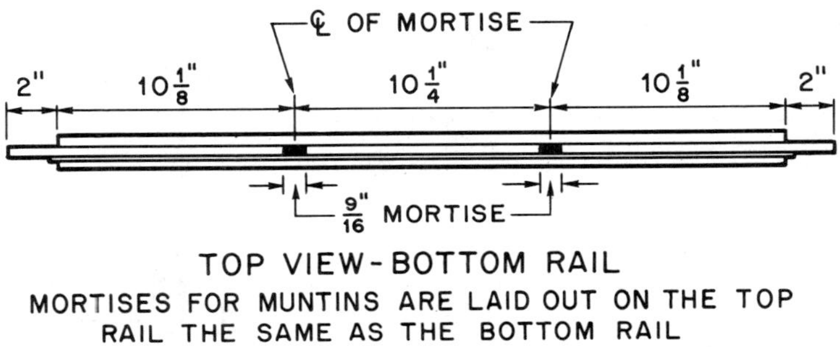

Fig. 39-6 Location of the mortises in the rails to receive the tenons of the muntins

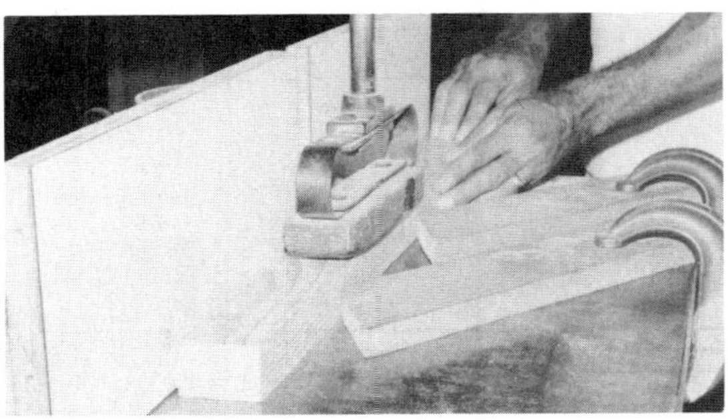

Fig. 39-7 The shaper may be used to stick sash parts.

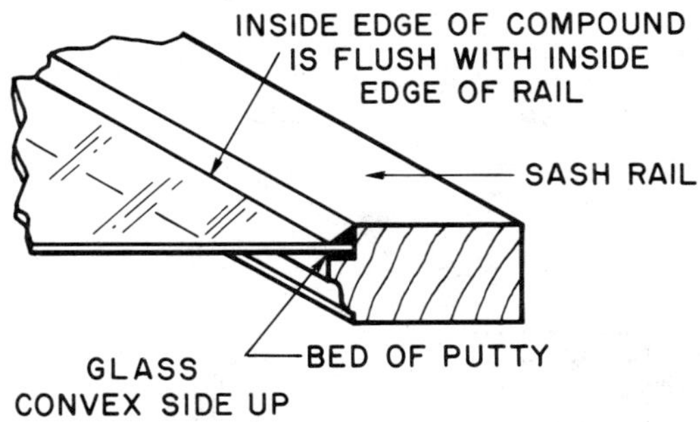

Fig. 39-8 Section of a glazed sash

Figure 39-9 shows the window frame needed for the 10" x 12", 3-light basement sash in figure 39-1. Figure 39-10 shows the layout and sizes for all the parts.

Cutting the Stock

Enough jamb stock to make the required number of frames is squared up to a size of 1 1/8" x 5 1/4". The 5 1/4-inch measurement may vary according to the wall thickness.

The head jambs are cut to a length of 2'-10 1/2". This is the same width as the sash. Both ends are cut square.

The side jambs are cut to a length of 22 inches. The horns left by this extra length enables a dado to be made instead of a rabbet. Dadoing

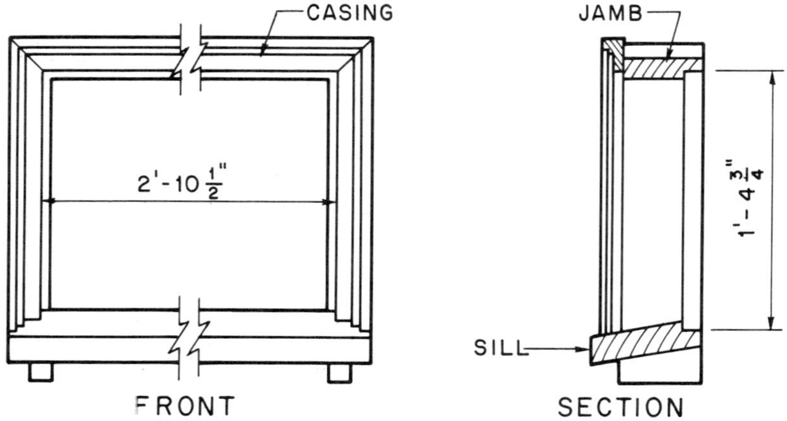

Fig. 39-9 Window frame for a 10" x 12", 3-light basement sash

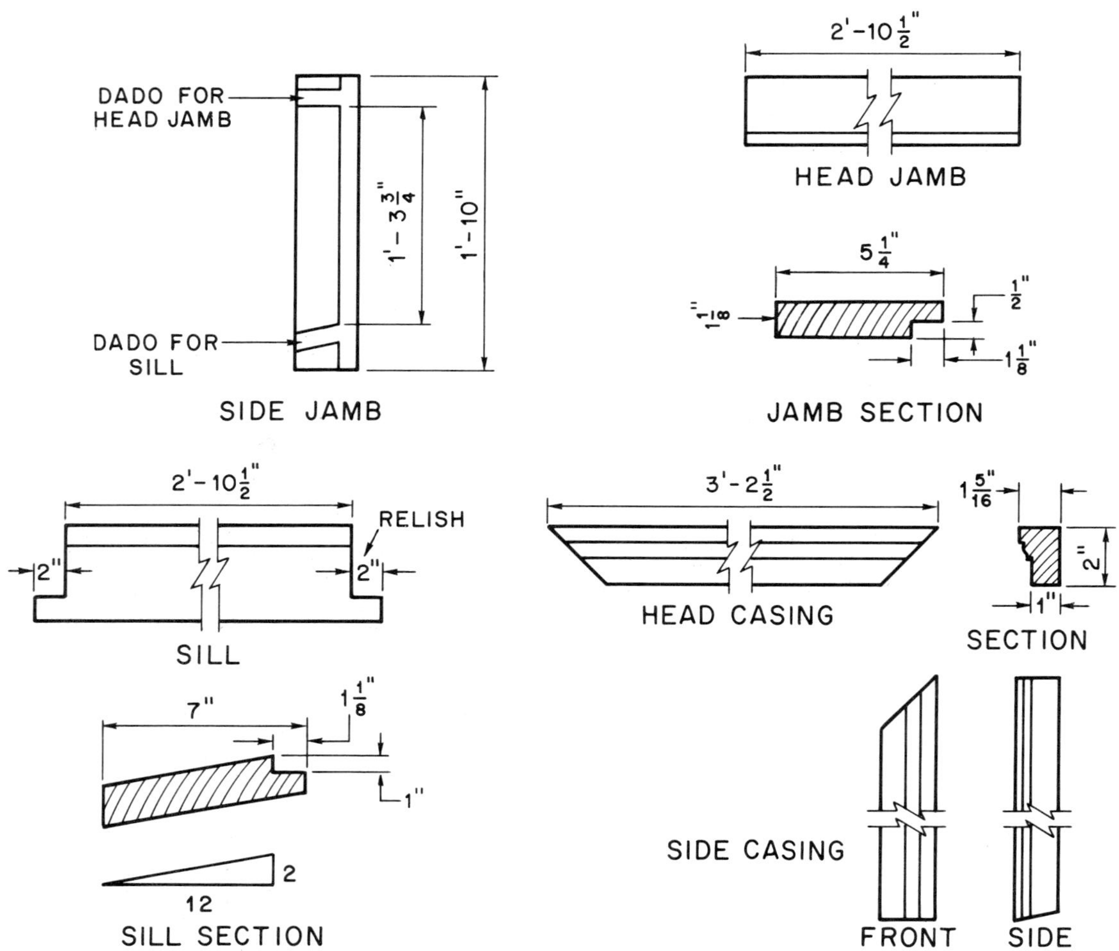

Fig. 39-10 Layout and sizes of the parts of the window frame

the side jambs makes assembly easier. The horns are usually cut off when the frame is installed in the building.

The sill stock is squared up to a size of 1 1/2" x 7 1/2".

Making the Frame

To make the window frame in figure 39-9:

1. Cut all stock to overall size.
2. Make a 1/2" x 1 1/8" rabbet on the jamb stock.
3. Dado the side jambs as shown in figure 39-10.
4. Bevel both edges of the sill and cut to width. The pitch of the sill is 2 inches in 12 inches.
5. Rabbet the sill the same size as the jambs.
6. Miter both ends of the head casings to an overall length of 38 1/2 inches.
7. Miter the side casings at a 45-degree angle on the top end. Bevel the bottom end to fit against the top of the sill. Remember to cut left and right casings and to reverse the directions of the mitering.
8. Fit the head jamb and sills into the dadoes of the side jambs. Wedge the head jamb and sill tightly against the inside shoulders of the dado if necessary.
9. Drive three fasteners of the proper length through the side jambs into the ends of the head jamb and sill. Keep the inside edges of the frame flush.
10. Apply the casings by driving fasteners into the edge of the jambs. Make sure the joint is tight between the side and head casings and the sill. Leave a 1/2 inch reveal on the jambs.

ACTIVITIES

1. Obtain a small basement sash. Disassemble it carefully, without breaking any parts. Observe the sticking, the joints made, and the fasteners used. Determine the pitch of the sill.
2. Make a list of rules for stating the sizes of sash and windows.
3. Make a display showing the many different kinds of windows and label them.
4. Determine the size of sash and windows in your classroom.
5. Make the sash for a double-hung window.
6. Cut glass to a specified size.
7. Make a window frame for the double-hung window sash constructed in #5.

UNIT REVIEW

Completion

1. A sash is the part of the window that holds the ________.
2. A sash is set in a ________.
3. The outside vertical members of a sash are called ________.
4. Small strips that divide the glass in a sash are called ________.
5. Sash are usually sized by the ________ and ________ of lights contained in the unit.
6. The stiles and rails of the sash are joined with ________ mortise-and-tenon joints.

7. The shoulders of the tenons on the rails are ________ so they fit snugly against the shaped edge of the stiles.

8. The cuts made in the rails of a sash to receive the muntins are called ________.

9. When glazing sash, the glass is laid in the opening with the ________ side out.

10. The glass is held in place in the sash with small triangular or diamond-shaped pieces of metal called ________.

Multiple Choice

1. Horizontal pieces of the sash frame are called
   a. rails. c. bars.
   b. stiles. d. mullions.

2. Openings in the sash that are filled with glass are called
   a. openings. c. lights.
   b. muntins. d. holes.

3. The width of a 10" x 12", 3-light sash is
   a. 30 inches. c. 34 1/2 inches.
   b. 32 inches. d. 36 1/2 inches.

4. The size of sash is stated in the following order
   a. height, width, number of lights. c. width, height, number of lights.
   b. number of lights, width, height. d. number of lights, height, width.

5. Standard finished thicknesses of sash are
   a. 1 1/8 and 1 3/8 inches. c. 1 1/2 and 1 3/4 inches.
   b. 1 1/4 and 1 1/2 inches. d. 1 3/4 and 2 inches.

6. Between lights of glass, muntins are usually made
   a. 3/16 inch thick. c. 3/8 inch thick.
   b. 1/4 inch thick. d. 1/2 inch thick.

7. The sides of a window frame are called
   a. side rails. c. side jambs.
   b. head jambs. d. muntins.

8. The bottom member of a window frame is called a
   a. jamb. c. header.
   b. casing. d. sill.

9. The cuts made in side jambs to receive the head jamb and sill are called
   a. rabbets. c. grooves.
   b. dadoes. d. miters.

10. The amount that the casings are set back from the inside edges of the jambs is called
   a. the reveal. c. the exposure.
   b. the setback. d. the offset.

unit 40 Stairways and Stair Parts

OBJECTIVES

After studying this unit, the student will be able to:

- locate and explain the purpose of the parts of a stairway.
- explain how to plan a stairway.
- lay out and make a housed stringer and open-mitered stringer.
- make a tread and riser.
- explain the manufacture of other stair parts.

STAIRWAYS

The art of laying out and building stairways and manufacturing stair parts is a specialized part of the woodworking trade. Millworkers who produce stair parts spend many years perfecting their skills.

Parts of a Stairway

Treads are the horizontal part of a stairway on which the feet are placed when climbing or descending the stairs, figure 40-1. *Risers* are vertical members that enclose and finish the space from tread to tread or from tread to floor.

A *nosing* is that part of a tread which extends beyond the riser. Its rounded edge is part of the tread. A *return nosing* is a separate piece mitered and fastened to the open end of a tread. Nosings finish off the edges and ends of treads.

Stringers are finish boards on each end of the risers and treads, figure 40-2. Stringers are called *closed* when they are placed against a wall into which the treads and risers are fitted. Stringers are *open* when they face the open end of a staircase. An open stringer is cut so the treads lay over and extend beyond it. It is also mitered where the risers join it.

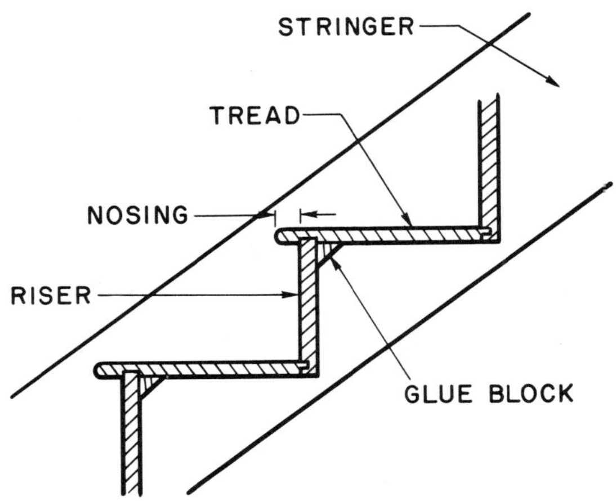

Fig. 40-1 Tread and riser construction

Housed stringers are dadoed and grooved out to receive the ends of treads and risers. Only closed stringers can be housed. Complete housed stringers are generally recommended and used in many areas of the country. They are self-supporting and are considered sturdier than plain stringers. However, plain stringers are often used in the construction of stairways in smaller homes. With the proper support, they provide a strong and safe stair.

A *starting step* is the first tread-and-riser unit in a stairway. Starting steps may be either rectangular or curved on one or both ends.

The *balustrade* is the total guard rail on the open side of the stairway. It includes all the rail details, including the handrail, balusters, and newel post, figure 40-3.

The *handrail* is the top finishing piece grasped by the hand when ascending or descending the stairs. Handrails come in various shapes. They are furnished by mills in straight lengths and in sections of special shapes. A *gooseneck* is generally used at the landing of a stairway. At the start of a stairway, both horizontal and vertical *volutes* and *turnouts* are used. Many other shapes are also available,

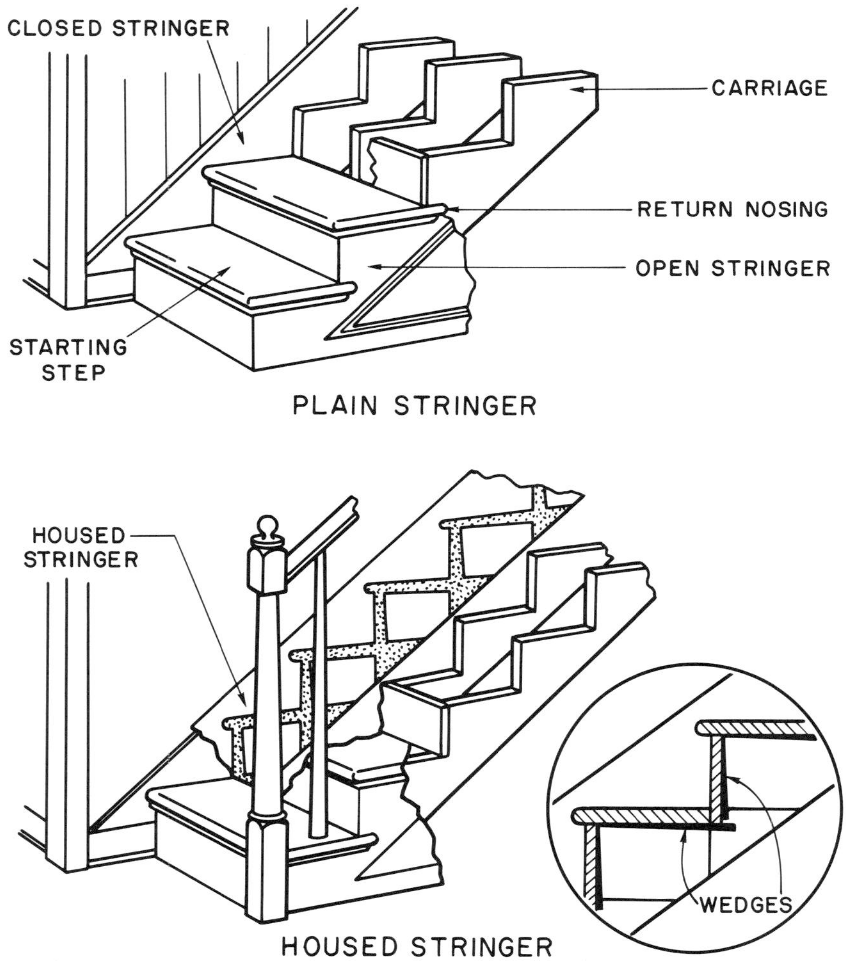

Fig. 40-2 Plain and housed stringers *(C.E. Morgan)*

such as *quarter* and *half-turns, concave,* and *convex easements,* figure 40-4.

Balusters are turned, often decorative pieces placed between the handrail and treads that close off the balustrade. The bottom end is turned smaller to fit into a hole bored in the treads. The top end is cut on a bevel to fit against the underside of the handrail. Balusters are furnished in lengths of 30, 33, 36, 39, and 42 inches so that they can be used in any part of the balustrade.

Newel posts are posts anchored to the staircase that support the ends of the handrail. *Starting*

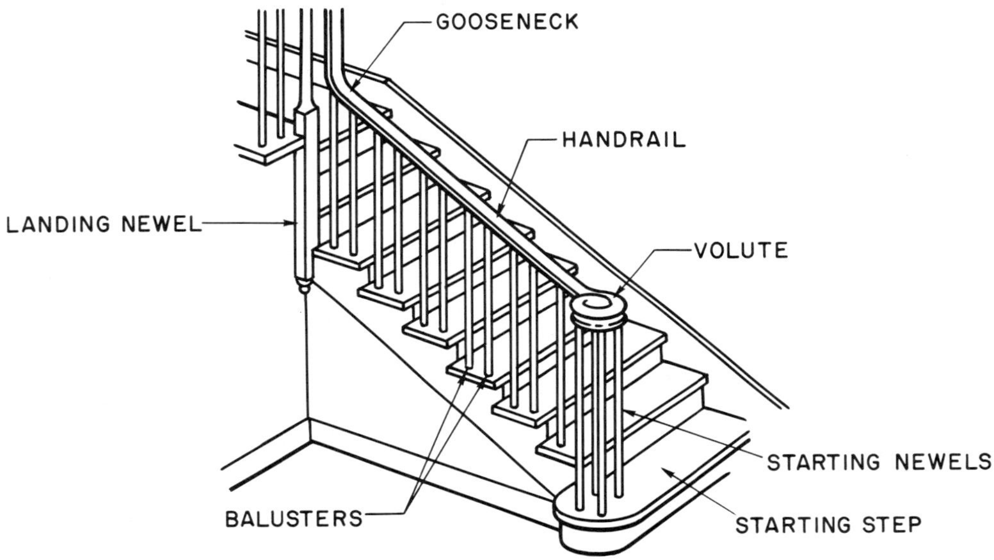

Fig. 40-3 Balustrade construction of an open staircase

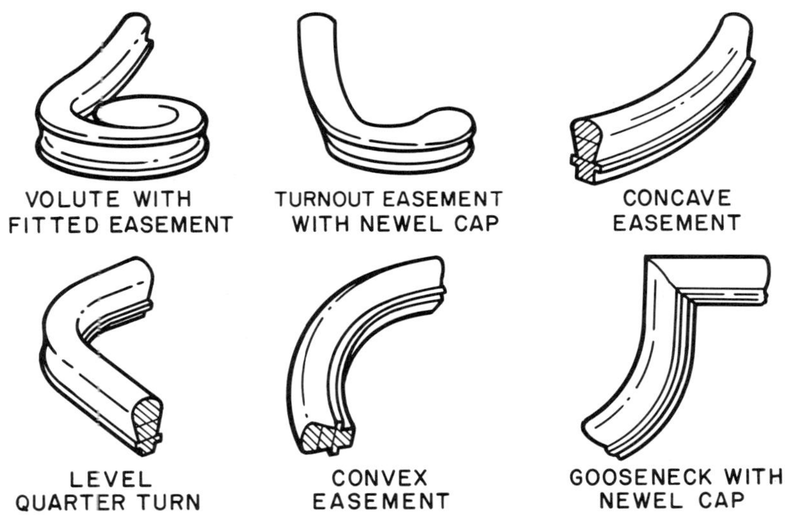

Fig. 40-4 Special handrail parts *(C.E. Morgan)*

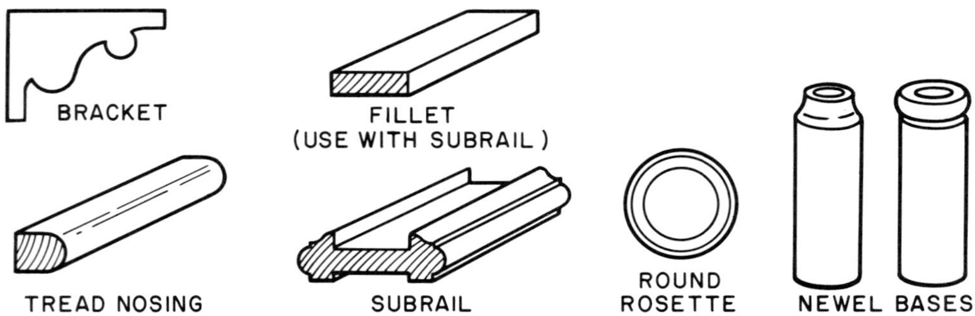

Fig. 40-5 Other stair parts *(C.E. Morgan)*

newels are used at the bottom of the staircase. *Landing newels* are used at landings or at the top of a staircase.

Other stair parts made in mills are *rosettes, tread nosings, brackets, subrails, fillets,* and *newel bases,* figure 40-5.

STAIRWAY PLANNING

Stair Widths

A staircase must be wide enough to allow two people to pass comfortably on the stairs. Furniture may also have to be carried up or down the stairs. Stairways are preferably made 3′-2″ to 3′-4″ wide. They should never be narrower than 3′-0″, figure 40-6.

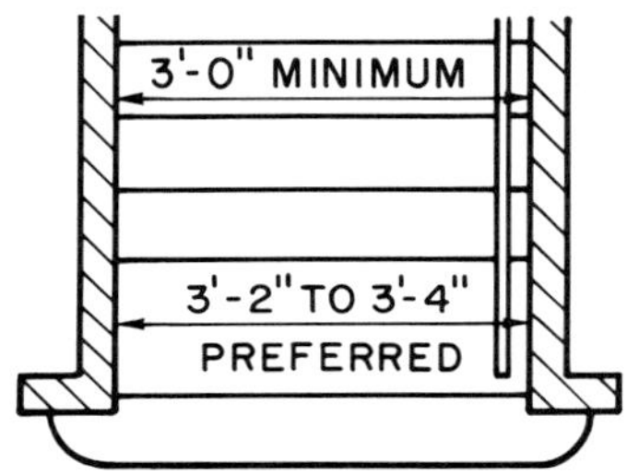

Fig. 40-6 Recommended stair widths *(C.E. Morgan)*

Rail Heights

Rail heights range from 30 to 36 inches along the stairs and from 34 to 42 inches at the landing. The higher rail assures a more secure landing rail installation. It is always desirable to use continuous handrails from floor to floor, figure 40-7.

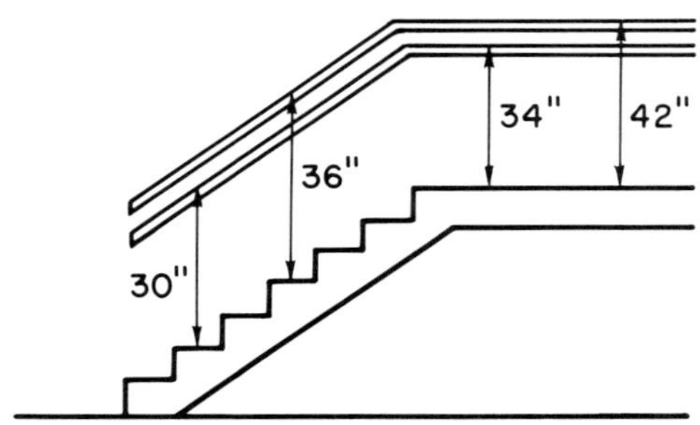

Fig. 40-7 Handrail heights *(C.E. Morgan)*

Rise and Run

The *total rise* of a stairway is the vertical height from floor to floor, figure 40-8. The *total run* of a stairway is the horizontal distance over which it travels. It is of vital importance that the stairway be built with the proper rise and run.

Stairways are built to provide maximum ease and safety in ascending and descending. If the stairway is too steep, it is difficult to climb and dangerous to descend. If the run is too long, much space and material is wasted.

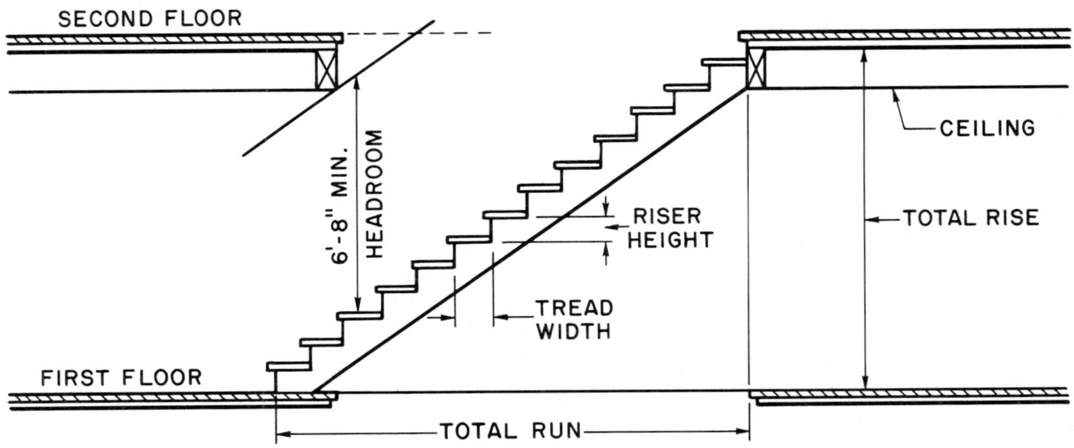

Fig. 40-8 Total rise and run

Determining Riser Height and Tread Width

To determine riser heights and tread widths, the total rise of the stairway must first be known. This height is divided by eight inches. The result is usually a whole number and a fraction. The next largest whole number is the total number of risers required.

For example, if the total rise is 9'-0" (108 inches), it is divided by 8 inches. The result is 13 1/2 rounded off to a total number of 14 risers.

Dividing the total rise by the number of risers gives the height of each riser. For example, dividing the total rise (108 inches) by the number of risers (14) gives a riser height of 7 5/8 inches.

Once the riser height is known, the tread width can be determined. The tread width is measured from the face of one riser to the next and does not include the nosing.

To find the tread width, apply the following rule:

- The sum of one riser and one tread should equal between 17 and 18.

For example, if the riser height is 7 5/8 inches, then the minimum tread width may be: 17 - 7 5/8" = 9 3/8". The maximum tread width may be: 18 - 7 5/8" = 10 3/8".

Figure 40-9 shows the recommended angle for stairs. Note that the number of treads in each flight of stairs is always one less than the number of risers. For a main staircase in a house, risers should not be higher than 7 5/8" and not less than 7 inches, combined with a tread width of 10 to 11 inches.

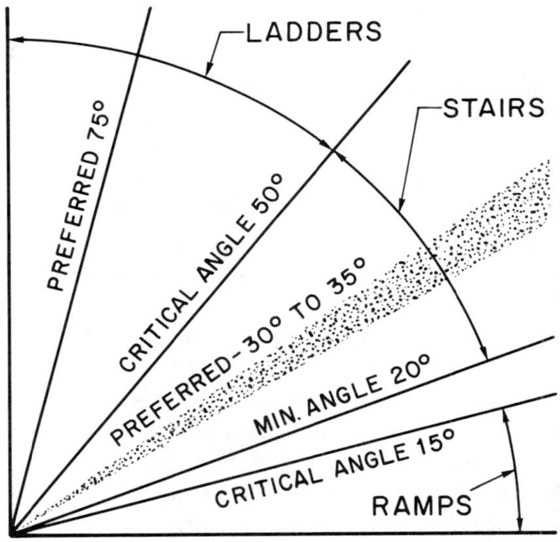

Fig. 40-9 Recommended angles for stairs

HOUSED STRINGERS

Staircases are either built on the job or made in a mill. A housed staircase differs from a job-built

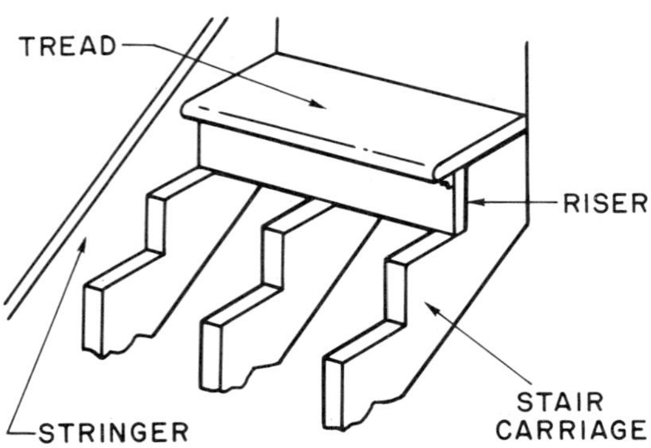

Fig. 40-10 Construction of a job-built staircase

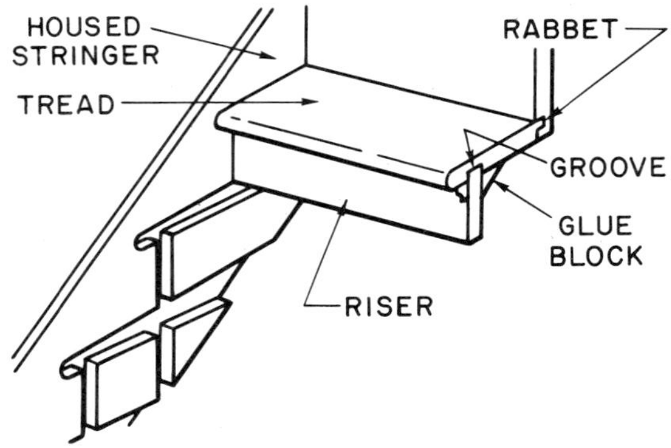

Fig. 40-11 Construction of a housed staircase

staircase mainly in the way the treads and risers are supported.

In a job-built staircase, the risers and treads are supported by cutout framing members, called *stair horses* or *carriages,* and are usually butted against a plain stringer, figure 40-10.

In a housed-stringer staircase, the risers and treads are supported by fitting into grooves routed into the stringer. The grooving of the stringer to "let-in" the risers and treads is called *housing the stringer.* This type of staircase is therefore called a housed-stringer staircase.

The back edge of the tread is rabbeted to fit into a groove made in the riser, figure 40-11. The underside of the tread is grooved to receive the top edge of the riser. Glue blocks are set in place on the underside of the stairway to reinforce the joint between riser and tread. At least three glue blocks are used for each step. Wedges are glued and driven in the housing on the back and underside of the riser and tread. These wedges drive the riser and tread against the shoulder of the housed stringer to make a tight joint.

Housed-stringer construction is considered to be best because it virtually eliminates squeaks. Also, this type of stair construction helps prevent joints from opening up due to shrinkage of the members and settling of the building. The housed staircase is a rigid, self-contained unit not affected by shrinkage or movement of the house frame.

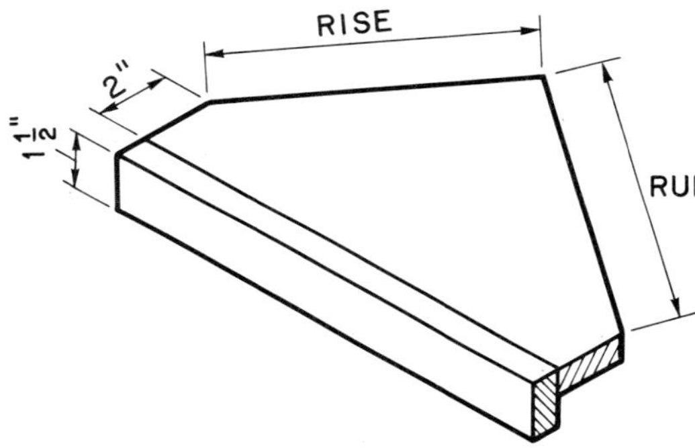

Fig. 40-12 Pitch board for a housed stringer

Pitch Board

The method of determining the riser height and tread width for a housed staircase is the same as for any staircase. After these have been determined, a pitch board is made.

A pitch board is often used instead of a framing square for laying out a stair stringer. A pitch board is a piece of stock, usually 3/4 inch thick, cut to the rise and run of the stairs. A strip of wood about 1 1/2 inches in width is fastened to the *rake* edge (long edge) of the pitch board. This is used to hold the pitch board against the edge while laying out the stringer, figure 40-12.

Making a Housed Stringer

On the face side of the stringer stock, draw a line parallel to and about 2 inches down from the

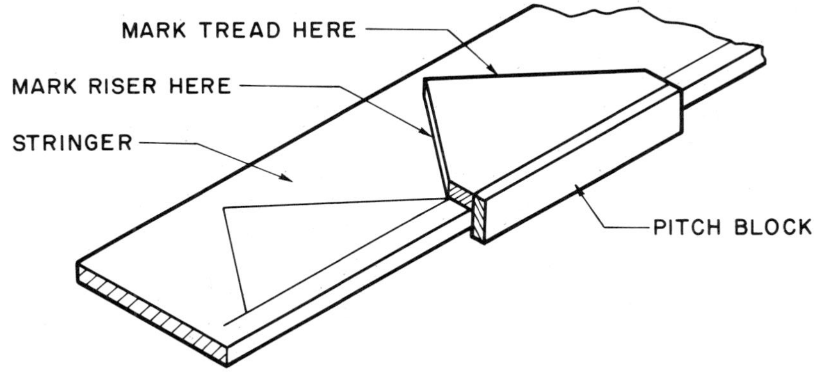

Fig. 40-13 Laying out a housed stringer with a pitch board

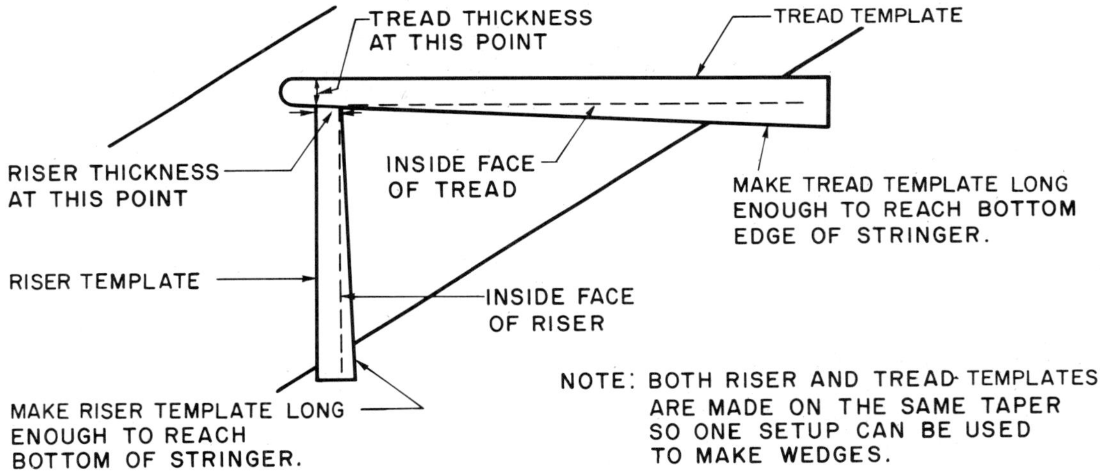

Fig. 40-14 Templates for laying out the inside edges of the stringer housing

top edge. This line is the intersection of the tread and riser faces. Using the pitch board, lay out the riser and treads for each step of the staircase. Use a sharp knife to scribe the lines. These lines show the location of the face side of each riser and tread and are the outside edges of the housing, figure 40-13.

To mark the inside edges of the housing, wedge-shaped templates may be used. The templates are tapered so the wedges will drive the risers and treads tight against the top edge of the housing. One template is made for the tread housing and one for the riser housing. Both templates are tapered the same to permit the use of stair wedges of the same pitch, figure 40-14.

To mark the inside edges of the housing, the edges of the templates are held to the lines previously scribed. The bottom edge of the template is marked outlining the cut for the bottom edge of the housing. The housing is now ready to cut.

Finished wall stringers may need extra length to provide sufficient stock for fitting against the baseboard at the top and bottom of the stairway.

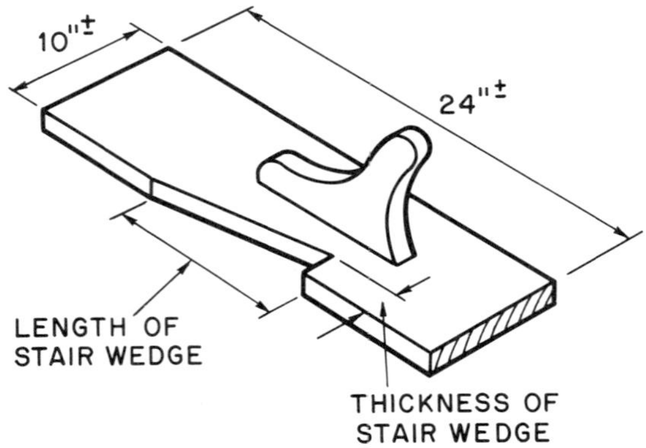

Fig. 40-15 Jig for making stair wedges

Fig. 40-16 Making stair wedges using a table saw and a wedge jig (guard removed for clarity)

A method of cutting a housed stringer is to use a routing template and router after the housing is laid out. Routing templates are adjustable for different rises and runs. Templates may also be made by cutting out thin plywood to the shape of the housing.

Making Stair Wedges

Stair wedges are made of dry hardwood, such as oak, ash, or maple. Wedge stock must be dry to prevent shrinkage. Hardwood is less likely to compress when the wedges are driven and after the stairs are installed and used.

To make stair wedges, first make a taper-ripping jig to be used on the table saw:

1. Cut a piece of stock the same thickness as the wedge about 10 inches wide and 24 inches long, figure 40-15.
2. In about the center of one edge, mark and cut out the shape of the wedge.
3. Fasten a thin strip of wood over the cutout. This strip will prevent the wedges from flying out and back toward the operator when being cut, and possibly causing serious injury.
4. Make and fasten a handle to the jig for pushing it safely through the table saw.

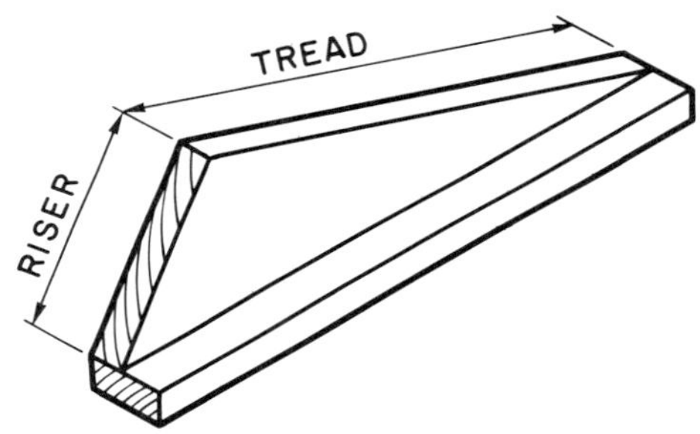

Fig. 40-17 Pitch block for an open mitered stringer

To make stair wedges:

1. Cut blocks of wood from hardwood the same length as the desired length of the wedges.
2. Set the fence of the table saw to the width of the jig.
3. Hold the block of hardwood into the cutout of the jig and run both pieces through the saw, figure 40-16.
4. Turn the block over. Place it back into the jig cutout and make another cut.
5. Continue making cuts, turning the hardwood block over with each cut. Use as many blocks and make as many cuts as necessary to make the number of wedges needed.

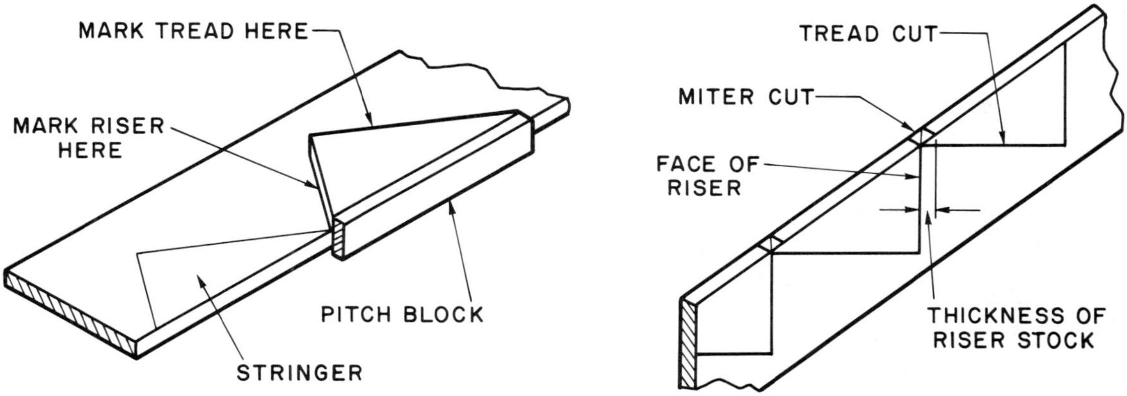

Fig. 40-18 Layout of an open-mitered stringer

OPEN-MITERED STRINGERS

The layout of an open stringer is similar to a housed stringer. The pitch block is held against the top edge of the stringer, however, instead of a line in from the edge. The risers and treads are marked lightly, figure 40-17.

To maɪk the mitered cut for the risers, figure 40-18, measure in from the riser face a distance equal to the thickness of the riser. Square both of these lines across the top edge of the stringer. Draw a diagonal line to mark the miter angle.

To mark the tread cut on the stringer, measure down from the mark representing the top of the tread a distance equal to the tread. Draw a line parallel to the line of the tread face. This is the cut on the stringer for the tread.

MACHINING A TREAD AND RISER

Treads are made of hardwood, usually oak or maple, to withstand the wear to which they are subjected. Risers receive relatively little wear and are usually made of less expensive softwood such as white pine.

Treads

Treads are made by gluing and clamping narrow strips of vertical grain hardwood. These narrow strips, usually 2 to 3 inches wide, provide a tread that is relatively warp-free. Rippings that otherwise might end up as waste may be used to glue together narrow strips. Using a vertical grain produces a tread that is less likely to cup and is able to withstand the wear to which it is subjected.

To make treads:

1. Glue and clamp strips of hardwood together using either butt or shaped edges.
2. Plane the glued-up treads to thickness and sand.
3. Joint one edge and cut the treads to width.
4. Rabbet the back edge of the tread, usually 3/8″ x 3/8″, to fit into the groove of the riser (see figure 40-1).
5. Nose the tread by shaping a half-round on its front edge, figure 40-19.
6. Cut the treads to standard lengths.

Risers

Risers are usually made from solid stock. To make risers:

1. Plane the stock to thickness and sand.
2. Joint one edge and rip the stock to width.
3. Make a 3/8″ x 3/8″ groove to receive the rabbeted back edge of the tread. Locate the top of the groove the thickness of the tread stock from the bottom edge of the riser.
4. Cut the risers to standard lengths.

OTHER STAIR PARTS

Handrails

Handrails are usually made of birch or maple. Straight sections of handrails are shaped by using a molder. Curved sections of handrails, such as volutes, gooosenecks, turnouts, and turns, are cut by machines to a rough shape. These roughed-out pieces are laboriously shaped by hand by skilled workers with many years of experience.

Fig. 40-19 Nosing the edge of a tread

Balusters and *newel posts* are usually turned. Automatic turning lathes are used to produce these stair parts in quantities. To duplicate or make a single piece, a wood lathe may be used. A worker must have a lot of experience using turning chisels to cut a turning with accuracy and good workmanship.

Other stair parts are produced on machines specially designed to make these parts in large quantities. Experience in the setup and operation of these machines is required to produce the desired parts.

ACTIVITIES

1. Make a scale model of a staircase. Label all the parts.
2. Calculate the riser height and tread width for a staircase with a total rise of 8′-9″. How many risers and treads will it have?
3. Lay out a housed stringer for the staircase in #2. Use a pitch board to lay it out. Make templates to mark the inside edges of the tread and riser housings.
4. Make stair wedges out of dry hardwood to reinforce the back and underside of risers and treads.
5. Lay out an open stringer for the staircase in #2.
6. Make a standard minimum size tread and riser for the staircase in #2.

UNIT REVIEW

Multiple Choice

1. A housed stringer differs from a job-built staircase mainly
   a. in the tools used.
   b. in the way treads and risers are supported.
   c. in the layout.
   d. in the design of the stair parts.
2. In quality stair construction, back edges of treads are
   a. splined into the risers.
   b. butted into the risers.
   c. rabbeted into the risers.
   d. glued against the risers.
3. The minimum number of glue blocks used in housed-stringer construction is
   a. one.
   b. two.
   c. three.
   d. four.

4. Risers and treads are driven tight against the shoulders of a housed stringer with
   a. wedges. c. screws.
   b. nails. d. a hammer.

5. A common method used to lay out risers and treads for a flight of stairs on a stringer is by using a
   a. pitch board. c. stair routing template.
   b. framing square. d. sliding T bevel.

6. Housed stringers are dadoed and grooved by using a
   a. dado head. c. routing template.
   b. shaper. d. mallet and chisel.

7. Stair wedges are made of
   a. hardwood. c. scrap wood.
   b. softwood. d. heartwood.

8. The molded front edge of a tread is called a
   a. balustrade. c. overhang.
   b. nosing. d. ovolo.

9. It is recommended that the rise of a flight of stairs from tread to tread is never less than
   a. 7 inches. c. 9 inches.
   b. 8 inches. d. 10 inches.

10. The sum of one riser and one tread should be between
   a. 7 and 8. c. 14 and 15.
   b. 10 and 11. d. 17 and 18.

Identification

Name each part.

A. ______________

B. ______________

C. ______________

D. ______________

E. ______________

F. ______________

G. ______________

H. ______________

I. ______________

J. ______________

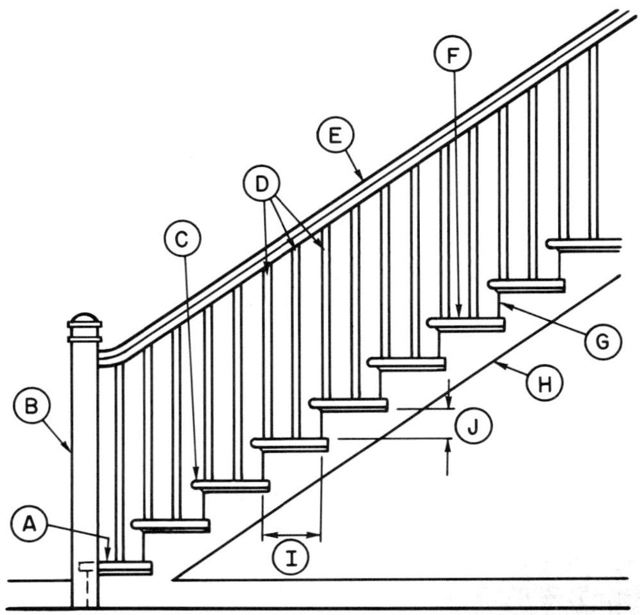

GLOSSARY

abrasive — a substance used for wearing away and polishing a surface by friction.
air-drying — method of drying lumber by exposing it to the air without artificial heat.
aluminum oxide — a type of abrasive used in sanding wood.
annular ring — the arrangement of the wood of a tree in concentric rings. There is one ring for each year of the tree growth.
apron — molding below the window sill or stool cap that finishes off the space between the bottom of the window and the wall.
arc — a part of the circumference of a circle.
astragal — a molding used to cover and make weathertight the joint between double doors.
backband — the outer molding applied to the casing of a window or door.
backset — the amount of wood left on between the gain of a hinge and the side of a door.
backsplash — a vertical board on the wall side of a countertop installed to protect the wall from abuse.
baluster — one of a series of small, decorative, vertical members of stair railing that supports the hand rail.
band — usually a molding that is applied around a flat casing.
barbed dowel pin — a short headless nail with barbs on the shank used largely for fastening the mortise-and-tenon joints in a window sash.
bed puttying — working putty into the rabbet of a sash before the glass is inserted.
bevel — an angle cut on the edge or end of a piece of stock that goes all the way through its thickness.
blind — a joint that does not go all the way through. The details of a blind joint are not seen when the pieces are jointed.
blind mortise — a mortise that does not go all the way through the stock.
board foot — a piece of stock 1 inch by 1 foot by 1 foot, or its equivalent.
bond — the permanent joining of stock.
bottom board — a board on which the flask is laid when making a sand mold.
bow — a defect in lumber consisting of a curve in the stock flatwise from end to end.
brad — a small finishing nail.
burnishing — turning the edge of a scraper blade.
cambium layer — layer of wood cells between the solid wood and bark of a tree. New growth takes place in this layer.
casework — rectangular, box-like pieces of cabinetwork that may contain doors and drawers, such as bookcases, desks, chests, dressers, etc.

casing — the decorative and finish trim around a door or window.
casting — an object made by pouring molten material into a form and allowed to cool.
chamfer — a bevel on the edge of stock that does not go all the way through the stock thickness.
check — a split across the annular rings in the end of wood stock. It is caused by more rapid drying of the end than the rest of the stock.
chord — a straight line joining the end points of an arc.
chuck — that part of a drill or bit brace that holds the drilling or boring tool.
close-grained wood — wood that has small and closely spaced pores that need not be filled to obtain a fine finish.
closed stringer — finished board usually fastened to the wall against which risers and treads butt.
collar — metal rings of different thicknesses and diameters used on the spindle of a shaper to hold cutting knives and/or to guide curved work.
concave — a hollowed out surface; a bowl-shape depression, like the inner surface of a circle or sphere.
convex — an outward curve such as the outer surface of a sphere.
cope — a joint where the end of one piece is cut and fitted to the irregular or molded surface of another.
core — the central part of a panel to which veneers are applied. Also a form placed in a sand mold to provide holes and interior passages in castings.
core box — a box made to form cores that are used when forming a mold for a casting.
core print — extensions on a pattern made to form impressions in the mold to support a core.
cove — a concave-shaped molding.
crook — a warp in wood in which the edge is either convex or concave from end to end.
cross-grain — a section of wood in which the grain runs more or less through its thickness or width instead of its length.
cup — a warp in lumber in which the surface is concave or convex from edge to edge.
cutting lips — in an auger bit, those edges that cut and lift the wood from the circle cut by the spurs.
dado — a wide cut or groove across the grain that does not go all the way through the piece.
dado head — cutting tool usually consisting of two circular blades and chippers of varying thicknesses used to cut dadoes and grooves.
diagonal — a straight line connecting the opposite corners of a rectangle or other polygon.
diameter — a straight line passing through the center of a circle and whose end points touch the circumference.
dovetail — an interlocking joint in which the tongue of one piece is wider or thicker on the end and fits into a mortise shaped to receive it.
dowel — a round pin of wood or metal used to strengthen joints.
draft — a taper given to the vertical surfaces of a pattern to facilitate drawing the pattern from the mold.
drag — the bottom part of a flask.
elliptical — oval or oblong in shape.
end grain — the grain of wood as seen from the end of the stock.
face — the finished or best appearing side of a piece of stock.
feed screw — the threaded end of an auger bit.
fence — a guide attached to wood-cutting tools and machines for guiding the work.
fiber saturation point — a point reached in drying lumber when the stock starts to shrink.
filler — a composition used for filling the pores of open-grained woods before finishing.

fillet — in patternmaking, material used for filling, rounding over, and strengthening inside corners of patterns.
fillet iron — a tool with metal balls of different diameters on each end used to apply fillets.
finish — boards, moldings, casings, and other finished pieces when applied to buildings are exposed and not covered.
finish — the final coats of protective material given to cabinets and other woodwork.
flask — a box-like form, made in two or more parts, used to hold sand in forming molds for castings.
flush — surfaces on the same level; pieces even with each other.
flutes — concave channels or grooves, usually in series, used to decorate columns or posts.
founding — the art of casting.
foundry — name given to the place where molds are formed and castings made.
gain — a shallow recess or mortise made to receive the leaf of a hinge or other thin material.
galvanized — a thin coating of zinc given to metal, usually iron or steel, to prevent rusting.
garnet — a reddish natural mineral used as an abrasive to sand and polish wood.
gate — a pathway in a sand mold by which the molten material reaches the molded sand made by the pattern.
gate cutter — a tool used by foundry workers used to cut the gate in a sand mold.
glazing — installing glass in windows and doors, including puttying.
glue block — a triangular block of wood usually glued in interior corners to strengthen joints.
gooseneck — curved part of a handrail in a staircase resembling a goose's neck.
green lumber — unseasoned lumber or lumber whose moisture content is too great to be used with satisfactory results.
green sand — a type of sand that contains enough clay content to retain its shape. It is used in sand molds for castings.
grit — the size of abrasive particles determining the coarseness of sandpaper.
groove — a wide cut made with the grain of lumber and partially through its thickness.
gullet — the valley between teeth in a saw blade.
gumming — the process of grinding and cleaning out the gullets of a circular saw blade.
half-lap — joining of two pieces by cutting out half the thickness of each piece so that the pieces fit together with flush surfaces.
handrail — the rail along the side of a stairway.
haunch — part of the tenon left on a rail of a paneled door to fit into the grooves of the stiles.
heartwood — wood near the center of the tree consisting of cells which have become inactive.
hexagonal — six-sided.
hone — polish a cutting tool to a keen edge by rubbing on a fine stone or leather strap.
horns — projections of the stiles of a door or frame beyond the top and bottom. It is left on to protect the door or frame during storage and transportation.
housed stringer — a wide finished board in a stairway that is routed to receive the ends of treads and risers.
indexing head — a round disc with spaced holes used for holding a round piece in position to make equally spaced cuts over its circumference.
interrupted mortise — a mortise made at intervals in the stile of a door to receive a wide rail and to preserve the strength of the stile.
jamb — the sides and top of a window or door frame.
jig — a device made to hold material and guide work.
jointing — the process of straightening an edge of stock.
jointing — the process of bringing teeth of a saw blade to an even height.

kerf – a saw cut.

kicker – a strip of wood placed in a drawer opening above the drawer to prevent the drawer from tilting down when opened.

kiln drying – method of drying lumber by placing it in ovens, called kilns, until it reaches a satisfactory moisture content.

knocked-down millwork – construction material delivered to the job site unassembled, but cut and ready for assembling.

laminating – the process of building up a piece with thin layers of wood.

layout rod – a strip of wood upon which lines showing the actual size and location of pieces of a cabinet are laid out.

light – a pane of glass in a door or window. Also, an opening for a pane of glass.

lignin – a mixture of substances which, when combined with cellulose, forms the essential part of wood tissue.

lipped – in reference to drawers and doors, the rabbeting of the four edges to fit inside and cover the drawer or door opening.

lock rail – a rail in a door at such a height to receive the door lock.

mantel – the ornamental facing usually of wood, around a fireplace, usually including a shelf.

match plate – a plate of wood or metal upon which multiple, small, fragile patterns are mounted to facilitate the molding process in a foundry.

medullary ray – band of wood cells extending from the pith of a tree toward the bark.

miter – an angle cut made across the width or through the thickness of lumber, usually made to fit with a corresponding piece.

miter gauge – a guide used on the table saw for holding pieces to make square or mitered cuts.

moisture content – the amount of moisture in wood; usually expressed as a percentage of the dry weight of wood.

mold – in foundry work, an impression made in sand by a pattern. Molten material is poured into it to make a casting.

mortise – a rectangular cavity cut in wood usually to receive the tenon projecting from another piece.

muntin – thin strips or bars of wood separating lights in a window or door.

newel post – an upright post supporting the handrail in a staircase. Newel posts are classified according to their location in a staircase, such as starting newel or landing newel.

nosing – the rounded edge of a stair tread projecting over the riser.

octagonal – eight-sided.

ogee – a molding with an S-shaped curve.

open-grained wood – wood having large pores which need to be filled before finishing.

open stringer – finished board on the open side of a staircase, usually mitered against the risers, where the treads overhang.

orange peel – a term given to a pebble-like effect in sprayed coats of lacquer or other finishes that resembles the peel of an orange.

overlay – a term given to doors or drawer fronts that overlap the door or drawer opening on all sides.

ovolo – a convex molding such as a quarter round.

parting – a straight line dividing the adjoining surfaces of the cope and drag of a flask.

pattern – a form, usually of wood, used to make impressions in sand for making castings.

pecky wood – small grooves running with the grain.

penny – a system of designating nail sizes.

pilot — a guide. On router bits, a knob or bearing on the end to control the sideways movement of the router.
pinholing — a defect in sprayed finishes caused by tiny holes or bubbles.
pith — the spongy center of a tree.
plain-sawing — method of sawing a log to produce pieces tangent to the annular rings.
plastic laminate — a tough, durable, thin material made in many colors and designs primarily used to cover kitchen cabinet countertops.
polygon — a plane figure with three or more sides.
primer — the first coat of finish given to wood or other material.
pumice — a finely-ground powdered abrasive used to rub down and polish finished surfaces.
quarter round — a molding whose end section is in the shape of a quarter circle.
quartersawing — a method of sawing a log to produce pieces with annular rings at right angles to the sides.
rabbet — a cut on the edge or end of lumber which takes out the corner.
radius — a straight line from the center of a circle to its circumference.
rail — horizontal pieces of a paneled door or sash.
raised panel — a door panel with tapered edges that gives the illusion that the center is raised or built-up.
rapping — the process of loosening the pattern in a mold.
reeds — a series of convex cuts usually made to decorate posts or columns.
regular polygon — a plane figure with three or more equal sides.
relief — a clearance given to the cutting edge of a twist drill.
return nosing — a piece with a curved edge installed on the open end of a tread in an open staircase.
reveal — the amount of jamb of a window or door frame exposed by backsetting the door or window casing.
riddle — a sifter used to sift sand over a pattern in a sand mold.
rise — the height from tread to tread in a stairway.
riser — the vertical finish board covering the space from tread to tread.
rosette — a piece of finish in a stairway fastened to a wall and against which the handrail butts.
rottenstone — a fine powder used in the polishing of finished surfaces.
run — the horizontal distance over which a flight of stairs travels.
sapwood — wood nearest the bark of a tree consisting of living cells.
sash — a framework in which lights of glass of a window are set.
scribe — to mark a fine line with a knife or sharp pencil.
scribe — to mark and fit cabinets or other trim to the irregularities of a wall or floor.
seasoning — the process of drying wood to a suitable moisture content.
set — the alternate bending of saw teeth to provide clearance in the cut for the saw blade.
shake — a defect in lumber identified by the separation of the annular rings.
shrink rule — rules used by patternmakers that are longer than the standard rule to make allowance for contraction of metals.
silex — a fine powder used in wood paste fillers for the purpose of filling open-grained wood.
silicon carbide — a man-made substance used as an abrasive for sanding wood or metal.
sill — the bottom member of a door or window frame.
skeleton frame — framework used in a cabinet instead of a solid piece to lighten the cabinet and to conserve material.
spline — a thin flat piece of wood or other material placed in grooves made in joints to strengthen them.

sprue — an opening in a sand mold through which molten material is poured when making castings.
spurs — sharp extensions on the end of an auger bit that cut the circle when holes are bored.
square — pieces at right angles to each other.
stave — long strips of wood with concave and convex sides.
sticking — the shape of the inside edges of the frame members of a paneled door or sash.
stile — the outside vertical members of a paneled door, sash, or other frame.
stool — used in some localities as another name for a sill in reference to door and window frames.
stop — usually a strip of wood, sometimes molded on one edge, against which a door or sash stops when closed.
straightedge — any strip of wood, metal, or other material with a straight edge used to test or make straight lines.
strickle — a straightedge used to screed or scrape sand even with the edges of a flask.
stringer — finished boards in a staircase.
surfaced lumber — lumber which is smoothed to size by running through a planer.
taper — gradual reduction in the size of a piece from one end to the other.
template — a pattern used as a guide to test or make duplicate pieces.
template guide — a round disc with lips inserted in the base of a router used as a guide when routing against templates.
tenon — a piece of lumber cut with an extension or tongue on the end usually made for fitting into a mortise.
throat — the distance from the saw blade to the framework of machines such as the band saw and scroll saw.
tongue-and-groove — a joint made by grooving the edge of one piece and making a tongue on the edge of the other piece that fits into the grooved edge.
tread — horizontal members of a staircase upon which the feet are placed when ascending or descending stairs.
true — a term that describes the straightness, squareness, alignment, and/or location of an object meaning that the object is in a manner that it ought to be.
turnout — a curved portion of a handrail, usually at the start of a staircase.
twist — a defect in lumber consisting of a distortion caused by turning or twist of the edges of a board; sometimes referred to as a wind.
veneer — thin pieces of wood or other material used for finishing purposes and usually covering an inferior piece of material.
volute — a spiral portion of a handrail at the start of a staircase, sometimes eliminating the need for a starting newel post.
wane — bark or missing wood on the edge of lumber.
warp — any distortion from straight in lumber. Examples of warp are cup, crook, bow and twist.
waste — excess material cut away from stock, but not necessarily to be disposed of.
whetting — the process of rubbing the cutting edge of a tool on an oilstone to produce a sharp edge.
wire edge — a burr formed on the cutting edge of a tool from grinding and which is removed by whetting and honing the edge.

APPENDIX

INCH/METRIC – EQUIVALENTS

Fraction	Decimal Equivalent Customary (in.)	Decimal Equivalent Metric (mm)	Fraction	Decimal Equivalent Customary (in.)	Decimal Equivalent Metric (mm)
1/64	.015625	0.3969	33/64	.515625	13.0969
1/32	.03125	0.7938	17/32	.53125	13.4938
3/64	.046875	1.1906	35/64	.546875	13.8906
1/16	.0625	1.5875	9/16	.5625	14.2875
5/64	.078125	1.9844	37/64	.578125	14.6844
3/32	.09375	2.3813	19/32	.59375	15.0813
7/64	.109375	2.7781	39/64	.609375	15.4781
1/8	.1250	3.1750	5/8	.6250	15.8750
9/64	.140625	3.5719	41/64	.640625	16.2719
5/32	.15625	3.9688	21/32	.65625	16.6688
11/64	.171875	4.3656	43/64	.671875	17.0656
3/16	.1875	4.7625	11/16	.6875	17.4625
13/64	.203125	5.1594	45/64	.703125	17.8594
7/32	.21875	5.5563	23/32	.71875	18.2563
15/64	.234375	5.9531	47/64	.734375	18.6531
1/4	.250	6.3500	3/4	.750	19.0500
17/64	.265625	6.7469	49/64	.765625	19.4469
9/32	.28125	7.1438	25/32	.78125	19.8438
19/64	.296875	7.5406	51/64	.796875	20.2406
5/16	.3125	7.9375	13/16	.8125	20.6375
21/64	.328125	8.3384	53/64	.828125	21.0344
11/32	.34375	8.7313	27/32	.84375	21.4313
23/64	.359375	9.1281	55/64	.859375	21.8281
3/8	.3750	9.5250	7/8	.8750	22.2250
25/64	.390625	9.9219	57/64	.890625	22.6219
13/32	.40625	10.3188	29/32	.90625	23.0188
27/64	.421875	10.7156	59/64	.921875	23.4156
7/16	.4375	11.1125	15/16	.9375	23.8125
29/64	.453125	11.5094	61/64	.953125	24.2094
15/32	.46875	11.9063	31/32	.96875	24.6063
31/64	.484375	12.3031	63/64	.984375	25.0031
1/2	.500	12.7000	1	1.000	25.4000

INDEX